ELEMENTARY ALGEBRA
A Straightforward Approach

Third Edition

Martin M. Zuckerman

City College of
The City University of New York

Ardsley House Publishers, Inc.
New York

Address orders and editorial
correspondence to:
Ardsley House Publishers, Inc.
320 Central Park West
New York, NY 10025

ISBN: 0-912675-69-1

Printed in the United States of America

10 9 8 7 6 5 4 3 2 1

In Memory
of
My Father

Contents

Preface

This book has been written for the multitudes of college students who have not had the good fortune to understand the basic algebraic and arithmetic concepts upon which all scientific studies depend. In clear, simple, yet precise language, *Elementary Algebra: A Straightforward Approach* presents basic material in an intuitive manner. Common learning difficulties are dealt with. Set-theoretic notation is avoided.

Major Improvements in the Third Edition

- A Diagnostic Test is given so that with their instructor's aid, students can initially assess their strengths and weaknesses.
- A second Practice Exam has been added at the end of each chapter.
- Throughout Chapter 1 more everyday applications of basic arithmetic concepts have been included.
- In Section 3.6 the notion of the *average rate* at which a vehicle travels is discussed so that distance problems conform more closely to reality. (Seldom do real objects travel at constant rates.)
- In Section 4.5 there is additional discussion of factoring binomials of the forms $x^4 - y^4$, $x^4 - y^2$, and $x^2 - y^4$.
- The Review Exercises to Chapter 4 include 50 new factoring examples in which the student must select the appropriate factoring technique.

- Section 6.3 begins with a discussion of conversions between percentages and decimals; this is preparatory for the presentation of interest problems.
- In Section 6.7 the relationship between inequalities involving a and b, $\frac{1}{a}$ and $\frac{1}{b}$, and $\frac{c}{a}$ and $\frac{c}{b}$ is discussed.

PRACTICE EXERCISES FOR TOPIC A

Answers
1. (a) ______
(b) ______

1. Multiply: (a) $r^8 \cdot r$ (b) $n^4 \cdot n^7 \cdot n^2$

Now try the exercises for Topic A on page 131.

B. Products of Monomials

Each section has been divided into several topics.

When multiplying monomials, use the Associative and Commutative Laws to rearrange the factors.

1. Group all coefficients at the beginning, and multiply them.
2. Group powers of the same variable together and multiply them, as previously explained.

Important points are highlighted in red to reinforce student understanding.

Example 4 ▶ Multiply $(4xy^2)(7x^2y^4)$
Solution:

$$(4xy^2)(7x^2y^4) = (4 \cdot 7)(xx^2)(y^2y^4) = 28x^3y^6$$

Group all coefficients at the beginning. Group powers of the same variable together. Multiply coefficients and powers of the same variable. ◀

Example 5 ▶ Multiply $(4ax^2y)(-2x^2y^3)(-3a^2bx)$
Solution:

$$(4ax^2y)(-2x^2y^3)(-3a^2bx) = [4(-2)(-3)](aa^2)b(x^2x^2x)(yy^3)$$
$$= 24a^3bx^5y^4$$ ◀

Lots of Illustrative Examples are provided.

PRACTICE EXERCISES FOR TOPIC B

Practice Exercises immediately follow each topic. Complete solutions to all the Practice Exercises are given in the back of the book.

Answers
2. (a) ______
(b) ______

2. Multiply: (a) $9a^3(-2a^5b^3)$ (b) $(-x^2y)(3xy^2)(-7x^3)$

Now try the exercises for Topic B on page 131.

A convenient answer column is provided for the Practice Exercises.

After mastering the Practice Exercises, the student should try the Exercises at the end of the section.

C. Quotients of Monomials

Recall that for every nonzero number a,

$$\frac{a}{a} = 1$$

This law also applies to variables a. Therefore, when dividing products, if the same factors appear in the dividend and divisor, we can divide these factors and replace them by 1's. In the following division process, observe that

- In Sections 7.3 and 7.4 a *slanting line* is defined as a line that is neither horizontal nor vertical. Properties of slanting lines are discussed. Equations in intercept form are presented so that given the graph, the student can immediately write an equation of a slanting line that does not pass through the origin. Then it is shown how to clear this equation of fractions.

- One or more *Pick-Out-the-Error* problems have been added to most Exercise Sets to enable the student to pinpoint common mistakes. Many new exercises have been added throughout the book.

Format

Users of earlier editions will note a completely new format designed to facilitate student understanding. Each section has been divided into several topics. The student should first read through Topic A and work through the illustrative examples presented. Immediately afterward, he or she should then attempt the Practice Exercises for Topic A which follow directly. *Complete Solutions to all Practice Exercises are provided in the back of the book* so that the student can master the techniques involved. When the student has grasped the pertinent notions, he or she should immediately try the Exercises for Topic A at the end of the section; these exercises are keyed to the illustrative examples and Practice Exercises of Topic A. The student can then procede to Topic B, followed by the Practice Exercises for Topic B, and then go on to the Exercises for Topic B at the end of the section, and so on.

At the end of each chapter there is a brief review of the most important concepts (What Have You Learned in Chapter __) as well as Review Exercises, including a number from previous chapters, and two Practice Exams. In this

A work-space column is provided.

Use this area for work space.

EXERCISES

A. In Exercises 1–7 multiply. (*See Examples 1–3.*)

1. $x^3 \cdot x$ **2.** $b^3 \cdot b^2$ **3.** $c^4 \cdot c^6$ **4.** $m^{10} \cdot m^8$

5. $c^4 \cdot c^2 \cdot c^3$ **6.** $y^9 \cdot y^4 \cdot y^7$ **7.** $z^2 \cdot z^5 \cdot z \cdot z^6$

8. Pick out the *error(s)*.

(a) $x^3 \cdot x^4 = x^{12}$ (b) $x^3 + x^4 = x^7$ (c) $x^2 \cdot x^6 = x^{2+6}$

(d) $x^2 \cdot x^6 = x^{2 \cdot 6}$ (e) $x^2 + x^6 = x^{2+6}$

***Pick-out-the-error* problems pinpoint common student mistakes.**

B. In Exercises 9–18 multiply. (*See Examples 4 and 5.*)

Exercises are arranged by topics and are keyed to the Illustrative Examples.

9. $4a^2 \cdot a^4$ **10.** $(-1)y^2 \cdot 5 \cdot y^2$

11. $(x^2y)(xy^2)$ **12.** $(a^2b)(a^3b^2)$

13. $(2c^2d)(5cd^2)$ **14.** $(3ab)(-2a^2b)$

15. $(5xy)(3x^2y)(2x^2)$ **16.** $(-4a^2b)(-ab)(abc)$

17. $(2xy^2z)(xyz^2)(3xz)(yz^2)$ **18.** $(-2abc)(bcd)(3abc^2)(4a^2d)$

19. Find the product of $2xy$ and $9xy^2$.

20. Find the product of $-2a^2bc$, $-5ac$, and abc^2.

way the student has several different opportunities to reinforce his or her knowledge of the techniques of the course.

Other Features

- The textual material is interspersed with **over 400 worked-out examples.**

Verbal problems are given in many Exercise Sets.

55. The sum of $2xy$ and $3x^2$ is multiplied by $3x^2y^3$. Find the result.

56. The sum of $4x^2y$ and $6xy^3$ is divided by $2xy$. Find the result.

57. Pick out the *error(s).*

(a) $\frac{x^2+4}{x} = x + 4$ (b) $\frac{x^2+4}{2} = x + 2$ (c) $\frac{x^4+4x^2}{x^2} = x^2 + 4$

(d) $\frac{2x^2+4}{2} = x^2 + 2$ (e) $\frac{x^2+y^2}{xy} = x + y$ (f) $\frac{x^2-y^2}{x} = x - y^2$

E. In Exercises 58–66 simplify each expression. (*See Example 13.*)

Sample Exercises with Solutions illustrate important techniques.

Sample: Simplify:	*Solution:*
$4x^2 - 3x(x + 2y)$	$4x^2 - 3x(x + 2y) = 4x^2 - 3x \cdot x - 3x \cdot 2y$
	$= 4x^2 - 3x^2 - 6xy$
	$= x^2 - 6xy$

The most important topics of the chapter.

What Have You Learned in Chapter 4?

REVIEW

You can find products such as $4x^2 \cdot y^2 \cdot 2xy^2$ and quotients such as $\frac{4x^2y^3}{2xy^2}$.

Also, you can find products such as $3ab(a + 5b)$ and quotients such as $\frac{12a^2b - 6ab}{6ab}$.

You know how to express an integer, such as 24, as the product of primes.
You know how to isolate the common factor of a polynomial.
You are able to multiply polynomials.
You can factor certain binomials and trinomials.
And you can solve some equations by factoring the left side.

There are Review Exercises at the end of each chapter.

Let's Review Chapter 4

4.1 *Multiplying Powers*

In Exercises 1 and 2 multiply.

1. $(3x^2y)(2xy^3)$ **2.** $(4m^2n)(-2mn^2)(-mn)$

In Exercises 3 and 4 divide.

3. $\frac{10x^3y^2}{5xy}$ **4.** $\frac{-32a^4b^3}{-4ab^3}$

In Exercises 5 and 6 multiply.

5. $x^2(x - 5y)$ **6.** $4a^2bc(2a^2b + 3abc)$

In Exercises 7 and 8 divide.

7. $\frac{6ab + 6b^2}{3b}$ **8.** $\frac{20x^2y^3 - 15xy^2}{5xy^2}$

9. Simplify $3a^3(a^2 - 2) + 2a^2(a^3 + 3a)$

- There are almost **4000 student exercises** ranging from simple drill exercises to a number of exercises that challenge the students to broaden their perspective. Each Exercise Set offers a large variety of problems, some of which deal with applications.
- A **second color** is used functionally to highlight major ideas and to reinforce important points.
- **Graphical techniques** are stressed. There are over 300 figures in the book.

At the end of each chapter there are exercises that review material from the preceding chapters.

And These from Chapters 1-3

91. Rearrange the following primes so that you can write "<" between any two of them: 7, 17, 41, 29, 23, 2.

92. Evaluate $(2t + 7)(t - 3)$ when $t = 4$.

93. Solve $4x - 3 + 5x = 12 - (5 - 7x)$

94. What number must be divided by 6 to obtain -12?

95. Find the value of $-4 - (-8)$.

96. How long must a rope be to enclose a rectangular area that is 54 feet long and 20 feet wide?

REVIEW

Practice Exam 1 on Chapter 4

In Questions 1–3 multiply:

1. $(4x^2y^5)(-3xy^3)$ **2.** $(4x - 3)(2x + 5)$ **3.** $(a^3 - 4a + 5)(a^2 - 2)$

4. Divide $\dfrac{15x^2y^2 - 25xy^4}{5xy^2}$

5. Express 108 as a product of primes.

In Questions 6–9 factor each polynomial.

6. $6x^2 - 8x$ **7.** $9a^2 - 16b^2$

8. $x^2 + 9x + 18$ **9.** $2y^2 - y - 1$

10. Solve and check the roots of $x^2 + 5x = 0$.

11. Solve $x^3 - 6x^2 + 8x = 0$

12. Solve $x^2 + 24x + 144 = 0$

Practice Exam 2 on Chapter 4

Two Practice Exams on each chapter are provided.

Complete Solutions to all Practice Exercises and to all Practice Exam questions appear at the back of the book.

CHAPTER 4

Practice Exercises for Section 4.1, pages 128–131

1. (a) r^9 (b) n^{13}

(a) $r^8 \cdot r = r^8 \cdot r^1 = r^{8+1} = r^9$

(b) $n^4 \cdot n^7 \cdot n^2 = n^{4+7+2} = n^{13}$

2. (a) $-18a^8b^3$ (b) $21x^6y^3$

Use the Commutative and Associative Laws of multiplication to rearrange and group the factors.

(a) $9a^3(-2a^5b^3) = 9(-2)(a^3a^5)(b^3) = -18a^8b^3$

(b) $(-x^2y)(3xy^2)(-7x^3) = (-1)(3)(-7)(x^2 \cdot x^1 \cdot x^3)(y^1 \cdot y^2)$
$= 21x^{2+1+3}y^{1+2}$
$= 21x^6y^3$

- **Complete solutions** are provided for *all* Practice Exercises and for *all* Practice Exam questions.

- **Answers** to *odd-numbered* Exercises at the end of each section and to *all* Review Exercises at the end of each chapter are provided at the back of the book.

Supplements

Answers to all the exercises (both odd- and even-numbered) appear in the accompanying **Instructor's Manual**. The Instructor's Manual also contains chapter-by-chapter comments, suggestions for presenting the material, five sample exams for each chapter, and three sample final exams.

A **microcomputer test bank** is available at no cost to instructors who adopt the book.

A **Student Workbook**, coordinated with the text, is available. On each left-hand page, examples are worked out that illustrate the techniques of the Key Problems of each section. On the corresponding right-hand page, the student must then solve several Key Exercises corresponding to each Key Problem. Additional Practice Exams are given for each chapter as well as Practice Final Exams. Also, Complete Solutions are given to about one-fourth of the exercises of the text.

Acknowledgments

I would like to thank each of the following reviewers of the various editions of this book for their many helpful suggestions. A number of these people gave unstintingly of their time in working on various aspects of this project.

Ignacio Bello, Hillsborough Community College
Ben P. Bockstege, Broward Community College
Thomas R. Butts, Michigan State University
Rodney Chase, Oakland Community College
Dorothy Crespin, Lower Columbia College
Paul R. Fallone, University of Connecticut at Hartford
Victor Klee, University of Washington
Ann Klein of Arlington, Massachusetts
The late Robert D. Klein, Northeastern University
F.J. Navarre, Widener College
Kevin Reilley of Utica, New York
David Segal, New York Institute of Technology
Ken Seydel, Skyline College
Fred Toxepeus, Kalamazoo Valley Community College
Rena Widener of Jerusalem, Israel
Frank Wright, Cerritos College

Finally, I would like to thank William Boot, Karen Curr, William Curr, Laura Jones, and Nancy Powers, who were involved in the production of the book, and Karen Berneath, who designed the cover.

Diagnostic Test

The purpose of this test is to help you understand which sections of the book you should emphasize in your studies. If in any section you are able to answer *all* parts of *every* question correctly, you may not need to study that section. But *if any part of your answer is wrong, you should study the section.* It is best to consult your instructor about your specific needs, but this test should help to pinpoint your strengths and weaknesses. When you have completed the test, turn to page A1 at the back of the book for the answers.

Chapter 1 Arithmetic and the Number Line

1.1 Integers and the Number Line

1. Indicate which integer corresponds to each of the points *A*, *B*, *C*, *D*, and *E* in Figure D1.

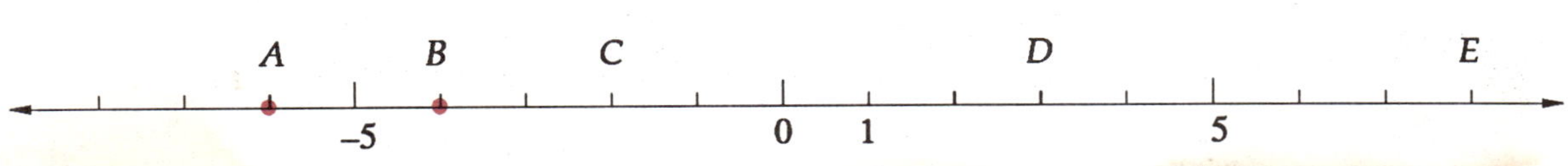

FIGURE D1

Use this area for work space.

2. Label the following integers on the number line of Figure D2.

(a) 6 (b) -4 (c) -1

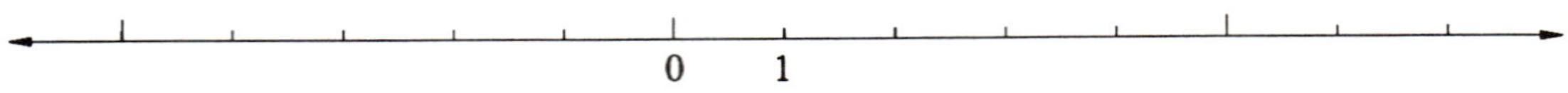

FIGURE D2

1.2 *Order*

3. Fill in the blanks with either "<" or ">".

(a) -5 ☐ -1 (b) 7 ☐ -10

4. Rearrange the following integers so that you can write "<" between any two of them.

5, -3, 0, -7, 1

1.3 *Inverse and Absolute Value*

5. Find the inverse of each number.

(a) 12 (b) -7 (c) 0

6. Find each absolute value.

(a) $|-16|$ (b) $|2|$

1.4 *Adding and Subtracting Signed Numbers*

7. Add $(-6) + (-10) + (-1)$

8. Subtract $(-15) - (-8)$

9. Add:

$$\begin{array}{r} -11 \\ 26 \\ -23 \\ -\ 4 \\ 10 \\ \hline \end{array}$$

1.5 *Multiplying and Dividing Signed Numbers*

10. Find each product.

(a) $8(-3)$ (b) $(-1)(-2)(5)(-7)(-1)$

11. Find the quotient, or indicate that division is undefined.

(a) $\dfrac{15}{-3}$ (b) $\dfrac{5(-8)}{-20}$ (c) $0 \div (-10)$ (d) $9 \div 0$

Use this area for work space.

1.6 Positive Integral Exponents

12. Find the value of each.

(a) 4^3 (b) $(-6)^2$ (c) $(-1)^5$

1.7 Combining Operations

13. Find the value of each.

(a) $7 + 3(-4)$ (b) $15 - 2(9 - 4)^2$

14. Divide the square of 6 by the product of 2 and 3.

Chapter 2 Polynomials

Use this area for work space.

2.1 Terms and Polynomials

1. Suppose x stands for some number. Express

five more than twice the number

in terms of x.

2. Use exponents to simplify $4 \cdot x \cdot x \cdot x \cdot y \cdot y$.

3. What is the (numerical) coefficient of the expression $-8a^3b^2$?

4. Which of the following are polynomials?

(a) $5x - 3$ (b) $\dfrac{2x^2 + x - 1}{x + 1}$ (c) $9x^2$

2.2 Evaluating Polynomials

5. Evaluate $x^2 + 5x + 8$ when $x = -3$.

6. Evaluate $x - 3xy + 2y$ when $x = 4$ and $y = -2$.

2.3 Adding Terms

7. Simplify by combining like terms.

(a) $x + 4x$ (b) $7a + 3b - b - 2a$ (c) $5x^2 - 5x + 3 - 3x + x^2$

2.4 Adding Polynomials

8. Classify each polynomial as (*i*) a monomial, (*ii*) a binomial, (*iii*) a trinomial, or (*iv*) none of these.

(a) $4x^3 - 3x^2 + 1$ (b) $5x - x^2 + 4x$ (c) $-xy$ (d) 0

Use this area for work space.

9. Add

$$\begin{array}{r} 10a - 10b - 1 \\ -3a + b - 6 \\ \hline \end{array}$$

10. Subtract $4x - y$ from $5x - 2y$.

11. Simplify $-[a - (5 - a) + 2]$

2.5 *Multiplying and Dividing by a Number*

12. Multiply: (a) $(7x)(-8y)$ (b) $-6(3a - b + 2)$

13. Divide $\dfrac{18x + 12y - 3}{3}$

14. Simplify $\dfrac{10x + 5y}{5} - 2(x - y)$

Chapter 3 Equations

Use this area for work space.

3.1 *Roots*

1. Check whether the given number is a root of the equation.

(a) $4x - 10 = 2$ $\boxed{2}$

(b) $3x = x - 2$ $\boxed{-1}$

2. Write an equation to express the following question. Use x as the variable. (You need not solve the equation.)

What number multiplied by 3 is 10 more than the number?

3.2 *Solving Simple Equations*

3. Solve each equation.

(a) $x + 8 = 5$ (b) $9y = 45$ (c) $\dfrac{z}{6} = 12$

3.3 *Isolating the Variable*

4. Solve each equation.

(a) $4x = 30 - x$ (b) $2x - 9 + x = 7x + 15 + 3x - x$ (c) $\dfrac{t}{2} - 1 = 13$

5. (a) Solve the equation. (b) Check the root.

$$11 + 7x = x - 7$$

3.4 *Equations with Parentheses*

6. Solve each equation.

(a) $9 - (3 - 2x) = 4(3x + 4)$ (b) $8(x + 7) = 16$ (c) $4t - [3 - (2t - 1)] = 8$

Use this area for work space.

3.5 Applied Problems

7. Set up an equation and solve it to answer the following.

A 22-foot rope must be cut in two pieces so that the longer piece is 1 foot longer than twice the shorter piece. How long must each piece be?

3.6 Distance Problems

8. An airplane travels at 150 miles per hour. Express in terms of t the distance it travels in t hours.

9. Two cars leave a toll booth at the same time and travel in the same direction along a straight road, one at 50 miles per hour, the other at 62 miles per hour. How far apart are they after 3 hours?

Chapter 4 Products and Factors

Use this area for work space.

4.1 Multiplying Powers

1. Multiply: (a) $(4x^2y^3)(-5xy^2)(-2x^3y)$ (b) $3x^2y(xy^2 - 2y + 1)$

2. Divide: (a) $\dfrac{28x^6y^3}{4x^3y^2}$ (b) $\dfrac{12x^4 - 9x^3 + 3x}{3x}$

3. Simplify $\dfrac{4a^3b + 2a^2b}{a} + 2a(a + b)$

4.2 Prime Factors

4. Express each integer as the product of primes.

(a) 165 (b) 84

5. Find the greatest common divisor of 18, 36, and 60.

4.3 Common Factors

6. Factor each polynomial by isolating the common factor.

(a) $10x - 8y + 6$ (b) $7x^2y + 21xy^2 - 14x^3y^3$ (c) $a^4 + 3a^3 - 6a^2$

4.4 Multiplying Polynomials

7. Multiply:

(a) $(3x - 8)(x + 1)$ (b) $(2x + 5)(2x - 5)$ (c) $\begin{array}{r} x^2 + x + 3 \\ x - 2 \\ \hline \end{array}$

Use this area for work space.

4.5 *Difference of Squares*

8. Factor each binomial.

(a) $x^2 - 100$ (b) $3a^2 - 27b^2$ (c) $16a - ax^4$

4.6 *Factoring Trinomials* $x^2 + Bx + C$

9. Factor each trinomial.

(a) $x^2 - 11x + 30$ (b) $n^2 + 4n - 32$ (c) $x^2 + 10xy + 9y^2$

4.7 *Factoring One Side of an Equation*

10. Solve each equation.

(a) $x^2 + 10x = 0$ (b) $x^2 - 6x - 16 = 0$ (c) $2x^2 - 50 = 0$

4.8 *Factoring Trinomials* $Ax^2 + Bx + C$

11. Factor each trinomial.

(a) $4x^2 + x - 3$ (b) $6x^2 - 9x - 6$ (c) $6x^2 + 11xy - 10y^2$

Chapter 5 Rational Expressions

Use this area for work space.

5.1 *Rational Numbers and Rational Expressions*

1. Cross multiply to show whether the given rational expressions are equivalent.

$$\frac{7x}{x - 2} \text{ and } \frac{14x^2}{2x^2 - 4x}$$

2. Simplify each rational number.

(a) $\frac{28}{42}$ (b) $\frac{36}{-16}$ (c) $-\frac{-8}{-72}$

3. Evaluate $\frac{x + 3y}{3xy}$ when $x = 2$ and $y = -1$.

5.2 *Dividing Powers*

4. Simplify each rational expression.

(a) $\frac{x^6y^2}{xy^3}$ (b) $\frac{25a^5}{40a^4}$ (c) $\frac{4x^3(x + 3)}{6x^2(x + 3)^2}$

Use this area for work space.

5.3 *Factoring and Simplifying*

5. Simplify:

(a) $\frac{x^2 + 3x}{3x^2}$ (b) $\frac{4x}{10x - 8}$ (c) $\frac{x^3 + 2x^2}{x^2 - 4}$

5.4 *Dividing Polynomials*

6. (a) Express in standard form: $3x^2 - x + 4x^3 - 10$

(b) What is the degree of this polynomial?

7. Divide $(2x^3 - 8x + 1) \div (2x - 4)$

(a) Find the quotient and (b) find the remainder.

5.5 *Multiplying and Dividing Rational Expressions*

8. Multiply:

(a) $\frac{5}{9} \cdot \frac{-1}{10} \cdot \frac{3}{4}$ (b) $\frac{8x + 4}{6x + 12} \cdot \frac{3}{4x^2 - 1} \cdot \frac{4x - 2}{9x + 18}$

9. Divide:

(a) $\frac{9}{10} \div \frac{3}{8}$ (b) $\frac{5a^2}{3a - 3} \div \frac{10a}{9}$

5.6 *Adding and Subtracting: Same Denominator*

10. Combine and simplify.

(a) $\frac{5}{12} - \frac{1}{12} + \frac{7}{12}$ (b) $\frac{5}{3x} + \frac{7}{3x}$ (c) $\frac{x - 1}{x + 2} + \frac{2x + 7}{x + 2}$

5.7 *Least Common Multiples*

11. Find the least common multiple of the given integers or polynomials.

(a) 16 and 24 (b) $6x^2$ and $9xy$ (c) $x^2 - 4x$, $x^2 - 3x - 4$, and $x^3 + x^2$

5.8 *Adding and Subtracting: Different Denominators*

12. Add or subtract, as indicated.

(a) $\frac{5}{6} + \frac{2}{3} - \frac{3}{4}$ (b) $\frac{5}{2x} + \frac{1}{x^2}$ (c) $\frac{4}{a} + \frac{a - 1}{a + 2} + \frac{3}{a^2 + 2a}$

Use this area for work space.

5.9 Complex Expressions

13. Simplify:

(a) $\dfrac{\frac{3}{4} + \frac{1}{2}}{\frac{1}{2}}$ (b) $\dfrac{1 + \frac{2}{x}}{\frac{x}{2} + 1}$ (c) $\dfrac{\frac{4}{x^2 - 4}}{\frac{x}{x - 2} - \frac{2}{x + 2}}$

Chapter 6 More Equations and Inequalities

Use this area for work space.

6.1 Proportions and Equations with Rational Expressions

1. Solve each proportion.

(a) $\dfrac{10}{x} = \dfrac{-5}{6}$ (b) $\dfrac{4}{x - 1} = \dfrac{5}{x + 1}$

2. A cafeteria needs 5 pounds of bread for every 28 customers it serves each day. How many pounds of bread should it order on a day in which 1400 customers are expected?

3. Solve each equation.

(a) $\dfrac{x}{4} - \dfrac{x}{6} = 1$ (b) $\dfrac{5}{8} + \dfrac{1}{x} = \dfrac{1}{2}$ (c) $\dfrac{4}{t - 2} - \dfrac{6}{t + 2} = \dfrac{12}{t^2 - 4}$

6.2 Decimals and Equations

4. (a) Express $\dfrac{13}{20}$ as a decimal.

(b) Express .44 as a rational number. Simplify.

5. Solve for x.

(a) $.2x + .04 = 2$ (b) $\dfrac{x}{.012} = 3$

6.3 Interest Problems

6. How much interest is earned on a principal of \$2000 invested for 1 year at an annual interest rate of $7\frac{1}{2}\%$?

7. How much money must be invested at a 6% annual interest rate in order to earn \$72 interest in a year?

8. How much interest is earned in a year by an investment of \$1000 if the interest rate is 8% compounded semiannually?

Use this area for work space.

6.4 *Commissions, Markups, and Discounts*

9. If a salesman earns \$180 commission on sales totalling \$1200, what is his rate of commission?

10. A retail store buys clocks at \$20 and marks them up by 75%. What is the retail price of a clock?

11. If a clothing store discounts its merchandise by 40%, what will be the sale price of a suit regularly priced at \$200?

6.5 *Literal Equations*

12. Solve for y in terms of x.

$$\frac{x - y}{3} = 5$$

13. Solve for a in terms of the other variables.

$$3a + b = c(b - a)$$

6.6 *Intervals*

14. Answer "true" or "false."

(a) 9 is in $[9, \infty)$.

(b) 0 is in $(-3, 0)$.

15. Describe in one of the forms

$$(a, b),\ [a, b],\ (a, \infty),\ [a, \infty),\ (-\infty, b),\ (-\infty, b]$$

the set of numbers satisfying each of the following.

(a) $0 \le x \le 12$

(b) $x > 6$

(c) $-9 < x < 1$

6.7 *Arithmetic of Inequalities*

16. Fill in "<" or ">".

(a) $c + 2 \ \square \ c + 1$

(b) If $c > 0$, then $\frac{5}{c} \ \square \ \frac{-6}{c}$

(c) If $c < 0$, then $-3c \ \square \ 2c$

6.8 *Solving Inequalities*

17. Solve each inequality.

(a) $10 - 3x > 1$

(b) $-5 \le 2x + 1 \le 5$

Chapter 7 Lines and Linear Systems

Use this area for work space.

7.1 *Rectangular Coordinates*

1. Plot the following ordered pairs on the rectangular coordinate system in Figure D3.

(a) $(4, -1)$ (b) $(-2, 1)$ (c) $(0, -4)$ (d) $(1, 3)$

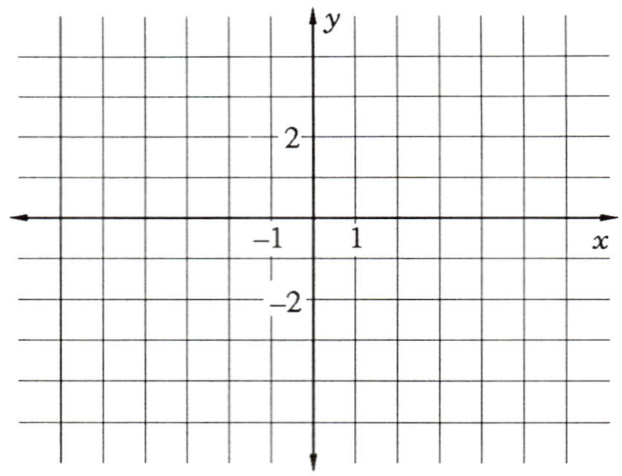

FIGURE D3

2. For each point in Figure D4 find

(a) its x-coordinate,

(b) its y-coordinate.

(c) Identify the point with its ordered pair of coordinates.

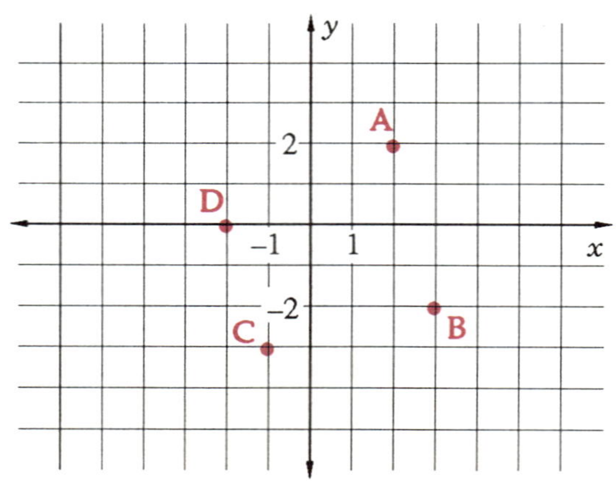

FIGURE D4

7.2 *Graphs of Equations*

3. Graph each equation.

(a) $y = x - 4$ (b) $2x + y = 6$ (c) $y = 2$

Use this area for work space.

7.3 *Slope*

4. In each case find the slope of the line passing through P_1 and P_2, or else indicate that the slope is undefined.

(a) $P_1 = (-6, 1)$, $P_2 = (4, -4)$

(b) $P_1 = (7, -2)$, $P_2 = (-4, -2)$

(c) $P_1 = (-3, 4)$, $P_2 = (-3, -1)$

5. Draw the line through (0, 3) with slope $-\frac{1}{2}$.

7.4 *Equation of a Line*

6. Find (a) the slope and (b) the *y*-intercept of the line with equation $3y = x - 3$.

7. Write the equation of the line with slope -1 and *y*-intercept 8.

8. (a) Find the equation in intercept form for the line with *x*-intercept 3 and *y*-intercept -5.

(b) Clear the equation of fractions.

7.5 *Intersection of Lines*

9. Find the intersection of the lines L_1 and L_2 graphically.

$$L_1: x + y = 6 \qquad L_2: y = \frac{x}{2}$$

10. Without graphing, determine which of the following lines are parallel.

$$L_1: y = 3x \qquad L_2: 3y = x + 3 \qquad L_3: y + 3x = 3 \qquad L_4: y = \frac{x}{3}$$

7.6 *Systems of Linear Equations*

11. Solve the following system of equations by substituting:

$$3x = y - 3$$
$$x + 2y = 13$$

12. Solve by the adding method:

$$3x - 4y = 1$$
$$2x - y = 9$$

7.7 *Direct Variation*

13. Suppose *y* varies directly as *x* and $y = 16$ when $x = 10$.

(a) Find the constant of variation.

(b) Find *y* when $x = 25$.

Chapter 8 Roots and Integral Exponents

Use this area for work space.

8.1 *Square Roots*

1. Find each square root.

(a) $\sqrt{81}$ (b) $-\sqrt{400}$ (c) $\sqrt{x^8}$ (d) $\sqrt{3^{16}}$

8.2 *Irrational Square Roots*

2. Between which two consecutive integers is $\sqrt{116}$ located?

3. Use $\sqrt{3} \approx 1.73$ to approximate $4 - 2\sqrt{3}$ to one decimal digit.

4. Find the length of side a in the right triangle shown in Figure D5.

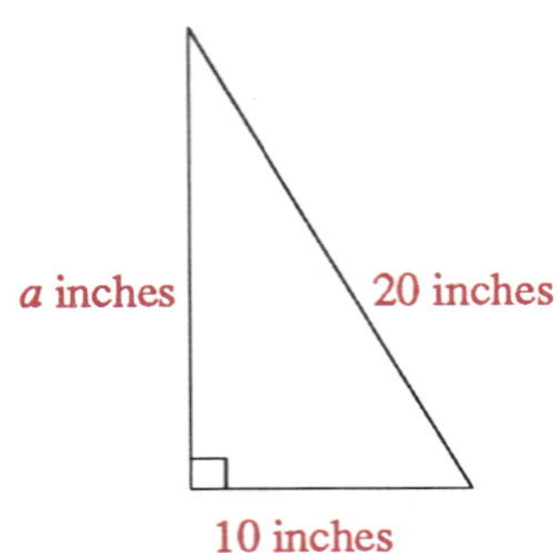

FIGURE D5

8.3 *Roots of Products and Quotients*

5. Find each square root.

(a) $\sqrt{36x^{16}y^4}$ (b) $\sqrt{\dfrac{16}{121}}$ (c) $\sqrt{\dfrac{x^6}{3^2}}$

6. Simplify each square root.

(a) $\sqrt{40}$ (b) $\sqrt{9x^3}$ (Assume $x > 0$.) (c) $\sqrt{\dfrac{2}{3}}$ (Rationalize the denominator.)

8.4 *Adding and Subtracting Roots*

7. Add or subtract, as indicated.

(a) $\sqrt{5} + 4\sqrt{5}$ (b) $\sqrt{60} - \sqrt{15}$ (c) $3x\sqrt{2} + 2\sqrt{2x^2}$ (Assume $x > 0$.)

(d) $\dfrac{\sqrt{3}}{2} - \dfrac{\sqrt{2}}{3}$

Use this area for work space.

8.5 *Multiplying and Dividing Roots*

8. Multiply:

(a) $\sqrt{2}\sqrt{72}$ (b) $\sqrt{3}(\sqrt{12} - 2)$ (c) $(2\sqrt{5} - 3\sqrt{2})(3\sqrt{5} + \sqrt{2})$

9. Simplify $\dfrac{\sqrt{75}}{\sqrt{3}}$

10. (a) Factor $12 - \sqrt{18}$ (b) Simplify $\dfrac{12 - \sqrt{18}}{6}$

8.6 *Zero and Negative Exponents*

11. Evaluate each.

(a) 5^{-3} (b) $(-4)^0$ (c) $\left(\dfrac{4}{3}\right)^{-2}$

12. Simplify. Express your answer without writing a fraction. Use negative exponents, if necessary.

(a) $x^5 \cdot x^{-3} \cdot x^{-4}$ (b) $\dfrac{x^2 \cdot x^5}{x^{-4} \cdot x^{-3}}$

8.7 *Scientific Notation*

13. Express in scientific notation.

(a) .000 631 (b) 725 000 000

14. Express the result of the following computation in ordinary decimal notation:

$$\frac{(5 \times 10^{-2})(6 \times 10^{2})}{2 \times 10^{-3}}$$

Chapter 9 Quadratic Equations

Use this area for work space.

9.1 *Reducible Equations*

1. Write each equation in the form

$$Ax^2 + Bx + C = 0, \text{ where } A \neq 0. \text{ (Do } \textit{not} \text{ solve.)}$$

(a) $3x^2 + 2 = 5x - x^2$

(b) $2x(x + 1) = (x + 4)^2$

2. Reduce the following to a first-degree equation and solve:

$$(x - 3)^2 = x(x + 3)$$

Use this area for work space.

9.2 *Factorable Equations*

3. Use the factoring method to solve each equation.

(a) $x^2 + 7x = 0$ (b) $2x^2 - 9x = 5$ (c) $\frac{6}{x} = \frac{x - 1}{x + 3}$

4. Find a quadratic equation with integral coefficients whose roots are 5 and -2.

9.3 *Equations of the Form* $x^2 = a$

5. Solve each equation by the rule that states that if $x^2 = a$, where $a > 0$, then $x = \pm\sqrt{a}$.

(a) $x^2 - 32 = 0$ (b) $(2y + 3)^2 = 25$

9.4 *The Quadratic Formula*

6. (a) Evaluate the discriminant of the equation

$$3x^2 + 2x - 1 = 0$$

(b) Without solving, indicate how many different (real) roots the equation has.

7. Solve each equation by using the quadratic formula.

(a) $x^2 - 3x + 1 = 0$ (b) $2x^2 + 4x - 3 = 0$

9.5 *Applications*

8. Find two positive integers such that the larger is 2 more than twice the smaller, and the product of these integers is 60.

9. Find the dimensions of a rectangle whose area is 18 square feet and whose perimeter is 22 feet.

ELEMENTARY ALGEBRA
A Straightforward Approach

Third Edition

Martin M. Zuckerman
City College of
The City University of New York

Ardsley House Publishers, Inc.
New York

1

Arithmetic and the Number Line

1.1 Integers and the Number Line

A. Picturing Positive Integers

DEFINITION

> The numbers
>
> $$1, 2, 3, 4, \ldots$$
>
> are known as **positive integers.** The three dots to the right are read "and so on."

Among the positive integers are 10, 17, 45, 101, and 3409.

There is a very convenient way of picturing 0 and the positive integers along a straight line—the so-called **number line.**

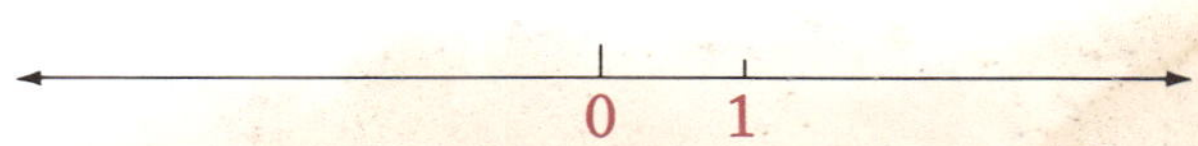

FIGURE 1.1. The number line.

Consider a *horizontal* line, as in Figure 1.1, and suppose that it extends indefinitely both to the left and to the right. Choose any point on the line and call this the **origin.** This point corresponds to the number 0. Now choose another point to the *right* of 0. This second point corresponds to the number 1.

A portion of a line that lies between two points (including these "end-points") is called a **line segment.** The *length* of the line segment between 0 and 1 determines one **unit of distance.** The number 2 corresponds to the point one (distance) unit to the right of 1. The number 3 corresponds to the point one unit further to the right. Continue this process to obtain the numbers 4, 5, 6, and so on. (See Figure 1.2.)

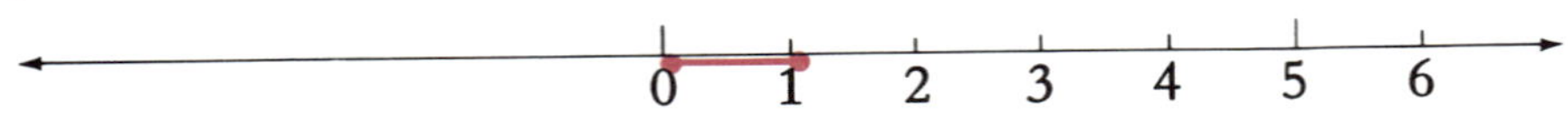

FIGURE 1.2. The line segment between 0 and 1 is in color.

Example 1 ▶ Consider the number line shown in Figure 1.3(a). Label the points that correspond to the following numbers.

(a) 4 (b) 7 (c) 8 (d) 10

FIGURE 1.3(a)

Solution:

FIGURE 1.3(b) ◀

Answers

2. ______

______

______

PRACTICE EXERCISES FOR TOPIC A

1. On the number line in Figure 1.4, label the points that correspond to the following numbers.

(a) 1 (b) 3 (c) 9 (d) 6

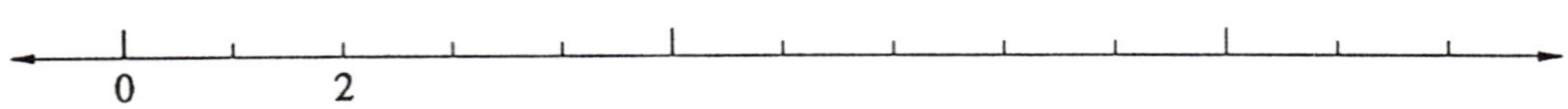

FIGURE 1.4

2. Indicate the integers that correspond to the following points on the number line in Figure 1.5.

FIGURE 1.5

Now try the exercises for Topic A on page 5.

B. Need for More Numbers

Example 2 ▶ Suppose a dress costs a store \$25. The store sells the dress for \$35. The profit is \$10.

\$35 ---- *sale price*
− \$25 ---- *cost*
\$10 ---- *profit*

However, at times it is necessary to sell an item *below cost*. The store *loses* money on the item. Thus, if it must sell the \$25 dress for \$15, it *loses* \$10. Its *profit* can be expressed by

−\$10 *Read: "minus ten dollars"* ◀

Example 2 illustrates that at times positive integers do not suffice. *Negative integers* must be introduced.

Example 3 ▶ When the temperature drops below 0°C (0 degrees Celsius), you express it in terms of a negative number. For example, the temperature in Milwaukee one January morning can be expressed by

−12°C ◀

Example 4 ▶ A football team *loses* 3 yards on a play. Its *gain* can be expressed by

−3 yards ◀

DEFINITION

> The numbers −1, −2, −3, −4, . . . are known as **negative integers.**

Example 5 ▶ Among the negative integers are the following.
(a) −10 (b) −53 (c) −4027 ◀

PRACTICE EXERCISES FOR TOPIC B

Answers
3. ______
4. ______
______
______
______
5. ______

3. Water will freeze when the temperature on a Celsius thermometer is 0° (zero degrees) or below. Which of the following Celsius measurements are *freezing* temperatures?

(a) −10°C (b) 30°C (c) 1°C (d) −2°C

4. The following words are often used in describing situations that are mathematical in nature. Separate those with positive connotations from those with negative connotations:

decrease, increase; profit, loss; diminishment, increment; grow, shrink

5. A stock *loses* 4 points. Which integer expresses its *gain*?

Now try the exercises for Topic B on page 5.

C. Picturing Negative Integers

Positive integers are pictured to the *right* of 0 along the number line. It seems reasonable that negative numbers should be pictured to the *left* of 0. Thus go one distance unit to the *left* of 0. This point corresponds to the integer -1. Go one unit further to the left to locate -2. Continue to the left to locate -3, -4, -5, and so on, as in Figure 1.6.

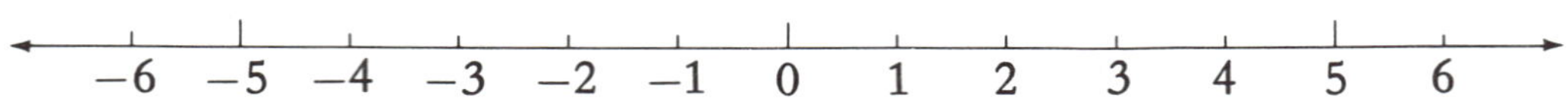

FIGURE 1.6

Example 6 ▶ Consider the number line shown in Figure 1.7(a). Label the points that correspond to the following numbers.

(a) -4 (b) -8 (c) -9 (d) -12

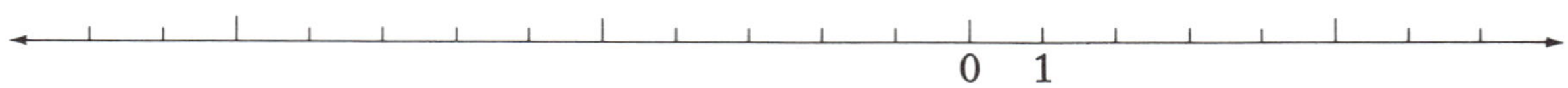

FIGURE 1.7(a)

Solution:

FIGURE 1.7(b) ◀

DEFINITION

> INTEGERS. The positive integers, 0, and the negative integers together are called the **integers.** Thus the integers are
>
> $$\ldots, -3, -2, -1, 0, 1, 2, 3, \ldots$$
>
> The three dots to the left and to the right indicate the pattern.

Example 7 ▶ (a) Which integer corresponds to the point 13 units to the right of 0?
(b) Which integer corresponds to the point 13 units to the left of 0?

Solution:

(a) 13 (b) -13 ◀

PRACTICE EXERCISES FOR TOPIC C

6. *Fill in the blanks.* On a number line positive integers are located to the _____ of zero; negative integers are located to the _____ of zero.

7. Indicate the integer that corresponds to each of the points labeled A, B, C, and D on the number line in Figure 1.8.

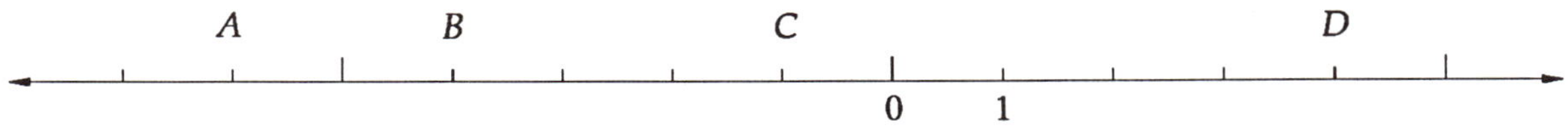

FIGURE 1.8

8. On the number line in Figure 1.9 label the points that correspond to the following integers: 2, 4, −1, −4, 6, −5

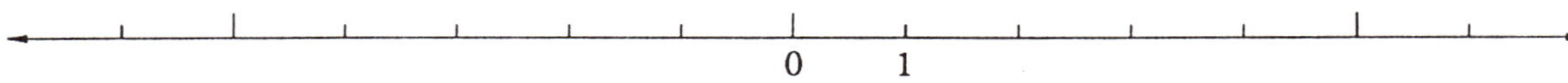

FIGURE 1.9

Now try the exercises for Topic C on page 6.

EXERCISES*

A. For Exercises 1–7, on the number line in Figure 1.10 label the points that correspond to the following numbers. (*See Example 1.*)

1. 4 **2.** 7 **3.** 8 **4.** 10

5. 9 **6.** 12 **7.** 13

FIGURE 1.10

8. In Figure 1.11 indicate the integer that corresponds to each of the points Q, R, S, T, and U.

FIGURE 1.11

Use this area for work space.

B. In Exercises 9–15 which of the following are negative integers? (*See Example 5.*)

9. 10 **10.** −2 **11.** 100 **12.** −100

13. 0 **14.** −1 **15.** −1 000 000

16. The following words are often used in describing situations that are mathematical in nature. Separate those with positive connotations from those with negative connotations:

rise, fall; lose, gain; below, above;
inflation, deflation; soar, plummet

*Answers to the odd-numbered exercises at the end of each section are given at the back of the book, beginning on page A5.

Use this area for work space.

17. A football team *loses* 5 yards on a play. Which integer expresses its *gain* (in yards)? (*See Examples 2–4.*)

18. During a clearance sale a shopkeeper sells a briefcase for $12 *below cost*. Which integer expresses his *profit* (in dollars)?

19. A merchant sells a pocketbook *at cost*. Which integer expresses her *profit* (in dollars)?

20. The temperature registers 15° below 0° (Fahrenheit). Which integer expresses this temperature reading?

21. The temperature *falls* 3° Celsius. Which integer expresses the *change* (in degrees Celsius)?

22. The Dead Sea lies 1292 feet below sea level. Which integer expresses this altitude?

C.

For Exercises 23–30, on the number line in Figure 1.12, label the points that correspond to the following numbers. (*See Example 6.*)

23. 3 **24.** 5 **25.** 9 **26.** −1

27. −3 **28.** −6 **29.** −9 **30.** −10

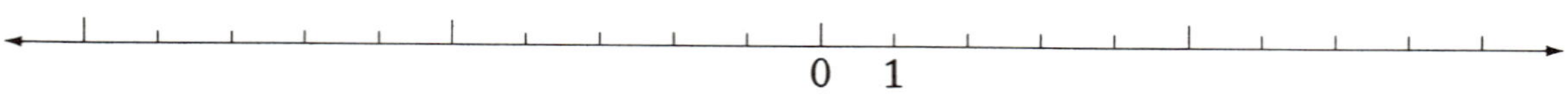

FIGURE 1.12

31. In Figure 1.13 indicate the integer that corresponds to each of the points *Q, R, S, T, U,* and *V.*

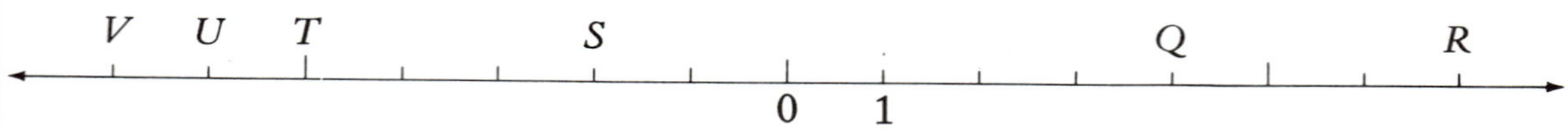

FIGURE 1.13

32. In Figure 1.14 indicate the integer that corresponds to each of the points *Q, R, S, T, U,* and *V.*

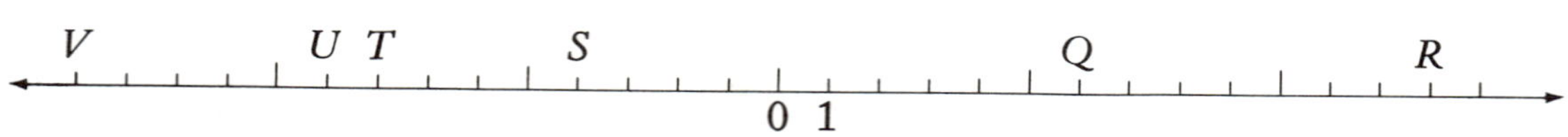

FIGURE 1.14

33. Which integer corresponds to the point 19 units to the right of 0? (*See Example 7.*)

34. Which integer corresponds to the point 19 units to the left of 0?

35. Which integer corresponds to the point 37 units to the right of 0?

36. Which integer corresponds to the point 61 units to the left of 0?

Use this area for work space.

37. Pick out the *error(s)*.

(a) 27 corresponds to the point 27 units to the right of 0.

(b) −32 corresponds to the point 32 units to the left of 0.

(c) 100 corresponds to the point 100 units to the left of 0.

(d) −200 corresponds to the point 200 units to the right of 0.

(e) Every integer either corresponds to a point to the left of 0 or to a point to the right of 0.

1.2 Order

A. Ordering Positive Integers

If you have 3 dollars, you have less money than if you have 5 dollars. In general, we say that 3 *is less than* 5. We indicate "is less than" by the symbol "<". Thus

$3 < 5$ *is read* *"3 is less than 5"*

Note that on the number line (Figure 1.15) 3 lies to the left of 5. For *any* two numbers, the *lesser* one will always be the one that lies to the left on the number line.

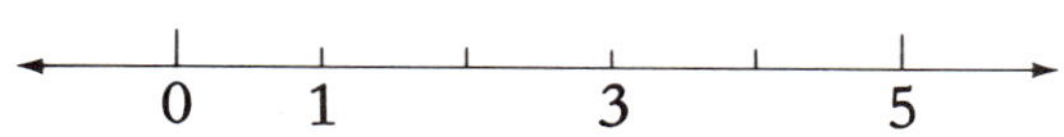

FIGURE 1.15

In mathematics, we often use letters to represent the numbers we are discussing. Instead of writing

$$\underbrace{\text{One integer}}_{a}\ \underbrace{\text{is less than}}_{<}\ \underbrace{\text{another integer,}}_{b}$$

write $a < b$

DEFINITION

> LESS THAN. Let a and b be any numbers.* If a lies to the **left** of b (on the number line), you say that ***a* is less than *b*.** In this case, you write
>
> $$a < b$$

When $a < b$, you can also say that ***b* is greater than *a*** and write

$$b > a$$

In this case b lies to the **right** of a.

*Although only integers have been introduced thus far, much of the discussion on order and on arithmetic applies to (real) numbers, in general. For this reason, many definitions and rules are stated in terms of numbers, rather than in terms of integers.

Example 1 ▶ $5 > 3$. Here, 5 lies to the right of 3. (See Figure 1.15.) ◀

Note that each of the symbols

$$<, \qquad >$$

points to the *lesser* number. Also

$$\text{if } a < b, \text{ then } b > a$$

Example 2 ▶ Fill in "<" or ">".

$$10 \ \square \ 8$$

Solution:

$$10 \ \boxed{>} \ 8$$ ◀

Example 3 ▶ Fill in "left" or "right."
5 lies to the $\square$ of 9 on the number line.
Solution:
5 lies to the $\boxed{\text{left}}$ of 9 on the number line. ◀

Answers

1. ______________

PRACTICE EXERCISES FOR TOPIC A

1. Express

4 is greater than 1

using one of the symbols "<" or ">".

2. Read each statement aloud and indicate why it is true in terms of the number line.

(a) $21 > 17$ (b) $0 < 2$

3. Fill in "<" or ">".

(a) $99 \ \square \ 100$ (b) $40 \ \square \ 37$

Now try the exercises for Topic A on page 11.

B. Ordering Negative Integers

You know that

$$2 < 5$$

But it is also true that

$$-5 < -2$$

In Figure 1.16 note that -5 lies to the left of -2. Thus -5 must be less than -2.

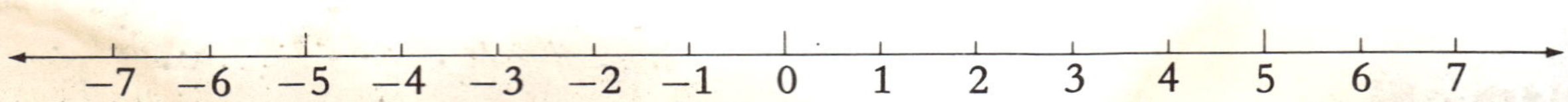

FIGURE 1.16

Example 4 ▶ Fill in "left" or "right."

-1 lies to the [ ] of -6 (on the number line).

Solution:

Refer to Figure 1.16.

-1 lies to the [right] of -6 (on the number line). ◀

Example 5 ▶ Fill in "$<$" or "$>$".

$$-7 \ \boxed{\phantom{<}}\ -3$$

Solution:

Refer to Figure 1.16.

$$-7 \ \boxed{<}\ -3$$ ◀

PRACTICE EXERCISES FOR TOPIC B

4. Fill in "left" or "right."

(a) On the number line -12 lies to the ______ of -9.

(b) On the number line -2 lies to the ______ of -3.

5. Fill in "$<$" or "$>$".

(a) $-12 \ \boxed{\phantom{<}}\ -9$ (b) $-2 \ \boxed{\phantom{<}}\ -3$

Now try the exercises for Topic B on page 12.

C. Positive Numbers and Negative Numbers

A number is *positive* if it lies to the *right* of 0, and *negative* if it lies to the *left* of 0. The number 0 is neither positive nor negative. Thus positive numbers p are numbers such that

$$p > 0$$

Negative numbers n are numbers such that

$$n < 0$$

For example,

$$7 > 0, \qquad \text{whereas } -4 < 0$$

Positive numbers and negative numbers together are called **signed numbers.** The **sign** of 5 is $+$ (*read:* "plus"); the sign of -5 is $-$ (*read:* "minus").

Every negative number n is less than every positive number p because a negative number lies to the left of a positive number (on the number line). (See Figure 1.17.)

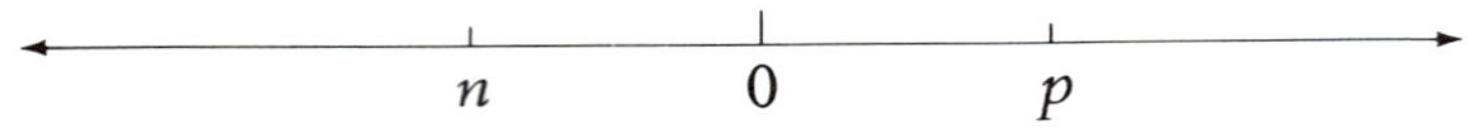

FIGURE 1.17

Example 6 ▶ (a) $-2 < 2$ (b) $-2 < 5$ (c) $-5 < 2$ (See Figure 1.18.)

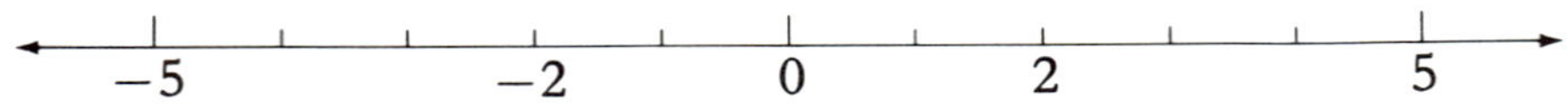

FIGURE 1.18 ◀

The number line extends indefinitely both to the right and to the left. Thus, given any number a, no matter how far to the right it lies, there is always a number b to the right of a. Hence b is greater than a. Similarly, no matter how far to the left a number c lies, there is always a number farther to the left, and thus less than c.

Let a and b be any numbers. Then according to the **Law of Trichotomy,** exactly one of the following is true:

$a = b$ (*a is equal to b.*)

$a < b$ (*a is less than b.*)

$a > b$ (*a is greater than b.*)

This is because on the number line,

a and b are the same point ($a = b$)

or a lies to the left of b ($a < b$)

or a lies to the right of b ($a > b$)

PRACTICE EXERCISES FOR TOPIC C

Answers

7. ______

______

8. ______

6. Fill in "<" or ">" and give the reason in terms of left or right on the number line.

(a) 0 ☐ −3 (b) −3 ☐ 0 (c) 7 ☐ −3 (d) −11 ☐ 10

7. Rearrange the integers so that you can write "<" between any two of them.

−8, −4, −10, −5

8. The temperature in Bangor was −5° Celsius at sunrise, and an hour later it was −10° Celsius. Did the temperature rise or fall over the course of that hour?

Now try the exercises for Topic C on page 12.

D. Applied Problems

Example 7 ► Jim owes $200; Jake owes $300. Who is wealthier?
Solution:
Jim is, because

$$-200 > -300$$ ◄

Answers

9. ______

PRACTICE EXERCISES FOR TOPIC D

9. Which is warmer? $-10°$ Celsius or $-15°$ Celsius?

Now try the exercises for Topic D on page 12.

Use this area for work space.

EXERCISES

A. In Exercises 1 and 2 express each sentence using one of the symbols "$<$" or "$>$". (*See Practice Exercise 1.*)

1. 5 is less than 7.

2. 10 is greater than 9.

In Exercises 3 and 4 express in words.

3. $8 > 5$

4. $2 < 6$

In Exercises 5–9 fill in "$<$" or "$>$". (*See Examples 1 and 2.*)

5. 2 ☐ 8

6. 3 ☐ 4

7. 9 ☐ 8

8. 16 ☐ 21

9. 53 ☐ 48

In Exercises 10–14 fill in "left" or "right." (*See Examples 1 and 3.*)

10. 4 lies to the ☐ of 9 (on the number line).

11. 10 lies to the ☐ of 3.

12. 1 lies to the ☐ of 3.

13. 12 lies to the ☐ of 9.

14. 30 lies to the ☐ of 40.

In Exercises 15–18 rearrange the integers so that you can write "$<$" between any two of them.

Sample: 2, 10, 1, 7	*Solution:* $1 < 2 < 7 < 10$

15. 5, 18, 12, 13

16. 40, 10, 19, 24

17. 18, 20, 14, 12, 8

18. 36, 24, 55, 48, 39

Use this area for work space.

B. In Exercises 19 and 20 express each sentence using one of the symbols "$<$ or "$>$".

19. -8 is less than -6.

20. -3 is greater than -4.

In Exercises 21–27 fill in "left" or "right." (*See Example 4.*)

21. -4 lies to the ______ of -5.

22. -5 lies to the ______ of -3.

23. -2 lies to the ______ of -6.

24. -10 lies to the ______ of -7.

25. -24 lies to the ______ of -26.

26. -3 lies to the ______ of -6.

27. -10 lies to the ______ of -20.

In Exercises 28–34 fill in "$<$" or "$>$". (*See Example 5.*)

28. -2 ______ -1

29. -8 ______ -3

30. -9 ______ -12

31. -10 ______ -5

32. -11 ______ -13

33. -20 ______ -6

34. -12 ______ -14

C. In Exercises 35–44 fill in "$<$" or "$>$" and give the reason in terms of left or right on the number line. (*See Example 6.*)

35. 4 ______ -4

36. 2 ______ -9

37. -14 ______ 12

38. -1 ______ 16

39. -16 ______ 1

40. -16 ______ -1

41. 0 ______ 13

42. 0 ______ -13

43. -7 ______ 0

44. 1 ______ -1000

In Exercises 45–51 rearrange the integers so that you can write "$<$" between any two of them.

45. $-1, -2, -3$

46. $-8, -9, -7, -10$

47. $-3, -6, 3, 6$

48. $-2, -4, 0, 4$

49. $-5, -3, 1, -7, 7$

50. $-12, 12, -3, 3, -4, 4$

51. $-15, -18, -3, 15, 18, -10, -4$

52. Pick out the *error(s)*.

(a) $-10 < 4$ (b) $-7 < 9$ (c) $-9 < -12$

(d) $1 > -4$ (e) $2 < -1$

D. **53.** Karen is 62 inches tall. Marilyn is 65 inches tall. Who is taller? (*See Example 7.*)

54. Alex is \$500 in debt. Sarah is \$250 in debt. Who is wealthier?

55. At noon the temperature in Duluth is $-4°$ Celsius. An hour later the temperature there is $-8°$ Celsius. **Has the temperature risen or fallen?**

56. After the first inning the Pirates trail the Reds by 3 runs. After the second inning the Pirates trail by 5 runs. Which team scored more runs in the second inning?

Use this area for work space.

57. In a mine Steve works 2300 feet below ground level and Joe works 1800 feet below ground level. Which miner works above the other?

58. Which date is earlier: 200 B.C. or 100 B.C.?

1.3 Inverse and Absolute Value

A. Inverse

The numbers 5 and -5 lie on opposite sides of the origin, 0, but are the same distance from 0. The *positive* number 5 lies 5 units to the *right* of the origin, whereas the *negative* number -5 lies 5 units to the *left* of the origin. (See Figure 1.19.)

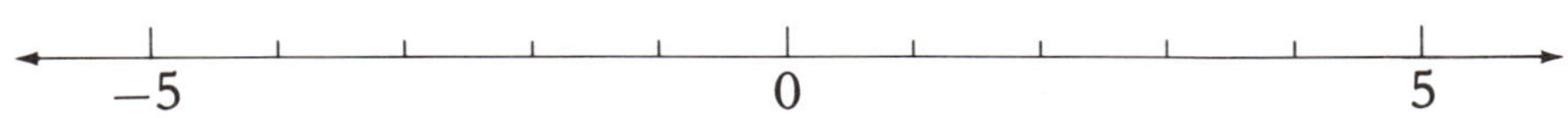

FIGURE 1.19

DEFINITION

INVERSE. Let a be a number other than 0. The **additive inverse of a,** or for short, the **inverse of a,** is the number that lies on the opposite side of the origin and is the same distance from the origin as is a. The **(additive) inverse of 0** is 0 itself.

For every number a, let

$$-a$$

denote the inverse of a. Thus -5 is the inverse of 5.

Example 1 ► (a) The inverse of 10 is -10.

(b) The inverse of -10 is 10. In order to denote the inverse of -10, we use parentheses, and write

$$\underbrace{-}_{\text{the inverse of}}(-10) = 10$$ ◄

In general, for any number a,

$$-(-a) = a$$

Observe that *the inverse of a positive number is a negative number. The inverse of a negative number is a positive number.*

Answers

1. (a) ______
(b) ______
(c) ______
(d) ______
(e) ______
2. ______

PRACTICE EXERCISES FOR TOPIC A

1. Write the additive inverse of each:

(a) 19 (b) -43 (c) 0 (d) $-(-9)$ (e) k, where k is negative

2. Does any integer have more than one additive inverse?

Now try the exercises for Topic A on page 16.

B. Absolute Value

DEFINITION

ABSOLUTE VALUE. Let a be any number. The **absolute value of a** is its distance from the origin.

Let

$$|a|$$

denote the absolute value of a.

Example 2 ▶ (See Figure 1.20.)
(a) $|7| = 7$ because 7 lies 7 units from the origin.
(b) $|-7| = 7$ because -7 lies 7 units from the origin.
(c) $|0| = 0$ ◀

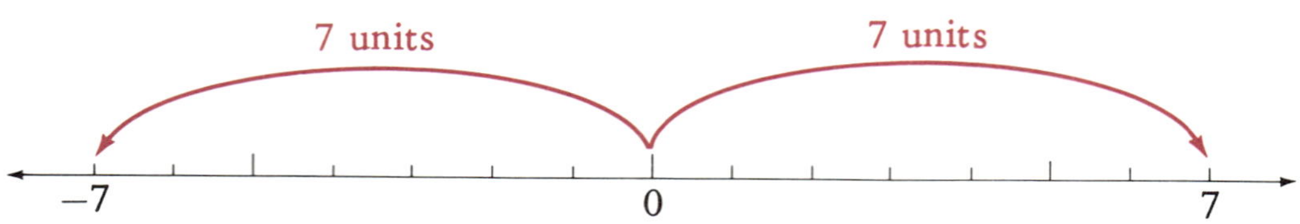

FIGURE 1.20

For every number a,

$$|a| = |-a|$$

Both numbers are the same distance from the origin, but on opposite sides of it. Thus *absolute value measures distance from the origin, but neglects direction.*

Observe that

$$|a| = a \qquad \textit{if } a \textit{ is positive or } 0$$

$$|a| = -a \qquad \textit{if } a \textit{ is negative}$$

In fact, if a is *negative,* a lies to the *left* of the origin, and $-a$ lies to the *right* of the origin. Thus, $-a$, which is *positive,* is the distance from a to the origin.

Example 3 ▶ (a) $|4| = 4$
(b) $|-4| = -(-4) = 4$ ◀

Example 4 ▶ A Datsun gets 25 miles to the gallon. It uses the same gallon of gas to travel 25 miles east as it would to travel 25 miles west. (See Figure **1.21.**) Thus gas consumption is measured in terms of absolute value. You consider distance but neglect direction. ◀

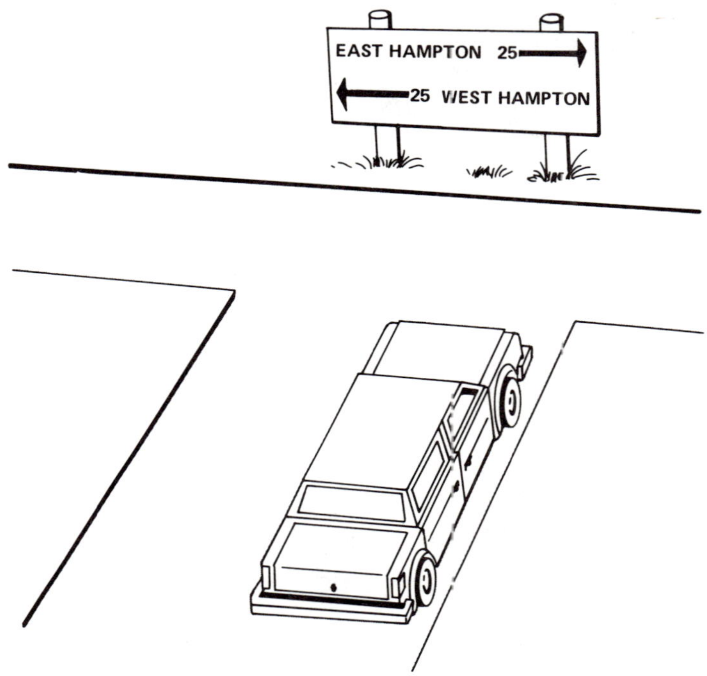

FIGURE 1.21

Example 5 ▶ Fill in "<" or ">".

$$|-2| \; \square \; |-5|$$

Solution:

$$|-2| = 2, \qquad |-5| = 5$$

$$|-2| < |-5| \qquad \text{because } 2 < 5$$ ◀

Answers

3. ____________

4. (a) ____________

(b) ____________

PRACTICE EXERCISES FOR TOPIC B

3. Which integers are 6 units from the origin?

4. Find each absolute value:

(a) $|10|$ **(b)** $|-36|$ **(c)** $|-7|$

(c) ______

5. (a) ______

(b) ______

(c) ______

5. Simplify:

(a) $-|-4|$ (b) $-(-13)$ (c) $-|13|$

6. Fill in "<" or ">".

(a) $4 \square |-5|$ (b) $|-5| \square |-4|$

Now try the exercises for Topic B on this page.

Use this area for work space.

EXERCISES

A. In Exercises 1–12 write the inverse of each number. (*See Example 1.*)

1. 2 **2.** 9 **3.** 12 **4.** 20

5. -2 **6.** -9 **7.** -12 **8.** -20

9. 0 **10.** 100 **11.** $-(-3)$ **12.** $-a$

13. Which integer is its own inverse?

14. Pick out the *errors(s)*.

(a) $-(-5) = 5$ (b) $-0 = 0$ (c) $-(-2) = -2$

(d) -7 is the inverse of 7. (e) 9 is the inverse of -9.

B. In Exercises 15–26 find each absolute value. (*See Examples 2 and 3.*)

15. $|8|$ **16.** $|1|$ **17.** $|17|$ **18.** $|32|$

19. $|-8|$ **20.** $|-1|$ **21.** $|-17|$ **22.** $|-32|$

23. $|0|$ **24.** $|100|$ **25.** $|-100|$ **26.** $|-5000|$

In Exercises 27–34 simplify.

Sample: $-\|-4\|$	*Solution:* $\|-4\| = 4$ and therefore $-\|-4\| = -4$

27. $-|7|$ **28.** $-|-7|$ **29.** $|-7|$ **30.** $-(-7)$

31. $|-0|$ **32.** $-|0|$ **33.** $-(-12)$ **34.** $|-(-12)|$

Sample: (For Exercises 35–38). Which integers are 5 units from the origin?	*Solution:* 5 and -5

35. Which integers are 8 units from the origin?

36. Which integers are 15 units from the origin?

37. Which positive integer is 11 units from the origin?

38. Which negative integer is 13 units from the origin?

Use this area for work space.

39. Which integer is further from the origin, 27 or -44?

40. Which integer is further from the origin, 32 or -29?

41. Which integers are less than 2 units from the origin?

42. Which integers are less than 10 units from the origin?

In Exercises 43–50 fill in "$<$" or "$>$". *(See Example 5.)*

43. $|-2|$ ☐ 0

44. $|-7|$ ☐ 6

45. 8 ☐ $|-3|$

46. -10 ☐ $|-12|$

47. $|-3|$ ☐ $|-5|$

48. $|-9|$ ☐ $|8|$

49. 100 ☐ $|-200|$

50. $-|10|$ ☐ $-|-5|$

51. Pick out the *error(s)*.

(a) $|-11| > -11$ (b) $|16| > 16$ (c) $|-3| = -(-3)$

(d) 17 is further from the origin than -13 is.

(e) -2 is closer to the origin than 4 is.

52. Betty is in a subway train 30 feet below ground level. At the same time Bill is in an elevated train 20 feet above ground level. Who is closer to ground level?

1.4 Adding and Subtracting Signed Numbers*

A. Adding and Subtracting on the Number Line

Let a and b be any numbers†—positive, negative, or 0. Then

$$a + b \qquad \text{read: "}a\text{ plus }b\text{"}$$

stands for the **sum** of a and b.

We assume that you are familiar with addition of positive integers.

Subtraction can be defined in terms of addition. Let a and c be positive numbers and, for the present, assume that either $a = c$ or $a > c$. Define $a - c$ (*a minus c*) as follows

$$a - c = b \quad \text{if} \quad a = b + c$$

For example,

$$5 - 3 = 2 \quad \text{because} \quad 5 = 2 + 3 \text{ (See Figure 1.22.)}$$

*For more examples and exercises on basic arithmetic, see M. Zuckerman, *Student Workbook to Accompany Elementary Algebra: A Straightforward Approach* (3rd ed.), Sections 1.4 and 1.5.

†The definitions and properties of arithmetic in the remainder of the chapter apply to real numbers, in general. However, the examples given will concern integers.

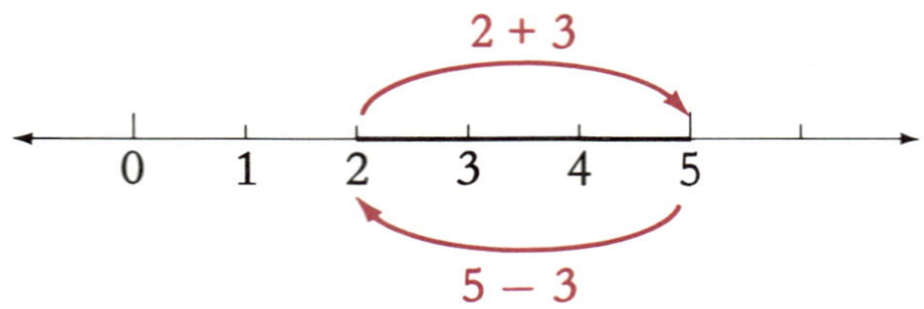

FIGURE 1.22

Also $$8 - 8 = 0 \quad \text{because} \quad 8 = 0 + 8$$

Thus subtraction "undoes" addition.

Observe that adding a *positive* number to a given number, a, yields a sum that is *greater* than a. For example, $2 + 3$, which equals 5, lies 3 units to the *right* of 2. (See Figure 1.22.) Thus, adding the *positive* number 3 to 2 yields a *greater* value than 2. The basic idea involved in extending addition to signed numbers will be that adding a *positive* number to a given number, a, results in a value *greater* than a, whereas adding a *negative* number to a results in a value *less* than a. Thus adding a *negative* number will correspond to moving to the *left* on the number line.

To *add negative numbers:*

1. Add their absolute values.
2. Prefix the sum with a minus sign.

For example,

$$(-5) + (-3) = -(5 + 3) = -8$$

In Figure 1.23 note that -8 lies 3 units to the *left* of -5.

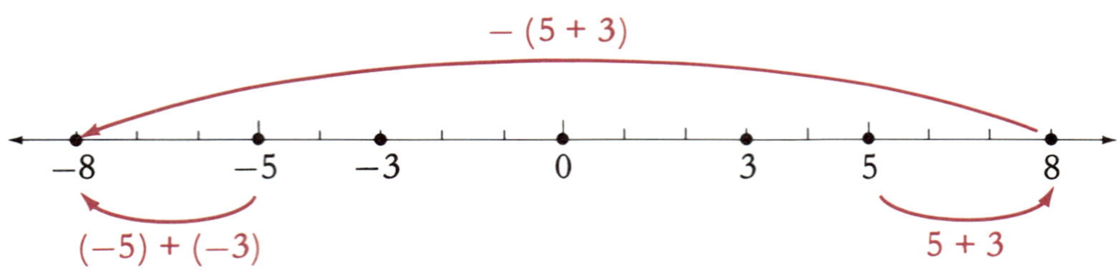

FIGURE 1.23

Example 1 ► (a) $(-6) + (-9) = -(6 + 9) = -15$
(b) $(-10) + (-20) = -(10 + 20) = -30$
(c) $(-4) + (-5) + (-3) = -(4 + 5 + 3) = -12$ ◄

To add $-4 + 4$, or $4 + (-4)$, recall that 4 and -4 are each the same distance from 0, but are on opposite sides of 0. Thus, to obtain -4 **+ 4,** move 4 units to the **right** of -4. This results in 0. Similarly 4 **+ (−4)** corresponds to moving 4 units to the **left** of 4. (See Figure 1.24.) Again the result is 0. In general, for any number a,

$$a + (-a) = (-a) + a = 0$$

Thus *the inverse of a is the number added to a to obtain* 0.

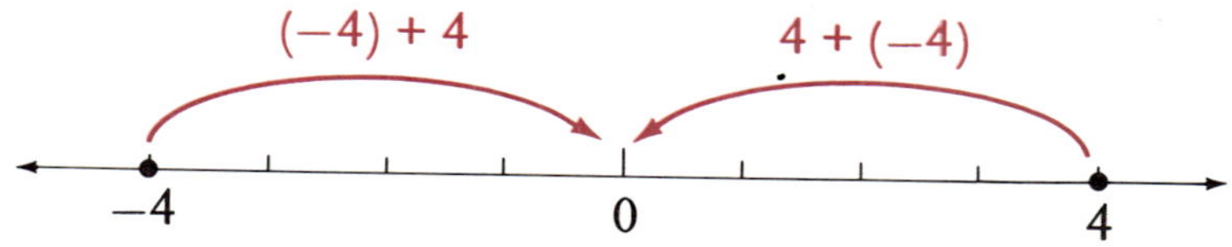

FIGURE 1.24

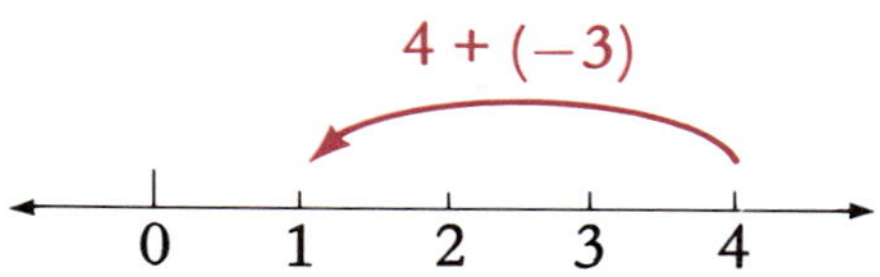

FIGURE 1.25(a)

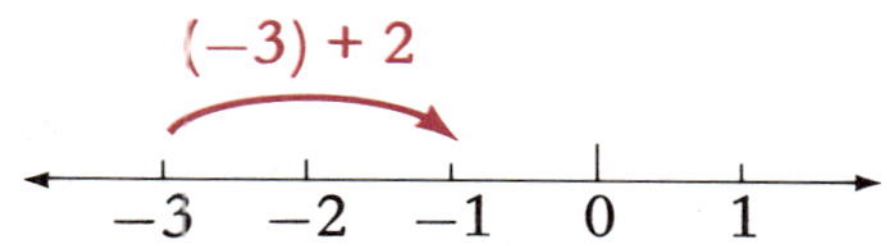

FIGURE 1.25(b)

To see how *to add numbers with different signs and different absolute values,* consider **Figure** 1.25.

Figure 1.25(a) indicates that 4 **+ (−3)** lies 3 units to the **left** of 4. Thus

$$4 + (-3) = 1$$

Note that

$$|4| > |-3|$$

Here the larger absolute value is that of the *positive* number and $4 + (-3)$ is *positive.*

Figure 1.25(b) shows that -3 **+ 2** lies 2 units to the **right** of -3. Thus

$$-3 + 2 = -1$$

Observe that

$$|-3| > 2$$

Here the larger absolute value is that of the *negative* number and $-3 + 2$ is *negative.*

To sum up, when adding numbers with *different* signs and *different* absolute values:

1. Subtract the smaller absolute value from the larger absolute value.
2. Prefix the difference with the sign of the number with the *larger* absolute value.

Thus, to obtain $4 + (-3)$ by this rule:

1. $\underbrace{|4|}_{4} - \underbrace{|-3|}_{3} = 4 - 3 = 1$
2. Because $|4| > |-3|$, prefix the difference, 1, with the sign of 4, which is +. (Usually, + is not written.) Thus

$$4 + (-3) = 4 - 3 = 1$$

To obtain $-3 + 2$ by this rule:

1. $\underbrace{|-3|}_{3} - \underbrace{|2|}_{2} = 3 - 2 = 1$
2. Because $|-3| > |2|$, prefix the difference, 1, with the sign of -3, which is $-$.

$$-3 + 2 = -(3 - 2) = -1$$

Example 2 ► Add.

(a) $9 + (-5)$ (b) $6 + (-12)$

Solution:

(a) $\underbrace{|9|}_{9} > \underbrace{|-5|}_{5}$

Prefix the difference, $9 - 5$, with the sign of 9. Thus

$$9 + (-5) = 9 - 5 = 4$$

(b) $\underbrace{|-12|}_{12} > \underbrace{|6|}_{6}$

Prefix the difference, $12 - 6$, with the sign of -12. Thus

$$6 + (-12) = -(12 - 6) = -6$$ ◄

Now that addition has been defined for any two numbers, subtraction can also be defined for any two numbers.

DEFINITION

> Let a and b be *any* two numbers. Define
>
> $$a - b = a + (-b)$$

Thus subtracting a number yields the same result as adding the inverse. For this reason, the same symbol $-$ is used for expressing $-b$, *the inverse of* b, as for expressing $a - b$, *the difference, a minus b.*

Example 3 ► (a) $5 - 3 = 5 + (-3) = 2$

Note that $5 - 3$ or 5 **+ (−3),** lies to the **left** of 5. (See Figure 1.26.)

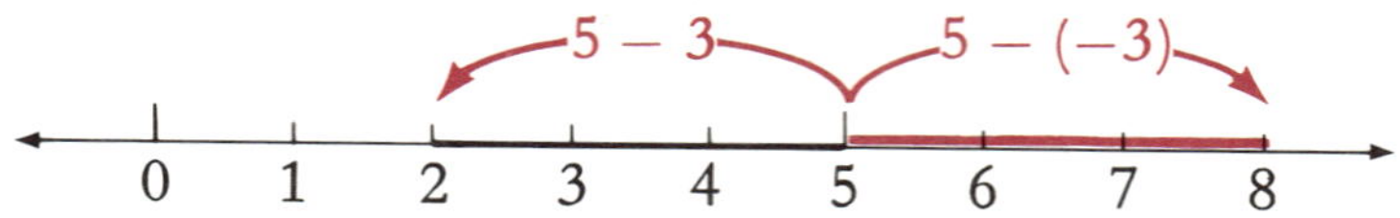

FIGURE 1.26

(b) $5 - (-3) = 5 + 3 = 8$

Note that $5 - (-3)$ or 5 **+ 3** lies to the **right** of 5. ◄

For every number a,

$$a + 0 = 0 + a = a \quad \text{and} \quad a - 0 = a$$

whereas

$$0 - a = 0 + (-a) = -a$$

Answers

1. (a) ________
   (b) ________
   (c) ________
   (d) ________
2. (a) ________
   (b) ________
   (c) ________
   (d) ________

PRACTICE EXERCISES FOR TOPIC A

1. Add each pair of signed numbers:

(a) $14 + 9$ (b) $(-14) + (-9)$ (c) $6 + (-13)$ (d) $-8 + 25$

2. Subtract as indicated:

(a) $-3 - (-7)$ (b) $-3 - 7$ (c) $12 - 20$ (d) $24 - (-6)$

3. Figure 1.27 illustrates:

☐ + ☐ = ☐

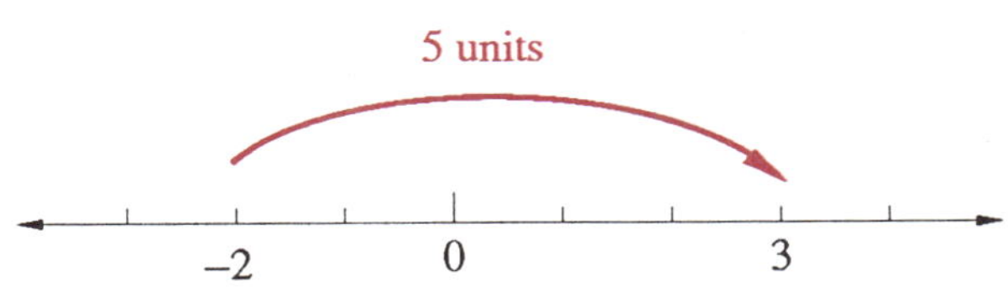

FIGURE 1.27

Now try the exercises for Topic A on page 23.

B. Commutative and Associative Laws

You can add two numbers in either order without changing the sum.

Let a and b be any numbers. Then

$$a + b = b + a$$

This is called the **Commutative Law of addition.**

Let a, b, and c be any numbers. When you first add a and b, and then add c, you obtain the same sum as when you first add b and c, and then add this to a:

$$(a + b) + c = a + (b + c)$$

This is called the **Associative Law of addition.** Because of this equality the final sum is often written

$$a + b + c$$

without parentheses.

Example 4 ► (a) $4 + 3 = 3 + 4 = 7$ by the Commutative Law of addition.

(b) $\underbrace{(3 + 6)}_{9} + 2 = 3 + \underbrace{(6 + 2)}_{8} = 11$ by the Associative Law (of addition).

(c) $[\underbrace{(-4) + (-2)}_{-6}] + (-3) = (-4) + [\underbrace{(-2) + (-3)}_{-5}] = -9$ by the Associative Law. ◄

To add three or more numbers, not all of the same sign, use the Commutative and Associative Laws, as in Examples 5 and 6, which follow:

1. Rearrange, and add the positive numbers and the negative numbers separately.
2. Let A be the sum of the positive numbers, and $-B$ be the sum of the negative numbers, where $A > 0$ and $B > 0$. Then the total sum is $A + (-B)$, or $A - B$.

Example 5 ► $2 + (-4) + 5 = 2 + [(-4) + 5]$

$= 2 + [5 + (-4)]$ by the Commutative Law

$= [2 + 5] + (-4)$ by the Associative Law

$= \underset{A}{7} + \underset{(-B)}{(-4)}$

$= 3$ ◄

Example 6 ► Add

$$\begin{array}{r} 24 \\ -49 \\ 29 \\ 37 \\ -30 \\ -42 \\ \hline \end{array}$$

Solution:
Rearrange, and add the positive numbers and the negative numbers separately.

$$\begin{array}{r} 24 \\ 29 \\ 37 \\ \hline 90 \end{array} \qquad \begin{array}{r} -49 \\ -30 \\ -42 \\ \hline -121 \end{array} \qquad \begin{array}{r} -121 \\ 90 \\ \hline -\ 31 \end{array}$$

The resulting sum, $90 + (-121)$, equals -31. ◄

Answers

4. (a) ______
(b) ______
5. (a) ______
(b) ______
(c) ______

PRACTICE EXERCISES FOR TOPIC B

4. Add to show that both sums are equal. Also indicate which law is being illustrated.

(a) $(-3) + 2 = 2 + (-3)$ (b) $[7 + (-4)] + 5 = 7 + [(-4) + 5]$

5. Find each sum.

(a) $7 + (-12) + 6$ (b) $(-22) + 30 + (-5)$ (c) $15 + (-8) + (-6) + 4 + (-10)$

Now try the exercises for Topic B on page 25.

C. Applied Problems

Example 7 ▶ The temperature in Detroit at 5 a.m. is $-13°$ Celsius. If the temperature rises 4° the next hour, what is the reading at 6 a.m.?

Solution:

A *rise* in temperature indicates adding 4 degrees to the original reading. Thus

$$-13 + 4 = -9$$

At 6 a.m. the reading is $-9°$ Celsius. ◀

Answers

6. ______

PRACTICE EXERCISES FOR TOPIC C

6. A stock rises 7 points on Monday, falls 10 points on Tuesday, and rises 8 points on Wednesday. In all, how much does it rise or fall for these three days?

Now try the exercises for Topic C on page 25.

Use this area for work space.

EXERCISES

A. In Exercises 1–20 add or subtract, as indicated. (*See Examples 1, 2, and 3.*)

1. $12 + 5$ **2.** $12 - 5$

3. $6 + 0$ **4.** $6 - 0$

5. $6 - 6$ **6.** $0 - 0$

7. $(-3) + (-7)$ **8.** $(-12) + (-6)$

9. $8 + (-5)$ **10.** $(-8) + 5$

11. $(-12) + 14$ **12.** $(-13) + 7$

13. $-30 + 50$ **14.** $30 + (-50)$

15. $3 - 6$ **16.** $-3 - 6$

17. $-7 - (-2)$ **18.** $-9 - (-9)$

19. $0 - 4$ **20.** $0 - (-4)$

In Exercises 21–26 fill in the integers. (See Figures 1.23–1.26 on pages 18–20 to see how addition and subtraction are represented on the number line.)

21. Figure 1.28 illustrates:

$$\square + \square = \square$$

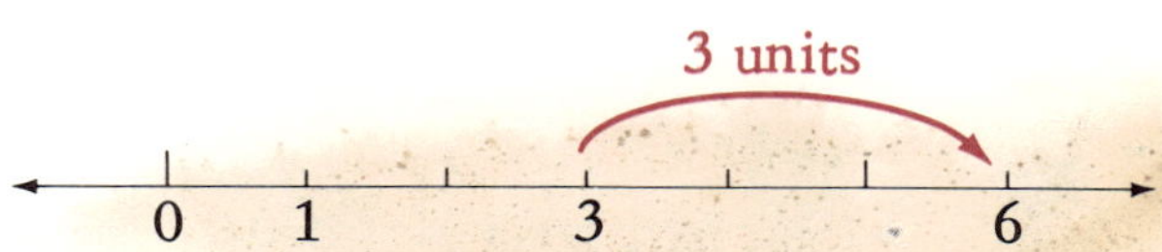

FIGURE 1.28

Use this area for work space.

22. Figure 1.29 illustrates:

□ + □ = □

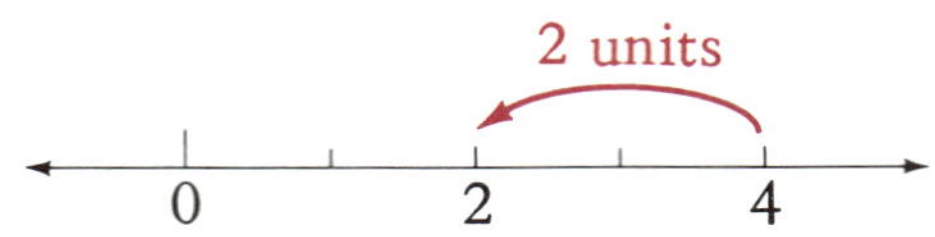

FIGURE 1.29

23. Figure 1.30 illustrates:

□ + □ = □

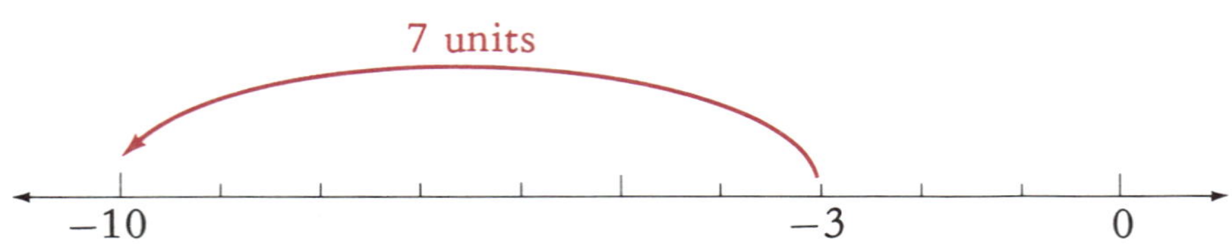

FIGURE 1.30

24. Figure 1.31 illustrates:

□ − □ = □

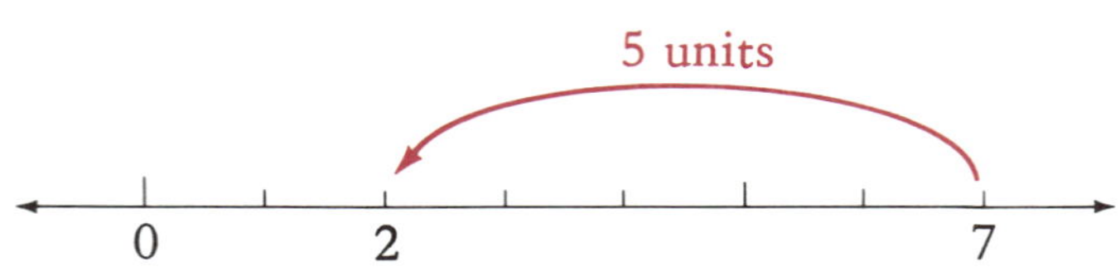

FIGURE 1.31

25. Figure 1.32 illustrates:

□ − □ = □

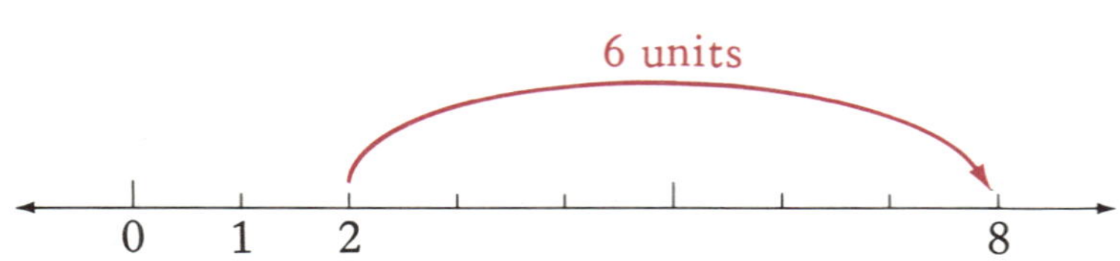

FIGURE 1.32

26. Figure 1.33 illustrates:

□ − □ = □

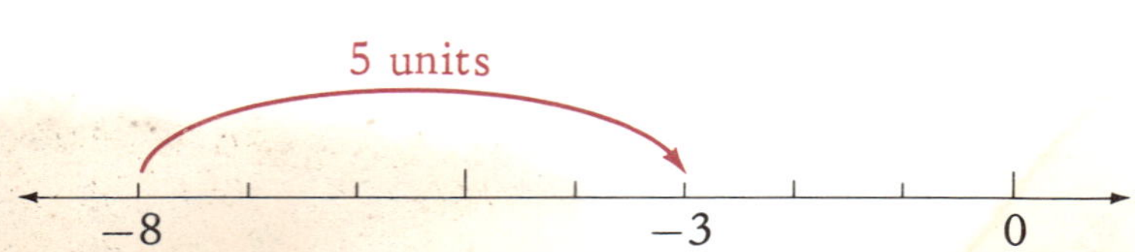

FIGURE 1.33

Use this area for work space.

27. Subtract 17 from -24.

28. Subtract -12 from 19.

B. In Exercises 29–34 add to show that both sums (a) and (b) are equal. Also indicate which law is being illustrated. (*See Example 4.*)

29. (a) $4 + 9$ (b) $9 + 4$

30. (a) $(-5) + (-6)$ (b) $(-6) + (-5)$

31. (a) $(-7) + 5$ (b) $5 + (-7)$

32. (a) $(8 + 3) + 6$ (b) $8 + (3 + 6)$

33. (a) $[(-2) + (-9)] + (-10)$ (b) $(-2) + [(-9) + (-10)]$

34. (a) $[5 + (-8)] + 7$ (b) $5 + [(-8) + 7]$

In Exercises 35–46 add or subtract, as indicated. (*See Examples 5 and 6.*)

35. $(-5) + (-3) + (-10) + (-2)$

36. $(-9) + (-12) + (-20) + (-5) + (-7)$

37. $9 + 8 + (-3) + (-7)$

38. $(-4) + (-3) + 8 + 12 + (-9)$

39. $(-40) + (-25) + 20 + (-60) + 30 + (-35)$

40. $-5 - 2 + 7 + 11 - 8$

41. $50 - 15 + 40 - 80 + 20$

42.	43.	44.	45.	46.
20	17	108	−209	−1062
−48	9	−205	−412	− 713
26	−25	99	−444	981
31	−24	429	389	1429
−19	−31	−308	117	−8101
−27	−39	−398	−604	

C. **47.** A carpenter earns \$27 on Monday, \$32 on Tuesday, \$29 on Wednesday, \$40 on Thursday, and \$36 on Friday. How much does he earn for this 5-day week? (*See Example 7.*)

48. Maria buys a dress for \$37, a hat for \$8, a pair of shoes for \$26, and a scarf for \$4. How much does she spend on her wardrobe?

49. Pamela's homework assignment is to read the first 30 pages of *Moby Dick*. She has already read 20 pages. How many more pages must she read?

50. At the track Ralph wins \$9 on one race and loses \$4 on the next race. Altogether, how much money has he gained?

51. The temperature in Omaha at 8 a.m. is $-10°$ Celsius. If the temperature drops 5 degrees the next hour, what is the reading at 9 a.m.?

52. The temperature in Chicago at noon is 12° Celsius. By midnight the temperature has dropped 20 degrees. What is the reading at midnight?

53. The temperature in Seattle rises from $-8°$ Celsius to $-2°$ Celsius. How many degrees does it rise?

Use this area for work space.

54. A car travels 40 miles north and then 50 miles south. How far from the starting point is the car? Is it north or is it south of the starting point?

55. A storekeeper is forced to sell a $55 item for $38. By how much has he lowered the price?

56. How much of a rise in altitude is there in climbing from 100 feet below sea level to 400 feet above sea level?

1.5 Multiplying and Dividing Signed Numbers

A. Multiplying

Let a and b be any numbers. The **product** of a and b, or a **times** b, can be written in various ways:

$$a \cdot b, \qquad ab, \qquad (a)(b), \qquad a(b), \qquad (a)b, \qquad a \times b$$

When two or more numbers are multiplied, they are each called **factors** of the product. Thus a and b are each factors of $a \cdot b$. When the factor b is a *positive integer*, multiplication by b amounts to repeated addition.

DEFINITION

Let a be any number and let b be a positive integer. Then

$$a \cdot b \quad \text{means} \quad \underbrace{a + a + a + \cdots + a}_{b \text{ times}}$$

Example 1 ► Refer to Figure 1.34.

(a) $2 \cdot 3 = 2 + 2 + 2 = 6$

(b) $(-4)2 = (-4) + (-4) = -8$ ◄

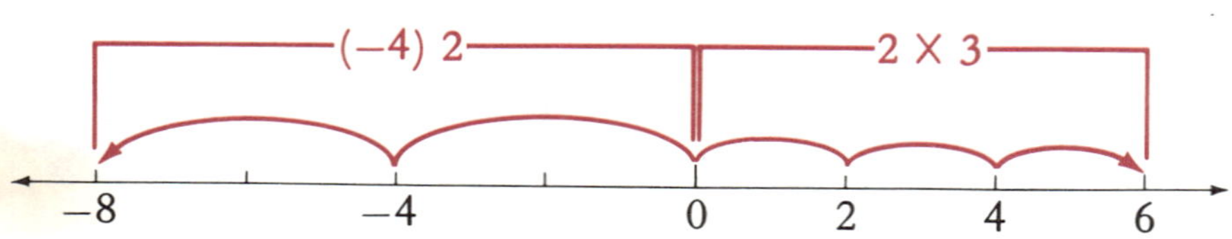

FIGURE 1.34

In order to discover the rules for multiplying by 0 and by a negative number, fill in the blanks in Table 1.1(a) on p. 27 by following the pattern in the right-hand column. Refer to Figure 1.35 on p. 27.

TABLE 1.1(a)

10×4	$=$	40
10×3	$=$	30
10×2	$=$	20
10×1	$=$	10
10×0	$=$	
$10 \times (-1)$	$=$	
$10 \times (-2)$	$=$	
$10 \times (-3)$	$=$	
$10 \times (-4)$	$=$	

Observe that when you multiply 10 by a positive integer, each time the right-hand factor decreases by 1, the product decreases by 10. It seems natural to complete the table as in Table 1.1(b).

TABLE 1.1(b)

10×4	$=$	40
10×3	$=$	30
10×2	$=$	20
10×1	$=$	10
10×0	$=$	0
$10 \times (-1)$	$=$	-10
$10 \times (-2)$	$=$	-20
$10 \times (-3)$	$=$	-30
$10 \times (-4)$	$=$	-40

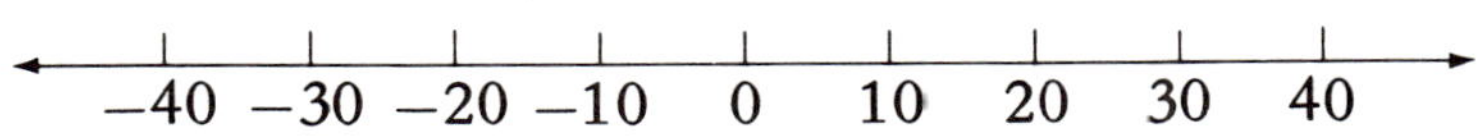

FIGURE 1.35

In general, if a and b are both positive, then

$$a(-b) = -(ab)$$

Moreover, for every number a,

$$a \cdot 1 = a \quad \text{and} \quad a \cdot 0 = 0$$

Example 2 ▶ (a) $5(-3) = -(5 \cdot 3) = -15$
(b) $12(-1) = -(12 \cdot 1) = -12$
(c) $9 \cdot 0 = 0$ ◀

In order to discover the rule for multiplying two negative numbers, fill in the blanks in Table 1.2(a) by considering the pattern in the right-hand column. Again refer to Figure 1.35.

TABLE 1.2(a)

$(-10) \times 4$	$= -40$
$(-10) \times 3$	$= -30$
$(-10) \times 2$	$= -20$
$(-10) \times 1$	$= -10$
$(-10) \times 0$	$= 0$
$(-10) \times (-1)$	$=$
$(-10) \times (-2)$	$=$
$(-10) \times (-3)$	$=$
$(-10) \times (-4)$	$=$

When multiplying -10 by a positive integer or by 0, each time the factor on the right *decreases* by 1, the product *increases* by 10. It seems natural to complete the table as in Table 1.2(b).

TABLE 1.2(b)

$(-10) \times 4$	$= -40$
$(-10) \times 3$	$= -30$
$(-10) \times 2$	$= -20$
$(-10) \times 1$	$= -10$
$(-10) \times 0$	$= 0$
$(-10) \times (-1)$	$= 10$
$(-10) \times (-2)$	$= 20$
$(-10) \times (-3)$	$= 30$
$(-10) \times (-4)$	$= 40$

In general, if a and b are positive numbers, then

$$(-a)(-b) = ab$$

Example 3 ▶ (a) $(-5)(-2) = 5 \cdot 2 = 10$
(b) $(-4)(-1) = 4 \cdot 1 = 4$
(c) $(-12)(-6) = 12 \cdot 6 = 72$ ◀

Multiplication is both commutative and associative. Thus let a, b, and c be any numbers.

$$a \cdot b = b \cdot a$$ **(Commutative Law of multiplication)**

$$(a \cdot b) \cdot c = a \cdot (b \cdot c)$$ **(Associative Law of multiplication)**

Because of the Associative Law, the final product is often written

$$a \cdot b \cdot c$$

without parentheses.

Example 4 ▶ (a) $4 \cdot 3 = 3 \cdot 4 = 12$ by the Commutative Law of multiplication.

(b) $\underbrace{(5 \cdot 2)}_{10} \cdot 3 = 5 \cdot \underbrace{(2 \cdot 3)}_{6} = 30$ by the Associative Law (of multiplication).

(c) $[\underbrace{(-4) \cdot 5}_{-20}] \cdot (-1) = (-20)(-1) = 20;$

$(-4) \cdot [\underbrace{5(-1)}_{-5}] = (-4)(-5) = 20$

Thus $[(-4) \cdot 5] \cdot (-1) = (-4) \cdot [5(-1)]$ by the Associative Law. ◀

It is worth mentioning that

$$a(-b) = -ab, \qquad (-a)b = -ab, \qquad \text{and} \qquad (-a)(-b) = ab$$

for all numbers a and b, regardless of their signs.

Answers

1. (a) ______
(b) ______
(c) ______
2. (a) ______
(b) ______

PRACTICE EXERCISES FOR TOPIC A

1. Multiply.

(a) 3×32 (b) $9(-4)$ (c) $(-20)(-1)$

2. Multiply to show that both products are equal. Indicate which law is being illustrated.

(a) $(-5) \times 4 = 4 \times (-5)$ (b) $2 \times [7 \times (-3)] = (2 \times 7) \times (-3)$

Now try the exercises for Topic A on page 33.

B. Sign of a Product

Example 5 ▶ (a) $6 \cdot 4 = 24$ (b) $6(-4) = -24$
(c) $(-6)4 = -24$ (d) $(-6)(-4) = 24$ ◀

When two or more numbers are multiplied, *the product is zero if at least one of the factors is zero*. Otherwise, the product is nonzero. Suppose none of the factors is zero. You can multiply in any order. The product of two negative numbers is positive. A third negative factor yields a negative product; a fourth negative factor yields a positive product, and so on.

1. The product is *positive* if there is an *even* number (0, 2, 4, 6, . . .) of negative factors.
2. The product is *negative* if there is an *odd* number (1, 3, 5, . . .) of negative factors.

Example 6 ▶ (a) $\underbrace{\underbrace{(-2)(-1)}_{2}(-2)}_{-4} = -4$

There are three negative factors. The product is negative

(b) $\underbrace{\underbrace{\underbrace{(-3)(-1)}_{3}(-2)}_{-6}(-1)}_{6} = 6$

There are four negative factors. The product is positive.

(c) $8(-2)(-1) = 16$

$\quad -16 \quad$
$\qquad -16$

Two of the factors are negative. The product is positive.

(d) $(-2)(-5)0(-7)10 = 0$
One of the factors is zero. The product is therefore zero. ◀

Answers

3. (a) ____

(b) ____

(c) ____

(d) ____

PRACTICE EXERCISES FOR TOPIC B

3. Find each product.

(a) $10(-2)(3)$ (b) $(-10)(2)(-3)$ (c) $10(-2)(-3)(-1)$ (d) $(-10)(-2)(-3)(0)$

Now try the exercises for Topic B on page 34.

C. Dividing

Division "undoes" multiplication. Figure 1.36 illustrates that division is defined in terms of multiplication, just as subtraction was defined in terms of addition.

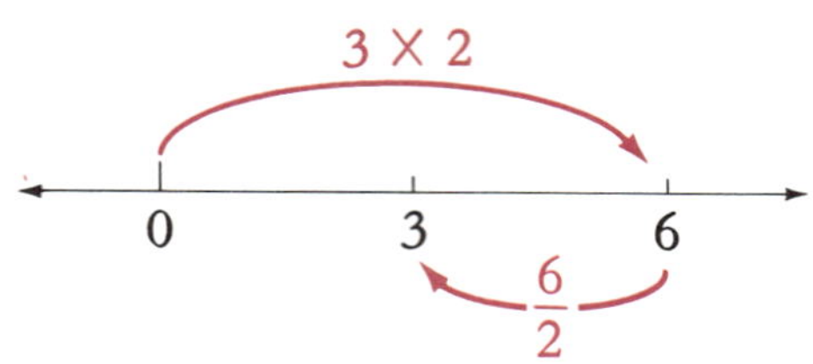

FIGURE 1.36

a divided by b is written:

$$\frac{a}{b}, \qquad a \div b, \qquad \text{or} \qquad b\overline{)a}$$

DEFINITION

Let a, b, and q be numbers and suppose that $b \neq 0$. (*Read:* "*b* is *not* equal to 0.") Define

$$\frac{a}{b} = q \qquad \textit{if} \qquad a = q \times b$$

Here a is called the **dividend** and b the **divisor;** q is called the **quotient.**

Thus

$$\frac{6}{2} = 3 \qquad \textit{because} \qquad 6 = 3 \times 2$$

(Here $a = 6$, $b = 2$, and $q = 3$.) To represent the number $\frac{\mathbf{6}}{\mathbf{2}}$ on the number line, divide the line segment between 0 and **6** into **2** equal parts, as in Figure 1.37. Observe that the point of division, which corresponds to $\frac{6}{2}$, is 3.

In general, if $\frac{a}{b} = q$, then both $\frac{a}{b}$ and q correspond to the same point on the number line.

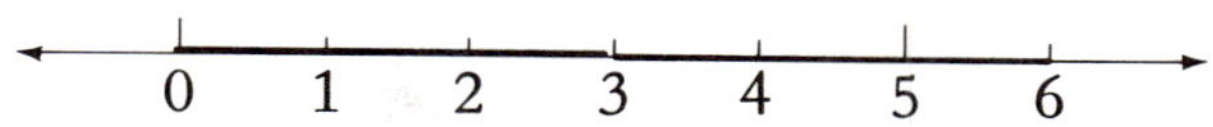

FIGURE 1.37. $\frac{6}{2} = 3$

Example 7 ▶ (a) $\frac{20}{4} = 5$ because $20 = 5 \times 4$

(b) $\frac{84}{7} = 12$ because $84 = 12 \times 7$ ◀

For every number a,

$$\frac{a}{1} = a$$

and for every *nonzero* number a,

$$\frac{a}{a} = 1$$

Both of these laws follow from the fact that

$$a \cdot 1 = 1 \cdot a = a$$

Answers

4. (a) ______

(b) ______

(c) ______

PRACTICE EXERCISES FOR TOPIC C

4. Divide. Check your answers by using multiplication.

(a) $42 \div 3$ (b) $\frac{91}{7}$ (c) $9\overline{)108}$

Now try the exercises for Topic C on page 34.

D. Sign of the Quotient

Let $\frac{a}{b} = q$. Then $a = q \times b$. Because division is defined in terms of multiplication, the quotient q is *positive* if a and b have the *same sign*; q is *negative* if a and b have *different signs.*

Example 8 ▶ (a) $\frac{15}{5} = 3$ (b) $\frac{-15}{5} = -3$

(c) $\frac{15}{-5} = -3$ (d) $\frac{-15}{-5} = 3$ ◀

When you divide products of numbers, count up the total number of negative factors in both dividend and divisor. If each factor is nonzero, then the quotient is

1. positive, if there is an even number of negative factors, possibly none at all, in dividend and divisor;
2. negative, if there is an odd number of negative factors in dividend and divisor.

Example 9 ▶ (a) $\frac{(-8)(-2)(-1)}{4(-4)} = \frac{-16}{-16} = 1$

Here there are four negative factors altogether, three in the dividend and one in the divisor. The quotient is positive.

(b) $\frac{(-8)(-2)}{4(-4)} = \frac{16}{-16} = -1$

Here there are three negative factors altogether, two in the dividend and one in the divisor. The quotient is negative. ◀

Answers

5. (a) ____

(b) ____

(c) ____

(d) ____

(e) ____

PRACTICE EXERCISES FOR TOPIC D

5. Find each quotient.

(a) $\frac{-36}{6}$ (b) $\frac{24}{-8}$ (c) $\frac{-55}{-5}$ (d) $\frac{100}{(-2)(-10)}$ (e) $\frac{6(-8)}{(-4)(-3)}$

Now try the exercises for Topic D on page 35.

E. Dividing by 0

Division by 0 is undefined. To understand why not, first try to divide a *nonzero* number, such as 4, by 0. Then

$$\frac{4}{0} = q \qquad \text{would mean that} \qquad 4 = q \times 0 = 0$$

This is obviously false. Next try to divide 0 by 0. Observe that

$$0 = 0 \times 0, \qquad 0 = 2 \times 0, \qquad 0 = 4 \times 0$$

There would be no *unique* quotient, q, as there is when the divisor is nonzero.

Although *division by* 0 *is undefined, you can divide* 0 *by any nonzero number b.* The quotient is then always 0.

Example 10 ► (a) $\frac{0}{6} = 0$ *because* $0 = 0 \times 6$

(b) $\frac{0}{-6} = 0$ *because* $0 = 0 \times (-6)$

(c) $\frac{6}{0}$ is *undefined.* ◄

Answers

6. (a) ______

(b) ______

PRACTICE EXERCISES FOR TOPIC E

6. Find each quotient, or state that the division is undefined.

(a) $\frac{0}{-8}$ (b) $\frac{-8}{0}$

Now try the exercises for Topic E on page 35.

F. Applied Problems

Example 11 ► A pair of socks costs three dollars. How much do six pairs of socks cost?
Solution:
Multiply three (dollars) by six.

$$\$3 \times 6 = \$18$$

Thus six pairs of socks cost $18. ◄

Answers

7. ______

PRACTICE EXERCISES FOR TOPIC F

7. If five pencils cost eighty-five cents, what is the cost per pencil?

Now try the exercises for Topic F on page 35.

Use this area for work space.

EXERCISES

A. In Exercises 1–10 find the product. (*See Examples 1, 2, 3, and 5.*)

1. 5×4 **2.** 10×7 **3.** 12×12

4. $7 \cdot 1$ **5.** $3(0)$ **6.** $0(-1)$

7. $5(-5)$ **8.** $(-7)(-2)$ **9.** $(-3)10$

10. $(-4)(-8)$

In Exercises 11–14 multiply to show that both products, (a) and (b), are equal. Also indicate which law is being illustrated. (*See Example 4.*)

11. (a) $3 \cdot 7$ (b) $7 \cdot 3$

12. (a) $(-10) \times 5$ (b) $5 \times (-10)$

Use this area for work space.

13. (a) $(2 \cdot 5) \cdot 4$ (b) $2 \cdot (5 \cdot 4)$

14. (a) $[(-1) \times (-4)] \times (-2)$ (b) $(-1) \times [(-4) \times (-2)]$

In Exercises 15 and 16 fill in the numbers. (See Figure 1.36 on page 30 to see how multiplication is represented on the number line.)

15. Figure 1.38 illustrates:

$\square \cdot \square = \square$

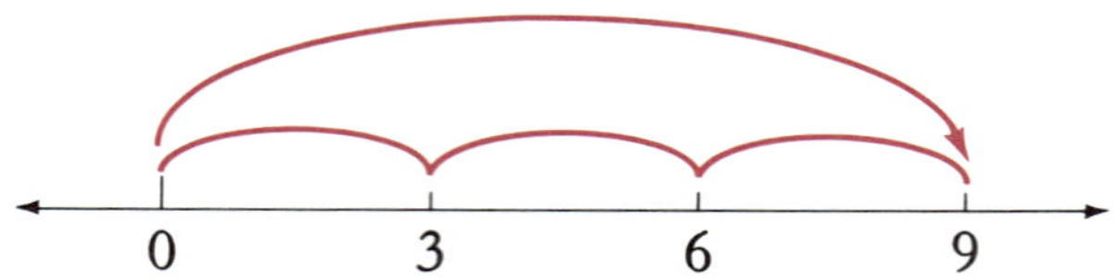

FIGURE 1.38

16. Figure 1.39 illustrates:

$\square \cdot \square = \square$

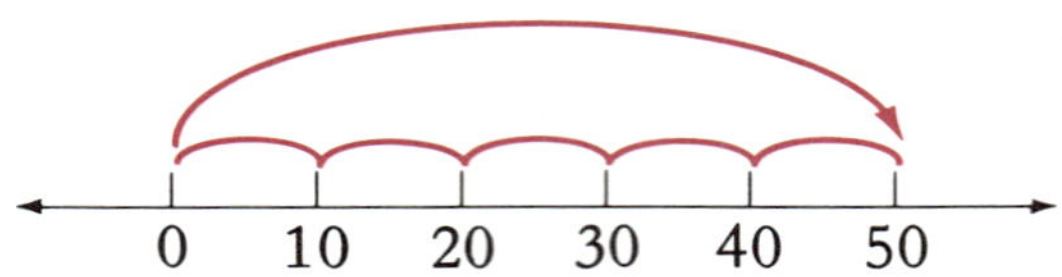

FIGURE 1.39

B. In Exercises 17–24 find the product. (*See Examples 5 and 6.*)

17. $(-1)(-2)(-3)$

18. $6(-2)(-1)(-2)$

19. $(-1)(-1)(-1)(-1)$

20. $(-1)(-1)(-1)(-1)(-1)$

21. $2 \times (-5) \times 0 \times (-3)$

22. $(-2)(-3)(-1)(-10)(-10)$

23. $(-10)(-10)(-10)(-10)(-10)(-10)$

24. $4 \times (-2) \times 3 \times (-5) \times (-10)$

C. In Exercises 25–30 find the quotient. (*See Example 7.*)

25. $\frac{8}{4}$

26. $\frac{36}{6}$

27. $\frac{120}{10}$

28. $64 \div 8$

29. $54 \div 9$

30. $7\overline{)84}$

In Exercises 31 and 32 fill in the numbers. (See Figure 1.36 on page 30 to see how division is represented on the number line.)

31. Figure 1.40 illustrates:

$\square \div \square = \square$

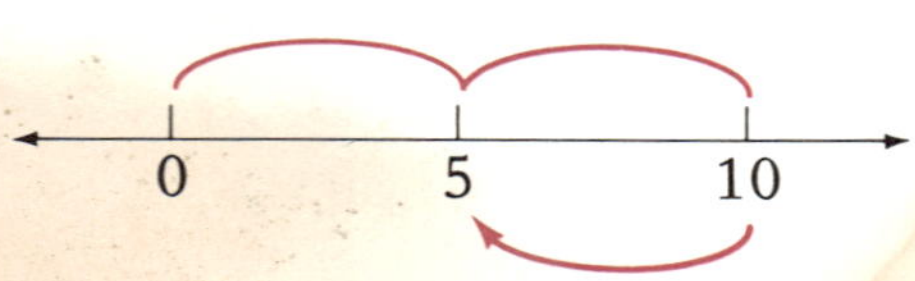

FIGURE 1.40

Use this area for work space.

32. Figure 1.41 illustrates:

$\square \div \square = \square$

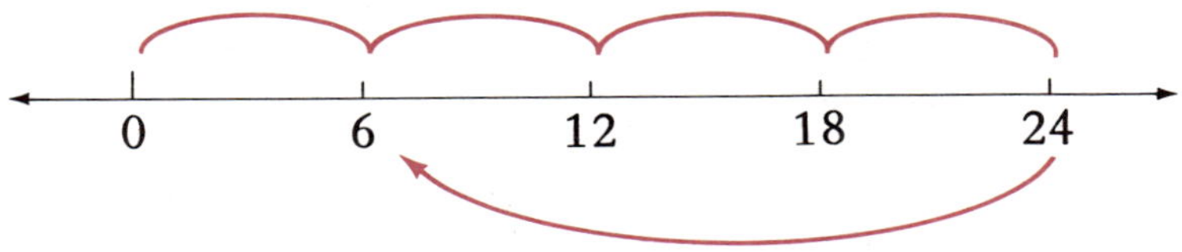

FIGURE 1.41

D. In Exercises 33–40 find the quotient. (*See Examples 8 and 9.*)

33. $\frac{-16}{2}$ **34.** $\frac{16}{-2}$ **35.** $\frac{-16}{-2}$

36. $\frac{(-7)(-2)}{(-1)(-14)}$ **37.** $\frac{3(-6)}{(-2)(-9)}$

38. $\frac{(-4)(-1)(-2)(-2)}{(-2)(-2)(-2)}$ **39.** $\frac{5 \times (-30) \times (-2)}{(-15) \times 4 \times (-1)}$

40. $\frac{(-30) \times 70 \times (-10) \times (-5)}{(-6) \times 10 \times (-5) \times 14}$

E. In Exercises 41–45 find the quotient, or state that division is undefined. (*See Example 10.*)

41. $\frac{0}{5}$ **42.** $\frac{-5}{0}$ **43.** $\frac{0}{0}$

44. $\frac{12 \times 0}{-4}$ **45.** $\frac{3 \times (-8)}{(-2) \times 0}$

46. Pick out the *error(s).*

(a) $\frac{0}{8} = 0$ (b) $\frac{8}{0} = 0$ (c) $\frac{8 \times 0}{0} = 0$

(d) $\frac{0}{0} = 1$ (e) $\frac{0}{-0} = -1$

F. **47.** There are 12 inches in a foot. How many inches are there in 5 feet? (*See Example 11.*)

48. There are 3 feet in a yard. A football field is 100 yards long. What is its length (a) in feet? (b) in inches?

49. There are 100 centimeters in a meter. How many centimeters are there in 8 meters?

50. There are 1000 meters in a kilometer. How many centimeters are there in 1 kilometer? (See Exercise 43.)

51. There are 60 minutes in one hour. An elevator operator works an 8-hour shift. How many minutes does he work?

52. There are 7 days in a week. How many days are there in 12 weeks?

53. Each of the 8 buildings on a campus contains 40 classrooms. Each classroom contains 50 seats. How many seats are there altogether?

Use this area for work space.

54. Each of three partners shares equally a total profit of $27. How much is each partner's share?

55. A dozen eggs cost 96 cents. How much does one egg cost?

56. An 11-floor high-rise contains 121 apartments. Each floor contains the same number of apartments. How many apartments are there on a floor?

57. A 9-ounce bottle of diet soda contains 63 calories. How many calories are there per ounce?

58. Five women each buy four tickets to a play at $15 per ticket. What is the total amount paid to the theater?

59. A man agrees to pay $84 for a radio. He does not have to put any money down. The payments are to last over a 7-week period. How much must he pay per week?

60. Suppose that the temperature change over the last 8 hours is $-16°$ Celsius. Assuming a constant rate of change, how many degrees Celsius has the temperature changed each hour?

1.6 Positive Integral Exponents

A. Squares

For any number a,

$$a^2 \text{ means } a \cdot a$$

Call a^2 the **square of *a*,** or ***a* squared.**

Example 1 ► (a) $4^2 = 4 \cdot 4 = 16$ (b) $9^2 = 9 \cdot 9 = 81$ (c) $0^2 = 0 \cdot 0 = 0$ ◄

For negative numbers, parentheses are used to clarify the intended meaning.

Example 2 ► (a) $(-2)^2 = (-2)(-2) = 4$
(b) $-2^2 = -4$ because, *by convention,* $-2^2 = -(2^2) = -4$
In part (a), square the negative number -2. In part (b), first square 2, and then consider the inverse of 2^2. ◄

PRACTICE EXERCISES FOR TOPIC A

1. (a) What is the square of 7?

(b) What is -5 squared?

2. Find the value of each.

(a) 9^2 (b) $(-6)^2$ (c) -4^2

Now try the exercises for Topic A on page 39.

Answers

1. (a) ______
(b) ______
2. (a) ______
(b) ______
(c) ______

B. Cubes

For any number a,

$$a^3 \text{ means } a \cdot a \cdot a$$

Call a^3 the **cube of a** or **a cubed.**

Example 3 ▶ (a) $3^3 = 3 \cdot 3 \cdot 3 = 27$
(b) $1^3 = 1 \cdot 1 \cdot 1 = 1$
(c) $(-2)^3 = (-2)(-2)(-2) = -(2 \cdot 2 \cdot 2) = -8$
(d) $-2^3 = -(2^3) = -8$ ◀

Answers

3. (a) ______
(b) ______
(c) ______
(d) ______

PRACTICE EXERCISES FOR TOPIC B

3. (a) Find the cube of -3.
(b) Find the value of 5 cubed.
(c) Find $(-6)^3$.
(d) Find -10^3.

Now try the exercises for Topic B on page 40.

C. *n*th Powers

Let a be any number.

$$a^1 \text{ means } a$$

Recall that

$$a^2 \text{ means } \underbrace{a \cdot a}_{\text{2 factors}} \qquad \text{and} \qquad a^3 \text{ means } \underbrace{a \cdot a \cdot a}_{\text{3 factors}}$$

Let n be a positive integer, where $n > 1$. Then

$$a^n \text{ means } \underbrace{a \cdot a \cdot a \ldots a}_{n \text{ factors}}$$

Thus, when considering a^n, a appears as a factor n times. Call a^n the **nth power of a,** or **a to the nth.**

Example 4 ▶ Find (a) 2^4 (b) 2^5 (c) 2^6

Solution:
(a) $2^4 = 2 \cdot 2 \cdot 2 \cdot 2 = 16$
(b) $2^5 = 2 \cdot 2 \cdot 2 \cdot 2 \cdot 2 = 32$
(c) $2^6 = 2 \cdot 2 \cdot 2 \cdot 2 \cdot 2 \cdot 2 = 64$ ◀

Example 5 ▶ Find (a) 10^4 (b) 10^5 (c) 10^8

Solution:

(a) $10^4 = 10\,000$ (with 4 zeros)
(b) $10^5 = 100\,000$ (with 5 zeros)
(c) $10^8 = 100\,000\,000$ (with 8 zeros) ◀

Answers

4. (a) ____
(b) ____
(c) ____
5. (a) ____
(b) ____

PRACTICE EXERCISES FOR TOPIC C

4. Evaluate each.

(a) $(-8)^3$ (b) 2^4 (c) 10^5

5. True or False. (a) $5^2 = 2^5$ (b) $4^2 = 2^4$

Now try the exercises for Topic C on page 40.

D. Base and Exponent

DEFINITION

> BASE, EXPONENT. In the expression a^n, a is called the **base** and n the **exponent (of a).** a^n is also called the **nth power of a.**

Example 6 ▶

$$4^3 = 64$$

Here 4 is the *base* and 3 the *exponent*. 64 is the *third power* (or *cube*) of 4. ◀

Answers

6. (a) ____
(b) ____
(c) ____

PRACTICE EXERCISES FOR TOPIC D

6. Consider the statement, $(-9)^3 = -729$.

(a) What is the base?
(b) What is the exponent?
(c) What is the cube of -9?

Now try the exercises for Topic D on page 40.

E. Powers of 1, 0, −1, and Negative Numbers

Example 7 ▶ (a) $1^n = 1$ for all positive integers n

(b) $0^n = 0$ for all positive integers n

(c) $(-1)^1 = -1$

$(-1)^2 = (-1)(-1) = 1$

$(-1)^3 = (-1)(-1)(-1) = -1$

$(-1)^4 = (-1)(-1)(-1)(-1) = 1$

Note the pattern:

$(-1)^n = -1$ for *odd* (positive) integers n, *that is, for* $n = 1, 3, 5, 7, \ldots$

$(-1)^n = 1$ for *even* (positive) integers n, *that is, for* $n = 2, 4, 6, 8, \ldots$ ◀

Example 7(c) can be generalized as follows.

1. *Odd powers of negative numbers are negative.*
2. *Even powers of negative numbers are positive.*

To see this, count the number of negative factors, as indicated by the exponent.

Example 8 ▶ (a) $(-2)^3 = (-2)(-2)(-2) = -8$

(b) $(-2)^4 = (-2)(-2)(-2)(-2) = 16$

(c) $(-2)^5 = (-2)(-2)(-2)(-2)(-2) = -32$

(d) $(-2)^6 = (-2)(-2)(-2)(-2)(-2)(-2) = 64$ ◀

Answers

7. (a) ______

(b) ______

(c) ______

(d) ______

PRACTICE EXERCISES FOR TOPIC E

7. Evaluate.

(a) $(-1)^{13}$ (b) $(-1)^{18}$ (c) $(-2)^7$ (d) 0^5

Now try the exercises for Topic E on page 41.

Use this area for work space.

EXERCISES

A.

1. What is the square of 10?

2. What is -9 squared?

In Exercises 3–14 find the values of the indicated squares or their inverses. (*See Examples 1 and 2.*)

3. 3^2 **4.** 5^2 **5.** 8^2 **6.** 11^2

7. $(-3)^2$ **8.** $(-7)^2$ **9.** $(-10)^2$ **10.** -10^2

11. 12^2 **12.** 20^2 **13.** $(-25)^2$ **14.** -100^2

Use this area for work space.

B.

15. Find the cube of 10.

16. What is the value of -4 cubed?

In Exercises 17–26 find the values of the indicated cubes or their inverses. (*See Example 3.*)

17. 2^3 **18.** 4^3 **19.** 5^3 **20.** 8^3

21. $(-1)^3$ **22.** -1^3 **23.** $(-5)^3$ **24.** $(-10)^3$

25. -20^3 **26.** $(-50)^3$

C. In Exercises 27–34 find the values of the indicated powers. (*See Examples 4, 5, 7, and 8.*)

27. 3^4 **28.** 3^5 **29.** 3^6 **30.** $(-3)^4$

31. 4^3 **32.** 4^4 **33.** 10^9 **34.** 10^{10}

35. *True or False.* $2^3 = 3^2$

36. (a) How many parents does a person have? (b) how many grandparents? (c) how many great-grandparents? (d) how many great-great-grandparents?

37. Suppose you start out with one dollar. Each month your money doubles. How much do you have after 5 months?

38. Suppose you start out with one dollar. Each month your money triples. In how many months will you have $81?

D. For Exercises 39–41, *see Example 6.* Consider the following statement.

$$5^2 = 25$$

39. What is the base? **40.** What is the exponent?

41. What power of 5 is 25?

In Exercises 42–44 consider the following statement.

$$3^4 = 81$$

42. What is the base? **43.** What is the exponent?

44. What power of 3 is 81?

45. What is the square of 6? **46.** What is the cube of 5?

47. What is 10 to the fourth power? **48.** What is 2 to the fifth power?

Sample: What power of 2 is 8?	*Solution:* $2^3 = 8$. Thus, 8 is the *third* power of 2.

Use this area for work space.

49. What power of 10 is 100?

50. What power of 3 is 27?

51. What power of (-2) is 4?

52. What power of 10 is 100 000?

53. What is the cube of the square of 2?

54. What is the square of the cube of -2?

E. In Exercises 55–68 find the values of the indicated powers.

55. 0^4 **56.** 1^{24} **57.** $(-1)^{10}$ **58.** $(-1)^{15}$

59. $(-1)^{20}$ **60.** $(-1)^{27}$ **61.** $(-2)^7$ **62.** $(-2)^8$

63. 10^7 **64.** $(-10)^{10}$ **65.** $(-3)^5$ **66.** $(-3)^6$

67. $(-5)^3$ **68.** $(-5)^4$

69. Pick out the *error(s)*.

(a) $(-2)^4 = (-4)^2$ (b) $3^5 = 5^3$ (c) $2^8 = 8^2$

(d) $1^{10} > 1^5$ (e) $(-1)^4 = (-1)^5$

1.7 Combining Operations

A. Adding and Subtracting

Addition and subtraction can be combined in the same example. Here *parentheses* or *brackets* are sometimes necessary to indicate the intended meaning.

Example 1 ▶ Find the value of $(4 - 8) - (3 + 7 - 2)$.
Solution:
First perform the operations within each pair of parentheses.

$$\underbrace{(4-8)}_{-4} - \underbrace{(3+7-2)}_{8} = -4 - 8 = -12$$ ◀

When pairs of parentheses or brackets are enclosed within other pairs, work from the innermost pairs outward.

Example 2 ▶ Find the value of $-[13 - (3 - 8 + 10)]$.
Solution:
Here the value of $3 - 8 + 10$ is subtracted from 13. You then determine the inverse of this difference.

$$-[13 - \underbrace{(3 - 8 + 10)}_{5}] = -\underbrace{[13 - 5]}_{8} = -8$$ ◀

Answers
1. ______
2. ______

PRACTICE EXERCISES FOR TOPIC A

1. Find the value of $(-18 + 3) - (7 - 16 + 5)$.

2. Find the value of $50 - [100 - (-14 + 9)]$.

Now try the exercises for Topic A on page 45.

B. Rules for Order of Operations

By convention, the expression

$4 \cdot 3^2$ means 4 times the square of 3

Thus $$4 \cdot 3^2 = 4 \cdot 9 = 36$$

Note that the exponent 2 only refers to the base 3. If you want to indicate the square of the product of 4 and 3, use parentheses. Thus

$$(4 \cdot 3)^2 = 12^2 = 144$$

Here the exponent applies to both factors 4 and 3.

The expression

$3 + 5 \cdot 2$ means add the product of 5 and 2 to 3.

Thus $$3 + 5 \cdot 2 = 3 + 10 = 13$$

If the sum of 3 and 5 is to be multiplied by 2, use parentheses. Thus

$$(3 + 5)2 = 8 \cdot 2 = 16$$

The order in which addition, subtraction, multiplication, division, and raising to a power are applied in an example is often crucial.

1. If parentheses are given, first perform the operations within parentheses. Following this:
2. Raise to a power.
3. Then multiply or divide from left to right.
4. Then add or subtract from left to right.

To remember the order of operations, think of:

Please rest in peace, my dear Aunt Sally.

Parentheses; raise to a power; multiply or divide; add or subtract.

Example 3 ▶ Find the value of $7(2 - 1)$.

Solution:

First perform operations within parentheses.

$$7(2 - 1) = 7 \cdot 1 = 7$$ ◀

Example 4 ▶ Find the value of $2 + 5^2 \cdot 3$.
Solution:
First raise to a power; then multiply; then add.

$$\begin{aligned} 2 + 5^2 \cdot 3 &= 2 + 25 \cdot 3 && \text{because } 5^2 = 25 \\ &= 2 + 75 && \text{because } 25 \cdot 3 = 75 \\ &= 77 \end{aligned}$$

◀

Example 5 ▶ Find the value of $\frac{-20}{5} - 1 + 2$.

Solution:
First divide; then subtract; then add.

$$\frac{-20}{5} - 1 + 2 = \underbrace{-4 - 1}_{-5} + 2 = -5 + 2 = -3$$

◀

Example 6 ▶ (a) $(4 + 2)^2 = 6^2 = 36$ (b) $4 + 2^2 = 4 + 4 = 8$
(c) $4^2 + 2^2 = 16 + 4 = 20$ (d) $(4 \cdot 2)^2 = 8^2 = 64$
(e) $4 \cdot 2^2 = 4 \cdot 4 = 16$ ◀

Example 7 ▶ Find the value of $\left(\frac{11 - 5}{3} - 1\right)^2$.

Solution:
Think of this as

$$\left([(11 - 5) \div 3] - 1\right)^2$$

and work from the innermost pair of parentheses outward. Thus first find $11 - 5$, then divide $11 - 5$ by 3, then subtract 1, and finally square the result. Therefore

$$\left(\frac{11 - 5}{3} - 1\right)^2 = \left(\frac{6}{3} - 1\right)^2 = (2 - 1)^2 = 1^2 = 1$$

◀

Example 8 ▶ Find the value of $3 \cdot 4^3 + (2 \cdot 3)^2$.
Solution:
Only the 4 is cubed; the product of 2 and 3 is squared. Thus

$$\begin{aligned} 3 \cdot 4^3 + (2 \cdot 3)^2 &= 3(4 \cdot 4 \cdot 4) + 6^2 \\ &= 3 \cdot 64 + 36 \\ &= 192 + 36 \\ &= 228 \end{aligned}$$

◀

PRACTICE EXERCISES FOR TOPIC B

Answers	
3. (a)	(b)
(c)	(d)
4.	

3. Find the value of each.

(a) $6 + 4(-3 + 10)$ (b) $19 - 5 \cdot 2^2$ (c) $2\left(\frac{35 - 2}{3} - 4\right)^2$ (d) $4 \cdot 3^3 + [5(-2)]^2$

4. Subtract 5 from the square of 3.

Now try the exercises for Topic B on page 45.

C. Distributive Laws

Observe that

$$4 \cdot \underbrace{(3 + 2)}_{5} = 4 \cdot 5 = 20$$

But you could also calculate as follows:

$$\begin{aligned} 4 \cdot (3 + 2) &= 4 \cdot 3 + 4 \cdot 2 \\ &= 12 + 8 \\ &= 20 \end{aligned}$$

The second method, though longer, has an important use in algebra. This method illustrates the following **Distributive Laws.** Let a, b, and c be any numbers.

$$a(b + c) = a \cdot b + a \cdot c, \qquad (b + c)a = b \cdot a + c \cdot a$$

Similarly

$$a(b - c) = a \cdot b - a \cdot c, \qquad (b - c)a = b \cdot a - c \cdot a$$

Answers

5. (a) ______

(b) ______

6. ______

PRACTICE EXERCISES FOR TOPIC C

5. Show that each of the following statements is true by evaluating both sides of the equals sign:

(a) $5(20 - 3) = 5 \cdot 20 - 5 \cdot 3$ (b) $(6 + 5) \cdot 9 = 6 \cdot 9 + 5 \cdot 9$

6. Use a Distributive Law to evaluate $7(10 + 8)$.

Now try the exercises for Topic C on page 46.

D. Applied Problems

Example 9 ▶ Peaches cost 79 cents per pound and plums cost 90 cents per pound. What is the total cost of 3 pounds of peaches and 5 pounds of plums?

Solution:

Multiply 79 by 3 and add to this the product of 90 and 5.

$$(79 \times 3) + (90 \times 5) = 237 + 450 = 687$$

This represents the cost in cents. Converting to dollars and cents, we find that the total cost is \$6.87. ◀

Answers

7. ______

PRACTICE EXERCISES FOR TOPIC D

7. How much change do you receive if you pay for twelve 25-cent stamps and 5 22-cent stamps with a five-dollar bill?

Now try the exercises for Topic D on page 46.

Use this area for work space

EXERCISES

A. In Exercises 1–8 find each value. (*See Examples 1 and 2.*)

1. $(4 - 2) - (2 - 4)$

2. $(3 - 6) - (10 + 4 - 2)$

3. $(8 - 5 - 3) - (2 - 7 + 10)$

4. $(6 - 3) - (3 - 9) - (7 - 2)$

5. $-[14 - (3 - 8)]$

6. $[6 - (2 - 8)] - [(-10) - 5 - (-20)]$

7. $-[7 - (-12 + 4)] - [(4 + 3 - 5) - 2 - (8 - 10)]$

8. $-[6 - (-10)] - (5 - 12) + [(-8) - (-4)] - (3 - 9 + 5 - 12)$

9. Find the value of:
(a) $4 - (3 - 2)$
(b) $(4 - 3) - 2$

10. Find the value of:
(a) $10 - (3 + 4)$
(b) $(10 - 3) + 4$

B. In Exercises 11–45 find each value. (*See Examples 3–8.*)

11. $3 + 5 \cdot 4$

12. $(3 + 5)4$

13. $4 \cdot 2^2$

14. $(4 \cdot 2)^2$

15. $\frac{4 + 2}{2}$

16. $\frac{4}{2} + 2$

17. $4 + \frac{2}{2} - 3 \cdot 7$

18. $4^2 - 2^2$

19. $4 - 2^2$

20. $(4 - 2)^2$

21. $(2 - 4)^2$

22. $2 - 4^2$

23. $(2 + 5)^2 - 1$

24. $2 + 5^2 - 1$

25. $\frac{(8 - 2)^2}{2}$

26. $\frac{8 - 2^2}{2}$

27. $5(3 - 1) + 2^2$

28. $[5(3 - 1) + 2]^2$

29. $\left(\frac{6 + 5}{11} - 2\right)^2$

30. $\frac{8 + 7}{5} - \frac{3 + 4}{-7}$

31. $\left[\frac{3 - (4 + 5)}{2}\right]^3$

32. $2\left(\frac{3 - 4 + 5}{2}\right)^3$

33. $\frac{(6 - 2)^2}{2} - 3(5 - 3)^3$

34. $\frac{(5 - 2)^2 - (-1 + 3)^2}{5}$

35. $6 \cdot 6 - 6 \div 6$

36. $6(6 - 6 \div 6)$

37. $6(6 - 6) \div 6$

38. $(6 \cdot 6 - 5) \div 6$

39. $48 \div 12 \cdot 2^2$

40. $48 \div (12 \cdot 2^2)$

41. $-(2 - 3)^3$

42. $2 + 3 \cdot 4^2 - 1$

43. $(2^4 - 12) \cdot 2 + 1$

44. $2^5 - 12 \cdot 2 + 1$

45. $3^3 - 6 \div 2 + 5$

46. Pick out the *error(s).*

(a) $3(4 + 5) = 3(4) + 5$

(b) $7(6 - 2) = 7(6) - 7(2)$

(c) $4 + 5 \cdot 8 = (4 + 5)8$

(d) $1^2 + 2^2 = 3^2$

(e) $(2 + 3)^2 = 2^2 + 3^2$

(f) $\frac{8 + 3}{2 + 3} = \frac{4 + 3}{1 + 3}$

Use this area for work space.

Sample:	Solution:
Sample: Divide the sum of 8 and 2 by 5.	*Solution:* $\frac{8+2}{5} = \frac{10}{5} = 2$

47. Add 3 to the square of 2.

48. Multiply the sum of 7 and 2 by 6.

49. Add 8 to the product of 4 and 3.

50. Divide the sum of 4 and 5 by 3.

51. Subtract 4 from the cube of 2.

52. Divide the square of (-4) by the cube of (-2).

C. In Exercises 53 and 54 show that each statement is true by evaluating both sides of the equals signs.

53. $6(29 - 13) = 6 \cdot 29 - 6 \cdot 13$

54. $(8 - 5) \times (-3) = 8 \times (-3) - 5 \times (-3)$

In Exercises 55–62 use the Distributive Laws in each case to find the value.

55. $4 \cdot (6 + 1)$

56. $(3 + 5) \cdot 2$

57. $8 \cdot (5 + 3)$

58. $(7 + 2) \cdot 3$

59. $9 \cdot (9 - 6)$

60. $(8 - 9) \cdot 9$

61. $17 \cdot 3 + 17 \cdot 7$

62. $86 \cdot 94 + 86 \cdot 6$

D. **63.** Onions cost 41 cents per pound and potatoes cost 37 cents per pound. What is the total cost of 2 pounds of onions and 3 pounds of potatoes? (*See Example 9.*)

64. A woman buys 9 pastries and pays for them with a $5 bill. She receives 2 quarters, 3 dimes, and 3 nickels in change. If each pastry costs the same, what is the price per pastry?

65. A business makes a profit of $40 000. Of this, $12 000 is reinvested. The rest of the profit is split equally among four partners. How much is each partner's share?

66. A typist charges $4 per page. If he types 5 pages per hour, how long does it take him to earn $100?

67. A football team must gain 10 yards to obtain a first down. In three plays it gains 5 yards, loses 7 yards, and gains 9 yards. How many yards does it still need for a first down?

68. Susan reads 30 pages per hour. If she has read 140 pages of a 350-page book, how much longer will it take her to finish the book?

69. A 250-pound man wishes to reduce to 200 pounds. If he is to lose 5 pounds per week, for how many weeks must he diet?

70. A publisher consigns 50 copies of a novel to a bookstore, which later returns 20 of these copies. If the bookstore pays $10 for each copy that it keeps, how much must it pay the publisher?

What Have You Learned in Chapter 1?

You can locate integers on the number line.

You know that a is less than b if a lies to the left of b on the number line. You can find the inverse as well as the absolute value of a number.

You can add, subtract, multiply, and divide integers.

You can work with positive integral exponents.

And you know in what order to apply these arithmetic operations in a given problem.

Let's Review Chapter 1*

1.1 Integers and the Number Line

1. In Figure 1.42 indicate the integer that corresponds to each of the points Q, R, S, T, U, and V.

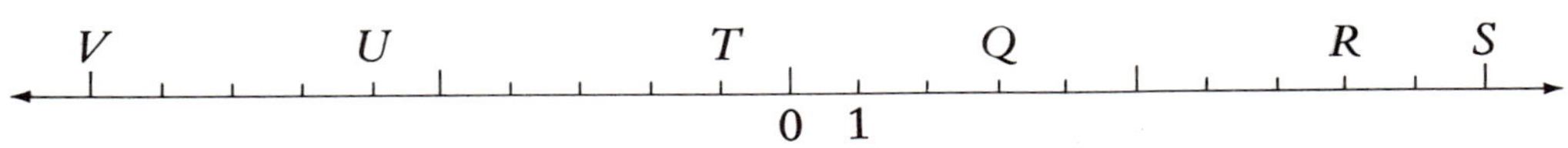

FIGURE 1.42

2. (a) Which integer corresponds to the point 12 units to the left of the origin?
(b) Which integer corresponds to the point 12 units to the right of the origin?

1.2 Order

3. Fill in "<" or ">".

$$5 \square 3$$

4. Fill in "left" or "right".

-8 lies to the ☐ of -4 (on the number line).

5. Rearrange the integers so that you can write "<" between any two of them.

$$0, -3, -10, -1, 4, 1$$

1.3 Inverse and Absolute Value

6. Write the inverse of:
(a) 7 (b) -3 (c) 0

7. Find each absolute value.
(a) $|5|$ (b) $|-2|$ (c) $|0|$

8. Which integers are 10 units from the origin?

---

*Answers to *all* review exercises are given at the back of the book, beginning on page A7.

REVIEW

1.4 *Adding and Subtracting Signed Numbers*

9. Add $(-4) + (-5) + (-7)$.

10. Subtract $(-10) - (-3)$.

11. Add
$$\begin{array}{r} 12 \\ -20 \\ -21 \\ 35 \\ \underline{-19} \end{array}$$

12. Add to show that both sums, (a) and (b), are equal. Also indicate which law is being illustrated.
(a) $[(-5) + (-7)] + (-3)$ (b) $(-5) + [(-7) + (-3)]$

13. The temperature in London at 6 a.m. is $-4°$ Celsius. If the temperature drops $2°$ the next hour, what is the reading at 7 a.m.?

1.5 *Multiplying and Dividing Signed Numbers*

14. Find the product.
$$(-2)(-1)(-3)(-2)$$

15. Find the quotient.
(a) $\dfrac{-36}{-12}$ (b) $\dfrac{(-4)(-6)}{2(-12)}$

16. Find the quotient, or state that division is undefined.
(a) $\dfrac{-6}{0}$ (b) $\dfrac{0}{-6}$

17. There are 60 minutes in an hour and 24 hours in a day. How many minutes are there in a day?

1.6 *Positive Integral Exponents*

18. Find each value.
(a) 7^2 (b) $(-3)^3$ (c) $(-2)^4$

19. What power of 5 is 125?

1.7 *Combining Operations*

20. Find the value of $4 \cdot 3^2 - 1$

21. Find the value of $\left(\dfrac{6}{2}\right)^3 + \dfrac{8}{2^2}$

22. Divide the square of 8 by the cube of 2.

23. Use the Distributive Laws to find $92 \cdot 7 + 8 \cdot 7$

Practice Exam 1 on Chapter 1

1. Which integer corresponds to the point 6 units to the left of the origin?

2. Fill in "<" or ">": $-4 \;\square\; -2$

3. Rearrange the following integers so that you can write "<" between any two integers. $-2,\ 0,\ -1,\ 12,\ -12,\ 1$

4. Write the inverse of -19.

5. Find $|-19|$.

6. Add $(-8) + (-2) + (-9)$.

7. Subtract $(-5) - (-3)$.

8. Multiply $(-2)(-4)(-1)(-3)$.

9. Find $\dfrac{-60}{(-2)(-3)}$.

10. Find $(-3)^4$.

11. What power of -2 is -8?

12. Find the value of $(-2 + 4)^2 - 2 \times 3$.

13. Find the value of $\dfrac{(8 - 2)^2}{9} + 4$.

14. Multiply the sum of 7 and -3 by the cube of 2.

Practice Exam 2 on Chapter 1

1. On the number line which integer corresponds to the point 23 units to the left of the origin?

2. Fill in "<" or ">".

   (a) $3 \;\square\; -5$ (b) $|-9| \;\square\; |8|$

3. Rearrange the following integers so that you can write "<" between any two of them.

   $7,\ 0,\ 4,\ -7,\ -3$

4. What is the additive inverse of -15?

REVIEW

5. Find the value of each.

(a) $-(-3)$ (b) $|-20|$ (c) $-|4|$

In Questions 6–13 perform the indicated operations.

6. $-20 + 8$

7. $-9 + (-4) + (-1)$

8. $17 - (-4)$

9. $-22 + (-3) - (-5)$

10. $15(-6)$

11. $(-4)(3)(-25)$

12. $21 \div (-3)$

13. $\dfrac{(-10)(-6)}{-12}$

14. Find the value of $(-4)^3$.

15. Evaluate $-10 + 3(2 - 6)^2$

16. Add the cube of -2 to the product of 5 and -8.

2

Polynomials

2.1 Terms and Polynomials

A. Variables

In arithmetic you add, subtract, multiply, and divide *individual* numbers. Sometimes the same arithmetic process can apply to *any number* under discussion. It is important to have symbols that can stand for any one of several numbers. For example, if you want to add 10 to a number (whatever that number may be — 1, 5, 0, −3, and so on) you can express this by

$$x + 10$$

If you want to express twice a number, write

$$2x$$

In both cases x stands for *any number* and is called a *variable*.

DEFINITION

> VARIABLE. A **variable** is a symbol that stands for any one of the numbers being discussed.

Most often, the letters x, y, and z will be the ones used as variables. Other letters, such as a, b, c, t, u, and v, will also be used at times.

A variable can be used to designate *any* number. However, a variable can also be used to stand for only *certain* numbers. For example, if you are only discussing positive integers, then the variable used will represent only positive integers.

Variables make it possible to apply arithmetic concepts to many numbers at one time. You can add, subtract, multiply, and divide these variables just as you can combine numbers. Powers of variables can also be considered. The arithmetic rules of Chapter 1 can be adapted to variables. Algebraic expressions containing variables, such as

$$x + 1$$

$$x - y$$

$$5uv$$

$$x^2$$

$$x^2 + 3x$$

will be used throughout the book. Most of this chapter will be a preparation for using variables in later applications.

Example 1 ► Suppose x stands for some number. Express each of the following in terms of x.

(a) 6 more than the number
(b) 3 less than the number
(c) 4 times the number
(d) the number divided by 7

Solution:

(a) Note that 9 is 6 more than the number 3, and $9 = 3 + 6$.
Also, 10 is 6 more than the number 4, and $10 = 4 + 6$. Thus

6 *more than* the number means the number *plus* 6

$$\underbrace{6}\;\underbrace{+}\;\underbrace{x},\quad \text{or}\quad \underbrace{x}\;\underbrace{+}\;\underbrace{6}$$

(b) Note that 4 is 3 less than the number 7, and $4 = 7 - 3$.
Also, 5 is 3 less than the number 8, and $5 = 8 - 3$. Thus

3 *less than* the number means the number *minus* 3

$$x - 3$$

(c) 4 times the number

$$4 \cdot x, \quad \text{or } 4x$$

(d) the number divided by 7

$x \div 7$, or $\frac{x}{7}$ ◀

Example 2 ▶ Suppose y stands for some number. Express each of the following in terms of y.

(a) the number increased by 5
(b) the number decreased by 1
(c) the square of the number
(d) 8 times the square of the number

Solution:

(a) Increasing a number by 5 means adding 5 to the number. Thus

the number increased by 5

$y + 5$

(b) Decreasing a number by 1 means subtracting 1 from the number. Thus

the number decreased by 1

$y - 1$

(c) the square of the number

y^2

(d) 8 times the square of the number

$8 \cdot y^2$, or $8y^2$ ◀

PRACTICE EXERCISES FOR TOPIC A

Answers
1. (a) ______
(b) ______
(c) ______

1. Suppose z stands for some number. Express the following in terms of z.

(a) 5 less than the number
(b) 7 times the square of the number
(c) 10 more than the cube of the number

Now try the exercises for Topic A on page 56.

B. What Is a Term?

The building blocks of algebraic expressions are *terms*.

DEFINITION

> TERM. A **term** is a product of numbers and variables. A number, by itself, is considered to be a term and is often called a **numerical term.** A variable, by itself, is also a term.

Example 3 ▶ Show that each of the following expressions is a term.

(a) $2x$
(b) xyz
(c) 5
(d) t

Solution:

(a) $2x$ is the product of 2 and the variable x.
(b) xyz is the product of the variables x, y, and z.
(c) 5 is a number. As such, it is a (numerical) term.
(d) A variable, such as t, is also a term. ◀

Example 4 ▶ Show that each of the following is a term.

(a) x^2
(b) $3x^2y^3$
(c) $(5x)(2y)$

Solution:

(a) x^2 stands for $x \cdot x$. Thus it is the product of variables.
(b) $3x^2y^3 = 3 \cdot x \cdot x \cdot y \cdot y \cdot y$
Thus $3x^2y^3$ is the product of a number and variables.
(c) $(5x)(2y)$ is the product of the numbers 5 and 2 and the variables x and y. The Associative and Commutative Laws of multiplication will enable you to rewrite this expression as $5 \cdot 2 \cdot x \cdot y$, or $10xy$. ◀

Example 5 ▶ (a) $x + 10$ is *not* a term because the definition of a term does not permit adding numbers and variables.
(b) $x + y$ is *not* a term because the definition of a term does not permit adding different variables. ◀

Answers

2. ________
3. (a) ________
(b) ________
(c) ________

PRACTICE EXERCISES FOR TOPIC B

2. Which of the following expressions are terms?

(a) $10x^2y$ (b) $3 + y^2$ (c) -7 (d) $x(-5)$ (e) $\frac{2}{x}$

3. Express each term using exponents.

(a) $x \cdot x \cdot x$ (b) $x \cdot x \cdot y \cdot y \cdot y \cdot y$ (c) $(3 \cdot a)(2 \cdot b \cdot b \cdot b)$

Now try the exercises for Topic B on page 57.

C. Coefficients

In the term $6xy$, the numerical factor 6 is known as the coefficient; and in the term $-6xy$, the numerical factor, -6, is the coefficient.

DEFINITION

> COEFFICIENT. The numerical factor (or the product of numerical factors) in a term is called the **numerical coefficient,** or simply, the **coefficient,** of the term. If only variables appear, the coefficient of the term is understood to be 1 or -1. (Thus the coefficient of xy is 1, and of $-xy$ is -1.)

A numerical term is its own coefficient. For example, the coefficient of the term 2 is 2, itself.

Example 6 ► Determine the coefficient of each term.

(a) $3x$ (b) $-4x^2y$ (c) $(5x)(-3z)$
(d) 10 (e) $-x$

Solution:

(a) 3
(b) -4
(c) -15 Note that $-15 = 5(-3)$.
(d) 10 The numerical term 10 is its own coefficient.
(e) -1 Note that $-x = (-1)x$. ◄

When you multiply by 0, the product is 0. Thus *any term,* such as $0xy$, *that has 0 as its coefficient reduces to* 0.

Do not confuse a coefficient with an exponent. For example, in the term

$$2x^3$$

2 is the coefficient; 3 is the exponent (of x).

Answers

4. (a) ______
(b) ______
(c) ______

PRACTICE EXERCISES FOR TOPIC C

4. Determine the coefficient of each term.

(a) $9x^3y$ (b) $-xz^2$ (c) $(-4a)(-6b)$

Now try the exercises for Topic C on page 57.

D. Polynomials

Terms can be added to form polynomials.

DEFINITION

> A **polynomial** is defined to be either a term or a sum of terms.

Example 7 ▶ (a) x^2 is a term. Therefore, it is also a polynomial.
(b) $x^2 + 3x$ is the sum of the terms x^2 and $3x$. Therefore it is a polynomial.
(c) $x^2 + xy - 4$ is the sum of the terms x^2, xy, and -4. It is a polynomial in two variables, x and y.
(d) $t^4 + 2t^3 - t^2 + 6t + 1$ is the sum of the terms t^4, $2t^3$, $-t^2$, $6t$, and 1. Thus it is a polynomial. ◀

Example 8 ▶ The expressions

$$\frac{x}{y} \quad \text{and} \quad \frac{x+1}{x+2}$$

are *not* polynomials. Such expressions involve division of polynomials, and will be discussed in Chapter 5. It will be seen that

$$\frac{x}{4} \quad \text{and} \quad \frac{x+1}{4}$$

are polynomials. These expressions involve division by numbers. ◀

Answers

5. ______

PRACTICE EXERCISES FOR TOPIC D

5. Which of the following expressions are polynomials?

(a) $-12xy$ (b) $3 + 2x^2$ (c) $1 + \dfrac{2}{x}$ (d) $\dfrac{5}{x+1}$

6. Determine the terms of each polynomial.

(a) $3x + 2$ is the sum of the terms ☐ and ☐.

(b) $y^3 - y^2 + 5y - 3$ is the sum of the terms ☐, ☐, ☐, and ☐.

Now try the exercises for Topic D on page 58.

Use this area for work space.

EXERCISES

A. In Exercises 1–10 suppose x stands for some number. Express each of the following in terms of x. (*See Examples 1 and 2.*)

1. the number plus 5

2. 5 times the number

3. the number minus 5

4. the number divided by 5

5. 3 more than the number

6. 8 less than the number

7. the number increased by 12

8. the number decreased by 6

9. twice the number

10. twice the cube of the number

In Exercises 11–16 suppose x stands for one number and y for another number. Express each of the following in terms of x and y.

Use this area for work space.

Sample:	*Solution:*
x increased by y	$x + y$

11. x minus y

12. x times y

13. y divided by x

14. twice the product of x and y

15. 1 more than the product of x and y

16. the sum of x and y divided by 10

B. In Exercises 17–22, which of the following expressions are terms? (*See Examples 3–5.*)

17. -12

18. x^3

19. $y + 3$

20. $3xy$

21. $\frac{x}{y}$

22. $0 \cdot x$

In Exercises 23–30 write each term as a product of a number and variables. (*See Example 4.*)

Sample:	*Solution:*
$(3x^4)(2y^2)$	$6 \cdot x \cdot x \cdot x \cdot x \cdot y \cdot y$

23. t^2

24. x^3

25. $5xy^2$

26. $12x^2yz^5$

27. $-xy$

28. $(4x)(3y)$

29. $(-2x^2)(-4y)$

30. $(-x)(-y)$

In Exercises 31–38 express each term using exponents.

Sample:	*Solution:*
$(4 \cdot x \cdot x \cdot x)(5 \cdot y \cdot y)$	$20x^3y^2$

31. $u \cdot u$

32. $v \cdot v \cdot v \cdot v$

33. $x \cdot x \cdot y \cdot y$

34. $3 \cdot a \cdot a \cdot a$

35. $a \cdot a \cdot b \cdot c \cdot c \cdot c$

36. $(2 \cdot t)(4 \cdot u \cdot u)$

37. $(2 \cdot x)(y \cdot y)(-3 \cdot z)$

38. $(-4 \cdot a)(2 \cdot b \cdot b)(-5 \cdot c \cdot c \cdot c \cdot c)$

C. In Exercises 39–51 determine the coefficient of each term. (*See Example 6.*)

39. $7x$

40. $-3y$

41. $5x^2$

42. $2x^5$

43. $-3u^2vw$

44. $-xy$

45. $(2x)(3y)$

46. $(3x)(-2y)$

47. $(-4y)(-5z)$

48. $(-x)(-yz)$

49. 14

50. -2

51. $0x^2y^3$

Use this area for work space.

52. Pick out the *error(s)*.

(a) The coefficient of $3x^4$ is 4.

(b) The coefficient of $4x^3$ is 4.

(c) The coefficient of $(-3x)(-4y)$ is 12.

(d) x^2y has no coefficient.

D. In Exercises 53–58, which of the following expressions are polynomials? (*See Examples 7 and 8.*)

53. 4 **54.** $5a^2bc^3$ **55.** $x^2 - x + 1$ **56.** $\frac{x+1}{y-2}$ **57.** $\frac{1}{x}$ **58.** $1 - \frac{1}{x}$

In Exercises 59–65 determine the terms of each polynomial. (*See Example 7.*)

Sample:	*Solution:*
$x^2 - 4x + 1$ is the sum of the terms ☐, ☐, and ☐	$x^2 - 4x + 1$ is the sum of the terms $\boxed{x^2}$, $\boxed{-4x}$, and $\boxed{1}$

59. $x + 4$ is the sum of the terms ☐ and ☐.

60. $x^2 - 3x$ is the sum of the terms ☐ and ☐.

61. $x + 3y - 2z$ is the sum of the terms ☐, ☐, and ☐.

62. $x^3 + 6x^2 + 5x + 10$ is the sum of the terms ☐, ☐, ☐, and ☐.

63. $t^8 + 2t^5 - 3t^3 + t^2 - t + 4$ is the sum of the terms ☐, ☐, ☐, ☐, ☐, and ☐.

64. The polynomial x^2 is the same as the term ☐.

65. The polynomial 10 is the same as the term ☐.

66. Pick out the *error(s)*.

(a) Every term is a polynomial.

(b) Every polynomial is a term.

(c) $\frac{x+1}{2}$ is a polynomial.

(d) $\frac{2}{x+1}$ is a polynomial.

2.2 Evaluating Polynomials

A. Substituting Numbers for Variables

Recall that a variable designates any number under discussion. Polynomials are constructed from numbers and variables by multiplication and addition. When you are given numbers that the variables of the polynomial represent, you can evaluate the polynomial by substituting the numbers for the variables.

Example 1 ▶ Evaluate $5x$ when $x = 6$.
Solution:
Substitute 6 for x in the expression $5x$, and obtain

$$5 \cdot 6 \quad \text{or} \quad 30$$ ◀

Example 2 ▶ Evaluate $3y - 1$ (a) when $y = -2$, (b) when $y = 0$.
Solution:
(a) Replace y by -2:

$$3(-2) - 1 = -6 - 1 = -7$$

(b) Replace y by 0:

$$3(0) - 1 = 0 - 1 = -1$$ ◀

When a variable occurs more than once in a polynomial, *each time the variable occurs*, it must be *replaced by the same number.*

Example 3 ▶ Evaluate $t^2 + 4t + 3$ when $t = 5$.
Solution:
Replace *each occurrence of* t by 5:

$$5^2 + 4 \cdot 5 + 3 = 25 + 20 + 3 = 48$$ ◀

If a polynomial contains two or more variables, replace each variable by a specified number.

Example 4 ▶ Evaluate $x^2 - 2y$ when $x = 3$ and $y = -4$.
Solution:
Substitute 3 for x and -4 for y, and obtain

$$3^2 - 2(-4) = 9 + 8 = 17$$ ◀

Example 5 ▶ Evaluate $x^2 + 3xy + yz$ when $x = 2$, $y = 1$, $z = 4$.
Solution:
Replace *each occurrence of* x by 2, *of* y by 1, *and of* z by 4:

$$2^2 + 3 \cdot 2 \cdot 1 + 1 \cdot 4 = 4 + 6 + 4 = 14$$ ◀

PRACTICE EXERCISES FOR TOPIC A

Answers

1. ______
2. ______
3. ______
4. ______

1. Evaluate $7x + 3$ when $x = -1$.

2. Evaluate $y^2 - 5y$ when $y = 3$.

3. Evaluate $2x^2 + 4xy - y^2$ when $x = 5$ and $y = -2$.

4. Evaluate $a^2 + 2bc - 10abc$ when $a = 2$, $b = 1$, and $c = 5$.

Now try the exercises for Topic A on page 61.

B. Formulas

Formulas are algebraic equations that indicate the relationship between various quantities. Geometric formulas are often expressed in terms of polynomials.

Example 6 ▶ The area, A, of a square is given by the formula

$$A = s^2$$

where s is the length of each side. Find the area if each side is of length (a) 10 inches, (b) 20 inches. (See Figure 2.1.)

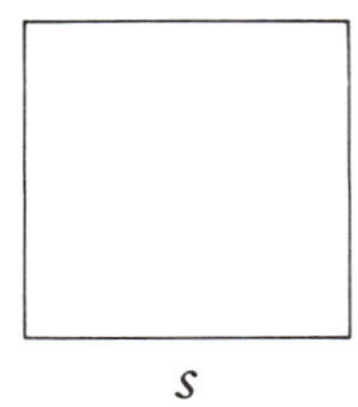

FIGURE 2.1. The area of a square is s^2.

Solution:

(a) $A = 10^2 = 100$

The area is 100 square inches.

(b) $A = 20^2 = 400$

The area is 400 square inches. ◀

Example 7 ▶ The area, A, of a triangle is given by the formula

$$A = \frac{bh}{2}$$

where b is the length of the base and h is the length of the altitude. Find the area if (a) $b = 6$ feet and $h = 10$ feet; (b) $b = 12$ feet and $h = 9$ feet. (See Figure 2.2.)

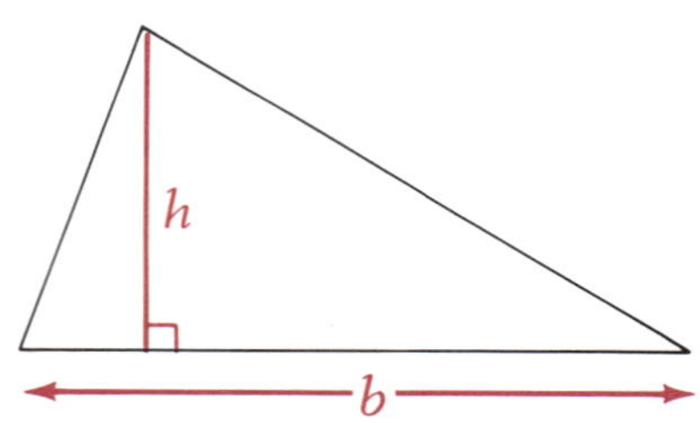

FIGURE 2.2. The area of a triangle is $\frac{bh}{2}$.

Solution:

(a) Substitute 6 for b and 10 for h in the formula $A = \frac{bh}{2}$, and obtain

$$A = \frac{6 \cdot \mathbf{10}}{2} = 30$$

Thus the area is 30 square feet.

(b) Substitute 12 for b and 9 for h:

$$A = \frac{12 \cdot \mathbf{9}}{2} = 54$$

Thus the area is 54 square feet. ◀

Example 8 ▶ A car travels at the rate of 50 miles per hour. The distance d (in miles) that it travels in t hours is given by the formula

$$d = 50t$$

Determine the distance it travels (a) in 1 hour; (b) in 4 hours.
Solution:
(a) Substitute 1 for t in the formula $d = 50t$, and obtain

$$d = 50 \cdot 1 = 50$$

Thus in 1 hour the car travels 50 miles.
(b) Substitute 4 for t:

$$d = 50 \cdot 4 = 200$$

In 4 hours the car travels 200 miles. ◀

Answers

5. ____________

6. ____________

PRACTICE EXERCISES FOR TOPIC B

5. Use the formula $A = \dfrac{bh}{2}$ to find the area of a triangle in which b, the length of the base is 15 meters and h, the length of the altitude is 8 meters.

6. The total number of degrees, d, for the angles of an n-sided polygon is given by the formula $d = (n - 2) \cdot 180$. Find the total number of degrees for the angles of a 5-sided polygon.

Now try the exercises for Topic B on page 62.

Use this area for work space.

EXERCISES

A. In Exercises 1–16 evaluate each polynomial for the specified number. (*See Examples 1 and 3.*)

1. $3x$; $\boxed{x = 4}$ **2.** $t + 5$; $\boxed{t = 2}$ **3.** $6x + 1$; $\boxed{x = 1}$

4. $2x - 3$; $\boxed{x = 1}$ **5.** $7y + 9$; $\boxed{y = 0}$ **6.** $5z - 3$; $\boxed{z = -1}$

7. z^2; $\boxed{z = 6}$ **8.** u^3; $\boxed{u = -3}$ **9.** $x^2 - 2$; $\boxed{x = 4}$

10. $x^2 + 4x$; $\boxed{x = -1}$ **11.** $x^2 + x + 1$; $\boxed{x = 3}$ **12.** $x^2 + 5x + 1$; $\boxed{x = -5}$

13. $t^4 - t^2 + 1$; $\boxed{t = 1}$ **14.** $t^5 - 4t^2 - 2t + 3$; $\boxed{t = -1}$

15. $x^3 - 7x^2 + 14x - 3$; $\boxed{x = 3}$ **16.** $z^4 - z^2 + 5$; $\boxed{z = 10}$

In Exercises 17–30 evaluate each polynomial for the specified numbers. (*See Examples 4 and 5.*)

17. xy; $\boxed{x = 4, y = -3}$ **18.** $2xy + 1$; $\boxed{x = 10, y = 4}$

19. $x + 2y$; $\boxed{x = -6, y = 8}$ **20.** $x^2 - y^2$; $\boxed{x = 1, y = -1}$

21. $s^2 - 3t^2$; $\boxed{s = 10, t = 5}$ **22.** $5mn + 2m - n$; $\boxed{m = 8, n = -2}$

Use this area for work space.

23. $3a^2 - 2b^2 + c$; $a = 4, b = 3, c = 2$

24. $x^2 + y^2 - z^2 + xyz$; $x = 4, y = -3, z = -2$

25. $2a - 7b + 9c - 2$; $a = b = c = 3$

26. $a + 2b - c + 5d$; $a = 10, b = 6, c = -1, d = 2$

27. $a^3 - bc^2 + 2ab$; $a = 2, b = 5, c = -2$

28. $x^2y - yx^2 + x - y$; $x = 10, y = -1$

29. $x^3 + x^2 - 2y + 3xy$; $x = 100, y = -1$

30. $2u^2 - 3v^2 + w^2$; $u = 1, v = 2, w = 3$

In Exercises 31–40 evaluate each polynomial for the three specified numbers. (*See Example 2.*)

Sample: Evaluate $x^2 - 2x + 5$ when (a) $x = 2$, (b) $x = -3$, (c) $x = 10$.	*Solution:* (a) $2^2 - 2 \cdot 2 + 5 = 4 - 4 + 5$ $= 5$ (b) $(-3)^2 - 2(-3) + 5$ $= 9 + 6 + 5$ $= 20$ (c) $10^2 - 2 \cdot 10 + 5$ $= 100 - 20 + 5$ $= 85$

31. $x + 10$; (a) $x = 2$, (b) $x = -2$, (c) $x = -10$

32. $2x - 8$; (a) $x = 6$, (b) $x = 0$, (c) $x = 4$

33. $5y + 9$; (a) $y = 0$, (b) $y = 2$, (c) $y = 10$

34. $z^2 + 1$; (a) $z = 0$, (b) $z = 1$, (c) $z = -1$

35. $x^2 + 2x$; (a) $x = 0$, (b) $x = 4$, (c) $x = -4$

36. $x^2 + 5x + 2$; (a) $x = 2$, (b) $x = 3$, (c) $x = 4$

37. $2x^2 - 3x + 1$; (a) $x = 4$, (b) $x = -4$, (c) $x = 6$

38. $u^3 - 2u + 6$; (a) $u = 0$, (b) $u = -1$, (c) $u = 2$

39. $b^4 - b^2 + 1$; (a) $b = 1$, (b) $b = -2$, (c) $b = 3$

40. $m^3 - 2m^2 + m - 4$; (a) $m = -1$, (b) $m = 2$, (c) $m = -3$

B.

41. Find the area of a square if each side is of length (a) 11 inches, (b) 15 inches, (c) 30 inches. (*See Example 6.*)

42. Find the area of a triangle if
(a) $b = 20$ inches and $h = 30$ inches;
(b) $b = 18$ inches and $h = 32$ inches. (*See Example 7.*)

Use this area for work space.

43. The area of a rectangle is given by the formula $A = lw$, where l is the length and w the width. Find the area if $l = 40$ feet and $w = 22$ feet. (*See Figure 2.3.*)

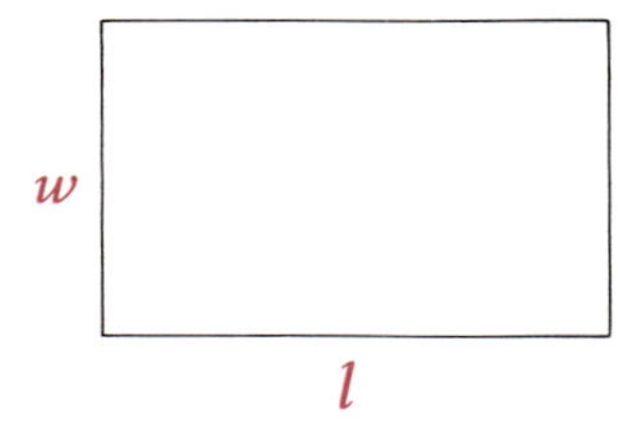

***FIGURE 2.3.* The area of a rectangle is lw. The perimeter is 2l + 2w.**

44. The perimeter of (distance around) a rectangle is given by the formula $P = 2l + 2w$, where l and w are as in Exercise 43. Find the perimeter if $l = 40$ inches and $w = 22$ inches. (*See Figure 2.3.*)

45. How long must a rope be to enclose a rectangular playground that is 100 meters long and 50 meters wide? (*Refer to Figure 2.3.*)

46. The volume of a rectangular box is given by the formula $V = lwh$, where l is the length, w the width, and h the height. Find the volume if
(a) $l = 10$ inches, $w = 6$ inches, and $h = 8$ inches;
(b) $l = 20$ inches, $w = 10$ inches, and $h = 12$ inches. (*See Figure 2.4.*)

47. The surface area of a rectangular box is given by the formula $S = 2lw + 2lh + 2wh$, where l, w, and h are as in Exercise 46. Find the surface area if
(a) $l = 10$ feet, $w = 6$ feet, and $h = 8$ feet;
(b) $l = 20$ feet, $w = 10$ feet, and $h = 12$ feet. (*See Figure 2.4.*)

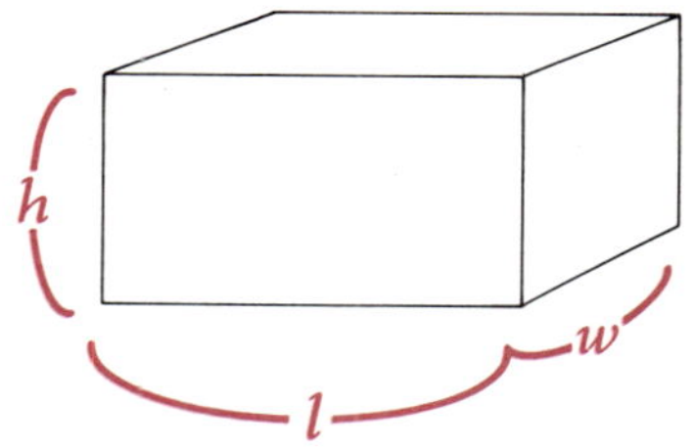

***FIGURE 2.4.* The volume of a rectangular box is lwh. The surface area is 2lw + 2lh + 2wh.**

48. The inside of a cigar box is of length 10 inches, width 6 inches, and height 4 inches.
(a) What is the volume of the inside of this box?
(b) How much paper is needed to line the inside? (*See Exercises 46 and 47.*)

49. An airplane travels at the rate of 300 miles per hour. The distance d (in miles) that it travels in t hours is given by the formula

$$d = 300t$$

Determine the distance it travels
(a) in 1 hour (b) in 3 hours (c) in 10 hours (*See Example 8.*)

50. A piece of material cost \$4 per square meter. The cost C (in dollars) for x square meters of this material is given by the formula

$$C = 4x$$

Determine the cost of
(a) 2 square meters (b) 5 square meters (c) 12 square meters
of this material.

Use this area for work space.

51. A ball is thrown upward from ground level with an initial velocity of 96 feet per second. Its height h (in feet) after t seconds is given by the formula

$$h = 96t - 16t^2$$

Determine its height (a) after 1 second (b) after 2 seconds (c) after 6 seconds.

52. The total value v (in dollars) of a number of coins is given by the formula

$$v = \frac{p + 5n + 10d + 25q + 50h}{100}$$

where p, n, d, q, and h are, respectively, the number of pennies, nickels, dimes, quarters, and half-dollars. Find the total value of 10 pennies, 12 nickels, 13 dimes, 14 quarters, and 9 half-dollars.

2.3 Adding Terms

A. Like Terms

Adding $\underbrace{5x + 3x}_{8x}$ is like adding $\underbrace{\text{5 apples and 3 apples}}_{\text{8 apples}}$, but adding $5x + 3y$ is like adding 5 apples and 3 oranges. For every value of x,

$$5x + 3x = 8x$$

For example, when $x = 2$,

$$5x = 5(2) = 10, \qquad 3x = 3(2) = 6, \qquad 8x = 8(2) = 16$$

and $$10 + 6 = 16$$

so that $$5(2) + 3(2) = 8(2)$$

When $x = -3$,

$$5x = 5(-3) = -15, \qquad 3x = 3(-3) = -9, \qquad 8x = 8(-3) = -24$$

and $$(-15) + (-9) = -24$$

so that $$5(-3) + 3(-3) = 8(-3)$$

You can add and thereby simplify terms that are alike in their variables and in the exponent of each variable.

DEFINITION

LIKE TERMS. Two or more nonzero terms are said to be **like terms** if

1. they contain exactly the same variables, and
2. each variable has the same exponent in each term.

Thus *like terms can differ only in their numerical coefficients or in the order of their variables.*

Example 1 ▶ (a) $5x$ and $3x$ are like terms.
(b) $9x^2$ and $-4x^2$ are like terms.
(c) $6xy$, $10xy$, and $-xy$ are like terms.
(d) $2uv$ and $5vu$ are like terms. They contain the same variables, u and v, and each variable has exponent 1. Note that like terms can differ in the *order* of their variables.
(e) $12x^2yz^3$, $25x^2yz^3$, and $-9x^2yz^3$ are like terms.
(f) 7, 2, and -1 are like (numerical) terms. None of these terms contains a variable. For that matter, any group of (nonzero) numerical terms are like terms. ◀

Here now are some examples of *unlike terms*.

Example 2 ▶ (a) $5x$ and $3y$ are unlike terms. They differ in their variables.
(b) x^2 and x^3 are unlike terms. They differ in the exponents of x.
(c) tu and tv are unlike terms. Each term contains a variable that is not in the other term.
(d) x^2y and xy^2 are unlike terms. They differ in the exponents of x as well as in the exponents of y. ◀

Answers

1. ______

2. ______

______

______

______

PRACTICE EXERCISES FOR TOPIC A

1. Indicate which pairs are like terms.

(a) $10x$ and $-x$ (b) $3x^2$ and $3x^3$ (c) $4x$ and $4y$ (d) $5xy^2$ and y^2x

2. Separate the following terms into groups of like terms:

$$2xy,\ -2,\ 3yx,\ -2yx^2,\ 5x^2,\ 5,\ 5x^2y,\ -xy$$

Now try the exercises for Topic A on page 68.

B. Adding Like Terms

To add or subtract *like* terms, extend the *Distributive Laws* from arithmetic to algebra. For example,

$$3x + 2x = (3 + 2)x = 5x$$

and

$$3x - 2x = (3 - 2)x = 1x = x$$

Let a and b be any numbers, and let x be a variable.

$$ax + bx = (a + b)x$$

and

$$ax - bx = (a - b)x$$

The Distributive Laws also apply to like terms that contain two or more variables.

Example 3 ▶ $5xy + 4xy = (5 + 4)xy = 9xy$ ◀

Example 4 ▶ $3x^2z^3 - x^2z^3 = (3 - 1)x^2z^3 = 2x^2z^3$ ◀

As with numbers, the Distributive Laws apply to adding or subtracting more than two like terms.

Example 5 ▶ $xyz + 4xyz - 2xyz = (1 + 4 - 2)xyz = 3xyz$ ◀

Answers

3. (a) ______

(b) ______

(c) ______

PRACTICE EXERCISES FOR TOPIC B

3. Add or subtract, as indicated.

(a) $9x + x$ (b) $-3xy + 5xy$ (c) $3z^2 - 7z^2$

Now try the exercises for Topic B on page 68.

C. General Addition of Terms

The Distributive Laws do not apply to unlike terms. Simply combine unlike terms with + or − signs.

Example 6 ▶ Add xy and $3z$.
Solution:
xy and $3z$ are unlike terms. Write

$$xy + 3z$$

This expression cannot be simplified. ◀

Example 7 ▶ Subtract $2t$ from the sum of r and $3s$.
Solution:
$2t$, r, and $3s$ are unlike terms. You obtain the expression

$$(r + 3s) - 2t$$

You can omit the parentheses here, and thereby simplify this to

$$r + 3s - 2t$$ ◀

To add or subtract several terms, arrange them so that like terms are grouped together. You may either group like terms in columns, as in Example 8, or you may group like terms by means of parentheses, as in Example 9. In rearranging the terms, you are extending the Commutative Law

$$a + b = b + a$$

and the Associative Law

$$(a + b) + c = a + (b + c)$$

of addition.

Example 8 ▶ Add and subtract as indicated.

$$2x + 3y - z + 4x - 4y + z$$

Solution:
Arrange like terms in columns.

$$\begin{array}{l} 2x + 3y - z \\ 4x - 4y + z \\ \hline 6x - y \end{array}$$ ◀

Example 9 ▶ Add and subtract as indicated.

$$3a + 2b - c + 4a + 2c - 7b + 9c$$

Solution:
Group like terms together within parentheses.

$$(3a + 4a) + (2b - 7b) + (-c + 2c + 9c) = 7a - 5b + 10c$$ ◀

↖Insert this sign.

Recall that for any *number* a,

$$a + 0 = 0 + a = a \quad \text{and} \quad a - 0 = a$$

These rules are true if a is any *term*. Thus

$$x - 0 = x$$

$$3t - 0 = 3t$$

$$5x^2y + 0 = 5x^2y$$

PRACTICE EXERCISES FOR TOPIC C

Answers
4. ______
5. ______
6. (a) ______
(b) ______
(c) ______
(d) ______

4. Write an expression for the sum of $3x$, xy, and $-y$.

5. Express the result of subtracting 8 from the sum of $9x$ and x^2.

6. Simplify the following expressions by combining like terms, wherever possible.

(a) $6x - 4 - 5x + 1$ (b) $9y - 7 + 3y - 8$ (c) $x^2 + 3x - 3x^2 + 2x$

(d) $ab - b - 2a + ab + b - a$

Now try the exercises for Topic C on page 68.

Use this area for work space.

EXERCISES

A. In Exercises 1–10 indicate which pairs are like terms. (*See Examples 1 and 2.*)

1. $2x$ and $4x$

2. $3y$ and $-3y$

3. $15xy$ and $5yx$

4. $13xy$ and $13xz$

5. $-2x^2$ and $2x^3$

6. 15 and 30

7. x^3y and xy^3

8. 1 and -1

9. x^2 and $4 \cdot x \cdot x$

10. $3x^2$ and $2x^3$

In Exercises 11–14 separate the terms into groups of like terms. (*See Practice Exercise 2.*)

11. $5, 5x, 3x, 5y, 3y$

12. $a^2, a, 4a^2, 2a^4, 2a, 4a, a \cdot a$

13. $x^2y, xy^2, -xy^2, 3xy, 3x^2y, x^2y^2, 2xy, -3xy^2$

14. $-1, ab, abc, -ab, 12, -abc, a^{12}bc, 12abc$

B. In Exercises 15–32 add or subtract, as indicated. (*See Examples 3–5.*)

15. $2x + x$

16. $5y + 2y$

17. $2x - x$

18. $5y + (-2)y$

19. $xy + 3xy$

20. $uv - 2uv$

21. $abc + 4abc$

22. $x^2y + 5x^2y$

23. $10xy - 4yx$

24. $2x^2yz + zyx^2$

25. $3x + x + x$

26. $5t - 2t + t$

27. $2s - s - s$

28. $5r - 3r + 6r$

29. $4w + w + 2w + w$

30. $3x - x + 2x - x$

31. $2xyz - xyz + 5xyz$

32. $5r + (-5r)$

33. Add $3x$, $5x$, and $6x$.

34. Subtract $2yz$ from the sum of $7yz$ and $-3yz$.

C. **35.** Write an expression for the sum of $5a$, $3b$, and 6. (*See Practice Exercise 4.*)

36. Express the result of subtracting $2x$ from the sum of $10y$ and 5. (*See Practice Exercise 5.*)

37. Express the result of subtracting the sum of $3c$ and $-4d$ from 12.

In Exercises 38–58 simplify the following expressions by combining like terms, wherever possible. In some cases you will not be able to simplify the given expression. (*See Examples 6–9.*)

38. $xy + 3x^2$

39. $6u - 3v$

40. $4s - 2t + s$

41. $2x - y + 3x + y$

42. $5x + 3$

43. $2a + 6b + 5a + 6b - a$

Use this area for work space.

44. $2m - 4n + 7n - 3m$

45. $2ab - 3cd + 5ab - cd$

46. $3xy + 5xy - 2z + 6z^2$

47. $4a + b - c + 2a - b - 2c$

48. $3x - 2y + 5x - z + 6x - 2z$

49. $3m - 2m + n - 3m + 2n$

50. $3x^2y + 5z^2 + 3x^2y + 5z^2$

51. $4xyz - 3abc + 2xyz - 5abc + 7xyz$

52. $2a + 5b - c + d + 6a - d + 7a$

53. $5x^2 + 3y^2 - z^2 + 6x^2 - z^2 - 11x^2$

54. $6x + y + 1 - 2x - 3 + y + 10$

55. $9 - 4u^2 - v^2 + 5 + 4u^2 + v^2$

56. $7 - 2u + 5v - 3 + 2u + 5v$

57. $9m + 2n - 6p + 2n - 4p - 9m$

58. $2x - 3y + 5z - 3y - 5z + 6y - 2x$

59. Add $2a$, $5b$, $3a$, and $-5b$.

60. Subtract $3y$ from the sum of $4y$, $2x$, and $-y$.

61. Add $3x$ to the sum of $5x$, $2y$, $3z$, $-4x$, and $6y$.

62. Subtract $2a$ from the sum of $3a$, $-4b$, $6c$, $-a$, and $7c$.

63. A carpenter earns x dollars per hour. On Monday he works 7 hours, on Tuesday 8 hours, on Wednesday 10 hours, on Thursday 8 hours, and on Friday 5 hours.
(a) Express his earnings on Monday in terms of x.
(b) Express his earnings for all 5 days in terms of x.

64. A wire is bent as indicated in Figure 2.5. Express the length of the wire in terms of l.

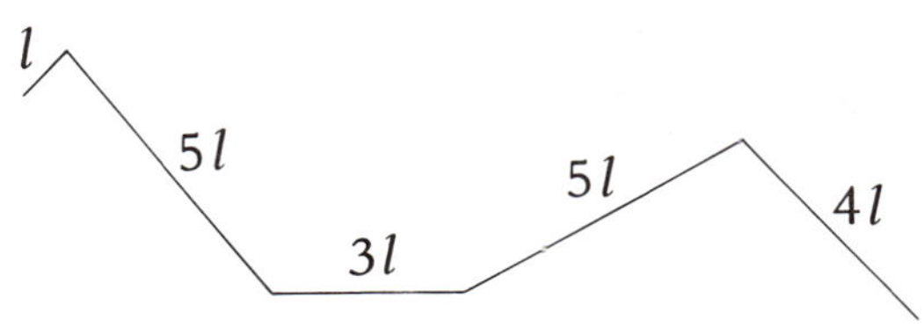

FIGURE 2.5

65. Which of the following equal $4xy$?
(a) $2x + 2y$ (b) $3xy + xy$ (c) $6x - 2y$
(d) $x + y + 2xy$ (e) $9xy - 5xy$

2.4 Adding Polynomials

A. Monomials, Binomials, Trinomials

For the most part, you will be working with polynomials that have either one, two, or three unlike terms.

DEFINITION

> A nonzero term is also called a **monomial.** A polynomial that is the sum (or difference) of two unlike *monomials* is called a **binomial,** and of three unlike *monomials,* a **trinomial.**

Example 1 ▶ (a) $3xy$ is a nonzero term, and hence a monomial.
(b) $2x + 5y$ is the sum of two unlike monomials. It is a binomial.
(c) $5x - y + z$ is the sum of three unlike monomials. It is a trinomial. ◀

Example 2 ▶ $2x + 5x + x$ is a monomial because $2x$, $5x$, and x are like terms. Thus

$$2x + 5x + x = 8x$$ ◀

A *nonzero* numerical term, such as 5 or -3, is classified as a monomial. But for technical reasons, the numerical term 0 is *not* classified as a monomial. It is, nevertheless, a particularly important polynomial, known as the **zero polynomial.**

Answers

1. (a) ____
(b) ____
(c) ____
(d) ____

PRACTICE EXERCISES FOR TOPIC A

1. Classify each polynomial as (*i*) a monomial, (*ii*) a binomial, (*iii*) a trinomial, or (*iv*) none of these.

(a) $2x - 3y$ (b) $3x^2 + 5x - 1$ (c) $x^2 - x^2$ (d) 2

Now try the exercises for Topic A on page 72.

B. Adding Polynomials

To add or subtract polynomials, group like terms together, as in the preceding section.

Example 3 ▶ Add the polynomials $2x + y$, $3x + y - z$, and $5y + z$.
Solution:
Rearrange the terms so that like terms are in the same column.

$$\begin{array}{r} 2x + y \phantom{{}+z} \\ 3x + y - z \\ \underline{\phantom{3x+{}} 5y + z} \\ 5x + 7y \phantom{{}+z} \end{array}$$ ◀

For any polynomial P,

$$P + 0 = 0 + P = P$$

Answers

2. ____

PRACTICE EXERCISES FOR TOPIC B

2. Add the polynomials:

$$\begin{array}{r} x - 9xy + 2y \\ 4x - xy - 3y \\ \underline{\phantom{4x-{}} 10xy - y} \end{array}$$

3. ______

3. Add the polynomials $3x^2 + 8$, $x^2 - 2x - 10$, and $8x + 1$

Now try the exercises for Topic B on page 73.

C. Inverse of a Polynomial

Recall that the inverse of a number a, written $-a$, is the number added to a to obtain 0. Thus the inverse of 7 is -7, and the inverse of -7 is 7. Note that

$$7 + (-7) = (-7) + 7 = 0$$

DEFINITION

> The **inverse of a polynomial** P is the polynomial added to P to obtain 0.

For any polynomial P, let $-P$ denote the inverse of P. Thus

$$P + (-P) = (-P) + P = 0$$

Example 4 ► The inverse of $x - 2y + 3z$ is $-x + 2y - 3z$.

$$\begin{array}{r} x - 2y + 3z \\ + \; -x + 2y - 3z \\ \hline 0 \end{array}$$
◄

Note that the inverse of $x - 2y - 3z$ is obtained by changing the sign of each of its terms. In general, *the inverse of a polynomial P is obtained by changing the sign of each term of P.*

Answers

4. (a) ______

(b) ______

(c) ______

PRACTICE EXERCISES FOR TOPIC C

4. Determine the inverse of each of the following polynomials.

(a) $-x + 4y + 3z$ (b) $2x^3 - x^2 + 8x - 4$ (c) $3x^2 - 3xy + y^2$

Now try the exercises for Topic C on page 73.

D. Subtracting Polynomials

Recall that subtraction of numbers was defined in terms of addition. *To subtract the polynomial Q from P, add −Q.*

$$P - Q = P + (-Q)$$

Example 5 ▶ *Subtract* the bottom polynomial from the top one.

$$\begin{array}{r} 4a - 3b + c \\ \underline{2a - 3b - c} \end{array}$$

Solution:
Change the sign of each term of the bottom polynomial and add.

$$\begin{array}{rrrr} & 4a & - 3b & + \ c \\ - & & + & + \\ & +2a & - 3b & - \ c \\ \hline & 2a & & + 2c \end{array}$$ ◀

Example 6 ▶ Simplify $10a - [4b - (3a - b)]$.
Solution:
Begin with the innermost parentheses and work outward, as you do when adding and subtracting numbers.

$$\begin{aligned} 10a - [4b - (3a - b)] &= 10a - [4b - 3a + b] \\ &= 10a - 4b + 3a - b \\ &= (10a + 3a) + (-4b - b) \\ &= 13a - 5b \end{aligned}$$ ◀

Answers

5. ____________

6. ____________

PRACTICE EXERCISES FOR TOPIC D

5. Subtract the bottom polynomial from the top one.

$$\begin{array}{r} 9m + mn - n \\ \underline{7m - 2mn + n} \end{array}$$

6. Simplify $6x + 1 - [12 - (3x + 4)]$.

Now try the exercises for Topic D on page 74.

Use this area for work space.

EXERCISES

A. In Exercises 1–12 classify each polynomial as (*i*) a monomial, (*ii*) a binomial, (*iii*) a trinomial, or (*iv*) none of these. (*See Examples 1 and 2.*)

1. $2x$ **2.** $3x + 5y$ **3.** 5 **4.** -5

5. 0 **6.** $-x$ **7.** $4x^2 + 3x + 1$

8. $x^3 + x^2 + x + 1$ **9.** $4x + 3x + 2x + x$ **10.** $2x + 3y + 2x + y$

11. $2x + y + 2x - y$ **12.** $2x + y - 2x - y$

Use this area for work space.

B. In Exercises 13–28 add the polynomials: (*See Example* 3.)

13.
$$\begin{array}{r} 2x + y \\ x + 4y \\ \hline \end{array}$$

14.
$$\begin{array}{r} 2r - s \\ 3r - 2s \\ \hline \end{array}$$

15.
$$\begin{array}{r} 8a - 5b + 2 \\ -7a + 5b + 1 \\ \hline \end{array}$$

16.
$$\begin{array}{r} 6x - 3y + 7 \\ 2x + 3y - 5 \\ \hline \end{array}$$

17.
$$\begin{array}{r} a + b + 2c \\ 2a + b + c \\ \hline \end{array}$$

18.
$$\begin{array}{l} 5r + s - t \\ 2r \quad\quad + t \\ \hline \end{array}$$

19.
$$\begin{array}{r} 3x + 5y - 2z \\ 2x - 5y + 2z \\ \hline \end{array}$$

20.
$$\begin{array}{r} 6r^2 + s^2 - 3t^2 \\ r^2 - s^2 + t^2 \\ \hline \end{array}$$

21.
$$\begin{array}{r} a + b - c \\ 2a + b - 2c \\ 4a - 2b + c \\ \hline \end{array}$$

22.
$$\begin{array}{r} 5x - 2y + z \\ 6x - 3y + z \\ x - y - z \\ \hline \end{array}$$

23.
$$\begin{array}{l} 3r + 2s - t - u \\ 2r - s + t \\ \quad\quad s \quad\quad + u \\ \hline \end{array}$$

24.
$$\begin{array}{r} w + x - y + z \\ w - x + y - z \\ 2w + x - y \quad\quad \\ x + y + z \\ \hline \end{array}$$

25.
$$\begin{array}{rrrr} 10a & + 7b & - 3c & + d \\ 5a & - 2b & + 3c & + 9d \\ 7a & - 2b & + 5c & \\ & 2b & & + d \\ \hline \end{array}$$

26.
$$\begin{array}{llll} 9t & + 5u & + 6v & + 12w \\ 3t & + 8u & - 2v & + 7w \\ 2t & & - 7v & + 2w \\ 3t & - 6u & & + 12w \\ \hline \end{array}$$

27.
$$\begin{array}{rrrrr} a & + b & & - d & + 2e \\ & - b & + 4c & & + e \\ 3a & - b & & + 7d & \\ 5a & & - 2c & + d & \\ -2a & & & - d & - e \\ \hline \end{array}$$

28.
$$\begin{array}{rrrr} 16w & + 22x & - 17y & + 9z \\ 12w & - 17x & + 9y & - 8z \\ 32w & - 25x & & + 10z \\ 28w & + 7x & - 13y & \\ 12w & - 5x & + 7y & \\ \hline \end{array}$$

29. Add the polynomials $6m + 9n$ and $5m + 5n$.

30. Add $3a + b$, $5a - 2b$, and $2a$.

31. Add $2x - y + 4z$, $10x - 5z$, and $-2x + 6y$.

32. Suppose x represents some number. Express each of the following numbers in terms of x.
(a) two more than the number
(b) three times the number
(c) one less than twice the number
(d) the sum of expressions (a), (b), and (c)

33. In order to tape up a rectangular box of length l, width w, and height h, Alice cuts off strips of tape that measure $l + 5$, $w + 3$, $h + 4$, $2l$, $w + 2$, and $2h - 3$. Express the total length of tape used in terms of l, w, and h.

34. Bob has x nickels and y dimes. Jane has x pennies and y quarters. How much money do they have together (in cents)?

C. In Exercises 35–38 determine the inverse of each polynomial. (*See Example* 4.)

35. $x + 2y + 5z$

36. $a - b + 4c - d$

37. $x^2 - 4x + 7$

38. $-5x^3 + 10x^2 - x - 1$

Use this area for work space.

D. In Exercises 39–47 subtract the bottom polynomial from the top one. (*See Example 5.*)

39. $\begin{array}{r} 4a + 2b \\ \underline{2a + b} \end{array}$

40. $\begin{array}{r} 7c - 6d \\ \underline{2c + 5d} \end{array}$

41. $\begin{array}{r} 2m + 5n \\ \underline{-m + 5n} \end{array}$

42. $\begin{array}{r} -7s + 8t \\ \underline{-7s - 8t} \end{array}$

43. $\begin{array}{r} 12y - 15z \\ \underline{-9y + 18z} \end{array}$

44. $\begin{array}{r} 6a - 3b + 2c \\ \underline{a - c} \end{array}$

45. $\begin{array}{r} 9x - 12y + 7z \\ \underline{2x + y - z} \end{array}$

46. $\begin{array}{r} 2a - b + c - d \\ \underline{-a + 2b - c + d} \end{array}$

47. $\begin{array}{r} r - s + t - 2u + v \\ \underline{3r - s - t + 2u - v} \end{array}$

48. Pick out the *error(s)*.

(a) $-(2x - 3y + 4z) = -2x - 3y - 4z$

(b) $-(4 - 5y - 9z) = 9z + 5y - 4$

(c) The sum of a polynomial and its inverse is always 0.

(d) The inverse of the inverse of polynomial P is P itself.

In Exercises 49–56 simplify each polynomial. (*See Example 6.*)

49. $(4a - 2b) - (3a - 2b)$

50. $(5x - 2y + z) - (2x - y - z) + (4x - y)$

51. $(r - s + t) - (3r + s - t) - (2r - 3s)$

52. $3x - y - [2x - y - (x + y)]$

53. $6x - 2y - [-x - (y - x)]$

54. $5a - 2b + c - [(3a - b) - (2a - b + c)]$

55. $7 - 2a - 3b + [5a - (3b - c)]$

56. $2x - 3y - [3x - (z - 2y) - (x - y + z)]$

57. Subtract $2x + 9y$ from $6x + 10y$.

58. Subtract $x - y$ from $x + y$.

59. Subtract $a - 2b$ from the sum of $3a + b$ and $5a - b$.

60. Subtract $2x^2$ from 0.

2.5 Multiplying and Dividing by a Number

A. Commutative and Associative Laws

Numbers can be multiplied in any order and in any grouping. Thus for any *numbers a, b,* and *c,*

$ab = ba$ *Commutative Law of multiplication*

$(ab)c = a(bc)$ *Associative Law of multiplication*

These laws also apply to multiplication of *terms* and of *polynomials.* For example,

$$3(2) = 2(3)$$
$$3(5) = 5(3)$$
$$3(-1) = (-1)(3), \qquad \text{and so on}$$

In general, for any number that x represents,

$$3x = x \cdot 3$$

Example 1 ▶

$$2(5x) = (2 \cdot 5)x \qquad \textit{by the Associative Law}$$
$$= 10x$$

Similarly, for any numbers x and y,

$$xy = yx \qquad \textit{by the Commutative Law}$$

and $(5x)(4y) = (5 \cdot 4)(xy) = 20xy$ ◀

Example 2 ▶ Multiply $(6a) \cdot (3b) \cdot (2c)$.

Solution:

Use the Commutative and Associative Laws to bring numerical coefficients to the left.

$$(6a) \cdot (3b) \cdot (2c) = (6 \cdot 3 \cdot 2) \cdot (a \cdot b \cdot c) = 36abc$$ ◀

Answers

1. (a) ____

(b) ____

(c) ____

(d) ____

PRACTICE EXERCISES FOR TOPIC A

1. Use the Commutative and Associative Laws of multiplication to multiply.

(a) $8(3x)$ (b) $(7x)(5)$ (c) $(6x)(-2y)$ (d) $(-10a)(-9b)(-2c)$

Now try the exercises for Topic A on page 78.

B. Distributive Laws

The perimeter P of a rectangle is given by the formula

$$P = 2l + 2w$$

where l is the length and w the width. (See Figure 2.6.) But observe in Figure 2.6 that

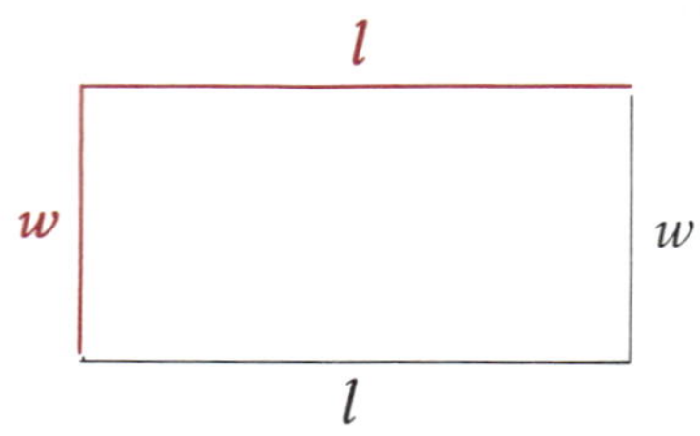

FIGURE 2.6. $\boldsymbol{P = 2l + 2w = 2(l + w)}$

the part of the boundary in black measures the same as the part in color. Each measures $l + w$. Thus the perimeter can also be given by

$$P = 2(l + w)$$

Therefore $$2(l + w) = 2l + 2w$$

This illustrates that the Distributive Laws, p. 44, apply to algebraic expressions as well as to numbers.

Let P, Q, and R be *terms* or *polynomials*. Then, according to the Distributive Laws,

$$P(Q + R) = PQ + PR$$

$$(Q + R)P = QP + RP$$

Similarly

$$P(Q - R) = PQ - PR$$

$$(Q - R)P = QP - RP$$

For the present, the factor P will be an integer.

Example 3 ▶ $$6(x + y) = 6x + 6y$$ ◀

Example 4 ▶ $$-2(x - y) = -2x \underbrace{-\,(-2)}_{+}y = -2x + 2y$$ ◀

The Distributive Laws also apply to sums of more than two terms.

Example 5 ▶
$$5(2x - y + 4z - 2) = 5(2x) - 5(y) + 5(4z) - 5(2)$$
$$= 10x - 5y + 20z - 10$$ ◀

Answers

2. (a) ______

(b) ______

(c) ______

PRACTICE EXERCISES FOR TOPIC B

2. Use the Distributive Laws to simplify.

(a) $15(2x + y - 3z)$ (b) $-8(-a + 5b)$ (c) $(7x - 2)(-3)$

Now try the exercises for Topic B on page 79.

C. Dividing

Recall that division of numbers is defined in terms of multiplication. Thus

$$\frac{12}{3} = 4 \qquad \textit{because} \qquad 12 = 4 \cdot 3$$

Similarly, division of algebraic expressions is defined in terms of multiplication. Thus,

$$\frac{12x}{3} = 4x$$

because, by the Commutative and Associative Laws of multiplication,

$$(4x) \cdot 3 = 3 \cdot (4x) = (3 \cdot 4)x = 12x$$

Observe that

$$\frac{3x + 3y}{3} = x + y$$

by the definition of division because

$$(x + y)3 = 3(x + y) = 3x + 3y$$

by the Commutative Law of multiplication. Note that *both* terms of the polynomial $3x + 3y$ are divided by 3. In fact,

$$\frac{3x + 3y}{3} = \frac{3x}{3} + \frac{3y}{3} = x + y$$

Example 6 ▶ Divide $\frac{2x + 4}{2}$

Solution:

$$\frac{2x + 4}{2} = \frac{2x}{2} + \frac{4}{2} = x + 2$$ ◀

Example 7 ▶ Divide $\frac{8x + 12y - 4}{-4}$

Solution:

$$\frac{8x + 12y - 4}{-4} = \frac{8x}{-4} + \frac{12y}{-4} + \frac{-4}{-4} = -2x - 3y + 1$$ ◀

PRACTICE EXERCISES FOR TOPIC C

3. Divide.

(a) $\frac{6x^2}{2}$ (b) $\frac{15x - 9}{3}$ (c) $\frac{20x - 30y + 10}{-10}$

Answers
3. (a)
(b)
(c)

Now try the exercises for Topic C on page 79.

D. Several Operations

Example 8 ▶ Simplify $3(2x + 3y) - 4(3x - 2y)$
Solution:

$$\begin{aligned} 3(2x + 3y) - 4(3x - 2y) &= (6x + 9y) - (12x - 8y) \\ &= 6x + 9y - 12x + 8y \\ &= (6x - 12x) + (9y + 8y) \\ &= -6x + 17y \end{aligned}$$ ◀

Example 9 ▶ Simplify $-2[(a - b) - 3(a + b)]$
Solution:
Work from inner parentheses outward.

$$\begin{aligned} -2[(a - b) - 3(a + b)] &= -2[a - b - 3a - 3b] \\ &= -2[-2a - 4b] \\ &= -2(-2a) - (-2)4b \\ &= 4a + 8b \end{aligned}$$ ◀

Answers

4. (a) ______
(b) ______
(c) ______
(d) ______
5. ______

PRACTICE EXERCISES FOR TOPIC D

4. Combine and simplify.

(a) $6(x - 4y) + 2(y - x)$ (b) $5(3a - b) - 2(a + 2b)$ (c) $4 - 2[a - (3a + 1)]$

(d) $\dfrac{12x + 8}{4} - 3(3 - x)$

5. Multiply the sum of $3a + b$ and $2a - 3b$ by 2.

Now try the exercises for Topic D on page 79.

Use this area for work space.

EXERCISES

A. In Exercises 1–12 multiply. (*See Examples 1 and 2.*)

1. $5(3a)$ **2.** $10(2a^2)$

3. $-7(4x)$ **4.** $(-3)(-2x)$

5. $(2x)(8y)$ **6.** $(5a)(7b)$

7. $(-3m)(4n)$ **8.** $(-10x)(-6y)$

9. $(2x)(3y)(4z)$ **10.** $(-2a)(-b)(4c)$

11. $(-3l)(-2m)(-6n)$ **12.** $(x^2)(-4y)(-2z)$

Use this area for work space.

B. In Exercises 13–30 use the Distributive Laws to simplify. (*See Examples 3–5.*)

13. $4(x + y)$ **14.** $3(y + z)$ **15.** $-10(a + b)$ **16.** $9(x^2 + y)$

17. $2(a - b)$ **18.** $12(a - b)$ **19.** $-3(m - n)$ **20.** $-2(x^2 - y^2)$

21. $5(x + y)$ **22.** $2(x - y)$ **23.** $2(ax + 3y)$ **24.** $-3(3x + 2y^2)$

25. $4(a + b + c)$ **26.** $5(x + 2y - 3z)$

27. $-2(5l - 4m + 3n)$ **28.** $-(x - y + z)$

29. $-4(3w - 2x + y - z)$ **30.** $-2(6v - w + 3x - 7y - 10z)$

C. In Exercises 31–44 divide. (*See Examples 6 and 7.*)

31. $\frac{8a}{4}$ **32.** $\frac{25x^2}{5}$

33. $\frac{-18x}{6}$ **34.** $\frac{-18x}{-6}$

35. $\frac{30y}{-10}$ **36.** $\frac{-132x^2y}{-11}$

37. $\frac{8x + 8y}{8}$ **38.** $\frac{4x + 6}{2}$

39. $\frac{9x - 12y}{3}$ **40.** $\frac{5x^2 + 10}{-5}$

41. $\frac{2a + 4b + 6c}{2}$ **42.** $\frac{12x - 16y - 20}{4}$

43. $\frac{6x - 9y + 3z - 12}{3}$ **44.** $\frac{25a - 30b + 10c - 40d}{-5}$

D. In Exercises 45–54 combine and simplify. (*See Examples 8 and 9.*)

45. $3(2x + y) + 5x$ **46.** $2(x + y) - (2x + y)$

47. $4(a + 3b) + 5(2a + b)$ **48.** $3(5y - z) - 2(y - z)$

49. $2(x^2 - 3x + 1) - (x^2 + 4x + 2)$

50. $4x - 3(2x + y - z) - 3(x - y + z)$

51. $-3[x - (y - 2z) - 3(x - y)]$

52. $-2(x + 2y - 4) - 4(3y - 2) + \frac{2x - 4}{2}$

53. $\frac{6a + 9b - 3c}{3} + \frac{5a + 10b - 15c}{5}$

54. $\frac{2a - 8b - 4c}{-2} - \frac{3a - 6b - 9c}{-3}$

55. Multiply the sum of $x + 2y$ and $3x - 5y$ by 4.

56. Add the product of 3 and $2y$ to $4y + 1$.

57. Divide the sum of $8x + 9y$ and $6x - 2y$ by 7.

58. Subtract the product of $4a$ and $3x$ from the sum of $5ax$ and $2ax$.

Use this area for work space.

59. Ann has x dimes and y quarters. She divides her money equally among her five nephews. How many cents does each nephew receive?

60. Pick out the *error(s)*.

(a) $5(a + b) = 5a + 5b$

(b) $2(x - 2y) = 2x - 2y$

(c) $-(a - 2b) = 2b - a$

(d) $-3(x - 2y + 5z) = -3x + 6y - 15z$

61. Pick out the *error(s)*.

(a) $\dfrac{5x + 5y}{5} = x + y$

(b) $\dfrac{12a - 4b}{4} = 3a - b$

(c) $\dfrac{2x - 9}{3} = 2x - 3$

(d) $\dfrac{10u - 20v + 100w}{10} = u - 2v + 10w$

What Have You Learned in Chapter 2?

You have learned that a variable designates any one of the numbers being discussed, that a term is a number or a variable or a product of numbers and variables, and that a polynomial is a term or a sum of terms.

You can evaluate polynomials for specified values of their variables.

You can add and subtract polynomials, and you can multiply and divide polynomials by numbers.

REVIEW

Let's Review Chapter 2

2.1 *Terms and Polynomials*

1. Suppose x stands for some number. Express

seven more than the number

in terms of x.

2. Write $-10x^2y^3$ as a product of a number and variables.

3. Determine the coefficient of $-5x^4$.

4. $x^2 - 2x + 5$ is the sum of the terms ☐, ☐, and ☐

2.2 *Evaluating Polynomials*

5. Evaluate $x^2 + 3x + 10$ when $x = 5$.

6. Evaluate $x^2 - y^2$ when $x = 9$ and $y = 6$.

7. Evaluate $x^3 - 2x^2 + 4$ when (a) $x = 2$, (b) $x = -2$.

8. Find the area of a square if each side is of length 20 inches.

2.3 *Adding Terms*

9. Are $3x^2y$ and $-6x^2y$ like terms?

10. Are $10xy^3$ and $10x^3y$ like terms?

In Exercises 11–14 add and subtract, as indicated.

11. $7x + 3x$

12. $2uv + 3uv - uv$

13. $3m + 5n - m + 10n$

14. $4x + 3y - z + x - y - 2z$

2.4 *Adding Polynomials*

In Exercises 15 and 16 add the polynomials.

15. $\begin{array}{r} 3a + b \\ \underline{5a + 6b} \end{array}$

16. $\begin{array}{rrrr} w + & x + & 2y - & z \\ w - & x + & 4y & \\ & x - & y + & z \\ \underline{2w} & \underline{-} & \underline{y -} & \underline{z} \end{array}$

17. Subtract the bottom polynomial from the top one:

$$\begin{array}{l} 3a + b - c \\ \underline{2a - 2c} \end{array}$$

18. Simplify $(5x + y - 2z) - (3x - z) + (x - y + 4z)$.

19. Simplify $15a + b - [4a - (2a - b)]$.

20. Subtract $2m + 5n$ from $5m - 2n$.

2.5 *Multiplying and Dividing by a Number*

In Exercises 21–23 multiply.

21. $(-2a)(-3b)(5c)$

22. $8(x^2 + 3y)$

23. $-4(3x - 2y + 5z)$

24. Divide $\dfrac{25a - 10b + 15c}{5}$.

25. Combine and simplify:

$$-2(x^2 - 4x + 3) - \frac{4x^2 + 8x - 16}{4}$$

26. Multiply the sum of $2a - 3b$ and $5a + 6b$ by -2.

And These from Chapter 1

27. Fill in "<" or ">". $-7 \;\square\; -8$

28. Fill in "left" or "right."

-3 lies to the $\square$ of -12 (on the number line).

29. Write the inverse of (a) 2, (b) -2.

30. Which integers are 5 units from the origin?

31. Fill in "<" or ">". $-|10| \;\square\; |-10|$

32. Find the value of $8 - [(5 - 3) - (4 - 9)]$.

33. Find the indicated product: $4(-7)0(-1)$

34. Use the Distributive Laws to find the value of $3(6 + 2)$.

35. A car travels 50 miles east, then 80 miles west, and finally 20 miles east. How far from its starting point is the car? Is it east or west of its starting point?

Practice Exam 1 on Chapter 2

REVIEW

1. Find the coefficient of $-4x^2y^3$.
2. Evaluate $x^2 + 8x - 3$ when $x = 4$.
3. Evaluate $2x^2 + 5y^2$ when $x = 10$ and $y = 5$.
4. Add and subtract, as indicated: $2a - 3b + 5a + 3b - 7a$.
5. Add.

$$\begin{array}{rrrr} a + & 7b - & 10c + & d \\ 5a - & b - & 3c + & 9d \\ 6a & - & c & \\ & 5b + & c + & d \\ \hline \end{array}$$

6. Subtract the bottom polynomial from the top one:

$$\begin{array}{r} 5x - 3y + 9z \\ 2x - 6y + 10z \\ \hline \end{array}$$

7. Simplify $4m - (3n + 2p) - [6m - (2n - p)]$.
8. Multiply $(-5)(2a - 4b + 3c)$.
9. Divide $\dfrac{12x^2 - 20x + 36}{-4}$.
10. Combine and simplify $-2(3x - 2y + 4z) - \dfrac{4x - 2y + 8z}{2}$.
11. Multiply the sum of $a + b - 2$ and $2a + 4$ by -5.
12. The volume of a rectangular box is given by $V = lwh$. Find the volume if $l = 8$ inches, $w = 2$ inches, and $h = 5$ inches.

Practice Exam 2 on Chapter 2

1. Suppose z stands for some number. Write an expression for
   5 more than twice the number
2. What is the coefficient of the term $-8a^2b^3$?
3. Evaluate $2x^2 + 7x$ when $x = -1$.
4. Evaluate $3a - ab - b^2$ when $a = 2$ and $b = -3$.
5. Are $6ab^2$ and $-6ba^2$ like terms?
6. Simplify by combining like terms:

$$x^2 - 3x + 2x^2 + 4x + 2$$

REVIEW

7. Add:

$$\begin{array}{rrr} -\ x & +\ 4xy & -\ 2y \\ 5x & -\ \ xy & \\ -\ 4x & -\ 2xy & +\ \ y \\ \hline \end{array}$$

8. Subtract the bottom polynomial from the top one.

$$\begin{array}{rrr} -\ 15a^2 & +\ \ a & -\ 3 \\ 5a^2 & -\ 2a & -\ 1 \\ \hline \end{array}$$

9. Multiply:

(a) $(-3x)(-6y)$ (b) $-7(2a - b)$

10. Divide $\dfrac{30x^2 - 6x + 18}{-6}$

11. Simplify $\dfrac{8x - 12}{4} - 3(x + 1)$

3

Equations

3.1 Roots

A. What is an Equation?

Suppose you are asked:

What number plus 2 is equal to 5?

It is convenient to write this as follows:

What number plus 2 is equal to 5?

$$\underbrace{\text{What number}}_{x} \; \underset{+}{\text{plus}} \; \underset{2}{2} \; \underbrace{\text{is equal to}}_{=} \; \underset{5}{5?}$$

Here the variable x represents the unknown number. The statement

$$x + 2 = 5$$

is known as an *equation*. You are going to learn how to "solve" equations such as this one.

DEFINITION

> An **equation** is a statement of equality. Thus
>
> $$x + 2 = 5$$
>
> is an equation. Here $x + 2$ is the **left side** of the equation and 5 is the **right side.**

$$3 + 2 = 5$$

is also an equation because it is a statement of equality.

Example 1 ▶ Consider the equation

$$5x = 30 - x$$

(a) What is the left side of this equation?
(b) What is the right side?
Solution:
(a) $5x$ is the left side.
(b) $30 - x$ is the right side. ◀

Example 2 ▶ Express the question

What number divided by 5 is equal to 4?

as an equation (in the variable x). You need not "solve" the equation.
Solution:

What number	divided by	5	is equal to	4?
x	$\div$	5	$=$	4

or more conveniently,

$$\frac{x}{5} = 4$$ ◀

PRACTICE EXERCISES FOR TOPIC A

Answers

1. (a) ________
   (b) ________
2. ________

1. Consider the equation

$$3x - 2 = 19$$

(a) What is the left side of this equation?
(b) What is the right side?

2. Express the question

What number subtracted from 10 is equal to 3?

as an equation (in the variable x). You need not "solve" the equation.

3. ____________________

3. Express the equation

$$x + 3 = 9$$

in words.

Now try the exercises for Topic A on page 88.

B. Roots of Equations

Thus far you have learned that an equation is a statement of equality. And you have learned to express various questions as equations. Now we consider how to recognize "roots" or "solutions" of equations.

DEFINITION

ROOT. A number is a **root** of an equation (in one variable) if a true statement results when the number replaces the variable. A root of an equation is also called a **solution** of the equation.

Example 3 ▶ (a) 3 is a root of the equation

$$x + 2 = 5$$

because when you replace x by 3, you obtain the *true* statement

$$3 + 2 = 5$$

(b) 4 is *not* a root of the equation because

$$4 + 2 = 5$$

is a *false* statement. ◀

Example 4 ▶ Is 9 a root of the equation

$$y - 3 = 6?$$

Solution:
Yes, 9 is a root of this equation because when you replace y by 9, you obtain the true statement

$$9 - 3 = 6$$ ◀

In this section, you will *check* whether or not a given number is a root of an equation. *To check this, replace the variable of the equation by the number to see whether a true statement results. If the variable occurs more than once in the equation, each time it occurs, replace it by the same number.* (In Section 3.2 you will begin to study the techniques of actually solving equations.)

When you check a root of an equation, write $\stackrel{?}{=}$ instead of $=$. At the end of the check, write $\stackrel{\checkmark}{=}$ if the number is a root; if not, write $\stackrel{\times}{=}$.

Example 5 ▶ Is 10 a root of the equation

$$2x - 7 = x + 3?$$

Solution:
Replace each occurrence of x by 10:

$$2(10) - 7 \stackrel{?}{=} 10 + 3$$

$$20 - 7 \stackrel{?}{=} 13$$

$$13 \stackrel{\checkmark}{=} 13$$

Thus 10 is a root of the given equation. ◀

Example 6 ▶ Is -1 a root of the equation

$$1 - 3y = 5 - y?$$

Solution:
Replace each occurrence of the variable y by -1:

$$1 - 3(-1) \stackrel{?}{=} 5 - (-1)$$

$$1 + 3 \stackrel{?}{=} 5 + 1$$

$$4 \stackrel{\times}{=} 6$$

Thus -1 is *not* a root of the given equation. ◀

Answers

4. (a) ______

(b) ______

(c) ______

PRACTICE EXERCISES FOR TOPIC B

4. Check whether the given number is a root of the equation.

(a) $x - 9 = -3$ [12]

(b) $4y + 1 = y - 8$ [-3]

(c) $\dfrac{z - 6}{5} = z$ [1]

Now try the exercises for Topic B on page 89.

Use this area for work space.

EXERCISES

A. In Exercises 1–4 for each given equation:
(a) What is its left side?
(b) What is its right side?
(*See Example 1.*)

1. $7x = 28$

2. $3y + 1 = 43$

3. $12 - 2x = 4$

4. $5t - 1 = t + 7$

Use this area for work space.

In Exercises 5–12 express each question as an equation (in the variable x). You need not "solve" the equation. (*See Example 2.*)

5. What number plus 5 is equal to 8?
6. What number minus 3 is equal to 7?
7. Twice a number equals -8. What is the number?
8. What number divided by 4 is equal to 5?
9. Six less than a number equals 10. What is the given number?
10. When a number is increased by 4, the result is 12. What is the given number?
11. When a number is decreased by 7, the result is 2. What is the given number?
12. The product of a number and 3 equals -6. What is the given number?

In Exercises 13–20 express each equation in words. You need not "solve" the equation.

13. $x + 1 = 5$
14. $6y = 18$
15. $10t = 40$
16. $\frac{z}{4} = 7$
17. $2x - 3 = 13$
18. $1 - x = 4$
19. $\frac{x - 3}{4} = 5$
20. $2x - 7 = 9 - 6x$

B. In Exercises 21–52 check whether the given number is a root of the equation. (*See Examples 3–6.*)

Sample: $2x + 1 = 11;\ \boxed{5}$	*Solution:* $2(5) + 1 \stackrel{?}{=} 11$ $10 + 1 \stackrel{?}{=} 11$ $11 \stackrel{\checkmark}{=} 11$ 5 is a root.

21. $x + 4 = 10;\ \boxed{6}$
22. $3x = 6;\ \boxed{2}$
23. $y - 2 = 5;\ \boxed{7}$
24. $4 - z = 2;\ \boxed{6}$
25. $x + 5 = 2;\ \boxed{-3}$
26. $y - 10 = -1;\ \boxed{9}$
27. $5t = 30;\ \boxed{-6}$
28. $-5t = -30;\ \boxed{6}$
29. $\frac{x}{2} = 5;\ \boxed{10}$
30. $\frac{y}{3} = -3;\ \boxed{-9}$
31. $x + 1 = -6;\ \boxed{-7}$
32. $z - 4 = -2;\ \boxed{-6}$
33. $2x + 1 = 7;\ \boxed{3}$
34. $5t + 4 = 14;\ \boxed{2}$
35. $3x + 1 = 4;\ \boxed{1}$
36. $5x - 2 = 12;\ \boxed{2}$
37. $1 - t = -1;\ \boxed{2}$
38. $\frac{x}{4} = 3;\ \boxed{12}$
39. $\frac{x + 1}{5} = 2;\ \boxed{9}$
40. $-3y = 9;\ \boxed{3}$

Use this area for work space.

41. $2x + 7 = 1$; $\boxed{-3}$

42. $1 - 5x = 1$; $\boxed{0}$

43. $-2x + 1 = 3$; $\boxed{2}$

44. $\frac{2}{x} = -2$; $\boxed{-1}$

45. $x + 1 = 2x$; $\boxed{1}$

46. $4x - 2 = 3x$; $\boxed{4}$

47. $1 + x = \frac{x}{2}$; $\boxed{0}$

48. $\frac{x - 1}{2} = x$; $\boxed{3}$

49. $\frac{1 - 5x}{3} = -3$; $\boxed{2}$

50. $x = 5x$; $\boxed{0}$

51. $x = -x$; $\boxed{-1}$

52. $\frac{x}{3} = x$; $\boxed{0}$

53. Pick out the *error(s)*.

(a) 1 is a root of $x = -x$

(b) 0 is a root of $x = -x$

(c) 0 is a root of $x = 2x$

(d) Every number is a root of $x = x$.

(e) No number is a root of $x + 1 = x + 2$.

3.2 Solving Simple Equations

A. Equivalent Equations

Up to now, you have *checked* whether or not a given number is the root of an equation. If you are given an equation, you can find its root(s) by simplifying the equation.

DEFINITION

> EQUIVALENT EQUATIONS. Equations are said to be **equivalent** if they have the same root (or roots).

Example 1 ▶

$$2x = 8$$

$$x + 1 = 5$$

and

$$x = 4$$

are all equivalent. Each has the single root 4. Clearly the last of these is the simplest equation because the variable is on one side all by itself. ◀

DEFINITION

> To **solve an equation** means to find its roots.

Each equation you will be solving in this chapter will have a single root.

You solve an equation by transforming it into simpler equivalent equations. Your goal is to obtain an equation in which one side is a variable and the other a number. For example, if after simplifying you obtain

$$x = 2$$

then 2 is the root of the given equation.

Answers

1.

PRACTICE EXERCISES FOR TOPIC A

1. Which of the following equations are equivalent to the equation $x = 20$?

(a) $x + 1 = 21$ (b) $2x = 20$ (c) $3x = 60$ (d) $x - 2 = 18$ (e) $\frac{x}{2} = 10$

Now try the exercises for Topic A on page 95.

B. Balanced Scales

You can think of an equation, such as

$$x - 2 = 4$$

as a balanced scale. (See Figure 3.1.)

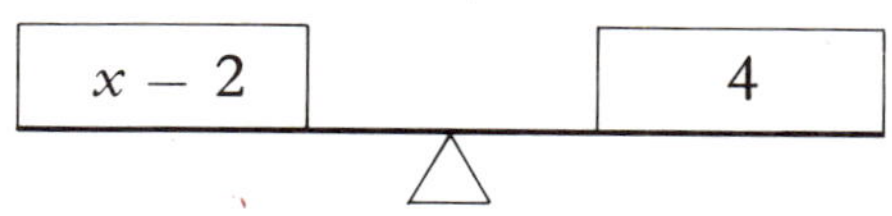

FIGURE 3.1. Balanced scale.

Your object is to isolate x on one side of the equation. What can you do to $x - 2$ to get x alone? *Addition undoes subtraction*. Because 2 *is subtracted from* x on the left side, you *add* 2 to undo subtraction. Therefore add 2 to the left side, as in Figure 3.2(a).

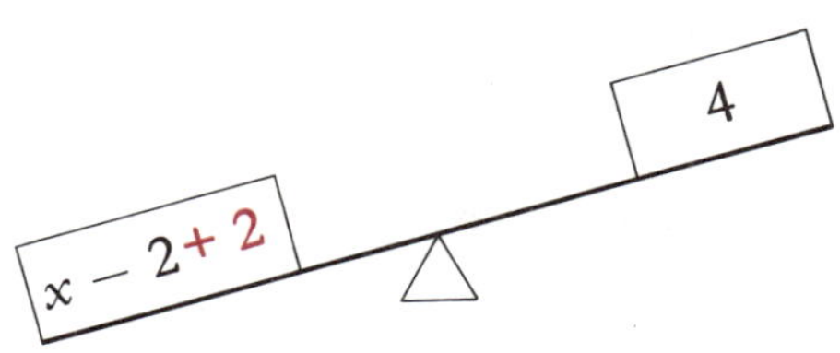

FIGURE 3.2(a). Unbalanced scale.

What you do to one side of the equation, you must also do to the other side. Thus add 2 to the right side to preserve the balance, as in Figure 3.2(b).

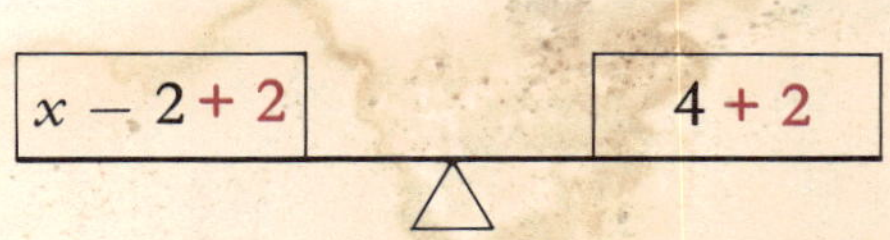

FIGURE 3.2(b). Balanced scale.

In symbols:

$$x - 2 + 2 = 4 + 2$$

or

$$x = 6$$

Thus, 6 is the root of the equation

$$x - 2 = 4$$

Example 2 ▶ Given the equation

$$2x = 8$$

Does the equation

$$\frac{2x}{2} = 8$$

preserve the balance?

Solution:

No. The left side of the given equation was divided by 2. Therefore, the right side should have been divided by 2 as well, that is,

$$\frac{2x}{2} = \frac{8}{2}$$

preserves the balance. ◀

Answers

2. ____________

PRACTICE EXERCISES FOR TOPIC B

2. Given the equation

$$\frac{x}{4} = 3$$

Does the equation

$$\frac{x}{4} \cdot 4 = 3 \cdot 4$$

preserve the balance?

Now try the exercises for Topic B on page 96.

C. Addition Property

When you add the same expression to both sides of an equation, you obtain an equivalent equation.

Example 3 ▶ Solve $x - 4 = 6$. *Check* your answer.

Solution:

What can you do to $x - 4$ to obtain x all by itself?

$$x - 4 = 6 \qquad \textit{Add 4 to both sides.}$$

$$x - 4 + 4 = 6 + 4$$

$$x = 10$$

The root is 10.

Check:
Replace x by 10 in the given equation.

$$10 - 4 \stackrel{?}{=} 6$$
$$6 \stackrel{\checkmark}{=} 6$$

◄

Subtracting a number *a* yields the same as adding $-a$. Thus *when you subtract the same expression from both sides of an equation, you obtain an equivalent equation.*

Example 4 ► Solve and *check* $x + 8 = 13$

Solution:

What can you do to $x + 8$ to obtain x by itself?

$$x + 8 = 13 \qquad \textit{Subtract 8 from both sides.}$$
$$x + 8 - 8 = 13 - 8$$
$$x = 5$$

The root is 5.
Check:

$$5 + 8 \stackrel{?}{=} 13$$
$$13 \stackrel{\checkmark}{=} 13$$

◄

Answers
3. (a) ______
(b) ______

PRACTICE EXERCISES FOR TOPIC C

3. Solve each equation and check your answers.

(a) $x - 9 = 15$ (b) $x + 7 = 2$

Now try the exercises for Topic C on page 96.

D. Multiplication Property

When you multiply or divide both sides of an equation by the same nonzero number, you obtain an equivalent equation.

Example 5 ► Solve $\frac{x}{3} = 6$

Solution:

What can you do to $\frac{x}{3}$ to obtain x alone?

$$\frac{x}{3} = 6 \qquad \textit{Multiply both sides by 3.}$$
$$\frac{x}{3} \cdot 3 = 6 \cdot 3$$
$$x = 18$$

The root is 18.

◄

Example 6 ▶ Solve $5y = 20$

Solution:

What can you do to $5y$ to obtain y by itself?

$$5y = 20 \qquad \textit{Divide both sides by 5.}$$

$$\frac{5y}{5} = \frac{20}{5}$$

$$y = 4$$

The root is 4. ◀

The Multiplication Property enables you to multiply or divide both sides of an equation by a *nonzero* number. You cannot multiply or divide both sides by 0. Also complications may arise if you multiply or divide by a *variable*. Example 7 illustrates some of these difficulties.

Example 7 ▶ The equation

$$1 = 2$$

is obviously a *false* statement.

(a) If you multiply both sides of the equation by 0, you obtain the *true* statement:

$$1 \cdot 0 = 2 \cdot 0$$

or

$$0 = 0$$

Thus if you multiply both sides of an equation by 0, you always obtain a true statement, even if the original statement is false.

(b) If you multiply both sides of the equation

$$1 = 2$$

by x, you obtain

$$1x = 2x$$

Note that 0 is a root of this equation because

$$1 \cdot 0 = 2 \cdot 0$$

But 0 was *not* a root of the original equation

$$1 = 2$$

Thus multiplying both sides of an equation by a variable may add roots that were not originally there. ◀

PRACTICE EXERCISES FOR TOPIC D

Answers
4. (a) ______
(b) ______
(c) ______

4. Solve each equation.

(a) $\frac{y}{7} = 4$ (b) $10k = 90$ (c) $\frac{x}{8} = -5$

Now try the exercises for Topic D on page 96.

E. Word Problems

Often a problem that is stated in words leads to an equation.

Here is the general procedure for solving a "word problem."

1. Your first task is to translate the words into algebraic symbols.
2. Then solve the resulting equation.
3. Finally, show how the solution of the equation yields a solution of the stated problem.

Example 8 ► What number must be added to 3 to obtain 5?

Solution:

1. Translate into algebraic symbols. Begin by letting x be the number in question. Note that when you *obtain* 5, the result *is equal to* 5.

$$\underbrace{\text{What number}}_{x}\ \underbrace{\text{must be added to}}_{+}\ 3\ \underbrace{\text{to obtain}}_{=}\ 5?$$

2. Solve the resulting equation.

$$x + 3 - 3 = 5 - 3$$

$$x = 2$$

3. Thus 2 is the number that is added to 3 to obtain 5. ◄

Answers

5. ________

6. ________

7. ________

PRACTICE EXERCISES FOR TOPIC E

Write an equation for each of the following and then solve it.

5. What number must be added to 12 to obtain 21?

6. When a certain number is diminished by 8, the result is 6. Find the number.

7. When 12 is multiplied by a certain number, the result is 60. What is the number?

Now try the exercises for Topic E on page 97.

Use this area for work space.

EXERCISES

A. In Exercises 1–4 are all of the equations in each group equivalent? (*See Example 1.*)

1. $x = 6$
$x - 3 = 3$
$4x = 24$

2. $-5x = 25$
$x + 8 = 3$
$x = 5$

3. $10 - x = 3$
$x + 7 = 14$
$x - 4 = 3$
$x = 7$

4. $\frac{x}{2} = 5$
$x = 10$
$10x = 100$
$\frac{-x}{10} = -1$

Use this area for work space.

5. Which of the following equations are equivalent to $x = 9$?

(a) $9x = 1$ (b) $\frac{x}{9} = 1$

(c) $9 - x = 0$ (d) $x - 3 = 6$ (e) $x - 6 = 3$

6. Which of the following equations are equivalent to $x = 1$?

(a) $6x = 6$ (b) $x - 4 = 4$

(c) $-x = -1$ (d) $5x - 2 = 3$ (e) $1 - x = -1$

7. Which of the following equations are equivalent to $x = -2$?

(a) $2x = 4$ (b) $x + 2 = 0$ (c) $x - 2 = 0$

(d) $-5x = -10$ (e) $x - 8 = -10$

8. Which of the following equations are equivalent to $x = 0$?

(a) $6x = 6$ (b) $1 - x = 1$ (c) $\frac{x}{4} = 0$

(d) $10 - x = 10$ (e) $-3x = 0$

B.

9. Given the equation
$$x - 5 = 7$$
Does the equation
$$x - 5 + 5 = 7 + 5$$
preserve the balance? (*See Example 2.*)

10. Given the equation
$$y + 2 = 3$$
Does the equation
$$y + 2 - 2 = 3 - 2$$
preserve the balance?

11. Given the equation
$$x + 3 = -3$$
Does the equation
$$x + 3 - 3 = -3 + 3$$
preserve the balance?

12. Given the equation
$$5x = 20$$
Does the equation
$$\frac{5x}{5} = \frac{20}{5}$$
preserve the balance?

C.

In Exercises 13–24 solve each equation. (*See Examples 3 and 4.*)

13. $x - 1 = 4$ **14.** $x - 2 = 2$ **15.** $y - 3 = 6$ **16.** $z - 4 = 0$

17. $t - 5 = -5$ **18.** $x - 4 = -2$ **19.** $x + 1 = 3$ **20.** $x + 3 = 9$

21. $y + 2 = 12$ **22.** $z + 10 = 11$ **23.** $t + 8 = 4$ **24.** $t + 6 = -9$

In Exercises 25–30 (a) solve each equation and (b) *check* the root. (*See Examples 3 and 4.*)

25. $x - 3 = 4$ **26.** $x - 10 = 10$ **27.** $y - 5 = 7$ **28.** $x + 3 = 10$

29. $y + 8 = 9$ **30.** $t + 12 = 10$

D.

In Exercises 31–52 solve each equation. (*See Examples 5 and 6.*)

31. $\frac{x}{2} = 7$ **32.** $\frac{x}{5} = 4$ **33.** $\frac{z}{9} = 9$ **34.** $\frac{t}{10} = 8$

35. $\frac{x}{11} = 5$ **36.** $\frac{y}{12} = 7$ **37.** $2x = 6$ **38.** $3y = 21$

Use this area for work space.

39. $5y = 45$ **40.** $7t = 49$ **41.** $12x = 60$ **42.** $\frac{x}{4} = -5$

43. $\frac{x}{5} = 0$ **44.** $\frac{x}{7} = -1$ **45.** $\frac{t}{-2} = 5$ **46.** $\frac{t}{-2} = -5$

47. $3x = -18$ **48.** $4y = 0$ **49.** $7t = -14$ **50.** $12t = -36$

51. $-2t = 8$ **52.** $-2t = -8$

In Exercises 53–60 (a) solve each equation and (b) *check* the root.

53. $\frac{x}{3} = 5$ **54.** $\frac{z}{6} = 9$ **55.** $2x = 16$ **56.** $5x = 60$

57. $\frac{x}{2} = -5$ **58.** $\frac{y}{3} = 0$ **59.** $2t = -24$ **60.** $-5x = 25$

E. In Exercises 61–70 write an equation and then solve it. (*See Example 8.*)

61. When 4 is added to a number, the sum is 6. Find the number.

62. When 4 is subtracted from a number, the difference is 6. Find the number.

63. What number must be added to 5 to obtain 9?

64. What number must be added to 7 to obtain 2?

65. What number must be divided by 2 to obtain 6?

66. What number must be multiplied by 2 to obtain 6?

67. What number must be divided by -5 to obtain 10?

68. What number must be multiplied by -5 to obtain -10?

69. When a number is increased by 3, the result is 7. What is the given number?

70. When a number is decreased by 8, the result is 2. What is the given number?

3.3 Isolating the Variable

A. Adding the Variable

Often the easiest way to solve an equation is by adding the variable to both sides, as in Example 1.

Example 1 ▶ Solve $6 - x = 0$

Solution:

$$6 - x = 0 \quad \text{Add } x \text{ to both sides.}$$

$$6 - x + x = 0 + x$$

$$6 = x$$

The root is 6. ◀

Note that in the final equation of Example 1, the variable, x, is on the right side, and the root, 6, is on the left side.

$$6 = x \qquad \text{means the same as} \qquad x = 6$$

In simplifying an equation, bring variables to one side, numerical terms to the other.

Example 2 ▶ Solve $3y - 5 = 2y + 10$

Solution:

The variables will be brought to the left side, and the numerical terms to the right side.

$$3y - 5 = 2y + 10 \qquad \textit{Add 5 to both sides.}$$

$$3y - 5 + 5 = 2y + 10 + 5$$

$$3y = 2y + 15 \qquad \textit{Now subtract 2y from both sides.}$$

$$3y - 2y = 2y - 2y + 15$$

$$y = 15$$

The root is 15. ◀

PRACTICE EXERCISES FOR TOPIC A

Answers
1. ______
2. ______
3. ______

1. Solve $-x + 15 = 0$

2. Solve and check $8y = 7y - 4$

3. Solve $5z + 3 = 6z - 2$

Now try the exercises for Topic A on page 101.

B. Using Both Addition and Multiplication Properties

Both the Addition and the Multiplication Properties frequently apply in the same example.

Example 3 ▶ Solve $4z + 1 = 2z + 9$

Solution:

Because the coefficient of $4z$ is larger than that of $2z$, bring the variables to the left side. This will result in a *positive* coefficient for the z-term.

$$4z + 1 = 2z + 9 \qquad \textit{Subtract 1 from both sides.}$$

$$4z + 1 - 1 = 2z + 9 - 1$$

$$4z = 2z + 8 \qquad \textit{Now subtract 2z from both sides.}$$

$$4z - 2z = 2z - 2z + 8$$

$$2z = 8 \qquad \textit{Finally, divide both sides by 2.}$$

$$\frac{2z}{2} = \frac{8}{2}$$

$$z = 4$$

The root is 4. ◀

Example 4 ▶ Solve and check $\frac{t}{4} - 1 = -5$

Solution:

$$\frac{t}{4} - 1 = -5 \qquad \textit{Add 1 to both sides.}$$

$$\frac{t}{4} - 1 + 1 = -5 + 1$$

$$\frac{t}{4} = -4 \qquad \textit{Multiply both sides by 4.}$$

$$\frac{t}{4} \cdot 4 = (-4) \cdot 4$$

$$t = -16$$

The root is -16.
Check:
Replace t by -16 in the given equation

$$\frac{t}{4} - 1 = -5$$

$$\frac{-16}{4} - 1 \stackrel{?}{=} -5$$

$$-4 - 1 \stackrel{?}{=} -5$$

$$-5 \stackrel{\checkmark}{=} -5$$

◀

Answers

4. ____________

5. ____________

PRACTICE EXERCISES FOR TOPIC B

4. Solve $9n - 7 = 5n + 17$

5. Solve and check $\frac{m}{5} + 8 = 2$

Now try the exercises for Topic B on page 102.

C. Collecting Terms

In Examples 5 and 6, like terms must first be combined on either side of the equation.

Example 5 ▶ Solve $2x + 5x = 20 + 8$

Solution:

Combine the variables of the left side and the numerical terms of the right side.

$$2x + 5x = 20 + 8$$

$$7x = 28 \quad \textit{Divide both sides by 7.}$$

$$\frac{7x}{7} = \frac{28}{7}$$

$$x = 4$$ ◀

Example 6 ▶ Solve $2y - 4 + 3y + 12 = 7y - 4$

Solution:

Combine the like terms of the left side.

$$2y - 4 + 3y + 12 = 7y - 4$$

$$5y + 8 = 7y - 4 \quad \textit{Add 4 to both sides.}$$

$$5y + 8 + 4 = 7y - 4 + 4$$

$$5y + 12 = 7y \quad \textit{Subtract 5y from both sides.}$$

$$5y + 12 - 5y = 7y - 5y$$

$$12 = 2y \quad \textit{Divide both sides by 2.}$$

$$\frac{12}{2} = \frac{2y}{2}$$

$$6 = y$$ ◀

Here then is a procedure that is often used for solving equations:

1. Combine like terms on either side of the equation.
2. Apply the Addition Property in order to bring variables to one side, numerical terms to the other.
3. Apply the Multiplication Property in order to obtain the variable on one side with coefficient 1.

Answers

6. ______

7. ______

PRACTICE EXERCISES FOR TOPIC C

Solve for x.

6. $7 - x + 1 = x - 5x - 1$

7. $x + 3 - x + 4 = 18 - x - 5$

Now try the exercises for Topic C on page 102.

D. Combining Steps

In Example 6, you **added 4** to both sides of the equation

$$5y + 8 = 7y - 4$$

and then **subtracted 5y** from both sides of the resulting equation. You can combine both steps by **adding 4 − 5y** to both sides. Thus

$$5y + 8 = 7y - 4 \qquad \textit{Add } 4 - 5y \textit{ to both sides.}$$

$$5y + 8 + 4 - 5y = 7y - 4 + 4 - 5y$$

$$12 = 2y$$

$$6 = y$$

Here is another example in which, in one step, several terms are added to both sides of an equation.

Example 7 ▶ Solve $6x + 7 = 15 - 2x$

Solution:

You want to **add 2x** to both sides and **subtract 7** from both sides. Thus **add 2x − 7** to both sides.

$$6x + 7 = 15 - 2x$$

$$6x + 7 + 2x - 7 = 15 - 2x + 2x - 7$$

$$8x = 8 \qquad \textit{Divide both sides by 8.}$$

$$\frac{8x}{8} = \frac{8}{8}$$

$$x = 1$$ ◀

Answers

8. ______

9. ______

PRACTICE EXERCISES FOR TOPIC D

Try to solve these equations in as few steps as possible.

8. $7x - 14 = 16 - 3x$

9. $x - 21 = 8x - 7$ (*Check.*)

Now try the exercises for Topic D on page 103.

Use this area for work space.

EXERCISES

A. In Exercises 1–8 solve each equation. (*See Examples 1 and 2.*)

1. $5 - x = 0$

2. $5 + x = 0$

3. $4t + 9 = 3t + 14$

4. $8z + 10 = 9z + 7$

5. $4x - 3 = 3x + 4$

6. $6u + 4 = 5u + 2$

7. $4 - 3x = 7 - 4x$

8. $9 - 10y = 10 - 9y$

In Exercises 9 and 10 (a) solve each equation and (b) *check* the root.

9. $-7 - x = 0$

10. $3t - 4 = -10 + 2t$

Use this area for work space.

In Exercises 11 and 12 write an equation and then solve it.

11. What number must be subtracted from 7 to obtain 0?

12. What number must be subtracted from -8 to obtain 0?

B. In Exercises 13–24 solve each equation. (*See Examples 3 and 4.*)

13. $3x - 5 = x + 3$

14. $6x + 2 = 4x - 6$

15. $2z - 9 = 5z - 15$

16. $4t + 7 = t + 7$

17. $6u + 9 = 2u - 7$

18. $8x - 7 = 5x - 4$

19. $1 - t = 4 + 2t$

20. $9 - 2x = 5 - 6x$

21. $\frac{y}{2} + 1 = 7$

22. $\frac{x}{5} - 3 = 4$

23. $\frac{t}{6} + 1 = 3$

24. $\frac{u}{4} - 2 = 0$

In Exercises 25 and 26 (a) solve each equation and (b) *check* the root. (*See Example 4.*)

25. $3y + 9 = 6y - 3$

26. $\frac{t}{4} - 3 = 7$

In Exercises 27 and 28 write an equation and then solve it.

27. Five more than twice a number equals the number minus 3. Find this number.

28. When 7 is subtracted from 3 times a number, the result equals the number increased by 13. Find this number.

29. An encyclopedia saleswoman receives a monthly salary of \$500 plus \$20 for each set of encyclopedias she sells. Thus her monthly earnings are given by the formula

$$y = 500 + 20x$$

where x is the number of sets she sells that month. How many sets must she sell in order to earn \$2000 for a month's work?

30. A ball is thrown upward. Its velocity (in feet per second) is given by the formula

$$v = 80 - 32t$$

where t is the number of seconds after the ball was thrown. After how many seconds will the velocity be 16 feet per second?

C. In Exercises 31–38 solve each equation. (*See Examples 5 and 6.*)

31. $4x + 3x = 14$

32. $5y - 2y + y = 8 + 3y + 4 - 3y$

33. $9x + 1 - 4x + 3 = 14$

34. $2u - 3 + 9u + 8 = 6u - 15$

35. $3x + 4x - 1 + 9 = 2x - 5 + x - 3$

36. $2t + 5 - 7t + 3 = 9t - 8 + 4t + 16$

37. $3x + x - 4x - 8x = 1 - 9x + 5$

38. $6y - 2y + 1 - 13 = 7y - 8 + 9y - 4$

In Exercises 39 and 40 (a) solve each equation and (b) *check* the root. (*See Example 4.*)

39. $2y + 4y = 27 + 9$

40. $6x + 3x - 1 + 2x = 11 - x$

Use this area for work space.

D. In Exercises 41–52 solve each equation in as few steps as possible. (*See Example 7.*)

41. $5t - 3 = 2t + 9$

42. $8x - 4 = 4x + 4$

43. $10 + 2x = 4x + 8$

44. $9 - 2t = 5 + 2t$

45. $7 + 5y = -28 - 2y$

46. $6z - 5 = 13 - 3z$

47. $10 - 3u = 40 + 7u$

48. $20 - 3x = -30 - 5x$

49. $2x - 34 = 6x + 6$

50. $5t - 7 = -28 + 12t$

51. $2x - 9 + 5x = 2 + 3x - 3$

52. $4y - 1 + 7y = 10y - 2$

In Exercises 53–56 (a) solve each equation in as few steps as possible and (b) *check* the root.

53. $4x - 3 = 7x + 3$

54. $12 - 3y = 8y + 1$

55. $6 - 2z = 9 - 5z$

56. $4x + 10 = 9x + 25$

In Exercises 57 and 58 write an equation and then solve it.

57. Six more than twice a number equals six less than six times the number. Find this number.

58. Three times a number is subtracted from 12. The result equals two more than seven times the number. Find this number.

3.4 Equations with Parentheses

A. Use of the Distributive Laws

Parentheses clarify the meaning of arithmetic and algebraic expressions. Sometimes one or both sides of an equation contain parentheses. The Distributive Laws are often used to remove parentheses.

Example 1 ▶ Solve $3(2x + 5) = 4x + 23$

Solution:

$$3(2x + 5) = 4x + 23$$ *Use the Distributive Laws on the left side.*

$$6x + 15 = 4x + 23$$ *Add $-4x - 15$ to both sides.*

$$6x + 15 - 4x - 15 = 4x + 23 - 4x - 15$$

$$2x = 8$$ *Divide both sides by 2.*

$$\frac{2x}{2} = \frac{8}{2}$$

$$x = 4$$

The root is 4. ◀

Example 2 ▶ Solve $3(2y + 9) = 7(y + 2)$
Solution:
Use the Distributive Laws to remove parentheses on both sides.

$$3(2y + 9) = 7(y + 2)$$

$$6y + 27 = 7y + 14 \qquad \text{Add } -6y - 14 \text{ to both sides.}$$

$$6y + 27 - 6y - 14 = 7y + 14 - 6y - 14$$

$$13 = y$$

The root is 13. ◀

Example 3 ▶ Solve $4 - (2x - 8) = x$
Solution:
Recall that

$$-(2x - 8) = (-1)(2x - 8) = -2x + 8$$

so that the given equation

$$4 - (2x - 8) = x$$

becomes

$$4 - 2x + 8 = x \qquad \text{Next, combine the numerical terms on the left side.}$$

$$12 - 2x = x$$

$$12 - 2x + 2x = x + 2x$$

$$12 = 3x$$

$$\frac{12}{3} = \frac{3x}{3}$$

$$4 = x$$

The root is 4. ◀

Answers
1. ______
2. ______

PRACTICE EXERCISES FOR TOPIC A

1. Solve $10(a + 7) = 4(3a - 4)$

2. Solve $5y + 9 = y - (6 - y)$

Now try the exercises for Topic A on page 105.

B. Dividing by a Factor

Sometimes both sides of an equation can be divided by the same factor. Although the Distributive Laws can also be used, the division method may be quicker.

Example 4 ▶ Solve and *check* $4(x + 2) = -8$

Solution:

$$4(x + 2) = -8 \qquad \textit{Divide both sides by 4.}$$

$$\frac{4(x + 2)}{4} = \frac{-8}{4}$$

$$x + 2 = -2 \qquad \textit{Add } -2 \textit{ to both sides.}$$

$$x + 2 - 2 = -2 - 2$$

$$x = -4$$

The root is -4.

Check:

$$4(-4 + 2) \stackrel{?}{=} -8$$

$$4(-2) \stackrel{?}{=} -8$$

$$-8 \stackrel{\checkmark}{=} -8$$

◀

PRACTICE EXERCISES FOR TOPIC B

Answers

3. ______

3. Solve and *check* $-8(x - 16) = 80$

Now try the exercises for Topic B on page 106.

Use this area for work space.

EXERCISES

A. In Exercises 1–24 solve each equation. (*See Examples 1–3.*)

1. $3(t - 4) = 2t + 1$

2. $6(x + 2) = 7x - 3$

3. $2(3x + 1) = 5x - 1$

4. $5(2y + 3) = 8y - 7$

5. $9(t - 1) = 10(t + 1)$

6. $3(z + 2) = 2(z - 7)$

7. $5(x - 4) = 4(x + 3)$

8. $7(x + 1) = 2(x - 3) - 2$

9. $4(5 - 2x) = 9(1 - x)$

10. $6(3x + 4) = 5(4x + 1) + 1$

11. $7 - (x + 2) = 1$

12. $4 - (y - 2) = 8$

13. $15 - 3(9 - x) = x$

14. $6(t + 1) = 12$

15. $8 - 2(z - 2) = 4$

16. $5(t + 2) = t + 2$

17. $y + 6 = 10(y + 2) - 5$

18. $2(x + 3) = 3x + 5$

19. $-7(t + 5) = t - 3$

20. $1 - [x - (1 - x)] = 0$

21. $2[4 - (3z - 5)] = 3z$

22. $-2[x - (4x - 5)] = 5x$

23. $2y - (1 - [y - (3 - 2y)]) = 1$

24. $5(x - 2) - 3(10 - x) = 0$

In Exercises 25–28 (a) solve and (b) *check*.

25. $x - (12 - 2x) = 0$

26. $10 - (y - 6) = y + 2$

27. $-3(y + 2) = y - 10$

28. $2x + 3(x - 2) = 19$

Use this area for work space.

29. In 1981 first-class domestic postage cost 18 cents for the first ounce and 17 cents for each additional ounce (or part of an ounce). The cost of mailing a letter weighing x ounces was therefore given by

$$C = 18 + 17(x - 1)$$

If a letter cost 86 cents to mail, what was the most it could weigh?

30. The charge for a certain long-distance telephone call is 40 cents for the first 3 minutes and 10 cents for each additional minute (or part of a minute). The cost for x minutes is thus given by

$$C = 40 + 10(x - 3)$$

For 90 cents, how long can a person speak on this call?

B. In Exercises 31–40 solve each equation. (*See Example 4.*)

31. $3(t + 2) = 6$

32. $7(x + 8) = -14$

33. $5(4y - 3) = 15y$

34. $100(11t - 7) = 400$

35. $4(2t - 1) = 12$

36. $6(5 - 3t) = 12t$

37. $2(3x - 5) = 4(1 - 2x)$

38. $7(5x + 3) = -7(4x + 15)$

39. $9(z + 12) = 0$

40. $-4(3x - 9) = 0$

In Exercises 41 and 42 (a) solve and (b) *check*. (*See Example 4.*)

41. $4(y + 3) = 20$

42. $-3(2t - 1) = -3t$

3.5 Applied Problems

A. From Words to Symbols

Many practical problems are solved by algebraic methods. Indeed, one of the main goals of this book is to teach you how to apply mathematical methods to situations that arise in everyday life.

Let us recall the procedure for solving a problem that is stated in words.

1. First translate the problem into algebraic symbols.
2. Then solve the resulting equation.
3. Show how the solution of the equation yields a solution of the stated problem.

Some problems directly or indirectly concern the integers:

$$\ldots, -3, -2, -1, 0, 1, 2, 3, \ldots$$

Of course, you do not *directly* worry about integers in everyday conversation. But you constantly *use* integers in describing other concepts, such as money, time, distance, and temperature. For example, you use

100 to describe *earning* \$100

−100 to describe *losing* \$100

32° Fahrenheit to describe the freezing point of water

−10° Fahrenheit to describe the temperature in Minneapolis on a January morning

You have already had practice in translating words into symbols. For example, if x represents an integer, then each of the following phrases can be expressed in terms of x.

$$\underbrace{\text{two more than}}_{2\,+}\ \underbrace{\text{an integer}}_{x}, \qquad \text{or } x + 2$$

$$\underbrace{\text{five times}}_{5\,\cdot}\ \underbrace{\text{an integer}}_{x}, \qquad \text{or } 5x$$

$$\underbrace{\text{an integer}}_{x}\ \underbrace{\text{decreased by}}_{-}\ \underset{10}{10}$$

$$\underbrace{\text{twice an integer}}_{2x}\ \underbrace{\text{increased by}}_{+}\ \underset{3}{3}$$

Example 1 ▶ (a) The sum of two integers is 20. If x represents one of these integers, express the other in terms of x.

(b) The sum of two integers is represented by s. If one of these integers is 10, express the other in terms of s.

(c) The larger of two integers is 4 more than the smaller. If y represents the smaller integer, express the sum of the integers in terms of y.

Solution:

(a) Let x represent one integer. Let $\square$ represent the other integer.

$$\underbrace{\text{The sum of two integers}}\ \text{is}\ 20.$$

$$x + \square = 20 \qquad \textit{Subtract } x \textit{ from each side.}$$

$$x - x + \square = 20 - x$$

$$\square = 20 - x$$

Thus $20 - x$ represents the second integer.

(b) Here one of the integers is given as 10. Let $\square$ represent the other integer.

$$\underbrace{\textbf{The sum of two integers}}\ \underbrace{\textbf{is represented by}}\ s.$$

$$10 + \square = s$$

$$10 - 10 + \square = s - 10$$

$$\square = s - 10$$

Thus here, $s - 10$ represents the second integer.

(c) Let y represent the smaller integer. You are told that the larger integer is

$$\underbrace{\text{4 more than } y}_{4 + y}$$

Thus $$y + 4$$

represents the larger integer. Because y represents the smaller integer, the sum of the integers is

$$y + (y + 4) \quad \text{or} \quad 2y + 4$$ ◀

PRACTICE EXERCISES FOR TOPIC A

Answers
1. (a) ______
(b) ______
2. ______
3. ______

1. If x represents a certain integer, express the following in terms of x.

(a) twice the integer (b) eight less than twice the integer

2. The difference between two integers is 50. If y represents the smaller integer, express the larger one in terms of y.

3. Suppose the sum of two integers is 70. If n represents one of these integers, how can you express the other in terms of n?

Now try the exercises for Topic A on page 111.

B. Solving Applied Problems

Now you will utilize the translations of Example 1 in solving stated problems (Examples 2 and 3).

Example 2 ▶ A 20-meter rope is cut into two pieces. One piece is three times as long as the other. Find the length of each piece.

Solution:

1. *Translate the problem into algebraic symbols.*

Let x be the length (in meters) of the *shorter* piece. Because the sum of the lengths of the two pieces is 20 meters, the longer piece must measure $(20 - x)$ meters. But the longer piece is three times as long as the shorter one. Thus the longer piece also measures $3x$ meters. Set up the equation:

$$\underbrace{\text{The length of the longer piece}}_{20 - x} \ \underset{=}{\text{is}} \ \underbrace{\text{three times that of the shorter one.}}_{3x}$$

2. *Solve this equation.*

$$20 - x = 3x \qquad \textit{Add x to each side.}$$

$$20 - x + x = 3x + x$$

$$20 = 4x \qquad \textit{Divide both sides by 4.}$$

$$5 = x \qquad \textit{You must also find 3x.}$$

$$15 = 3x$$

3. *Interpret the solution of the equation in terms of the stated problem:*

The shorter piece is 5 meters long. The longer piece is 15 meters long. ◀

Example 3 ▶ In a two-candidate election Helen received 4 more votes than Bob. If 48 people voted, how many votes did each candidate receive?

Solution:

Let y be the number of votes Bob received. Then Helen received $(y + 4)$ votes, and together they received $y + (y + 4)$, or $(2y + 4)$, votes.

$$\underbrace{\text{Bob's votes}}_{y} + \underbrace{\text{Helen's votes}}_{(y+4)} = \underbrace{48 \text{ votes}}_{48}$$

$$2y + 4 = 48 \qquad \textit{Subtract 4 from each side.}$$

$$2y + 4 - 4 = 48 - 4$$

$$2y = 44$$

$$y = 22 \qquad \textit{Bob received 22 votes.}$$

$$y + 4 = 26 \qquad \textit{Helen received 26 votes.}$$

◀

Example 4 ▶ Walter's boss earns \$1000 more than three times as much as Walter earns. If the boss earns \$40 000 per year, what is Walter's yearly salary?

Solution:

Let x represent Walter's yearly salary. Then the boss earns

$$\underbrace{\$1000}_{1000}\ \underbrace{\text{more than}}_{+}\ \underbrace{\text{three times as much as}}_{3\cdot}\ \underbrace{\text{Walter earns.}}_{x}$$

$$\underbrace{\text{The boss's earnings}}_{3x + 1000}\ \text{are}\ \underbrace{\$40\,000 \text{ per year.}}_{40\,000}$$

$$3x + 1000 = 40\,000 \qquad \textit{Subtract 1000 from each side.}$$

$$3x = 39\,000 \qquad \textit{Divide both sides by 3.}$$

$$x = 13\,000$$

Walter earns \$13 000 per year. ◀

Example 5 ▶ Six more than an integer is four times this integer. Find the integer.

Solution:

Let x represent this integer.

Translate the problem:

$$\underbrace{\text{Six more than an integer}}_{x + 6}\ \text{is}\ \underbrace{\text{four times this integer.}}_{4x}$$

$$x + 6 = 4x$$

Solve the equation:

$$6 = 3x$$

$$2 = x$$

Thus 2 is the integer you seek. ◀

Example 6 ▶ The sum of the measures of the angles of any triangle is 180°. One angle of a certain triangle measures 20° more than the smallest angle. The third angle measures twice as much as the smallest angle. Find the measures of the three angles.

Solution:

Let x be the measure of the smallest angle in degrees. Then the second angle measures (x + 20) degrees and the third angle measures 2x degrees. Because the sum of the three angles is 180°, set up the equation as follows.

$$\underbrace{\text{Measure of smallest angle}}_{x} + \underbrace{\text{Measure of second angle}}_{(x+20)} + \underbrace{\text{Measure of third angle}}_{2x} = \underbrace{180°}_{180}$$

$$4x + 20 = 180$$

$$4x + 20 - 20 = 180 - 20$$

$$4x = 160$$

$$x = 40$$

$$x + 20 = 60$$

$$2x = 80$$

The three angles measure 40°, 60°, and 80°, respectively. ◀

Example 7 ▶ A man has two more \$10 bills than \$5 bills in his wallet. If altogether he has \$95 in \$5 and \$10 bills, how many \$5 bills does he have?

Check your result.

Solution:

Let x be the number of \$5 bills. Then x + 2 is the number of \$10 bills. The following table is useful.

	Dollars per Bill ·	*Number of Bills* =	*Total Amount in Dollars*
\$5 bills	5	x	5x
\$10 bills	10	x + 2	10(x + 2)

$$\underbrace{\text{Altogether}}_{5x + 10(x+2)} \ \underbrace{\text{he has}}_{=} \ \underbrace{\$95.}_{95}$$

$$5x + 10x + 20 = 95$$

$$15x + 20 - 20 = 95 - 20$$

$$15x = 75$$

$$x = 5$$

He has five \$5 bills (and seven \$10 bills).

Check:

$$5(5) + 10(7) \stackrel{?}{=} 95$$

$$25 + 70 \stackrel{?}{=} 95$$

$$95 \stackrel{\checkmark}{=} 95$$

◀

Answers

4. ____________

5. ____________

PRACTICE EXERCISES FOR TOPIC B

4. Jack bought a doughnut and a cup of coffee for a total of 90 cents. If the doughnut cost 20 cents more than the cup of coffee, find the cost of each.

5. Three roommates earn salaries that total $400 per week. Barbara earns 40 dollars per week more than Amy, and Chris earns twice as much as Amy. How much does each earn per week?

Now try the exercises for Topic B on this page.

Use this area for work space.

EXERCISES

A.

1. Let x represent an integer. Express each of the following in terms of x. (*See Example 1.*)
   (a) two more than the integer
   (b) six less than the integer
   (c) triple the integer
   (d) the integer decreased by five

2. Let y represent an integer. Express each of the following in terms of y.
   (a) twice the integer
   (b) one less than twice the integer
   (c) four more than twice the integer
   (d) the integer, increased by four, then doubled

3. Let t represent an integer. Express each of the following in terms of t.
   (a) half of the integer
   (b) an integer that is four units larger than t
   (c) an integer that is ten units less than t
   (d) five times the sum of the integers found in parts (a), (b), and (c).

4. Suppose the sum of two integers is 50. If x represents one of these integers, express the other in terms of x.

5. Suppose the sum of two integers is represented by x. If one of these integers is 50, express the other in terms of x.

6. The difference between two integers is 19. If x represents the larger integer, express the smaller one in terms of x.

7. The larger of two integers is 8 more than the smaller. If x represents the smaller integer, express the larger one in terms of x.

8. The larger of two integers is 5 more than the smaller. If x represents the larger integer, express the sum of these integers in terms of x.

B.

In Exercises 9–16 write an equation and then solve it. When called for, *check* your result. (*See Example 5.*)

9. Ten more than an integer is 23. Find this integer.

10. Three times an integer is -27. Find this integer.

11. A certain integer is decreased by 7, and the result is 25. Find this integer.

Use this area for work space.

12. One more than twice an integer is 15. Find this integer. *Check your result.*

13. One less than three times an integer is 23. Find this integer.

14. Seven more than four times an integer is 51. Find this integer. *Check your result.*

15. Two less than five times an integer is 28. Find this integer.

16. The sum of two integers is 50. The larger is six more than the smaller. Find these integers. *Check your result.*

In Exercises 17–32 write an equation and then solve it.

17. A rope 16 feet long is cut into two pieces. If one piece is x feet long, express the length of the other piece using x. (*See Example 2.*)

18. A rope that is y feet long is cut into two pieces. If one piece is 10 feet long, express the length of the other piece using y.

19. A 36-meter rope is cut into two pieces in such a way that one piece is 6 meters longer than the other. How long is each piece?

20. A 12-meter board is cut into two pieces. The longer piece is twice as long as the shorter one. Find the length of the shorter piece.

21. A 35-foot rope is cut into three pieces. If the largest piece is twice as long as the smallest piece and 5 feet longer than the remaining piece, how long is each piece?

22. A 24-foot board is cut into three pieces. One piece is 3 feet longer than another piece. If the remaining piece is 5 feet long, how long are the other pieces?

23. In a school election 2420 votes are cast for two candidates. One candidate wins by 10 votes. How many votes does this candidate receive? (*See Example 3.*)

24. In a class election Jill receives 6 more votes than Jack and 5 fewer than Jerry. If Jill receives x votes,
(a) how many votes does Jack receive?
(b) how many votes does Jerry receive?
(c) If altogether 41 votes are cast for the three candidates, how many votes does each receive?

25. In a town election the Democratic candidate receives 22 more votes than the Republican candidate. An Independent write-in candidate receives as many votes as the Republican and Democrat combined. If altogether 2004 votes are cast for these three candidates, how many votes does each candidate receive?

26. Eric and Doug together have \$60. Eric has \$10 more than Doug. How much does Doug have?

27. Joe's wife, Judy, earns \$40 more per week than Joe does. Together they earn \$160 less weekly than Joe's father, who earns three times as much as Joe. What is Joe's weekly salary? (*See Example 4.*)

28. It costs a manufacturer \$10 to produce each sled. In addition, there is a general overhead cost of \$20 000. She decides to produce 5000 sleds. If she receives \$20 per sled from a retailer,
(a) how many sleds must she sell to break even?
(b) how many must she sell to make a profit of \$10 000?

Use this area for work space.

29. A manufacturer finds that it costs him \$3 to produce a novelty item. There is also a general overhead cost of \$500. If the manufacturer has \$5000 to spend, how many items can be produced?

30. The perimeter of a rectangle is given by the formula

$$P = 2(l + w)$$

where P is the perimeter, l the length, and w the width. (See Figure 3.3.) Suppose the perimeter of a rectangle is 50 inches and the length is 3 inches longer than the width. Find the dimensions of the rectangle.

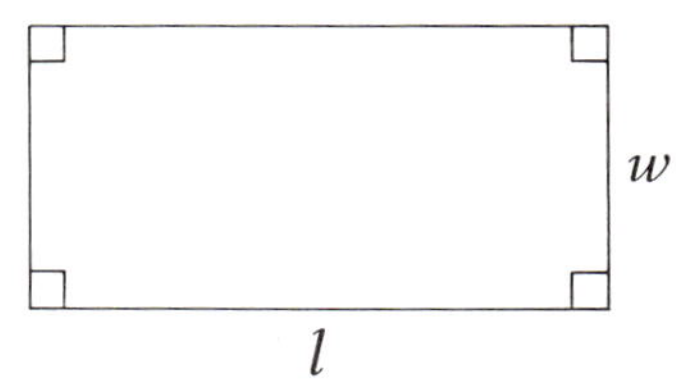

FIGURE 3.3. $\boldsymbol{P = 2l + 2w = 2(l + w)}$
$\boldsymbol{A = lw}$

31. The area of a rectangle is given by the formula

$$A = lw$$

where A is the area, l the length, and w the width. (See Figure 3.3.) Suppose that the length of a rectangle is twice the width. If the perimeter is 60 meters, find (a) the width, (b) the area.

32. The area of a triangle is given by the formula

$$A = \frac{bh}{2}$$

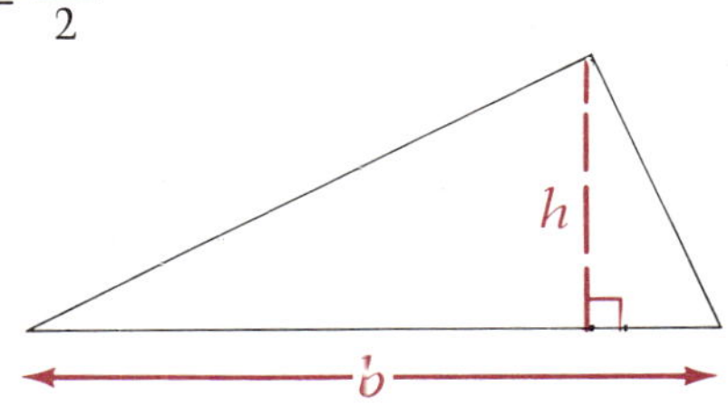

FIGURE 3.4. $\boldsymbol{A = \frac{bh}{2}}$

where b is the length of the base of the triangle and h is the length of the corresponding altitude. (See Figure 3.4.) Suppose the area of a triangle is 60 square centimeters and the base is 10 centimeters long. Find the length of the corresponding altitude.

For Exercises 33–36 note that the *complement* of $\angle A$ (*angle* A) measures 90° minus the measure of $\angle A$. (See Figure 3.5.) The *supplement* of $\angle A$ measures 180° minus the measure of $\angle A$. (See Figure 3.6.).

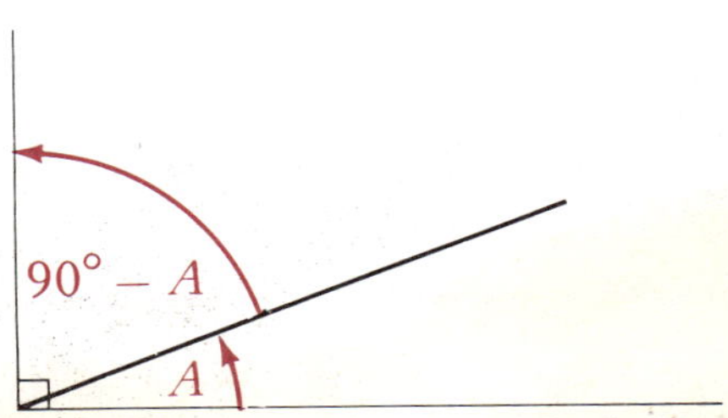

FIGURE 3.5

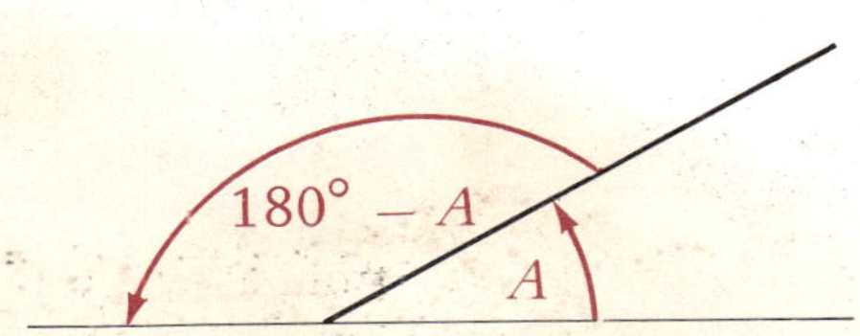

FIGURE 3.6

Use this area for work space.

33. An angle measures twice as much as its complement. Find the measure of the angle.

34. An angle measures 30° more than its supplement. Find the measure of the angle.

35. The supplement of angle A measures three times as much as the complement of angle A. Find the measure of angle A.

36. The supplement of angle B measures 20° less than three times the complement of angle B. Find the measure of angle B.

For Exercises 37–40 note that the sum of the measures of the angles of a triangle is 180°. Also, a right triangle contains a right (90°) angle. Each of the other angles measures less than 90° and is called an *acute angle*. (*See Example* 6.)

37. In a right triangle one acute angle measures twice as much as the other acute angle. Find the measure of each acute angle.

38. In a right triangle one acute angle measures 10° more than the other acute angle. Find the measure of each acute angle.

39. The largest angle of a triangle measures 20° more than the second largest angle and 25° more than the smallest angle. Find the measure of each angle.

40. One angle of a triangle measures 30° less than each of the other two angles. Find the measure of each angle.

In Exercises 41–46 write an equation and then solve it. Check Exercise 41. (*See Example* 7.)

41. Jean has twice as many dimes as nickels in her purse. If altogether she has 75 cents in nickels and dimes, how many nickels does she have? *Check your result.*

42. Frank has $97 in $1 bills, $5 bills, and $10 bills. If he has four more $5 bills than $10 bills and three more $10 bills than $1 bills, how many of each does he have?

43. It costs Sheila 87 cents to mail her letter. She uses one more of the 15-cent stamps than she uses of the 3-cent stamps. Assuming these are the only stamps she has, how many of each does she use?

44. A store buys lamps for $14 apiece. Two of the lamps get broken, and the rest are sold for $25 apiece. The store makes a profit of $632 on this item. How many lamps were originally purchased by the store?

45. Mozart wrote five more than four times the number of symphonies that Beethoven wrote. Altogether they wrote 50 symphonies. How many did each write?

46. A television channel has 12 minutes of programming for every 3-minute commercial. Assuming these are the only commercials, how many commercials are there on a 3-hour show?

3.6 Distance Problems

A. The Distance Formula

The distance that can be traveled at a *constant rate* equals the rate multiplied by the time spent traveling. For example, if a car travels at the constant rate of 50 miles per hour, in 4 hours it will travel

$$50 \cdot 4, \text{ or } 200, \text{ miles}$$

Let

$$r = \text{rate}, \qquad t = \text{time}, \qquad \text{and} \qquad d = \text{distance}$$

Then

$$r \cdot t = d$$

This is known as the **distance formula.** It is convenient to express this formula by means of a table. Thus for the above example, write

r	$\cdot\ t$	$=\ d$
50	4	200

Next, suppose an airplane flies at the constant rate of 400 miles per hour. In 6 hours it travels 2400 miles.

r	$\cdot\ t$	$=\ d$
400	6	2400

The units of measurement must match. Thus, when distance is measured in miles and time in hours, then rate is given in *miles per hour,* which can be written as

$$\frac{\text{miles}}{\text{hours}}$$

Corresponding to the formula

$$r \cdot t = d$$

you have

$$\frac{\text{miles}}{\text{hours}} \cdot \text{hours} = \text{miles}$$

In practice, cars, airplanes, and other vehicles travel at a *varying rate,* rather than at a *constant rate,* over a period of time. A car may start at a toll booth and gradually accelerate until it reaches a speed of 60 miles per hour, then slow down to 50 miles per hour, speed up to 65 miles per hour, and so on. This greatly complicates the picture, and the distance formula no longer applies. (Indeed, the techniques of calculus are generally required to calculate distance.) However, if we assume an *average rate* at which a vehicle travels, this average rate is, in fact, a constant rate, so that the distance formula does apply.

Throughout this section *all rates are assumed to be average rates.*

Other forms of the distance formula apply when you are given some other information. For example, suppose you are told that a man walks 12 miles at the average rate of 3 miles per hour. To find the time he walks, divide both sides of the distance formula

$$r \cdot t = d$$

by r to obtain

$$t = \frac{d}{r}$$

Thus
$$t = \frac{12}{3}$$
$$t = 4$$

He walks 4 hours.

Similarly, suppose you are told that a bus covers 180 miles of a highway in 3 hours. To find the rate, divide both sides of the distance formula by t to obtain

$$r = \frac{d}{t}$$

Then the bus travels at the rate of

$$\frac{180}{3}, \text{ or } 60, \text{ miles per hour}$$

Example 1 ► A train traveling at the rate of 90 miles per hour goes 450 miles. How long does this take?

Solution:

Let t be the number of hours.

r	$t = \frac{d}{r}$	d
90	t	450

$$t = \frac{d}{r}$$
$$t = \frac{450}{90}$$
$$t = 5$$

The trip takes 5 hours. ◄

PRACTICE EXERCISES FOR TOPIC A

Answers
1. (a)
(b)
(c)
2. (a)
(b)
(c)

1. If a car travels at an average rate of 55 miles per hour, how far will it travel in

(a) 3 hours (b) 5 hours (c) t hours

2. A bicyclist covers 22 miles in 2 hours.

(a) What is her average rate of speed (in miles per hour)?

(b) At this rate how long will it take her to travel 55 miles?

(c) How long will it take her to travel x miles?

Now try the exercises for Topic A on page 120.

B. Same Direction, Opposite Directions

Suppose two cars leave from the same place at the same time along a straight road, one traveling at 60 miles per hour, the other at 50 miles per hour.

1. If they travel in *the same direction,* then *each hour* the faster car goes 10 (or 60 − 50) miles further than the slower car. (*See Figure* 3.7(*a*).)

FIGURE 3.7(a). If the cars travel in the same direction, the faster car goes 10 miles further than the slower car in 1 hour.

2. If they travel in *opposite directions,* then *each hour* the cars separate 110 (or 60 + 50) miles. (*See Figure* 3.7(*b*).)

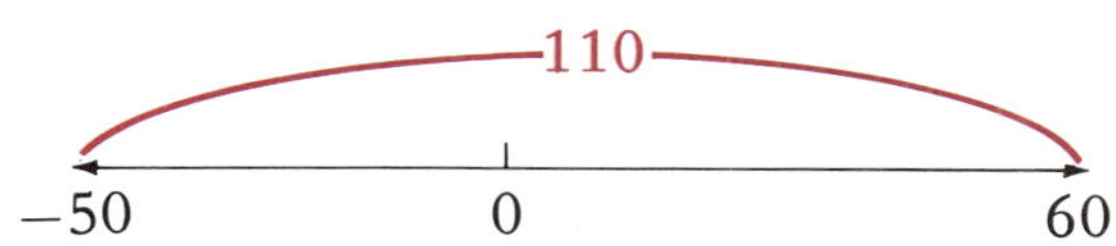

FIGURE 3.7(b). If the cars travel in opposite directions, the cars are 110 miles apart in 1 hour.

Example 2 ► Two cars leave a restaurant at the same time headed in the same direction along a straight road. One car travels at 60 miles per hour, the other at 40 miles per hour. How far apart are they after 4 hours?

Solution:

	r	$\cdot\ t\ =$	d
faster car	60	4	240
slower car	40	4	160

(*See Figure 3.8.*)

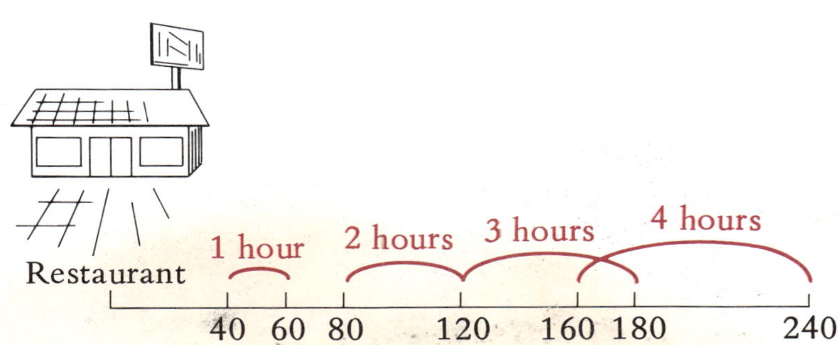

FIGURE 3.8

The cars travel in the same direction. In 1 hour the faster car is 20 (= 60 − 40) miles ahead of the slower car. In 4 hours it is 80 (= 4 · 20) [or 240 − 160] miles ahead of the slower vehicle. ◄

Example 3 ▶ Two cars leave a ball park at the same time. One car travels eastward at 55 miles per hour; the other car travels westward at 70 miles per hour. How far apart are they after 2 hours?

Solution:

	r ·	t =	d
eastward	55	2	110
westward	70	2	140

The cars travel in opposite directions. Their distance apart, d, is the sum of the distances each has traveled in 2 hours:

$$d = 110 + 140$$

$$d = 250$$

The distance apart is 250 miles. Each hour the cars move apart 55 + 70 (= 125) miles. Thus 250 = 125 · 2. (*See Figure 3.9.*) ◀

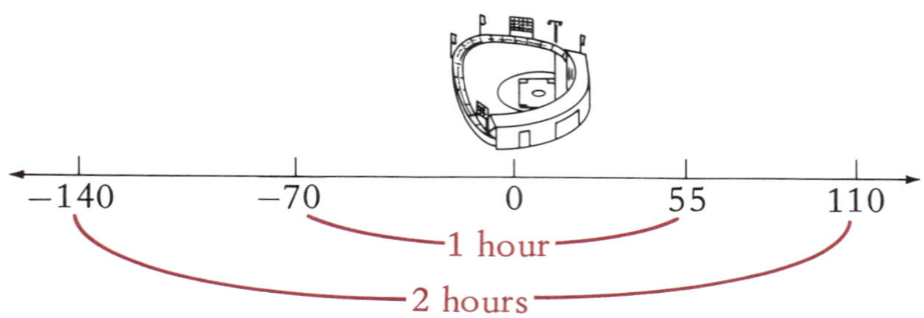

FIGURE 3.9

Answers

3. ____________

4. ____________

PRACTICE EXERCISES FOR TOPIC B

3. A car and a bus leave a toll booth at the same time and travel in the same direction. The car travels at the constant rate of 65 miles per hour and the bus travels at 50 miles per hour. What is the distance between them after 3 hours?

4. Two airplanes leave an airport at the same time, headed in opposite directions. One cruises at 230 miles per hour, the other at 150 miles per hour. How far apart are they after 2 hours?

Now try the exercises for Topic B on page 121.

C. Round Trip

Next, suppose a man runs to a mailbox and then walks back home. The round trip takes 10 minutes.

If he runs for t minutes, then he walks for $(10 - t)$ minutes. For example, if he runs for 4 minutes, then he walks for 10 − 4, or 6, minutes. Assume he takes the same route going and returning. Then *the distance each way is the same.* (*See Figure 3.10.*)

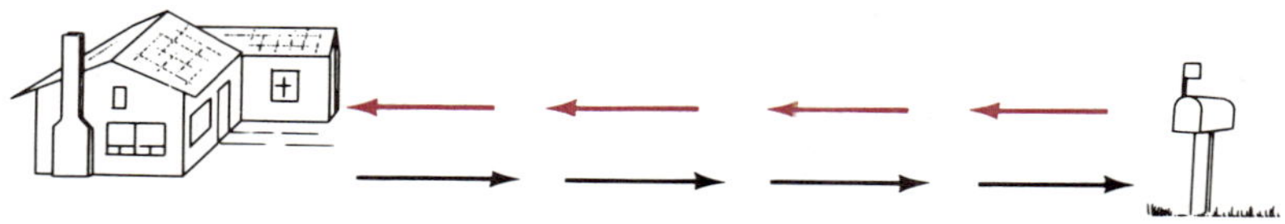

FIGURE 3.10

Example 4 ▶ A canoe goes upstream at the rate of 5 miles per hour. It returns downstream at the rate of 10 miles per hour. The round trip takes 3 hours.

(a) How long does the canoe travel upstream?

(b) How far upstream does it go?

(c) Check your results.

Solution:

(a) The canoe travels the same distance upstream as downstream. (*See Figure 3.11.*)

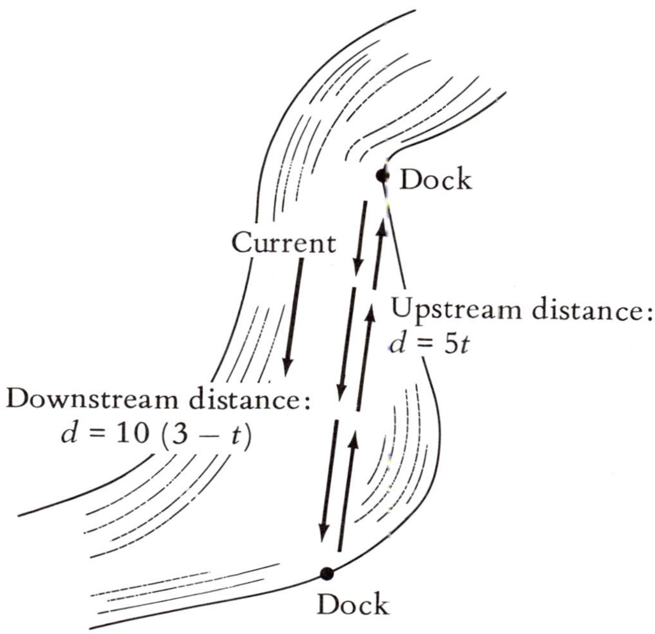

FIGURE 3.11

Let t be the number of hours it travels upstream. Because the round trip takes 3 hours, it travels for $(3 - t)$ hours downstream.

	r	$\cdot$ t	$=$ d
upstream	5	t	$5t$
downstream	10	$3 - t$	$10(3 - t)$

The distance upstream equals the distance downstream.

$$5t = 10(3 - t)$$

Solve this equation:

$$5t = 30 - 10t$$

$$15t = 30$$

$$t = 2$$

The canoe travels for 2 hours upstream (and for 1 hour downstream).

(b) To find the *distance* upstream, use the distance formula with this new piece of information from part (a).

	r ·	t =	d
upstream	5	2	10

The canoe travels 10 miles upstream.

(c) You can check both parts by showing that the distance downstream is also 10 miles. Recall that the time traveling downstream is 1, or (3 − 2), hour.

	r ·	t =	d
downstream	10	1	10

◀

Answers

5. ______

PRACTICE EXERCISES FOR TOPIC C

5. Sam drove from home to Fairfield at an average rate of 60 miles per hour. On the way back home the trip took exactly one hour longer because he drove at an average rate of 50 miles per hour all the way. Use an equation to find out the distance from home to Fairfield and how long it took to drive each way. [*Hint*: Let t = the time it took to get there at 60 miles per hour; fill in the table and write a meaningful equation about these quantities.]

	r ·	t =	d
going	60	t	
returning	50		

Now try the exercises for Topic C on page 121.

Use this area for work space.

EXERCISES*

A.

1. A car travels 30 miles per hour for 3 hours. How far does it travel?

2. How long does it take Bob to bicycle home from his office if he travels 12 miles per hour and lives 24 miles from his office? (*See Example 1.*)

3. An automobile travels for 2 hours at 60 miles per hour. It then slows down and travels 40 miles per hour in the same direction for the next 3 hours. How far does it travel?

*All rates are *average rates*.

Use this area for work space.

4. A train travels 280 miles in 4 hours. (a) At what rate does it travel? (b) At this rate, how far will it travel in 7 hours? (c) in y hours?

5. An airplane travels 1050 miles in 3 hours. (a) At what rate does it travel? (b) At this rate, how many hours does it take to travel 1750 miles? (c) to travel m miles?

6. A car covers a distance of 300 kilometers in six hours. A bus covers a distance of 420 kilometers in seven hours. Which vehicle travels at a faster rate?

7. A jogger can run at the rate of 8 miles per hour over level ground and at 4 miles per hour over hilly ground. Altogether it takes her 3 hours to cover 20 miles. How many of these miles are level?

8. A car travels for 4 hours at 60 miles per hour. At this speed it gets 16 miles to the gallon. Gasoline costs $1.20 per gallon.
(a) How many gallons of gas does the car use?
(b) What is the total cost of gasoline for this trip?

B.

9. A car traveling 70 miles per hour along a straight road passes another car cruising at 45 miles per hour. The two cars are headed in the same direction.
(a) How far apart are they after 1 hour?
(b) How far apart are they after 3 hours? (*See Example* 2.)

10. Two cars leave a toll booth at the same time. They travel in opposite directions along a straight road, one at 65 miles per hour, the other at 60 miles per hour.
(a) How far apart are they after 1 hour?
(b) How far apart are they after 4 hours? (*See Example* 3.)

11. Two cars leave town at the same time traveling in opposite directions along a straight road. One car goes 50 miles per hour, the other 65 miles per hour. How far apart are they after 6 hours? *Check your result.*

12. Two trains approach one another along (straight) parallel tracks, starting at the same time from stations 540 miles apart. One train travels 100 miles per hour, the second train 80 miles per hour. How long after they start do they pass each other?

13. Two joggers leave from the same place at the same time, headed in the same direction along a straight path. One goes 10 kilometers per hour, the other 7 kilometers per hour. How far apart are they after 2 hours? *Check your result.*

14. A car leaves a gas station an hour after a motor scooter does. They both travel along a straight highway in the same direction. If the car travels 20 miles per hour faster than the scooter and passes the scooter in 2 hours, how fast is the car traveling?

C.

15. A car travels 45 miles per hour for 4 hours. On the return trip it travels 60 miles per hour. How long does the return trip take? *Check your result.* (*See Example* 4.)

16. A motorcyclist travels 60 miles per hour for 6 hours along a straight highway. He then sells his motorcycle and travels back the same route by train in just 4 hours. Assuming the train tracks are alongside the highway, how fast does the train travel?

17. A canoe goes upstream at the rate of 6 miles per hour. It returns downstream at the rate of 9 miles per hour. If the round trip takes 5 hours, how far upstream does it go? *Check your result.*

18. A rowboat goes upstream at the rate of 4 kilometers per hour. It returns downstream at the rate of 8 kilometers per hour. If the round trip takes 6 hours, how far upstream does it go?

REVIEW

What Have You Learned in Chapter 3?

You have learned that an equation is a statement of equality, and that a number is a root of an equation (in one variable), if a true statement results when the number replaces the variable. To check whether a number is a root of the equation, replace each occurrence of the variable by this number. If a true statement results, the number is a root.

You can solve an equation by transforming it into simpler equivalent equations, that is, simpler equations with the same roots. When you add the same expression to, or subtract the same expression from, both sides of an equation, you obtain an equivalent equation. When you multiply or divide both sides of an equation by the same nonzero number, you obtain an equivalent equation. In simplifying an equation, you bring variables to one side, numerical terms to the other.

Many applied problems can be solved by first translating the problem into a mathematical equation, and then solving the equation by the methods you have learned.

Let's Review Chapter 3

3.1 Roots

In Exercises 1–4 check whether the given number is a root of the equation.

1. $3x + 2 = 14$; $\boxed{4}$

2. $-6y = 18$; $\boxed{-3}$

3. $\frac{5t - 5}{5} = 5$; $\boxed{5}$

4. $2x = -x$; $\boxed{0}$

5. Express the following question as an equation (in the variable x). You need not solve the equation.

What number divided by 3 is equal to 10?

3.2 *Solving Simple Equations*

In Exercises 6–8 solve each equation.

6. $x - 5 = 7$

7. $7y = -63$

8. $\frac{t}{4} = 12$

9. (a) Solve the following equation. (b) Check the root.

$$x + 12 = -6$$

In Exercise 10 write an equation and then solve it.

10. What number must be multiplied by -3 to obtain 18?

3.3 *Isolating the Variable*

In Exercises 11–14 solve each equation.

11. $12 + 4x = 0$

12. $3t + 8 = 5t - 6$

13. $\frac{u}{5} + 2 = 12$

14. $3y + 1 - 2y = 7 - y + 4$

15. (a) Solve the following equation. (b) Check the root.

$$7 - 3x = 9x - 5$$

16. A salesman earns a weekly salary of \$250 plus \$15 for each sale. His weekly earnings are given by

$$y = 250 + 15x$$

where x is the number of sales that week. How many sales does he have to make in order to earn \$400 for a week's work?

3.4 *Equations with Parentheses*

In Exercises 17–20 solve each equation.

17. $3(2t - 1) = 7t + 5$

18. $9(4u - 3) = 3(7u + 1)$

19. $6 - (3 - x) = 2x - 3$

20. $9(3t - 1) = 18$

3.5 *Applied Problems*

In Exercises 21–25 set up an equation and then solve it.

21. Five more than three times an integer is 80. Find this integer.

22. A 40-meter rope is divided into three pieces in such a way that the middle piece is 5 meters shorter than each of the end pieces. Find the length of the middle piece.

23. In an election, the winner receives twice as many votes as the other two candidates combined. The runner-up edges out the third candidate by 2 votes. If altogether 144 votes are cast for these three candidates, how many votes does each receive?

24. The supplement of angle *A* measures 10° less than three times the complement of angle *A*. Find the measure of angle *A*.

25. Judy has three more dimes than nickels and twice as many nickels as quarters. If altogether she has 85 cents in nickels, dimes, and quarters, how many of each coin does she have?

3.6 *Distance Problems*

26. Two cars leave a gas station at the same time traveling in opposite directions along a straight road. One car is going 45 miles per hour, the other 60 miles per hour. How far apart are they after 5 hours?

27. A motorcycle travels to its destination at the rate of 40 miles per hour for a period of 6 hours. At what rate does it have to travel if it is to make the return trip in 5 hours?

28. Patricia drives at an average rate of 50 miles per hour for 4 hours. She then drives for 2 hours at 60 miles per hour. How fast must she drive for the remainder of the trip if the entire 530-mile trip is to take 9 hours?

REVIEW

And These from Chapters 1 and 2

29. What power of -2 is -32?

30. Find the value of $\frac{(-5)(-1)(-2)40}{(-2)(-2)(-2)}$.

31. There are 24 hours in a day and 7 days in a week. How many hours are there in 4 weeks?

32. Find the value of $\frac{14 - 2}{3} - 3(5 - 1)$.

33. Find the value of $(-3)^2 - (-3)^3$.

34. Suppose x stands for some number. Express 1 less than the square of the number in terms of x.

35. Evaluate $2xy - 3$ when $x = 2$ and $y = -1$.

36. Evaluate $x^3 + 4x^2 - 1$ when $x = 2$.

37. Simplify $4x - [3y - (2x - y + 1)]$.

38. Multiply $(5a)(-4b)(-10c)$

39. Divide $\frac{12x - 8y + 20z}{-4}$

40. Multiply the sum of $x^2 + 2x + 4$ and $3x^2 - x - 2$ by -3.

Practice Exam 1 on Chapter 3

1. Is -2 a root of $5x - 2 = -8$?

2. Is 3 a root of $\frac{7t - 6}{3} = 5$?

3. Solve $y + 12 = 3$

4. Solve $\frac{x}{-3} = 2$

5. Solve $3t - 1 = t + 9$

6. Solve $\frac{z - 2}{4} = 6$

7. Solve $4x + 3(x - 1) = 25$

8. Solve and check $5(2t - 1) = 3(7 - t)$

In Questions 9–12 write an equation and then solve it.

9. What number must be subtracted from 10 to obtain 12?

10. A 38-foot rope is cut into three pieces. The longest piece is 1 foot longer than twice the shortest piece. The third piece is 4 feet shorter than the longest piece. Find the length of each piece.

11. In a right triangle one acute angle measures four times as much as the other. Find the measure of the smallest angle.

12. A Ford leaves a restaurant heading due north at 50 miles per hour. A Plymouth leaves the restaurant heading due south at 60 miles per hour 2 hours later. How far apart are the cars 3 hours after the Plymouth departs?

Practice Exam 2 on Chapter 3

1. Is -9 a root of the equation $4(x - 5) = -16$?
2. Is 8 a root of $3y - 10 = 22 - y$?
3. Solve $x + 15 = 10$
4. Solve $\frac{n}{6} = -4$
5. Solve $30 - \frac{a}{2} = 21$
6. Solve $y - 8 = 19 - 2y$
7. Solve and check $4(2x + 1) = 6 - (x + 2)$
8. A car travels 135 miles in 3 hours. At this rate, how far will it travel in 8 hours?

In Questions 9–12 write an equation and then solve it.

9. When 18 is added to a certain number, the result is 4 times the number. What is the number?
10. When a certain number is multiplied by 5, the result is 12 less than the original number. Find the number.
11. A class with 23 children has 5 more boys than girls. How many girls are in the class?
12. Betty flew to Chicago in a plane that traveled at 175 miles per hour. She returned by bus at 50 miles per hour, and the trip took 5 hours longer than the plane ride. How many hours did it take to fly to Chicago?

4

Products and Factors

4.1 Multiplying Powers

In Chapter 2 you learned how to multiply and divide a polynomial by a number. Thus, by the Distributive Laws,

$$4(x^2 + 2x + 3) = 4x^2 + 4(2x) + 4(3)$$
$$= 4x^2 + 8x + 12$$

and

$$\frac{9a^2 - 6a + 15}{3} = \frac{9a^2}{3} - \frac{6a}{3} + \frac{15}{3}$$
$$= 3a^2 - 2a + 5$$

A. $a^m \cdot a^n$

What happens when you multiply two powers of the same number? Observe that

$$2^2 \cdot 2^1 = (2 \cdot 2) \cdot 2 = 2 \cdot 2 \cdot 2 = 2^3$$

Thus $$2^2 \cdot 2^1 = 2^{2+1}$$

Next, consider what happens when you multiply two powers of the same variable.

Example 1 ▶ Multiply $x^3 \cdot x^2$
Solution:

$$x^3 \cdot x^2 = (xxx)(xx) = xxxxx = x^5$$

Thus $$x^3 \cdot x^2 = x^{3+2}$$ ◀

Let m and n be positive integers. To find the product

$$a^m \cdot a^n$$

observe that

$$a^m = \underbrace{a \cdot a \cdot a \ldots a}_{m \text{ factors}}$$

$$a^n = \underbrace{a \cdot a \cdot a \ldots a}_{n \text{ factors}}$$

Therefore $$a^m \cdot a^n = \underbrace{(a \cdot a \cdot a \ldots a)}_{m \text{ factors}}\underbrace{(a \cdot a \cdot a \ldots a)}_{n \text{ factors}}$$

$$= \underbrace{a \cdot a \cdot a \ldots a \cdot a \cdot a \cdot a \ldots a}_{(m+n) \text{ factors}}$$

$$= a^{m+n}$$

Thus $$a^m \cdot a^n = a^{m+n}$$

In other words, to multiply two powers of the same number (or variable) a, use the base a and add the exponents.

Example 2 ▶ Multiply $y^8 \cdot y^5$
Solution:

$$y^8 \cdot y^5 = y^{8+5} = y^{13}$$ ◀

Three or more powers of the same number (or variable) are multiplied by using the base and adding the exponents.

Example 3 ▶ Multiply $t^4 \cdot t^2 \cdot t$
Solution:
Recall that $t = t^1$. Thus

$$t^4 \cdot t^2 \cdot t = t^{4+2+1} = t^7$$ ◀

Answers

1. (a) ______
   (b) ______

PRACTICE EXERCISES FOR TOPIC A

1. Multiply: (a) $r^8 \cdot r$ (b) $n^4 \cdot n^7 \cdot n^2$

Now try the exercises for Topic A on page 131.

B. Products of Monomials

When multiplying monomials, use the Associative and Commutative Laws to rearrange the factors.

1. Group all coefficients at the beginning, and multiply them.
2. Group powers of the same variable together and multiply them, as previously explained.

Example 4 ▶ Multiply $(4xy^2)(7x^2y^4)$

Solution:

$$(4xy^2)(7x^2y^4) = (4 \cdot 7)(xx^2)(y^2y^4) = 28x^3y^6$$

Group all coefficients at the beginning. — Group powers of the same variable together. — Multiply coefficients and powers of the same variable. ◀

Example 5 ▶ Multiply $(4ax^2y)(-2x^2y^3)(-3a^2bx)$

Solution:

$$\begin{aligned}(4ax^2y)(-2x^2y^3)(-3a^2bx) &= [4(-2)(-3)](aa^2)b(x^2x^2x)(yy^3)\\ &= 24a^3bx^5y^4\end{aligned}$$ ◀

Answers

2. (a) ______
   (b) ______

PRACTICE EXERCISES FOR TOPIC B

2. Multiply: (a) $9a^3(-2a^5b^3)$ (b) $(-x^2y)(3xy^2)(-7x^3)$

Now try the exercises for Topic B on page 131.

C. Quotients of Monomials

Recall that for every nonzero number a,

$$\frac{a}{a} = 1$$

This law also applies to variables a. Therefore, when dividing products, if the same factors appear in the dividend and divisor, we can divide these factors and replace them by 1's. In the following division process, observe that

$$\frac{x^5}{x^2} = \frac{\overset{1}{\cancel{x}} \cdot \overset{1}{\cancel{x}} \cdot x \cdot x \cdot x}{\underset{1}{\cancel{x}} \cdot \underset{1}{\cancel{x}}} = \frac{x \cdot x \cdot x}{1} = x^3$$

Thus $$\frac{x^5}{x^2} = x^{5-2}$$

To divide powers of the same *nonzero* number (or variable) a, write down the base a and subtract the exponents:

$$\frac{a^m}{a^n} = a^{m-n}$$

where m and n are positive integers, and $m > n$.

Example 6 ▶ Divide $\frac{y^8}{y^3}$

Solution:

$$\frac{y^8}{y^3} = y^{8-3} = y^5$$ ◀

Powers of several variables may have to be divided.

Example 7 ▶ Divide $\frac{a^7b^4c^3}{a^4b^3c^2}$

Solution:

$$\frac{a^7b^4c^3}{a^4b^3c^2} = a^{7-4}b^{4-3}c^{3-2} = a^3bc$$ ◀

In the following example, first divide the coefficients. Also note that the powers of y divide out. In fact,

$$\frac{y^3}{y^3} = 1$$

Example 8 ▶ Divide $\frac{10x^6y^3}{5x^4y^3}$

Solution:

$$\frac{10x^6y^3}{5x^4y^3} = 2x^{6-4} = 2x^2$$ ◀

Answers

3. (a) ____

(b) ____

(c) ____

PRACTICE EXERCISES FOR TOPIC C

3. Divide:

(a) $\frac{c^9}{c^4}$ (b) $\frac{a^2b^4c}{ab^2c}$ (c) $\frac{28x^5y^2}{4x^2y}$

Now try the exercises for Topic C on page 131.

D. Using the Distributive Laws

Example 9 ▶ Use the Distributive Laws to find $x^2(x + y)$.

Solution:

$$x^2(x + y) = x^2 \cdot x + x^2 \cdot y$$
$$= x^3 + x^2y$$ ◀

Example 10 ▶ Multiply $4xyz^2(xy - 2z + 3x^2z)$

Solution:

$$4xyz^2(xy - 2z + 3x^2z) = 4xyz^2(xy) - 4xyz^2(2z) + 4xyz^2(3x^2z)$$
$$= 4x^2y^2z^2 - 8xyz^3 + 12x^3yz^3$$ ◀

Example 11 ▶ Divide $\frac{2x^4 - 7x^3}{x^2}$

Solution:

$$\frac{2x^4 - 7x^3}{x^2} = \frac{2x^4}{x^2} - \frac{7x^3}{x^2} = 2x^2 - 7x$$

To check this by multiplying, note that

$$(2x^2 - 7x) \cdot x^2 = 2x^2 \cdot x^2 - 7x \cdot x^2$$
$$= 2x^4 - 7x^3$$ ◀

Example 12 ▶ Divide $\frac{5x^2y + 10xy^2 - 15xy^3}{5xy}$

Solution:

$$\frac{5x^2y + 10xy^2 - 15xy^3}{5xy} = \frac{5x^2y}{5xy} + \frac{10xy^2}{5xy} - \frac{15xy^3}{5xy}$$
$$= x + 2y - 3y^2$$ ◀

PRACTICE EXERCISES FOR TOPIC D

Answers

4. ____________

5. ____________

4. Multiply $3x^3y(x^4y^2 - 5x^3y^3)$

5. Divide $\frac{10a^8b^3 - 5a^3b^2 + 15a^2b}{5a^2b}$

Now try the exercises for Topic D on page 132.

E. Several Operations

Example 13 ▶ Simplify $\frac{10a^3b + 20a^2b}{10ab} - \frac{4ab^2}{2b^2}$

Solution:

$$\frac{10a^3b + 20a^2b}{10ab} - \frac{4ab^2}{2b^2} = \frac{10a^3b}{10ab} + \frac{20a^2b}{10ab} - \frac{4ab^2}{2b^2}$$

$$= a^2 + 2a - 2a$$

$$= a^2$$

◄

Answers

6. ______

7. ______

PRACTICE EXERCISES FOR TOPIC E

6. Simplify $4a(a^2 + 2b) - 3ab$

7. Simplify $\dfrac{9a^2b - 6ab^2}{3ab} + \dfrac{10a^2b^2}{5ab^2}$

Now try the exercises for Topic E on page 132.

Use this area for work space.

EXERCISES

A. In Exercises 1–7 multiply. (*See Examples 1–3.*)

1. $x^3 \cdot x$ **2.** $b^3 \cdot b^2$ **3.** $c^4 \cdot c^6$ **4.** $m^{10} \cdot m^8$

5. $c^4 \cdot c^2 \cdot c^3$ **6.** $y^9 \cdot y^4 \cdot y^7$ **7.** $z^2 \cdot z^5 \cdot z \cdot z^4$

8. Pick out the *error(s)*.

(a) $x^3 \cdot x^4 = x^{12}$ (b) $x^3 + x^4 = x^7$ (c) $x^2 \cdot x^6 = x^{2+6}$

(d) $x^2 \cdot x^6 = x^{2 \cdot 6}$ (e) $x^2 + x^6 = x^{2+6}$

B. In Exercises 9–18 multiply. (*See Examples 4 and 5.*)

9. $4a^2 \cdot a^4$ **10.** $(-1)y^2 \cdot 5 \cdot y^2$

11. $(x^2y)(xy^2)$ **12.** $(a^2b)(a^3b^2)$

13. $(2c^2d)(5cd^2)$ **14.** $(3ab)(-2a^2b)$

15. $(5xy)(3x^2y)(2x^2)$ **16.** $(-4a^2b)(-ab)(abc)$

17. $(2xy^2z)(xyz^2)(3xz)(yz^2)$ **18.** $(-2abc)(bcd)(3abc^2)(4a^2d)$

19. Find the product of $2xy$ and $9xy^2$.

20. Find the product of $-2a^2bc$, $-5ac$, and abc^2.

C. In Exercises 21–30 divide. (*See Examples 6–8.*)

21. $\dfrac{x^4}{x^2}$ **22.** $\dfrac{a^{10}}{a^7}$ **23.** $\dfrac{a^3b^2}{cb}$ **24.** $\dfrac{s^5t^8}{s^4t^5}$

Use this area for work space.

25. $\dfrac{x^4y^3z^2}{xy^2z}$ 26. $\dfrac{15b^{10}}{5b^8}$ 27. $\dfrac{30ab^2}{-10ab}$ 28. $\dfrac{21x^3y^6}{7x^2y^4}$

29. $\dfrac{35s^8t^7}{-5s^7t}$ 30. $\dfrac{48x^3y^3z^7}{-12x^2yz^5}$

31. Divide $42a^4bc^3$ by $7a^3bc$.

32. Pick out the *error(s)*.

(a) $\dfrac{x^6}{x^3} = x^2$ (b) $\dfrac{x^8}{x^2} = x^6$ (c) $\dfrac{x^6}{x^2} = x^3$ (d) $\dfrac{x^5}{x} = x^4$ (e) $\dfrac{x^9}{x^9} = 0$

D. In Exercises 33–44 multiply. (*See Examples 9 and 10.*)

33. $a^2(a + b)$ 34. $2a(a^2 + a^3)$ 35. $-4y^2(y + yz)$

36. $2ab(3a^2 - 5b)$ 37. $3xyz(xz - y^2z)$ 38. $5s^2t(2s^3 - 4t^2)$

39. $10x^2yz^3(5x^4yz^3 + 8xy^2z^2)$ 40. $7ab^2c(4a^2bc^4 - 3a^7bc^2)$

41. $2ab(a^2 + ab + b^2)$ 42. $4xyz^2(x^2z - 4xy + 3xz)$

43. $-x^2yz^2(x + 2xy - 3xyz)$ 44. $-2a^2bc^2(-ab + 5bc - 2a^2bc^7)$

In Exercises 45–54 divide. (*See Examples 11 and 12.*)

45. $\dfrac{a^2 + a}{a}$ 46. $\dfrac{m^7 + m^4}{m^3}$ 47. $\dfrac{x^3 - x^2}{x^2}$

48. $\dfrac{a^3b^2 + ab}{ab}$ 49. $\dfrac{ax^2 - a^2x}{ax}$ 50. $\dfrac{3x^3y^2 + 9xy}{3x}$

51. $\dfrac{4ab - 12a}{4a}$ 52. $\dfrac{20x^4y^3 + 30x^3y^3}{5x^2y^3}$

53. $\dfrac{12a^2b + 18ab + 9ab^2}{3ab}$ 54. $\dfrac{16x^2y^2z^2 + 12xy^2z^2 - 4xyz}{4xyz}$

55. The sum of $2xy$ and $3x^2$ is multiplied by $3x^2y^3$. Find the result.

56. The sum of $4x^2y$ and $6xy^3$ is divided by $2xy$. Find the result.

57. Pick out the *error(s)*.

(a) $\dfrac{x^2 + 4}{x} = x + 4$ (b) $\dfrac{x^2 + 4}{2} = x + 2$ (c) $\dfrac{x^4 + 4x^2}{x^2} = x^2 + 4$

(d) $\dfrac{2x^2 + 4}{2} = x^2 + 2$ (e) $\dfrac{x^2 + y^2}{xy} = x + y$ (f) $\dfrac{x^2 - y^2}{x} = x - y^2$

E. In Exercises 58–66 simplify each expression. (*See Example 13.*)

Sample: Simplify: $4x^2 - 3x(x + 2y)$	*Solution:* $4x^2 - 3x(x + 2y) = 4x^2 - 3x \cdot x - 3x \cdot 2y$ $= 4x^2 - 3x^2 - 6xy$ $= x^2 - 6xy$

Use this area for work space.

58. $(5a + 2)b - 3ab$

59. $2a^2 + a(3a + b)$

60. $5ab - 2b(a - 3b)$

61. $4x^2(x + y) - 2x^2(x - y)$

62. $a(a + b) - a(a - b)$

63. $x(x^2 - 2) + x^2(x + 2)$

64. $\dfrac{4a^2 + ab}{a} + 5a - b$

65. $\dfrac{2a^2 + 4ab}{a} + \dfrac{9b^3 + 6ab^2}{3b^2}$

66. $2x(x^2y + x^3) - \dfrac{x^5 + x^4y}{x}$

4.2 Prime Factors

It is important to know how to write an integer as a product of other integers. This process is known as **factoring.** You will also learn how to factor a polynomial, that is, how to write it as a product of other polynomials. Factoring will play a role in solving equations (Section 4.7) and in combining fractions (Chapter 5).

A. Primes

Observe that

$$15 = 3 \cdot 5$$

Thus 15 can be expressed as the product of two integers, 3 and 5, which are called *factors* of 15.

DEFINITION

FACTOR. Let a and b be integers, where $b \neq 0$. Then b is called a **factor of *a*** if

$$a = b \cdot c$$

for some integer c. In this case, if $c \neq 0$, then c is also a factor of a, and a is said to be **divisible by *b* and by c.** For any integers a, b, and c, if

$$a = b \cdot c$$

then a is said to be a **multiple of *b* and of c.**

Example 1 ▶ Find all positive factors of 10.
Solution:

$$10 = 10 \cdot 1 \quad \text{and} \quad 10 = 5 \cdot 2$$

The *positive* factors of 10 are

1, 2, 5, and 10

Thus 10 is divisible by each of these numbers and is a multiple of each. ◀

Note that the only way you can express 5 as the product of positive integers is

$$5 = 5 \cdot 1 = 1 \cdot 5$$

Thus the only *positive* factors of 5 are 5 and 1.

DEFINITION

> PRIME; COMPOSITE. Let p be an integer, where $p > 1$. Then p is said to be a **prime** if the only positive factors of p are p itself and 1. An integer n, $n > 1$, that is not a prime is called a **composite.**

2, 3, 5, and 7 are primes, whereas 6 and 10 are composites.

$$6 = 2 \cdot 3 \quad \text{and} \quad 10 = 2 \cdot 5$$

4 is also a composite because

$$4 = 2 \cdot 2$$

0, 1, and negative integers are not classified as either primes or composites.

Example 2 ▶ Which of the following are primes? Which are composites?
(a) 11 (b) 15 (c) 23 (d) 33
Solution:
(a) 11 is a prime. (b) 15 is a composite because $15 = 3 \cdot 5$.
(c) 23 is a prime. No smaller positive integer other than 1 is a factor of 23.
(d) 33 is a composite because $33 = 3 \cdot 11$. ◀

Clearly
$$12 = 4 \cdot 3$$

But the factor 4 can be factored further:

$$4 = 2 \cdot 2$$

Thus
$$12 = 2 \cdot 2 \cdot 3$$

Every composite can be expressed as the product of primes. For example,

$$\begin{aligned} 40 &= 8 \cdot 5 \\ &= (2 \cdot 2 \cdot 2)5 \\ &= 2^3 \cdot 5 \end{aligned}$$

It is convenient to use exponents to express repeated prime factors. *Except for the order of the factors, there is only one way of expressing a composite as a product of primes.* No matter how you begin factoring, the *prime* factors are the same. Thus,

$$\begin{aligned} 40 &= 4 \cdot 10 \\ &= (2 \cdot 2)(2 \cdot 5) \\ &= 2^3 \cdot 5 \end{aligned}$$

Also you can factor the *negative* integer -40 as follows:

$$-40 = -(2^3 \cdot 5)$$

Example 3 ▶ Express each integer as the product of primes.
(a) 30 (b) 50 (c) 72 (d) 96

Solution:

(a) $30 = 6 \cdot 5 = 2 \cdot 3 \cdot 5$ (b) $50 = 5 \cdot 10 = 5 \cdot 2 \cdot 5 = 2 \cdot 5^2$
(c) $72 = 8 \cdot 9 = 2^3 \cdot 3^2$ (d) $96 = 12 \cdot 8 = 2^2 \cdot 3 \cdot 2^3 = 2^5 \cdot 3$ ◀

Answers

1. ____________
2. (a) ______ (b) ______
(c) ______ (d) ______
3. (a) ____________
(b) ____________
(c) ____________
(d) ____________

PRACTICE EXERCISES FOR TOPIC A

1. Find all positive factors of 8.

2. Which of the following integers are primes? Which are composites?

(a) 35 (b) 29 (c) 49 (d) 1

3. Express each integer as the product of primes.

(a) 87 (b) 200 (c) -81 (d) 117

Now try the exercises for Topic A on page 137.

B. Common Factors

Clearly 5 is a factor of both 10 and 15.

DEFINITION

> COMMON FACTOR. Let m and n be any integers—positive, negative or 0. Then the *positive* integer c is called a **common factor of *m* and *n*** if c is a factor of both m and n. The largest common factor of m and n is called the **greatest common divisor of *m* and *n*,** and is written
>
> $$gcd(m, n)$$

Example 4 ▶ Find (a) $gcd(10, 15)$ (b) $gcd(-10, -15)$ (c) $gcd(8, 9)$ (d) $gcd(6, 12)$

Solution:

(a) 1 and 5 are the common factors of 10 and 15. The larger of these is 5. Thus $gcd(10, 15) = 5$

(b) By definition, the common factors of *any* two integers—positive, negative, or 0—must be *positive*. Thus 1 and 5 are also the common factors of -10 and -15, and

$$gcd(-10, -15) = 5$$

(c) 1 is the only common factor of 8 and 9. Thus $gcd(8, 9) = 1$

(d) The common factors of 6 and 12 are 1, 2, 3, and 6. Thus $gcd(6, 12) = 6$ ◀

Part (d) of Example 4 illustrates the fact that *if m is a factor of n, then* $gcd(m, n) = m$. As a second example, $gcd(10, 100) = 10$

You can also speak of common factors and greatest common divisors of three or more integers. The *gcd* notation is again used.

When the integers are relatively simple, their *gcd* can be found at sight, as in Example 4. When difficulties arise, use the prime factorizations of each integer.

Example 5 ▶ Find $gcd(36, 54, 72)$

Solution:

$$36 = 4 \cdot 9 = 2^2 \cdot 3^2, \qquad 54 = 2 \cdot 27 = 2 \cdot 3^3, \qquad 72 = 8 \cdot 9 = 2^3 \cdot 3^2$$

Observe that 2 divides 36, 54, and 72, but 2^2 does not divide 54. Similarly, 3^2 divides 36, 54, and 72, but 3^3 divides neither 36 nor 72. Thus consider the *smallest* power of each *common* prime factor of 36, 54, and 72. *The product of these smallest powers* is then the largest common factor of 36, 54, and 72, and hence is the *gcd*.

Thus here the *common* prime factors are 2 and 3. The *smallest* power of 2 that occurs is 2 (in the factorization of 54). The *smallest* power of 3 is 3^2.

$$gcd(36, 54, 72) = 2 \cdot 3^2 = 18$$ ◀

PRACTICE EXERCISES FOR TOPIC B

Answers

4. (a) ______ (b) ______

(c) ______

4. Find the *gcd* of the given integers.

(a) 30 and 42 (b) 18 and 90 (c) 32, 48, and 60

Now try the exercises for Topic B on page 137.

C. Coefficients of Polynomials

When "factoring" a polynomial P, it is sometimes necessary to find the *gcd* of the coefficients of P.

Example 6 ▶ Find the *gcd* of the coefficients of $32x - 80$.

Solution:

Factor both coefficients.

$$32 = 4 \cdot 8 = 2^2 \cdot 2^3 = 2^5, \qquad -80 = -(16 \cdot 5) = -(2^4 \cdot 5)$$

The only common prime factor is 2, and the smallest power of 2 that occurs is 2^4. Thus

$$gcd(32, -80) = 2^4 = 16$$

and the *gcd* of the coefficients of $32x - 80$ is 16. ◀

PRACTICE EXERCISES FOR TOPIC C

Answers

5. (a) ______ (b) ______

5. Find the *gcd* of the coefficients of each polynomial.

(a) $25x + 30$ (b) $60x^2 - 45x + 15$

Now try the exercises for Topic C on page 137.

Use this area for work space.

EXERCISES

A. In Exercises 1–4 find all positive factors of the given integer. (*See Example 1.*)

1. 4 **2.** 12 **3.** 30 **4.** 32

In Exercises 5–12 determine which integers are primes and which are composites. (*See Example 2.*)

5. 9 **6.** 13 **7.** 17 **8.** 22

9. 27 **10.** 31 **11.** 37 **12.** 41

In Exercises 13–32 express each integer as the product of primes. (*See Example 3.*)

13. 14 **14.** 16 **15.** 18 **16.** 20 **17.** −24

18. 25 **19.** 28 **20.** 30 **21.** 36 **22.** −40

23. 42 **24.** −44 **25.** 65 **26.** 66 **27.** 70

28. −100 **29.** 108 **30.** 144 **31.** 169 **32.** 221

33. Find the smallest (positive) integer that is the product of two different primes.

34. Find the smallest (positive) integer that is the product of three different primes.

35. List all primes less than 50.

36. List all primes between 50 and 100.

37. Pick out the *error(s)*.

(a) 1 is a prime.

(b) If p is a prime, then p^2 is a prime.

(c) Every prime factor of a composite n is smaller than n.

(d) The product of two primes is also a prime.

(e) The product of two composites is also a composite.

B. In Exercises 38–52 find the *gcd* of the given integers. (*See Examples 4 and 5.*)

38. 9 and 15

39. 9 and 18

40. 25 and 35

41. 12 and −36

42. −24 and −36

43. 48 and 72

44. 60 and 84

45. 4, 6, and 8

46. 25, 30, and 50

47. −8, 16, and 32

48. −40, −60, and −80

49. 27, 36, and 45

50. 64, 72, and 144

51. 12, 18, 24, and 36

52. 30, 42, 60, and 96

C. In Exercises 53–62 find the *gcd* of the coefficients of each polynomial. (*See Example 6.*)

53. $3x + 6$

54. $4x - 8$

55. $10x + 15$

56. $12x + 20$

Use this area for work space.

57. $-x^2 + 5$

58. $48x^2 + 108x$

59. $12x^2 - 18x + 6$

60. $5x^2 - 10x + 25$

61. $24x^2 + 36x - 42$

62. $27x^3 - 18x^2 + 54x - 45$

4.3 Common Factors

A. Factors of Polynomials

Recall that a nonzero integer b is a factor of (the integer) a if

$$a = bc$$

for some integer c. The same notion applies to polynomials. You will want to call 2 and x *factors* of the polynomial 2x.

DEFINITION

> Let P and Q ($\neq 0$) be polynomials. Then Q is called a **factor of *P*** if
>
> $$P = Q \cdot R$$
>
> for some polynomial R.

Example 1 ▶ Show that 3 is a factor of $6x + 9$.

Solution:

First note that 3 is the *gcd* of the coefficients of $6x + 9$. Also

$$\frac{6x + 9}{3} = \frac{6x}{3} + \frac{9}{3} = 2x + 3$$

Thus

$$\underbrace{6x + 9}_{P} = \underset{Q}{3} \cdot \underbrace{(2x + 3)}_{R}$$ ◀

Example 2 ▶ Show that x^2 is a factor of $x^5 + 2x^2$.

Solution:

$$\frac{x^5 + 2x^2}{x^2} = \frac{x^5}{x^2} + \frac{2x^2}{x^2} = x^3 + 2$$

Thus

$$\underbrace{x^5 + 2x^2}_{P} = \underset{Q}{x^2} \cdot \underbrace{(x^3 + 2)}_{R}$$ ◀

Numbers, such as 3 (in Example 1), as well as powers of variables, such as x^2 (in Example 2), can be factors of a polynomial. In the next example, *the product of a number and of powers of several different variables is a factor of the given polynomial.*

Example 3 ▶ Let $P = 8x^3y + 12x^2y^2 + 4x^2yz$. First consider the coefficients of P, and observe that

$$gcd\ (8, 12, 4) = 4$$

Thus 4 is a factor of P. Similarly

$$x^2 \text{ and } y$$

are each factors of *all three terms of P*, and are therefore factors of P. The *product* of these,

$$4x^2y$$

is a factor of P because

$$\underbrace{8x^3y + 12x^2y^2 + 4x^2yz}_{P} = \underbrace{4x^2y}_{Q} \cdot \underbrace{(2x + 3y + z)}_{R}$$ ◀

DEFINITION

> Let P be a polynomial. The **greatest common monomial factor of *P*,** or for short, the **common factor of *P*,** is defined to be the product of
>
> **1.** the greatest common divisor of the coefficients of P and
> **2.** the smallest power of each variable that occurs in each term of P.

Thus in Example 3, $4x^2y$ is the common factor of P.

PRACTICE EXERCISES FOR TOPIC A

Answers
1. ______
2. ______
3. (a) ______
(b) ______
(c) ______
(d) ______

1. Show that 4 is a factor of $12a + 8b - 20$.

2. Show that $5x^2$ is a factor of $15x^3 + 10x^2$.

3. Show that each of the following is a factor of $9x^2y^3z^3 - 12xy^2z^4$

(a) 3 (b) x (c) y^2 (d) z^3

Now try the exercises for Topic A on page 141.

B. Isolating the Common Factor

DEFINITION

> To **isolate the common factor** Q of a polynomial P, write
>
> $$P = Q \cdot R$$
>
> where Q is the common factor of P and where R is another polynomial.

Apply the *Distributive Laws* to isolate the common factor of a polynomial.

Example 4 ▶ Isolate the common factor of $4a + 8b$.
Solution:

$$gcd(4, 8) = 4$$

Let

$$P = 4a + 8b$$

Then the common factor Q of P is 4. Isolate the common factor Q by writing

$$\underbrace{4a + 8b}_{P} = \underbrace{4}_{Q}(\underbrace{a + 2b}_{R})$$ ◀

Example 5 ▶ Isolate the common factor of $5x^2y^2 - 10xy^3$.
Solution:

$$gcd(5, -10) = 5$$

Let

$$P = 5x^2y^2 - 10xy^3$$

Then the common factor Q of P is $5xy^2$. Isolate the common factor Q by writing

$$\underbrace{5x^2y^2 - 10xy^3}_{P} = \underbrace{5xy^2}_{Q}(\underbrace{x - 2y}_{R})$$ ◀

Example 6 ▶ Isolate the common factor of $2a^2bc + 4a^2c + 5a^3b$.
Solution:

$$gcd(2, 4, 5) = 1$$

The only variable that occurs in all three terms is a; the smallest power to which it occurs is the second power, a^2. Thus a^2 is the common factor. Isolate the common factor by writing

$$\underbrace{2a^2bc + 4a^2c + 5a^3b}_{P} = \underbrace{a^2}_{Q}(\underbrace{2bc + 4c + 5ab}_{R})$$ ◀

Example 7 ▶ Let $P = -3y^4 - 6y^3 - 9y^2$. Note that the common factor Q would be given by

$$Q = 3y^2$$

But because all of the coefficients of P are negative, it is better to consider

$$-Q = -3y^2$$

and to write

$$\underbrace{-3y^4 - 6y^3 - 9y^2}_{P} = \underbrace{-3y^2}_{-Q}(\underbrace{y^2 + 2y + 3}_{R})$$ ◀

Example 8 ▶ Consider $2x + 3$. The common factor of $2x + 3$ is 1. It is not possible to isolate the common factor of this polynomial, $2x + 3$. ◀

Answers

4. (a) ____

(b) ____

PRACTICE EXERCISES FOR TOPIC B

4. Isolate the common factor of each polynomial.

(a) $21a^2b + 7ab^2$ (b) $36x^3 - 12x^2 + 6x$

Now try the exercises for Topic B on this page.

Use this area for work space.

EXERCISES

A.

1. Show that 2 is a factor of $4a + 10$. (*See Example 1.*)

2. Show that 5 is a factor of $20x^2 + 25$.

3. Show that 4 is a factor of $8y^2 + 12y - 20$.

4. Show that a is a factor of $7a^2 + 6a$. (*See Example 2.*)

5. Show that b^2 is a factor of $5b^3 - b^2$.

6. Show that x^4 is a factor of $x^7 - 3x^6 + 9x^5$.

7. Show that each of the following is a factor of $8a^2 + 12a$. (*See Example 3.*)

(a) 4 (b) a

8. Show that each of the following is a factor of $9x^4 - 6x^3 + 18x^2$.

(a) 3 (b) x^2

9. Show that each of the following is a factor of $25a^2b^3 - 15ab^3 + 40ab^2$.

(a) 5 (b) a (c) b^2

10. Show that each of the following is a factor of

$$36x^2y^4z^6 - 48xy^9z^4 + 24x^5y^3z^{10} - 144x^6y^5z^7$$

(a) 12 (b) x (c) y^3 (d) z^4

11. Which of the following is *not* a factor of $40a^2b^5c^3 - 32a^5b^4c^4 + 36a^6b^3c^5$?

(a) 4 (b) a^2 (c) b^5 (d) c^3

12. Which of the following is *not* a factor of

$$144x^3y^4z^5 - 96x^5y^3z^4 + 120x^4y^3z^7 - 156x^7y^3z^4?$$

(a) 24 (b) x^3 (c) y^3 (d) z^4

B. In Exercises 13–56 isolate the common factor of each polynomial. If the common factor is 1, leave the given polynomial unchanged. (*See Examples 4–8.*)

13. $2a + 2$ **14.** $3b - 3$ **15.** $-6x - 6$ **16.** $2y + 4$

17. $5x^2 + 10$ **18.** $3m^3 - 9$ **19.** $20x^2 + 50$ **20.** $12x^2 + 18$

21. $16a + 25$ **22.** $9a^2 - 6$ **23.** $t^2 + t$ **24.** $x^2 - x$

25. $2x^2 + 3x$ **26.** $5x^3 - 6x$ **27.** $y^3 - y^2$ **28.** $c^4 + c^2$

29. $a^{10} - a^7$ **30.** $2b^5 - 4b^2$ **31.** $10x^3 - 20x^2$ **32.** $12x^3 - 18x$

Use this area for work space.

33. $32a^4 + 12a^3$

34. $15x^{12} + 20x^7$

35. $x^2y + xy^2$

36. $a^2bc + abc$

37. $x^2yz^2 - xyz^2$

38. $abc^3 - a^3b^2c$

39. $2a^2bc + 4ab^2$

40. $6x^2yz + 15xz^2$

41. $16m^2np + 20m^4n^4$

42. $25a^2b^3c^2 + 20ab^2c^3$

43. $a^3b^4c^2 + a^2b^3c^2 + ab^3c^2$

44. $x^7y^6z^5 - x^5y^6z^4 + x^3z^4$

45. $2x^2 + 4x + 2$

46. $6x^2 + 8x + 3$

47. $10x^3 - 20x^2 + 50x$

48. $t^4 + 8t^2 + 16t$

49. $5a^4 - 10a^3 + 15a^2$

50. $12b^4 - 15b^3 + 18$

51. $20x^6y - 24x^4y + 30x^3y$

52. $36y^2z^5 + 60y^2z^3 - 24yz^2$

53. $-18m^2n^3 - 54mn^4 - 72mn^3$

54. $96x^2y^5z^3 - 144x^2y^{10}z^2 + 72x^3y^4z^5$

55. $2x^5y + 12x^4y^2 + 8x^3y^3 - 6x^2y^4 + 4xy^5$

56. $9x^5 + 15x^4y + 18x^3y^2 - 6x^2y^3 + 3xy^4 + y^5$

57. Pick out the *error(s)*.

(a) $4x + 4y = 4(x + y)$

(b) $3ax - ay = a(3x - y)$

(c) $a^2b - a^2 = a^2(b - 1)$

(d) $x^2y + x^3 = x^2(y + 1)$

(e) $a^4 + a^3b = a^3(a + b)$

4.4 Multiplying Polynomials

A. Using the Distributive Laws

Recall the *Distributive Laws:*

$$a(b + c) = ab + ac$$

$$(b + c)a = ba + ca$$

$$a(b - c) = ab - ac$$

$$(b - c)a = ba - ca$$

To multiply two polynomials (not both monomials), first use the Distributive Laws. *Every term of the first polynomial is multiplied by every term of the second polynomial.* The sum of the resulting terms is then the product of the two polynomials.

Example 1 ▶ Multiply $(x + 5)(x + 3)$

Solution:

According to the first Distributive Law,

$$\underbrace{(x + 5)}_{a}\underbrace{(x + 3)}_{(b+c)} = \underbrace{(x + 5)}_{a}\underbrace{x}_{b} + \underbrace{(x + 5)}_{a}\underbrace{3}_{c}$$

According to the second Distributive Law,

$$(x + 5)x + (x + 5)3 = x^2 + 5x + 3x + 15 = x^2 + 8x + 15$$

$$(b + c)a + (B + C)A = ba + ca + AB + CA$$

Observe that every term of the polynomial $x + 5$ is multiplied by every term of the polynomial $x + 3$. Thus you obtain

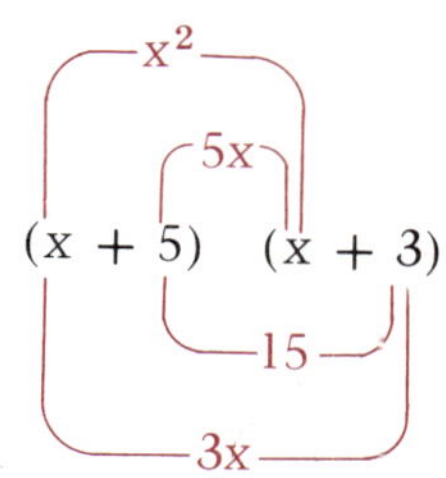

You then add

$$x^2 + 5x + 3x + 15$$

to obtain the polynomial

$$x^2 + 8x + 15$$ ◀

In practice it is probably easier to multiply by writing the factors one above the other. Write *like terms* in the *same column*. First multiply by x:

$$\begin{array}{r} x + 5 \\ x \\ \hline x^2 + 5x \end{array}$$

then by 3:

$$\begin{array}{r} x + 5 \\ x + 3 \\ \hline x^2 + 5x \qquad \\ 3x + 15 \\ \hline x^2 + 8x + 15 \end{array}$$

Answers

1. ____________

PRACTICE EXERCISES FOR TOPIC A

1. Multiply $(x + 1)(x + 9)$

Now try the exercises for Topic A on page 148.

B. Products of Binomials

To remember the various terms that are added, the FOIL method is sometimes used when multiplying *binomials*. Although the order in which you add the terms is different in the FOIL method, the resulting product is the same. FOIL stands for the first letters of:

First terms product	x^2	$(x + 5)(x + 3)$
Outer product	$3x$	
Inner product	$5x$	$(x + 5)(x + 3)$
Last terms product	15	

Thus $\quad (x + 5)(x + 3) = x^2 + 3x + 5x + 15 = x^2 + 8x + 15$

(We will multiply *in columns* rather than by the FOIL method.) The column method applies when you multiply polynomials with *any number* of terms. The FOIL method applies only when you multiply binomials.

Example 1 illustrates a product that is frequently encountered—namely,

$$(x + a)(x + b)$$

(In Example 1, $a = 5$ and $b = 3$.) In general, you obtain

$$\begin{array}{l} x \quad + a \\ \underline{x \quad + b} \\ x^2 + ax \\ \underline{\qquad\quad bx + ab} \\ x^2 + (a + b)x + ab \end{array}$$

Therefore $\quad (x + a)(x + b) = x^2 + (a + b)x + ab$

Example 2 ▶ Multiply $(x + 6)(x - 3)$

Solution:

$$\begin{array}{r} x + 6 \\ \underline{x - 3 } \\ x^2 + 6x \\ \underline{- 3x - 18} \\ x^2 + 3x - 18 \end{array}$$ ◀

Example 3 ▶ Multiply (a) $(x + 5)^2$ (b) $(x - 5)^2$

Solution:

(a) $(x + 5)^2 = (x + 5)(x + 5)$ (b) $(x - 5)^2 = (x - 5)(x - 5)$

$$\begin{array}{r} x + 5 \\ \underline{x + 5 } \\ x^2 + 5x \\ \underline{5x + 25} \\ x^2 + 10x + 25 \end{array} \qquad \begin{array}{r} x - 5 \\ \underline{x - 5 } \\ x^2 - 5x \\ \underline{- 5x + 25} \\ x^2 - 10x + 25 \end{array}$$ ◀

In general,

$$(x + a)^2 = (x + a)(x + a) = x^2 + 2ax + a^2$$

In Example 3, part (a), take $a = 5$ to obtain

$$(x + 5)^2 = x^2 + 2 \cdot 5x + 5^2 = x^2 + 10x + 25$$

In part (b), take $a = -5$ to obtain

$$(x - 5)^2 = x^2 + 2(-5)x + (-5)^2 = x^2 - 10x + 25$$

You can also consider $(x - 5)^2$ to be of the form $(x - a)^2$, by taking $a = 5$, instead of -5. In this case the general formula is

$$(x - a)^2 = (x - a)(x - a) = x^2 - 2ax + a^2$$

Example 4 ▶ Multiply $(x + 4)(x - 4)$

Solution:

$$\begin{array}{l} x + 4 \\ \underline{x - 4} \\ x^2 + 4x \\ \underline{\quad\;\; - 4x - 16} \\ x^2 \qquad\quad - 16 \end{array}$$

Note that the "cross terms" $4x$ and $-4x$ are inverses. Their sum is 0. ◀

In general,

$$(x + a)(x - a) = x^2 + \underbrace{(a - a)}_{0}x - a^2$$

Thus $$(x + a)(x - a) = x^2 - a^2$$

This formula will be the basis of the "difference of squares" factoring method discussed in the next section.

PRACTICE EXERCISES FOR TOPIC B

Answers
2. (a) ______
(b) ______
(c) ______

2. Multiply:

(a) $(x + 6)(x - 1)$ (b) $(x + 8)^2$ (c) $(x + 10)(x - 10)$

Now try the exercises for Topic B on page 148.

C. More Examples

Example 5 ▶

$$\begin{array}{r} 2x + 1 \\ \underline{2x + 3} \\ 4x^2 + 2x \quad\;\; \\ \underline{6x + 3} \\ 4x^2 + 8x + 3 \end{array}$$ ◀

Example 6 ▶

$$\begin{array}{r} 2a^2 + 3 \\ a + 2 \\ \hline 2a^3 + 3a \\ 4a^2 + 6 \\ \hline 2a^3 + 4a^2 + 3a + 6 \end{array}$$

Leave room for the a^2 term. ◀

Example 7 ▶

$$\begin{array}{r} 5a + b \\ a - b \\ \hline 5a^2 + ab \\ - 5ab - b^2 \\ \hline 5a^2 - 4ab - b^2 \end{array}$$

◀

Answers
3. (a)
(b)

PRACTICE EXERCISES FOR TOPIC C

3. Multiply:

(a) $(3x - 1)(x + 5)$ (b) $3y(y + 1)(y - 2)$

Now try the exercises for Topic C on page 148.

D. Multiplying Trinomials

Products of trinomials, or for that matter, of any two polynomials can be multiplied *in columns.*

Example 8 ▶ Multiply $(x^2 + 3x + 2)(x^2 + x - 2)$
Solution:
Multiply every term of $x^2 + 3x + 2$ first by x^2, then by x, then by -2. Then add in columns.

$$\begin{array}{r} x^2 + 3x + 2 \\ x^2 + x - 2 \\ \hline x^4 + 3x^3 + 2x^2 \\ x^3 + 3x^2 + 2x \\ - 2x^2 - 6x - 4 \\ \hline x^4 + 4x^3 + 3x^2 - 4x - 4 \end{array}$$

◀

Answers
4. (a)
(b)

PRACTICE EXERCISES FOR TOPIC D

4. Multiply:

(a) $(x - 4)(x^2 + 2x - 3)$ (b) $(a^2 - a + 2)(a^2 + a - 1)$

Now try the exercises for Topic D on page 149.

E. Combining Operations

Example 9 ▶ Combine $(x^2 + 4)(x + 3) + (x + 1)(x - 5)$

Solution:
First multiply, as indicated.

$$\begin{array}{rrrr} & & x^2 & +4 \\ & & x & +3 \\ \hline x^3 & & +4x & \\ & 3x^2 & & +12 \\ \hline x^3 & +3x^2 & +4x & +12 \end{array} \qquad \begin{array}{rrr} x & +1 & \\ x & -5 & \\ \hline x^2 & +x & \\ & -5x & -5 \\ \hline x^2 & -4x & -5 \end{array}$$

Then add the resulting polynomials.

$$\begin{array}{lrrrr} & x^3 & +3x^2 & +4x & +12 \\ + & & x^2 & -4x & -5 \\ \hline & x^3 & +4x^2 & & +7 \end{array}$$

The result is $x^3 + 4x^2 + 7$. ◀

Answers

5. ______

PRACTICE EXERCISES FOR TOPIC E

5. Simplify $(x - 4)^2 - (x + 2)^2$

Now try the exercises for Topic E on page 149.

F. A Geometric Application

Example 10 ▶ The total region (outlined in color) of Figure 4.1(a) consists of a square together with two rectangles. This region has the same area as that of the rectangle of Figure 4.1(b). Find the length indicated by ? in Figure 4.1(a).

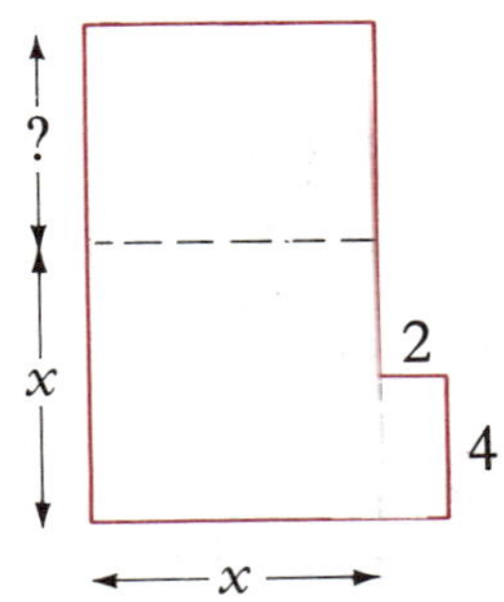

FIGURE 4.1(*a*)

Solution:
The rectangle of Figure 4.1(b) has length $x + 4$ and width $x + 2$. Its area is therefore $(x + 4)(x + 2)$, which equals $x^2 + 6x + 8$.

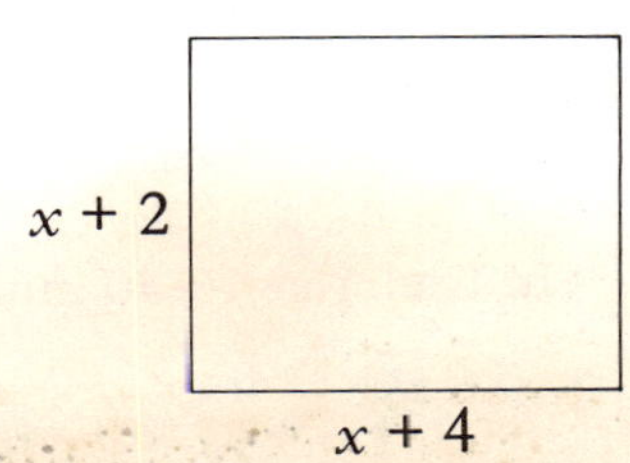

FIGURE 4.1(*b*)

The square in Figure 4.1(a) has area x^2, and the rectangle on the right has area 4×2, or 8. The combined area of these two figures is thus $x^2 + 8$. Because the total area of the region must be $x^2 + 6x + 8$, the rectangle on top must have area $6x$. Note that this rectangle shares a side with the square. Hence this common side has length x, and thus the side in question must be of length 6. ◀

Answers

6. ______

PRACTICE EXERCISES FOR TOPIC F

6. Show that the area of the rectangle in Figure 4.2(a) is equal to the area of the region in Figure 4.2(b) (a rectangle from which a smaller rectangle, measuring 1 by 2, has been cut away).

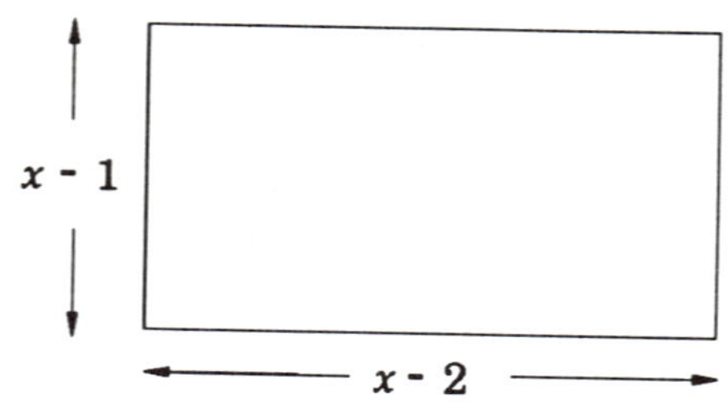

FIGURE 4.2(a)

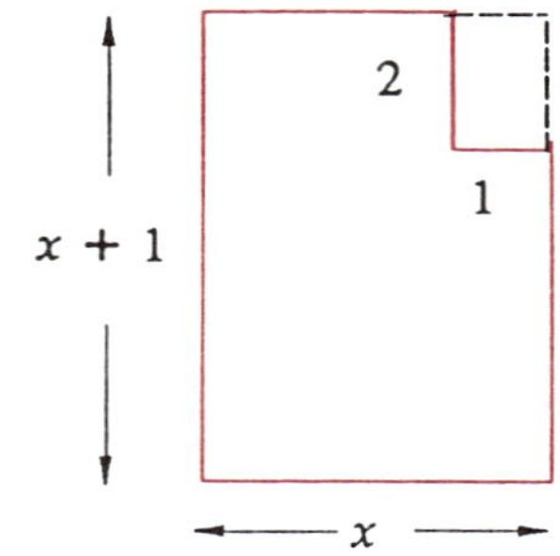

FIGURE 4.2(b)

Now try the exercises for Topic F on page 149.

Use this area for work space.

EXERCISES

A. In Exercises 1–4 multiply. *(See Example 1.)*

1. $(x + 1)(x + 6)$
2. $(y + 2)(y + 3)$
3. $(x + 4)(x + 2)$
4. $(a + 7)(a + 4)$

B. In Exercises 5–14 multiply. *(See Examples 2–4.)*

5. $(a + 3)(a - 2)$
6. $(b + 3)(b - 7)$
7. $(b - 2)(b - 4)$
8. $(x - 9)(x + 3)$
9. $(c + 6)^2$
10. $(b - 7)^2$
11. $(x + 8)(x - 8)$
12. $(y - 4)(y + 4)$
13. $(x + y)(x + y)$
14. $(b + c)(b - c)$
15. Pick out the *error(s)*.
    (a) $(x + 4)^2 = x^2 + 16$
    (b) $(x + 2)(x - 2) = x^2 - 4$
    (c) $(x - 3)^2 = x^2 - 9$
    (d) $(x - a)^2 = x^2 - a^2$
    (e) $(x + a)(x - a) = x^2 - a^2$

C. In Exercises 16–40 multiply. *(See Examples 5–7.)*

16. $(3n - 10)(n + 3)$
17. $(y - 5)(2y - 3)$
18. $(2m + 1)(3m + 2)$
19. $(3n - 2)(4n - 5)$
20. $(5x + 2)(2x - 7)$
21. $5(z + 2)(z - 1)$
22. $-2(a - 9)(a + 5)$
23. $-3(b - 1)(b - 2)$
24. $-(c - 3)(c + 6)$
25. $(x^2 + 4)^2$
26. $(x^2 + 3)(x^2 - 3)$
27. $(a^4 + 1)(a^4 - 1)$
28. $(u^4 + 1)(u + 2)$
29. $(x^3 + 1)^2$
30. $(x^3 - 3)^2$

Use this area for work space.

31. $(x^2 + 3)(x^2 + 2)$ **32.** $(b^2 + 5)(b^2 - 2)$ **33.** $(c^3 - 1)(c^3 - 4)$

34. $(c^3 - 1)(c^2 + 1)$ **35.** $(m + 2n)(m + n)$ **36.** $(x + 3a)(x + 2a)$

37. $(m - 4n)(m + 2n)$ **38.** $(x - 2y)(x + y)$ **39.** $(5a - 3b)(-2a + 6b)$

40. $(3a + b)(2a - 2b)$

D. In Exercises 41–48 multiply as indicated. (*See Example* 8.)

41. $\begin{array}{l} x^2 + x + 3 \\ \underline{x \; + 2} \end{array}$ **42.** $\begin{array}{l} y^2 + 4y + 1 \\ \underline{y \; - 1} \end{array}$

43. $\begin{array}{l} a^2 - 5a + 2 \\ \underline{a^2 - 3} \end{array}$ **44.** $\begin{array}{l} b^4 - b^2 + 1 \\ \underline{b^2 + 4} \end{array}$

45. $\begin{array}{r} 2x^2 + 4x + 3 \\ \underline{x^2 + 3x + 5} \end{array}$ **46.** $\begin{array}{r} 3y^2 - y + 1 \\ \underline{2y^2 + 2y + 3} \end{array}$

47. $\begin{array}{l} x^3 + 4x^2 - x + 3 \\ \underline{x \; + 4} \end{array}$ **48.** $\begin{array}{l} x^2 + 2xy + y^2 + 1 \\ \underline{x \; - y} \end{array}$

E. In Exercises 49–54 combine as indicated. (*See Example* 9.)

49. $(x + 2)(x + 3) + (x + 1)(x + 4)$ **50.** $(x + 4)(x - 2) - (x + 3)(x + 5)$

51. $(y + 7)(y - 2) + (y + 5)^2$ **52.** $(y + 3)^2 - (y - 1)^2$

53. $(a^2 + 5)(a - 2) - (a^2 + 4a + 1)(a - 1)$

54. $(b^2 + 3)(b - 1) + \dfrac{4b^4 + 2b^3 + 6b^2}{2b^2}$

55. Multiply the sum of $3x + 1$ and $-2x$ by $2x + 5$.

56. Add the product of $2x + 3$ and $3x - 1$ to $4x + 5$.

57. Divide the product of $x^2 + 3x$ and $x^3 - x^2$ by x^3.

58. Subtract $x^3 - x + 1$ from the product of $x^2 + 3$ and $x - 5$.

F. **59.** The region (outlined in color) of Figure **4.3(a)** consists of a square together with two rectangles. This region has the same total area as that of the rectangle in Figure **4.3(b)**. Find the length indicated by ? in Figure **4.3(a)**. (*See Example 10.*)

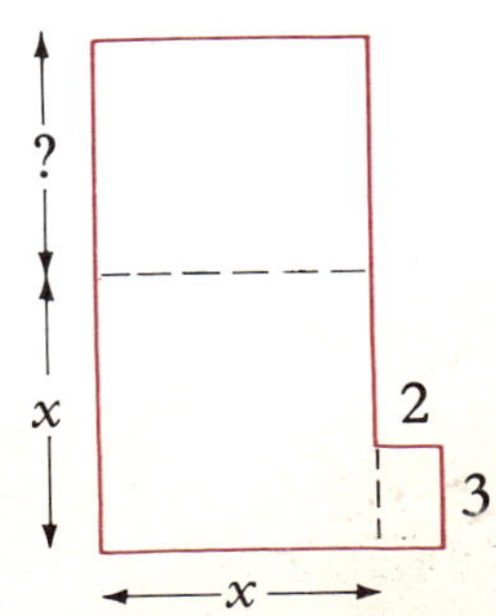

FIGURE 4.3(a)

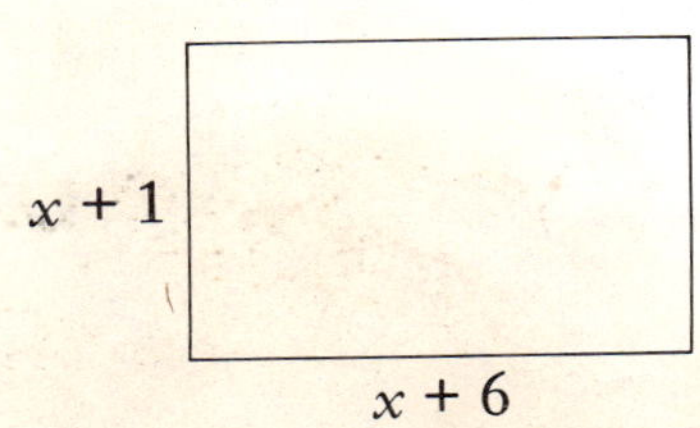

FIGURE 4.3(b)

Use this area for work space.

60. From the large rectangle of Figure 4.4(a), two smaller rectangles have been cut out. The resulting region (outlined in color) has the same area as that of the rectangle of Figure 4.4(b). Express the length indicated by ? in terms of x.

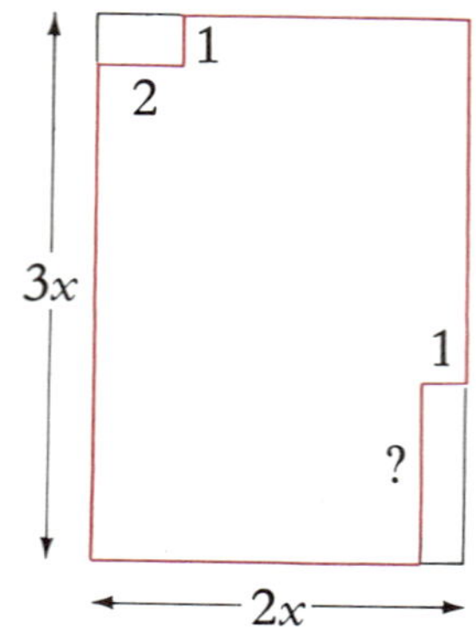

FIGURE 4.4(a)

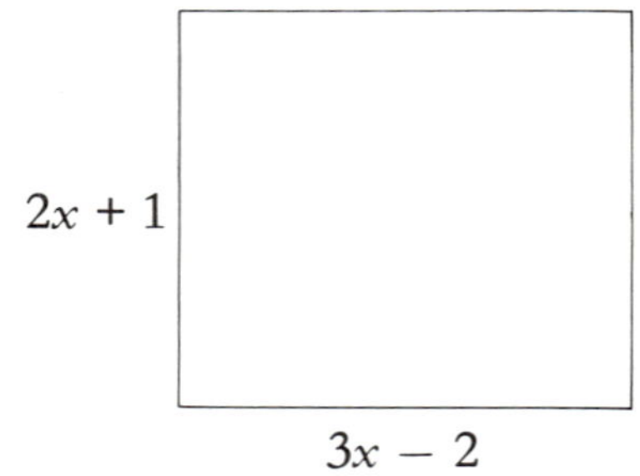

FIGURE 4.4(b)

61. Each side of a square photograph of side length s inches is enlarged by three inches. Express the area of the enlarged photograph as a polynomial in the variable s.

62. A (rectangular) Navajo rug has length l inches and width w inches. A Chinese rug is 5 inches longer, but 2 inches narrower. Express the area of the Chinese rug as a polynomial in the variables l and w.

4.5 Difference of Squares

Recall that you factor a polynomial P by writing P as a product of other polynomials. In Section 4.3 you learned how to isolate the common factor of a polynomial. In the remainder of the chapter you will learn other techniques for factoring polynomials. You will also apply these factoring techniques to solving equations.

A. $x^2 - a^2$

One of the easiest types of factoring to recognize is known as the **difference of squares.** Recall that

$$(x + a)(x - a) = x^2 - a^2$$

The difference-of-squares method applies to a binomial, such as

$$x^2 - a^2$$

whose two terms are *squares* that are separated by a *minus* sign.

Example 1 ▶ Factor $x^2 - 100$

Solution:

$$100 = 10^2$$

Thus $$x^2 - 100 = x^2 - 10^2 = (x + 10)(x - 10)$$ ◀

Example 2 ▶ Factor $4 - a^2$

Solution:

Although the number 4 is written first, the difference-of-squares method again applies.

$$4 = 2^2$$

Thus $$4 - a^2 = 2^2 - a^2 = (2 + a)(2 - a)$$ ◀

Example 3 ▶ Factor $s^2 - t^2$

Solution:

$$s^2 - t^2 = (s + t)(s - t)$$ ◀

A common error in applying this method is to factor the individual terms, but not the polynomial as a whole. For instance, in Example 3, do **not** write

$$s^2 - t^2 = s \cdot s - t \cdot t$$

but rather

$$s^2 - t^2 = (s + t)(s - t)$$

as above. (The first equality is correct, but is not the desired factored result.)

Answers

1. (a) ______

(b) ______

PRACTICE EXERCISES FOR TOPIC A

1. Factor each binomial.

(a) $k^2 - 64$ (b) $100 - y^2$

Now try the exercises for Topic A on page 154.

B. Squares of Products

You may be given a binomial such as

$$x^2 - a^2b^2$$

By the Associative and Commutative Laws of multiplication,

$$\begin{aligned} a^2b^2 &= (aa)(bb) \\ &= (ab)(ab) \\ &= (ab)^2 \end{aligned}$$

The product of the squares is the square of the product. Thus

$$x^2 - a^2b^2 = x - (ab)^2 = (x + ab)(x - ab)$$

Example 4 ▶ Factor $x^2 - 100y^2$
Solution:

$$100y^2 = 10^2y^2 = (10y)^2$$

Thus $$x^2 - 100y^2 = x^2 - (10y)^2 = (x + 10y)(x - 10y)$$ ◀

Example 5 ▶ Factor $16a^2 - 25b^2$
Solution:

$$16a^2 = (4a)^2; \qquad 25b^2 = (5b)^2$$

Thus $$16a^2 - 25b^2 = (4a)^2 - (5b)^2 = (4a + 5b)(4a - 5b)$$ ◀

Answers

2. (a) ______

(b) ______

PRACTICE EXERCISES FOR TOPIC B

2. Factor each binomial.

(a) $1 - 49a^2$ (b) $25y^2 - 9x^2$

Now try the exercises for Topic B on page 155.

C. Common Factors

Sometimes both common factoring and the difference-of-squares method apply to a polynomial. Always isolate the common factor *before* applying the difference-of-squares method.

Example 6 ▶ Factor $5a^2 - 20b^2$
Solution:

$$5a^2 - 20b^2 = 5(a^2 - 4b^2) = 5(a + 2b)(a - 2b)$$ ◀

Example 7 ▶ Factor $ax^2 - a^3$
Solution:

$$ax^2 - a^3 = a(x^2 - a^2) = a(x + a)(x - a)$$ ◀

A polynomial is considered to be in factored form only when it is *completely* factored. In Examples 6 and 7 and in many subsequent examples and exercises, more than one factoring method applies. Do not stop after one step, but rather, consider whether any further factoring is possible.

Answers

3. (a) ______

(b) ______

(c) ______

PRACTICE EXERCISES FOR TOPIC C

3. Factor. (Isolate the common factor first.)

(a) $2n^2 - 18$ (b) $40a - 10a^3$ (c) $12ab^3 - 3a^3b$

Now try the exercises for Topic C on page 155.

D. Sums of Squares

The above method applies to the difference of squares, but *not to the sum of squares.*

Example 8 ▶ Factor $a^2 + b^2$

Solution:

This binomial cannot be factored by the present method. Notice that the cross terms do not cancel in the following product:

$$\begin{array}{r} a + b \\ a + b \\ \hline a^2 + ab \\ ab + b^2 \\ \hline a^2 + 2ab + b^2 \end{array}$$

◀

Answers

4. ______

PRACTICE EXERCISES FOR TOPIC D

4. Which of the following binomials cannot be factored (by the present methods)?

(a) $4x^2 + 9$ (b) $4x^2 - 9$ (c) $9 - 4x^2$ (d) $-9 + 4x^2$ (e) $-9 - 4x^2$

Now try the exercises for Topic D on page 155.

E. Squares of Squares

The difference-of-squares method can apply more than once in factoring a polynomial. Thus when considering

$$b^4 - c^4$$

observe that

$$b^4 = b^2 \cdot b^2 = (b^2)^2 \quad \text{and} \quad c^4 = c^2 \cdot c^2 = (c^2)^2$$

Therefore b^4 and c^4 are each squares (of squares) and consequently,

$$b^4 - c^4 = \underbrace{(b^2)^2}_{x} - \underbrace{(c^2)^2}_{a}$$

is of the form

$$x^2 - a^2, \text{ where } x = b^2 \text{ and } a = c^2$$

The difference-of-squares method applies again. Thus

$$\begin{aligned} b^4 - c^4 &= (b^2)^2 - (c^2)^2 && \textit{difference of squares} \\ &= \underbrace{(b^2 + c^2)}_{\text{sum of squares}}\underbrace{(b^2 - c^2)}_{\text{difference of squares}} \\ &= (b^2 + c^2)(b + c)(b - c) \end{aligned}$$

Example 9 ▶ Factor $16 - t^4$

Solution:

Note that $16 = 4^2 = (2^2)^2$ and $t^4 = (t^2)^2$. Then

$$16 - t^4 = \underbrace{(4 + t^2)}_{\text{sum of squares}}\underbrace{(4 - t^2)}_{\text{difference of squares}}$$

$$= (4 + t^2)(2 + t)(2 - t)$$ ◀

Example 10 ▶ Factor $x^2 - y^4$

Solution:

Here only y^4, which equals $(y^2)^2$, is the square of a square. The difference-of-squares method applies only once. Thus

$$x^2 - y^4 = x^2 - (y^2)^2 = (x + y^2)(x - y^2)$$ ◀

Answers

5. (a) ______

(b) ______

PRACTICE EXERCISES FOR TOPIC E

5. Factor: (a) $a^4 - 1$ (b) $x^4 - a^2$

Now try the exercises for Topic E on page 155.

Use this area for work space.

EXERCISES

A. In Exercises 1–12 factor each binomial. (*See Examples 1–3.*)

1. $x^2 - 9$	**2.** $y^2 - 4$	**3.** $x^2 - 49$	**4.** $a^2 - 25$
5. $m^2 - 100$	**6.** $s^2 - 36$	**7.** $y^2 - 81$	**8.** $z^2 - 144$
9. $1 - x^2$	**10.** $121 - c^2$	**11.** $x^2 - y^2$	**12.** $m^2 - n^2$

13. A square of side length 3 is cut from a larger square of side length s, as in Figure 4.5(a). The resulting region (outlined in color) has the same area as does the rectangle of Figure 4.5(b). Express the length of this rectangle, indicated by ?, in terms of s.

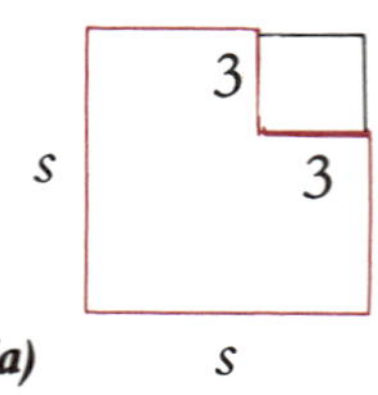

FIGURE 4.5(a)

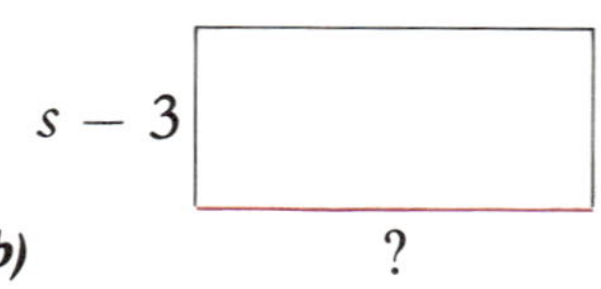

FIGURE 4.5(b)

14. A small square is cut from a larger square of side length s, as in Figure 4.6(a). The resulting region (outlined in color) has the same area as does the rectangle in Figure 4.6(b). Express the width of this rectangle, indicated by ?, in terms of s.

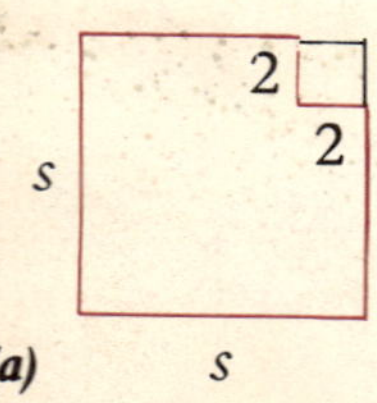

FIGURE 4.6(a)

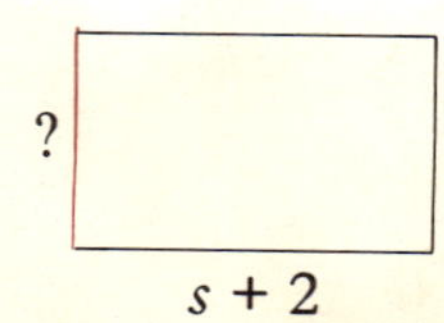

FIGURE 4.6(b)

Use this area for work space.

B. In Exercises 15–28 factor each binomial. (*See Examples 4 and 5.*)

15. $4x^2 - 1$ **16.** $9a^2 - 25$ **17.** $36x^2 - 49$ **18.** $100z^2 - 121$

19. $1 - 9u^2$ **20.** $25 - 81v^2$ **21.** $4x^2 - y^2$ **22.** $9x^2 - 16a^2$

23. $9m^2 - 49n^2$ **24.** $100z^2 - 81a^2$ **25.** $144y^2 - 25x^2$ **26.** $64r^2 - 169s^2$

27. $400t^2 - u^2$ **28.** $49a^2 - 225b^2$

In the remaining exercises *completely* factor the given binomials.

C. In Exercises 29–56 factor each binomial. (*See Examples 6 and 7.*)

29. $2x^2 - 2$ **30.** $5y^2 - 45$ **31.** $3s^2 - 75$ **32.** $7 - 7t^2$

33. $300 - 3x^2$ **34.** $180 - 5y^2$ **35.** $2x^2 - 32y^2$ **36.** $3y^2 - 27z^2$

37. $8a^2 - 200b^2$ **38.** $7x^2 - 175y^2$ **39.** $6m^2 - 54n^2$ **40.** $44a^2 - 99b^2$

41. $24x^2 - 294y^2$ **42.** $288c^2 - 242d^2$ **43.** $a^3 - a$ **44.** $25y - y^3$

45. $a^4 - a^2$ **46.** $x^3 - x^5$ **47.** $x^3 - xy^2$ **48.** $a^3 - 4ab^2$

49. $m^4 - 49m^2n^2$ **50.** $4s^4 - 9s^2t^2$ **51.** $2x^3 - 2x$ **52.** $5a^3 - 20a$

53. $3x^3 - 3xy^2$ **54.** $7m^2n - 7n^3$ **55.** $8x^3 - 50xy^2$ **56.** $12a^2b - 27b^3$

D. In Exercises 57–60 indicate which one binomial *cannot* be factored (by the methods you have learned). (*See Example 8.*)

57. (a) $x^2 - 9$ (b) $x^2 - 1$ (c) $x^2 + 1$ (d) $x^2 - a^2$

58. (a) $a^2 - x^2$ (b) $x^2 - a^2$ (c) $x^2 + a^2$ (d) $4x^2 + 4a^2$

59. (a) $y^2 - 16$ (b) $3y^2 - 48$ (c) $16y^2 + 16$ (d) $y^2 + 16$

60. (a) $ax^2 + ab^2$ (b) $a^2c + ad^2$ (c) $25a^2 + 49b^2$ (d) $ax^2 + a^2y$

61. Pick out the *error(s)*.

(a) $a^2 + 4 = (a + 2)^2$ (b) $x^2 - y^2 = (x + y)(x - y)$

(c) $x^4 + y^4 = (x^2 + y^2)^2$ (d) $a^2x^2 + b^2x^2 = (ax + bx)(ax - bx)$

E. In Exercises 62–69 factor each binomial. (*See Examples 9 and 10.*)

62. $m^4 - n^4$ **63.** $x^4 - 10\,000$ **64.** $a^4 - 1$

65. $81 - t^4$ **66.** $y^4 - z^2$ **67.** $5c^2 - 5d^4$

68. $20ac^4 - 20a$ **69.** $x^8 - 1$

4.6 Factoring Trinomials $x^2 + Bx + C$

A. The Basic Method

Now you will learn how to factor trinomials such as

$$x^2 + 3x + 2$$

and

$$x^2 - 5x + 4$$

In general, such trinomials are of the form

$$x^2 + Bx + C$$

where B and C are integers and $B \neq 0$.

Here is an example that illustrates the basic method.

Example 1 ▶ Factor $x^2 + 3x + 2$

Solution:

In order to obtain x^2, try to multiply:

$$\begin{array}{l} x + \_\_\_ \\ \underline{x + \_\_\_} \end{array}$$

Because the numerical term of $x^2 + 3x + \mathbf{2}$ is 2, fill in the blanks with integers whose product is 2. Try 1 and 2.

$$\begin{array}{r} x + 1 \\ \underline{x + 2} \\ x^2 + x \phantom{{}+2} \\ \underline{\phantom{x^2 + {}} 2x + 2} \\ x^2 + 3x + 2 \end{array}$$

Note that $1 \cdot 2 = 2$

(2 is the numerical term of $x^2 + 3x + \mathbf{2}$.) Also

$$1 + 2 = 3$$

(3 is the coefficient of the *middle* term $3x$ in $x^2 + \mathbf{3}x + 2$.) Thus

$$x^2 + 3x + 2 = (x + 1)(x + 2)$$ ◀

Answers

1. __________

PRACTICE EXERCISES FOR TOPIC A

1. Factor $t^2 + 12t + 11$

Now try the exercises for Topic A on page 161.

B. Analyzing the Basic Method

In Example 1 there was really only one choice for filling in the blanks. This is not true in general.

Example 2 ▶ Factor $x^2 + 5x + 4$

Solution:

Try

$$\begin{array}{l} x + \_\_\_ \\ \underline{x + \_\_\_} \end{array}$$

Fill in the blanks with integers whose product is 4. You could try $2 \cdot 2$ or $1 \cdot 4$:

$$\begin{array}{r} x + 2 \\ \underline{x + 2} \\ x^2 + 2x \qquad \\ \underline{\qquad 2x + 4} \\ x^2 + \mathbf{4}x + 4 \end{array}$$

The coefficient of the *middle* term should be 5, not 4.

$$\begin{array}{r} x + 1 \\ \underline{x + 4} \\ x^2 + x \qquad \\ \underline{\qquad 4x + 4} \\ x^2 + \mathbf{5}x + 4 \end{array}$$

This is the desired coefficient of 5x.

Thus $$x^2 + 5x + 4 = (x + 1)(x + 4)$$

Here $$1 \cdot 4 = 4$$

(4 is the numerical term of $x^2 + 5x + \mathbf{4}$); and

$$1 + 4 = 5$$

(5 is the coefficient of 5x in $x^2 + \mathbf{5}x + 4$.) ◀

As you see, you may have to try more than one pair of integers. To factor

$$x^2 + Bx + C$$

try to find integers m and n whose *product* is C and whose *sum* is B. In symbols:

$$mn = C$$

and $$m + n = B$$

Then

$$\begin{array}{l} x + m \\ \underline{x + n} \\ x^2 + mx \\ \underline{\qquad nx + mn \qquad\qquad\qquad\qquad} \\ x^2 + \underbrace{(m + n)}_{B}x + \underbrace{mn}_{C} = x^2 + Bx + C \end{array}$$

What happens when at least one of the coefficients B or C is negative? Note that B represents a *sum* and C a *product*.

The product of two positive integers, or of two negative integers, is positive. The product of a positive and a negative integer (in either order) is negative.

The sum of positive integers is positive; the sum of negative integers is negative. The sum of integers unlike in sign can be positive, negative, or 0.

Table 4.1 is useful in determining the signs of m and n when factoring

$$x^2 + Bx + C, \qquad \text{that is,} \qquad x^2 + (m + n)x + mn$$

TABLE 4.1

$B(= m + n)$	$C(= mn)$	m and n
positive	positive	both positive
negative	positive	both negative
positive	negative	one positive, one negative
negative	negative	one positive, one negative

Example 3 ▶ Factor $x^2 - 5x + 4$

Solution:

Here B, *the coefficient of* the term $-5x$, *is negative:*

$$x^2 - \mathbf{5}x + 4$$

Modify the factors of Example 2 by taking -1 and -4 (instead of 1 and 4).

$$\begin{array}{r} x - 1 \\ \underline{x - 4} \\ x^2 - x \quad\; \\ \underline{\; - 4x + 4} \\ x^2 - \mathbf{5}x + 4 \end{array}$$

Thus $$x^2 - 5x + 4 = (x - 1)(x - 4)$$ ◀

Example 4 ▶ Factor $y^2 + 2y - 3$

Solution:

Here C, *the numerical term, is negative:*

$$y^2 + 2y \mathbf{- 3}$$

One of the factors of -3 must be positive, the other negative (because the product, -3, is negative). Try $1(-3)$ and $3(-1)$.

$$\begin{array}{r} y + 1 \\ \underline{y - 3} \\ y^2 + y \quad\; \\ \underline{\; - 3y - 3} \\ y^2 - \mathbf{2}y - 3 \end{array}$$

Wrong! The coefficient of the middle term should be 2, not -2.

$$\begin{array}{r} y + 3 \\ \underline{y - 1} \\ y^2 + 3y \quad\; \\ \underline{\; - y - 3} \\ y^2 + \mathbf{2}y - 3 \end{array}$$

Right coefficient of the middle term! Thus

$$y^2 + 2y - 3 = (y + 3)(y - 1)$$ ◀

Example 5 ▶ Factor $x^2 + 8x + 12$

Solution:

Here $B = 8$ and $C = 12$. *Both B and C are positive.* Try *positive* integers m and n such that

$$mn = 12 \qquad \text{and} \qquad m + n = 8$$

The possibilities are

$$1 \cdot 12, \qquad 2 \cdot 6, \qquad \text{and} \qquad 3 \cdot 4$$

Try these:

$1 + 12 = 13$	$\mathbf{2 + 6 = 8}$	$3 + 4 = 7$
(Wrong)	**(Right)**	(Wrong)

(In practice, as soon as you find the right integral factors, immediately factor the given polynomial.)

Thus $$x^2 + 8x + 12 = (x + 2)(x + 6)$$ ◀

Example 6 ▶ Factor $x^2 + x + 3$

Solution:

The only positive factors of 3, 1 and 3, do not work because $1 + 3 = \mathbf{4}$, whereas the coefficient of x is **1.** Thus

$$x^2 + x + 3$$

cannot be factored by this method. ◀

PRACTICE EXERCISES FOR TOPIC B

Answers

2. (a) ______

(b) ______

(c) ______

2. Factor each trinomial.

(a) $a^2 + 6a + 8$ (b) $x^2 - 9x + 8$ (c) $y^2 - 2y - 8$

Now try the exercises for Topic B on page 161.

C. $(x + a)^2$

Note that

$$(x + 3)(x + 3) = x^2 + 6x + 9$$

that is,

$$(x + 3)^2 = x^2 + 2 \cdot 3x + 3^2$$

Now consider any number a in place of 3. Then in general,

$$x^2 + 2ax + a^2 = (x + a)^2$$

Thus $x^2 + 10x + 25 = (x + 5)^2$ *Here $a = 5$*

and $x^2 - 10x + 25 = (x - 5)^2$ *Here $a = -5$*

Answers
3. (a)
(b)

PRACTICE EXERCISES FOR TOPIC C

3. Factor: (a) $x^2 + 12x + 36$ (b) $x^2 - 22x + 121$

Now try the exercises for Topic C on page 162.

D. Common Factors

It is often best to isolate the common factor before applying the present method.

Example 7 ▶ Factor $3x^3 - 21x^2 + 30x$

Solution:

$$gcd\,(3, -21, 30) = 3$$

Thus $\quad 3x^3 - 21x^2 + 30x = 3x(x^2 \mathbf{-}\ \mathbf{7}x\ \mathbf{+}\ \mathbf{10})$

negative (pointing to −) positive (pointing to +)

Try $\quad (-1)(-10)$ and $(-2)(-5)$

$$(-1) + (-10) = -11 \quad (Wrong)$$

$$(-2) + (-5) = -7 \quad (Right)$$

Therefore $\quad 3x^3 - 21x^2 + 30x = 3x(x - 2)(x - 5)$ ◀

Answers
4. (a)
(b)

PRACTICE EXERCISES FOR TOPIC D

4. Factor: (a) $7x^2 - 35x + 42$ (b) $5x^3 - 5x^2 - 60x$

Now try the exercises for Topic D on page 162.

E. Two Variables

This method applies with very little modification to trinomials in two variables.

Example 8 ▶ Factor

(a) $x^2 - 5x - 14$ (b) $x^2 - 5xy - 14y^2$ (c) $x^3 - 5x^2y - 14xy^2$

Solution:

(a) $x^2 \mathbf{-}\ \mathbf{5}x\ \mathbf{-}\ \mathbf{14}$

negative negative

The possible factors of -14 are

1 and -14, $\quad 2$ and -7,

-1 and 14, $\quad -2$ and 7

Because the coefficient of $-5x$ is negative, the negative factor must have larger absolute value than the positive factor. Thus try only: $1(-14)$ and $2(-7)$

$$1 + (-14) = -13 \quad (\textit{Wrong})$$

$$2 + (-7) = -5 \quad (\textit{Right})$$

Therefore $$x^2 - 5x - 14 = (x + 2)(x - 7)$$

(b) The considerations of part (a) lead you to try:

$$\begin{array}{l} x + 2y \\ \underline{x - 7y} \\ x^2 + 2xy \\ \underline{\quad - 7xy - 14y^2} \\ x^2 - 5xy - 14y^2 \end{array}$$

(c) First isolate the common factor, x. Then use part (b):

$$x^3 - 5x^2y - 14xy^2 = x(x^2 - 5xy - 14y^2)$$
$$= x(x + 2y)(x - 7y)$$ ◀

PRACTICE EXERCISES FOR TOPIC E

Answers

5. (a) ______

(b) ______

5. Factor: (a) $x^2 + 15xy + 26y^2$ (b) $4x^2 + 20xy - 24y^2$

Now try the exercises for Topic E on page 162.

EXERCISES

Use this area for work space.

A. In Exercises 1–4 factor each trinomial. (*See Example 1.*)

1. $x^2 + 4x + 3$ **2.** $y^2 + 6y + 5$ **3.** $a^2 - 8a + 7$ **4.** $b^2 + 14b + 13$

B. In Exercises 5–28 factor each trinomial. (*See Examples 2–5.*)

5. $z^2 + 7z + 6$ **6.** $c^2 + 4c + 4$ **7.** $a^2 + a - 2$

8. $m^2 + 3m - 4$ **9.** $b^2 - b - 2$ **10.** $n^2 - 4n - 5$

11. $s^2 - s - 12$ **12.** $x^2 - x - 12$ **13.** $y^2 + 8y + 12$

14. $t^2 - 11t - 12$ **15.** $x^2 - 9x + 14$ **16.** $y^2 + 13y - 14$

17. $a^2 + 5a - 14$ **18.** $b^2 - 5b - 14$ **19.** $c^2 + 9c + 20$

20. $d^2 - 12d + 20$ **21.** $x^2 + 13x + 40$ **22.** $x^2 + 11x + 30$

23. $y^2 - 7y - 30$ **24.** $y^2 - 10y + 16$ **25.** $x^2 + 9x - 36$

26. $x^2 - 17x + 66$ **27.** $x^4 + 6x^2 + 8$ **28.** $y^4 + 10y^2 + 24$

In Exercises 29 and 30 indicate which one trinomial *cannot* be factored (by the present methods). (*See Example 6.*)

29. (a) $x^2 + 5x + 6$ (b) $x^2 - 5x + 6$ (c) $x^2 - 6x + 5$ (d) $x^2 + 6x - 5$

30. (a) $y^2 + 7y + 12$ (b) $y^2 + 8y + 12$ (c) $y^2 + 9y + 12$ (d) $y^2 + 13y + 12$

Use this area for work space.

C. In Exercises 31–36 factor.

31. $b^2 + 2b + 1$ **32.** $c^2 - 2c + 1$ **33.** $x^2 + 6x + 9$

34. $x^2 + 14x + 49$ **35.** $a^2 - 12a + 36$ **36.** $b^2 - 10b + 25$

D. In Exercises 37–48 factor. (*See Example* 7.)

37. $2x^2 + 6x + 4$ **38.** $2x^2 + 16x + 14$ **39.** $3x^2 + 6x - 24$

40. $4y^2 - 28y + 48$ **41.** $-y^2 + 9y - 20$ **42.** $30 + z - z^2$

43. $a^3 + 3a^2 - 10a$ **44.** $b^3 + 12b^2 + 32b$ **45.** $4m^2 + 40m + 100$

46. $-2n^2 + 24n - 72$ **47.** $72 + 6x - 3x^2$ **48.** $uv^2 - 16uv + 64u$

E. In Exercises 49–62 factor. (*See Example* 8.)

49. $2t^3 - 24t^2 + 70t$ **50.** $63x + 12x^2 - 3x^3$ **51.** $x^2 + 2ax + a^2$

52. $y^2 - 2ay + a^2$ **53.** $x^2 + 5ax + 4a^2$ **54.** $y^2 - 3ay - 4a^2$

55. $m^2 + 4mn + 4n^2$ **56.** $m^2 + 3mn - 4n^2$ **57.** $x^2 + 6xy + 9y^2$

58. $u^2 - 8uv + 16v^2$ **59.** $2x^2 + 24xy + 72y^2$ **60.** $3x^2 + 3xy - 36y^2$

61. $ax^2 + 6axy + 8ay^2$ **62.** $x^3 + 6x^2y + 5xy^2$

4.7 Factoring One Side of an Equation

Factoring techniques enable you to solve equations other than the types considered in Chapter 3.

A. When Is a Product 0?

To solve an equation such as

$$x^2 - x = 0$$

in which *the right side of the equation equals* 0, factor the left side to obtain

$$x(x - 1) = 0$$

Now the left side is expressed as the product of x and $x - 1$.

A product is 0 if at least one of the factors is 0. Moreover, if none of its factors is 0, then a product cannot be 0. For example,

$$0 \cdot 6 = 0$$

$$-7 \cdot 0 = 0$$

$$0 \cdot 0 = 0$$

$$4(-7)6 \neq 0$$

Thus in order that

$$x(x - 1) = 0$$

at least one of the factors x or x − 1 must be 0. Therefore you can solve the two simpler equations

$$x = 0, \qquad x - 1 = 0$$

separately:

$$x = 0 \quad \Big| \quad x - 1 = 0$$
$$x = 1$$

Thus the roots of the equation

$$x^2 - x = 0$$

are 0 and 1. To *check* these roots, replace x first by 0 and then by 1 in the given equation,

$$x^2 - x = 0$$

$$0^2 - 0 \stackrel{?}{=} 0 \quad \Big| \quad 1^2 - 1 \stackrel{?}{=} 0$$
$$0 \stackrel{\checkmark}{=} 0 \quad \Big| \quad 0 \stackrel{\checkmark}{=} 0$$

Answers

1. (a) ______
(b) ______

PRACTICE EXERCISES FOR TOPIC A

1. Solve each equation.

(a) $x(x + 3) = 0$ (b) $(x - 8)(x + 1) = 0$

Now try the exercises for Topic A on page 166.

B. Factoring the Left Side

If you can factor the left side of an equation of the form

$$x^2 + Bx + C = 0$$

you can apply the above method.

Example 1 ▶ Solve $x^2 - 3x + 2 = 0$
Solution:

$$x^2 - 3x + 2 = (x - 1)(x - 2)$$

Thus

$$x^2 - 3x + 2 = 0$$

can be written as

$$(x - 1)(x - 2) = 0$$

$$x - 1 = 0 \quad \Big| \quad x - 2 = 0$$
$$x = 1 \quad \Big| \quad x = 2$$

The roots are 1 and 2. ◀

Example 2 ▶ Solve $t^2 + 4t + 3 = 0$

Solution:

$$t^2 + 4t + 3 = (t + 1)(t + 3)$$

Thus

$$t^2 + 4t + 3 = 0$$

can be written as

$$(t + 1)(t + 3) = 0$$

$$t + 1 = 0 \quad \Big| \quad t + 3 = 0$$

$$t = -1 \quad \Big| \quad t = -3$$

The roots are -1 and -3. ◀

Answers

2. (a) ______

(b) ______

PRACTICE EXERCISES FOR TOPIC B

2. Solve each equation.

(a) $x^2 - 3x - 10 = 0$ (b) $x^2 + 10x + 9 = 0$

Now try the exercises for Topic B on page 166.

C. $a^2 = 0$

Observe that

$$a^2 = 0 \quad \text{means that} \quad a \cdot a = 0$$

and thus it follows that

$$a = 0$$

Example 3 ▶ Solve $x^2 + 9 = 6x$

Solution:

Rewrite this equation in the equivalent form

$$x^2 - 6x + 9 = 0$$

so that the right side equals 0. Consider the left side.

$$x^2 - 6x + 9 = (x - 3)(x - 3) = (x - 3)^2$$

Thus $x^2 - 6x + 9 = 0$ can be written as

$$(x - 3)^2 = 0$$

This is of the form $a^2 = 0$ in which $a = x - 3$. Therefore

$$x - 3 = 0$$

$$x = 3$$

The *only* root is 3. ◀

PRACTICE EXERCISES FOR TOPIC C

Answers
3. (a)
(b)

3. Solve each equation.

(a) $x^2 - 14x + 49 = 0$ (b) $t^2 + 4t = -4$

Now try the exercises for Topic C on page 167.

D. Difference of Squares

The difference-of-squares method can be used to factor the left side of an equation of the form

$$x^2 - a^2 = 0$$

Example 4 ► Solve $x^2 - 16 = 0$

Solution:

The left side is the difference of squares:

$$x^2 - 16 = (x + 4)(x - 4)$$

Thus $$(x + 4)(x - 4) = 0$$

$$x + 4 = 0 \quad | \quad x - 4 = 0$$

$$x = -4 \quad | \quad x = 4$$

The roots are 4 and -4. ◄

In Chapter 9 another technique will be employed to solve equations of the form

$$x^2 - a^2 = 0$$

PRACTICE EXERCISES FOR TOPIC D

Answers
4. (a)
(b)

4. Solve each equation.

(a) $x^2 - 100 = 0$ (b) $y^2 = 64$

Now try the exercises for Topic D on page 167.

E. Common Factors

Often the left side of an equation contains a common factor. Isolate the common factor before applying other factoring techniques.

Example 5 ► Solve $4x^3 - 100x = 0$

Solution:

Isolate the common factor on the left side.

$$4x^3 - 100x = 4x(x^2 - 25)$$

The second factor, $x^2 - 25$, is the difference of squares. Thus

$$4x^3 - 100x = 4x(x + 5)(x - 5)$$

Use this to solve the given equation.

$$4x^3 - 100x = 0$$

$$4x(x + 5)(x - 5) = 0 \qquad \textit{Divide both sides by 4.}$$

$$\frac{4x(x + 5)(x - 5)}{4} = \frac{0}{4} \qquad \textit{Recall that } \frac{0}{4} = 0$$

$$x(x + 5)(x - 5) = 0$$

$x = 0$	$x + 5 = 0$	$x - 5 = 0$
	$x = -5$	$x = 5$

The roots are 0, 5, and -5. ◀

Answers

5. (a) ______

(b) ______

PRACTICE EXERCISES FOR TOPIC E

5. Solve each equation.

(a) $3x^3 + 9x^2 + 6x = 0$ (b) $5u^2 = 80$

Now try the exercises for Topic E on page 167.

Use this area for work space.

EXERCISES

A. In Exercises 1–8 solve each equation. *Check* the roots, when indicated, by using the given equation. (*See Practice Exercise 1.*)

1. $x(x - 4) = 0$

2. $x(x + 6) = 0$

3. $(x - 2)(x - 3) = 0$

4. $(x - 1)(x + 4) = 0$

5. $2(y + 3)(y - 8) = 0$ (*Check.*)

6. $-3(z + 2)(z + 4) = 0$

7. $(x - 5)^2 = 0$

8. $(x + 2)^2 = 0$

B. In Exercises 9–24 solve each equation. *Check* the roots, when indicated, by using the given equation. (*See Examples 1 and 2.*)

9. $x^2 - 3x = 0$

10. $x^2 + 8x = 0$

11. $x^2 - 5x + 4 = 0$

12. $x^2 - 5x + 6 = 0$

13. $t^2 + 6t + 8 = 0$ (*Check.*)

14. $y^2 + y - 12 = 0$

15. $x^2 - 2x - 8 = 0$

16. $y^2 + 2y - 8 = 0$

17. $x^2 + 8x + 15 = 0$

18. $x^2 - 2x - 15 = 0$ (*Check.*)

19. $z^2 + 9z + 18 = 0$

20. $t^2 - t - 30 = 0$

21. $x^2 + 4x - 21 = 0$

22. $x^2 + 7x + 10 = 0$ (*Check.*)

23. $x^2 - 3x = 10$

24. $x^2 = 9x - 20$

Use this area for work space.

C. In Exercises 25–30 solve each equation. (*See Example 3.*)

25. $u^2 + 2u + 1 = 0$

26. $v^2 - 4v + 4 = 0$

27. $x^2 - 10x + 25 = 0$

28. $x^2 + 8x + 16 = 0$

29. $y^2 + 64 = 16y$

30. $z^2 + 20z = -100$

D. In Exercises 31–36 solve each equation. (*See Example 4.*)

31. $x^2 - 1 = 0$

32. $y^2 - 9 = 0$

33. $t^2 - 81 = 0$

34. $x^2 = 144$

35. $y^2 - 121 = 0$

36. $z^2 - 10\,000 = 0$

E. In Exercises 37–52 solve each equation. *Check* the roots, when indicated, by using the given equation. (*See Example 5.*)

37. $3x^2 + 24x - 27 = 0$

38. $10x^2 - 70x + 120 = 0$

39. $2x^2 + 2x - 4 = 0$ (*Check.*)

40. $3x^2 + 6x = 9$

41. $2x^2 - 72 = 0$

42. $7x^2 = 28$

43. $x^3 - 49x = 0$

44. $x^4 - 100x^2 = 0$

45. $3y^3 - 27y = 0$

46. $-6z^3 + 24z = 0$

47. $x^3 - 10x^2 + 9x = 0$

48. $t^3 - 6t^2 + 8t = 0$

49. $2x^3 - 8x^2 + 6x = 0$

50. $5x^3 - 5x^2 - 60x = 0$

51. $5u^2 = 20$

52. $6v^2 = 150$

4.8 Factoring Trinomials $Ax^2 + Bx + C$

A. The Basic Method

In this section you will factor trinomials, such as

$$2x^2 + 3x + 1 \qquad \text{and} \qquad 3x^2 + 8x + 4$$

in which the coefficient of the x^2 term is greater than 1. In general, these trinomials are of the form

$$Ax^2 + Bx + C, \qquad A > 1$$

Example 1 ► Factor $2x^2 + 3x + 1$

Solution:

In order to obtain $2x^2$, try:

$$\begin{array}{l} 2x + \underline{\quad} \\ \underline{x + \underline{\quad}} \end{array}$$

Because the numerical term of

$$2x^2 + 3x + \mathbf{1}$$

is **1,** and because the coefficient of the term $3x$ is *positive,* try $1 \cdot 1$.

$$\begin{array}{r} 2x + 1 \\ x + 1 \\ \hline 2x^2 + x \\ 2x + 1 \\ \hline 2x^2 + 3x + 1 \end{array}$$

This works. Therefore

$$2x^2 + 3x + 1 = (2x + 1)(x + 1)$$ ◀

Example 2 ▶ Factor $2y^2 - 5y + 2$

Solution:

Again try

$$\begin{array}{l} 2y + \underline{\quad} \\ y + \underline{\quad} \end{array}$$

But because of the *negative* coefficient of the middle term $-5y$

try $(-1)(-2)$

Now there are two combinations to try:

$$\begin{array}{r} 2y - 2 \\ y - 1 \\ \hline 2y^2 - 2y \\ -2y + 2 \\ \hline 2y^2 - \mathbf{4}y + 2 \end{array} \ (\textit{Wrong}) \qquad \text{and} \qquad \begin{array}{r} 2y - 1 \\ y - 2 \\ \hline 2y^2 - y \\ -4y + 2 \\ \hline 2y^2 - \mathbf{5}y + 2 \end{array} \ (\textit{Right})$$

Thus $$2y^2 - 5y + 2 = (2y - 1)(y - 2)$$ ◀

Example 3 ▶ Factor $3x^2 + 8x + 4$

Solution:

Consider

$$\begin{array}{l} 3x + \underline{\quad} \\ x + \underline{\quad} \end{array}$$

Here try: $1 \cdot 4$ and $2 \cdot 2$

You will have to try both combinations

$$\begin{array}{r} 3x + 1 \\ x + 4 \\ \hline 3x^2 + x \\ 12x + 4 \\ \hline 3x^2 + 13x + 4 \end{array} \ (\textit{Wrong}) \qquad \text{and} \qquad \begin{array}{r} 3x + 4 \\ x + 1 \\ \hline 3x^2 + 4x \\ 3x + 4 \\ \hline 3x^2 + 7x + 4 \end{array} \ (\textit{Wrong})$$

as well as

$$\begin{array}{l} 3x + 2 \\ \underline{x + 2} \\ 3x^2 + 2x \\ \underline{\qquad\quad 6x + 4} \\ 3x^2 + 8x + 4 \quad (\textit{Right}) \end{array}$$

Therefore $3x^2 + 8x + 4 = (3x + 2)(x + 2)$ ◀

Answers

1. (a) ______

(b) ______

PRACTICE EXERCISES FOR TOPIC A

1. Factor each trinomial.

(a) $5x^2 - 3x - 2$ (b) $6x^2 - 5x + 1$

Now try the exercises for Topic A on page 171.

B. Common Factors

You may have to isolate the common factor before applying this method.

Example 4 ▶ Factor $-8x^2 + 6x + 2$

Solution:

Because you want to consider

$$Ax^2 + Bx + C \qquad \textit{where } A > 0$$

instead of isolating the common factor 2, isolate -2:

$$-8x^2 + 6x + 2 = (-2)(4x^2 - 3x - 1)$$

Consider $4x^2 - 3x - 1$

Now there are two possibilities for $4x^2$:

$$\begin{array}{l} 2x + \_\_\_ \\ \underline{2x + \_\_\_} \end{array} \quad \text{and} \quad \begin{array}{l} 4x + \_\_\_ \\ \underline{x + \_\_\_} \end{array}$$

Because the numerical term of $4x^2 - 3x - 1$ is -1 (*negative*), try

$$1(-1)$$

Remember to try the integral factors *both* ways with 4x and x, if the first way doesn't work.

$$\begin{array}{l} 2x + 1 \\ \underline{2x - 1} \\ 4x^2 + 2x \\ \underline{\qquad - 2x - 1} \\ 4x^2 \qquad\quad - 1 \quad (\textit{Wrong}) \end{array} \qquad \begin{array}{l} 4x - 1 \\ \underline{x + 1} \\ 4x^2 - x \\ \underline{\qquad 4x - 1} \\ 4x^2 + 3x - 1 \quad (\textit{Wrong}) \end{array} \qquad \begin{array}{l} 4x - 1 \\ \underline{x - 1} \\ 4x^2 + x \\ \underline{\qquad - 4x - 1} \\ 4x^2 - 3x - 1 \quad (\textit{Right}) \end{array}$$

Therefore $$4x^2 - 3x - 1 = (4x + 1)(x - 1)$$

and $$-8x^2 + 6x + 2 = -2(4x + 1)(x - 1)$$ ◀

Answers

2. (a) ______

(b) ______

PRACTICE EXERCISES FOR TOPIC B

2. Factor each trinomial.

(a) $9x^2 - 3x - 6$ (b) $-2ax^2 - 3ax + 5a$

Now try the exercises for Topic B on page 171.

C. Two Variables

Now consider a polynomial in two variables, as in part (b) of the following example.

Example 5 ▶ Factor

(a) $6x^2 + 17x + 5$

(b) $6x^2 + 17xy + 5y^2$

Solution:

(a) Try all combinations for

$$\begin{array}{r} 6x + \_\_\_ \\ x + \_\_\_ \\ \hline \end{array} \quad \text{and for} \quad \begin{array}{r} 3x + \_\_\_ \\ 2x + \_\_\_ \\ \hline \end{array}$$

along with $5 \cdot 1$

$$\begin{array}{r} 6x + 5 \\ x + 1 \\ \hline 6x^2 + 5x \\ 6x + 5 \\ \hline 6x^2 + 11x + 5 \end{array} \quad (\textit{Wrong}) \qquad \begin{array}{r} 6x + 1 \\ x + 5 \\ \hline 6x^2 + x \\ 30x + 5 \\ \hline 6x^2 + 31x + 5 \end{array} \quad (\textit{Wrong})$$

$$\begin{array}{r} 3x + 5 \\ 2x + 1 \\ \hline 6x^2 + 10x \\ 3x + 5 \\ \hline 6x^2 + 13x + 5 \end{array} \quad (\textit{Wrong}) \qquad \begin{array}{r} 3x + 1 \\ 2x + 5 \\ \hline 6x^2 + 2x \\ 15x + 5 \\ \hline 6x^2 + 17x + 5 \end{array} \quad (\textit{Right})$$

Thus $$6x^2 + 17x + 5 = (3x + 1)(2x + 5)$$

(b) Modify the above factors:

$$\begin{array}{r} 3x + y \\ 2x + 5y \\ \hline 6x^2 + 2xy \\ 15xy + 5y^2 \\ \hline 6x^2 + 17xy + 5y^2 \end{array}$$

Thus $$6x^2 + 17xy + 5y^2 = (3x + y)(2x + 5y)$$ ◀

It is important to be able to recognize which factoring technique applies to a given polynomial. In Review Exercises 41–90 on pages 174 and 175 you will be given various polynomials to factor without being told which method applies. It is strongly recommended that you work through most of these exercises.

Equations of the form

$$Ax^2 + Bx + C = 0, \qquad A > 1$$

can often be solved by factoring the left side. Because "fractional" roots are involved, this will be discussed in Chapter 9.

Answers

3. ____________

PRACTICE EXERCISES FOR TOPIC C

3. Factor $14x^2 + 9xy + y^2$

Now try the exercises for Topic C on page 172.

Use this area for work space.

EXERCISES

A. In Exercises 1–40 factor each trinomial. (*See Examples 1–3.*)

1. $3x^2 + 4x + 1$

2. $5y^2 + 6y + 1$

3. $7a^2 + 8a + 1$

4. $7b^2 - 8b + 1$

5. $3x^2 + 2x - 1$

6. $3x^2 - 2x - 1$

7. $5y^2 - 6y + 1$

8. $5y^2 - 4y - 1$

9. $2x^2 + 7x + 3$

10. $2x^2 + 5x + 2$

11. $2y^2 + 5y + 3$

12. $2y^2 + 7y + 3$

13. $2a^2 + 7a + 5$

14. $2b^2 - 9b + 9$

15. $2m^2 - 5m + 2$

16. $2m^2 - 5m - 3$

17. $2n^2 - 7n + 3$

18. $2a^2 - 7a - 4$

19. $2a^2 + 7a - 4$

20. $2a^2 - 9a + 4$

21. $2a^2 + 15a + 25$

22. $4b^2 + 20b + 25$

23. $9y^2 - 6y + 1$

24. $9z^2 - 10z + 1$

25. $4a^2 + 4a + 1$

26. $4b^2 + 5b + 1$

27. $3y^2 - 4y - 7$

28. $3z^2 - 20z - 7$

29. $4m^2 + 8m + 3$

30. $4n^2 - 8n + 3$

31. $6x^2 + 5x - 6$

32. $6z^2 + 13z + 6$

33. $7a^2 + 9a + 2$

34. $7b^2 + 15b + 2$

35. $10x^2 + 3x - 4$

36. $10x^2 + 13x + 4$

37. $9x^2 - 36x - 13$

38. $9y^2 - 42y + 13$

39. $5a^2 + 17a - 12$

40. $5b^2 + 16b + 12$

B. In Exercises 41–50 factor each trinomial. (*See Example 4.*)

41. $4x^2 + 6x + 2$

42. $6y^2 + 8y + 2$

43. $2b^2 + 23b - 12$

44. $8x^2 + 16x + 6$

45. $6b^2 + 9b + 3$

46. $8c^2 + 20c + 12$

Use this area for work space.

47. $4a^2 - 6a + 2$ **48.** $5b^2 - 5b - 30$ **49.** $4x^3 + 10x^2 + 6x$

50. $10y^4 + 21y^3 + 2y^2$

C. In Exercises 51–60 factor each trinomial. (*See Example* 5.)

51. $4x^2 + 4xy + y^2$ **52.** $4s^2 - 4st + t^2$ **53.** $2y^2 + 3yz + z^2$

54. $6x^2 + 5ax + a^2$ **55.** $2x^2 + 5xy + 2y^2$ **56.** $u^2 - 5uv + 4v^2$

57. $9a^2 - 24ab - 20b^2$ **58.** $16x^2 + 2xy - 3y^2$ **59.** $6x^2 + 14xy + 4y^2$

60. $4u^3 - 4u^2v - 3uv^2$

What Have You Learned in Chapter 4?

You can find products such as $4x^2 \cdot y^2 \cdot 2xy^2$ and quotients such as $\frac{4x^2y^3}{2xy^2}$.

Also, you can find products such as $3ab(a + 5b)$ and quotients such as $\frac{12a^2b - 6ab}{6ab}$.

You know how to express an integer, such as 24, as the product of primes.

You know how to isolate the common factor of a polynomial.

You are able to multiply polynomials.

You can factor certain binomials and trinomials.

And you can solve some equations by factoring the left side.

REVIEW

Let's Review Chapter 4

4.1 *Multiplying Powers*

In Exercises 1 and 2 multiply.

1. $(3x^2y)(2xy^3)$

2. $(4m^2n)(-2mn^2)(-mn)$

In Exercises 3 and 4 divide.

3. $\frac{10x^3y^2}{5xy}$

4. $\frac{-32a^4b^3}{-4ab^3}$

In Exercises 5 and 6 multiply.

5. $x^2(x - 5y)$

6. $4a^2bc(2a^2b + 3abc)$

In Exercises 7 and 8 divide.

7. $\frac{6ab + 6b^2}{3b}$

8. $\frac{20x^2y^3 - 15xy^2}{5xy^2}$

9. Simplify $3a^3(a^2 - 2) + 2a^2(a^3 + 3a)$

4.2 *Prime Factors*

10. Find all primes less than 20.

11. Find the prime factorization of (a) 36 (b) 132

12. Find *gcd* (24, 60).

13. Find the *gcd* of the coefficients of $8x^2 + 12x - 16$.

4.3 *Common Factors*

In Exercises 14–17 isolate the common factor of each polynomial.

14. $4x + 6$

15. $x^3 + x^2 - 4x$

16. $9a^6 - 15a^4 + 12a^3$

17. $x^5y^4z^3 + x^3y^3z^3 + x^2y^3z$

REVIEW

4.4 *Multiplying Polynomials*

In Exercises 18–22 multiply.

18. $(x + 1)(x + 5)$

19. $(a + 2)(a - 7)$

20. $(z^2 + 1)(z^2 - 4)$

21. $(m + 2n)(m - 3n)$

22. $(x^2 - 3x + 2)(x^3 - 2x)$

23. Combine $(x^2 + 5)(x - 3) + (x - 2)^2$

4.5 *Difference of Squares*

In Exercises 24–27 factor each binomial.

24. $a^2 - 25$

25. $4y^2 - 49z^2$

26. $a^3 - a$

27. $3xy^2 - 12x$

4.6 *Factoring Trinomials $x^2 + Bx + C$*

In Exercises 28–31 factor each trinomial.

28. $c^2 + 7c + 12$

29. $m^2 - m - 20$

30. $t^3 + 6t^2 + 9t$

31. $x^2 - 5xy + 6y^2$

4.7 *Factoring One Side of an Equation*

In Exercises 32–36 solve each equation. *Check* the roots in Exercise 34.

32. $x^2 - 9x = 0$

33. $x^2 + 7x + 12 = 0$

34. $y^2 - 2y - 24 = 0$ (*Check.*)

35. $t^2 - 36 = 0$

36. $x^3 - 9x^2 + 18x = 0$

4.8 *Factoring Trinomials $Ax^2 + Bx + C$*

In Exercises 37–40 factor each trinomial.

37. $4a^2 + 5a + 1$

38. $2x^2 + 7x + 5$

39. $2y^2 - 7y + 3$

40. $4x^2 + xy - 3y^2$

Next, the different types of factoring are mixed up. In Exercises 41–90 factor each polynomial.

41. $5m - 10n$

42. $36 - a^2$

43. $x^2 + 9x + 14$

44. $7a^2 - 28b^2$

45. $x^3 - 4x$

46. $3x^3 - 5x^2 - 2x$

47. $m^2 + 8mn + 15n^2$

48. $4m^2 - 16n^4$

49. $s^3 + 6s^2 - 7s$

50. $t^4 - 16t^2$

51. $a^2 - 10a + 16$

52. $5x^2 - 8x - 4$

53. $x^2 + 17x - 18$

54. $k^2 + 6k - 40$

55. $7x^2 - 6x - 1$

56. $n^2 - 6n - 55$

57. $6x^2 + 17x + 10$

58. $49 - b^6$

59. $x^2 - 8xy - 20y^2$

60. $a^3 - 25a$

61. $72a^2 - 200$

62. $10x^2 + 8xy - 2y^2$

63. $18a^2 + 18a - 8ab$

64. $81x^2 - 4$

65. $x^2 - 7x - 30$

66. $b^2 - 7b + 10$

67. $4x^2 + 24x + 32$

68. $x^2 - y^2z^2$

69. $4x^4 - 100x^2$

70. $18a^2b^2 - 8b^2$

71. $3a^3 + 15a^2 + 12a$

72. $9x^2 + 8xy - y^2$

73. $24x^2 - 150y^2$

74. $9a^3 + 3a^2 - 12a$

75. $6x^2 - 11x + 5$

76. $120 - 30n^2$

77. $10a^2b + 25ab^2 - 5ab$

78. $11k - 44k^3$

79. $9a + 10ab - a$

80. $4x^2 - 28x + 49$

81. $a^2x^2 - 9a^4$

82. $10a^2 - 10a - 560$

83. $14x^3 + 7x^2 - 7x$

84. $54x^6 - 24y^4$

85. $-x^4 + 81$

86. $5r - 20rq^2$

87. $(14x^2 + 3x) - (5x^2 - 3x)$

88. $a^5 - a$

89. $a^2x^4 - a^4x^2$

90. $5a^2b^2 + 15ab^2 - 20b^2$

REVIEW

And These from Chapters 1-3

91. Rearrange the following primes so that you can write "$<$" between any two of them: 7, 17, 41, 29, 23, 2.

92. Evaluate $(2t + 7)(t - 3)$ when $t = 4$.

93. Solve $4x - 3 + 5x = 12 - (5 - 7x)$

94. What number must be divided by 6 to obtain -12?

95. Find the value of $|-4| - (-8)$.

96. How long must a rope be to enclose a rectangular area that is 54 feet long and 20 feet wide?

97. Subtract $b^2 - 4ab$ from the sum of $5b^2 + ab$ and $6a^2 - 2ab$.

98. Divide $\dfrac{6x + 27xy - 9y}{-3}$

99. Simplify $-2x(x^2 - y) - 3(x^3 - y)$.

100. Add the product of $5x$ and 3 to $x^2 - 10x$.

REVIEW

Practice Exam 1 on Chapter 4

In Questions 1–3 multiply:

1. $(4x^2y^5)(-3xy^3)$
2. $(4x - 3)(2x + 5)$
3. $(a^3 - 4a + 5)(a^2 - 2)$
4. Divide $\dfrac{15x^2y^2 - 25xy^4}{5xy^2}$
5. Express 108 as a product of primes.

In Questions 6–9 factor each polynomial.

6. $6x^2 - 8x$
7. $9a^2 - 16b^2$
8. $x^2 + 9x + 18$
9. $2y^2 - y - 1$
10. Solve and check the roots of $x^2 + 5x = 0$.
11. Solve $x^3 - 6x^2 + 8x = 0$
12. Solve $x^2 + 24x + 144 = 0$

Practice Exam 2 on Chapter 4

In Questions 1–5 multiply.

1. $k^3 \cdot k^9 \cdot k$
2. $3x^4y(-7x^2y^3)$
3. $5ab^2(a^4 - 8b^3)$
4. $(9x + 2)(2x - 1)$
5. $(y + 3)(y^2 - 4y + 2)$
6. Divide $\dfrac{18x^6y + 6x^3y - 12x^2}{6x^2}$
7. Express 350 as the product of primes.

In Questions 8–12 factor each polynomial.

8. $77a^3 + 11a^2$
9. $12x^2 - 27$
10. $x^2 - 12x + 35$
11. $x^2 + 3xy - 40y^2$
12. $2x^2 - 5x + 3$
13. Solve $x^2 + 19 = 20x$
14. Solve $x^3 - 4x = 0$

5

Rational Expressions

5.1 Rational Numbers and Rational Expressions

A. Rational Numbers

There are times when integers do not express common notions. Other numbers, known as *rational numbers,* must be introduced. *Rational numbers express division of integers.*

Example 1 ▶ A man divides a square tract of land among his 9 grandchildren. Each grandchild receives $\frac{1}{9}$ (one-ninth) of the land.

One grandson, Bob, buys his sister's share. Bob then owns $\frac{2}{9}$ (two-ninths) of the land. (*See Figures 5.1(a) and 5.1(b) on p. 178.*) ◀

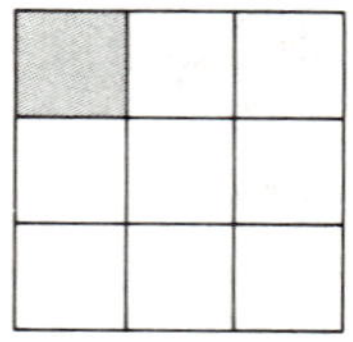

FIGURE 5.1(a). $\frac{1}{9}$

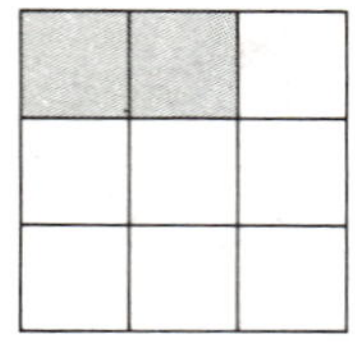

FIGURE 5.1(b). $\frac{2}{9}$

DEFINITION

> A **rational number** (or **rational fraction**) is a number that can be written in the form $\frac{N}{D}$, where N and D are integers and $D \neq 0$. Here N is called the **numerator** and D, the **denominator.**

Example 2 ▶ (a) $\frac{2}{5}$ is a rational number with numerator 2 and denominator 5. The rational number $\frac{2}{5}$ expresses dividing 2 into 5 equal parts.

(b) $\frac{-1}{3}$ is a rational number with numerator -1 and denominator 3.

(c) $\frac{5}{2}$ is a rational number with numerator 5 and denominator 2. Divide 5 by 2 to see that $\frac{5}{2}$ can also be expressed as the **mixed number** $2\frac{1}{2}$, in which 2 is the **integral part** and $\frac{1}{2}$ is the **fractional part.** ◀

An integer, such as 6, can be regarded as a rational number. In fact, $6 = \frac{6}{1}$, and thus 6 is of the form $\frac{N}{D}$, where $N = 6$ and $D = 1$.

Consider a rational number with a *positive* denominator. In order to represent this rational number on the number line, divide the line segments between integer points, as indicated in Example 3.

Example 3 ▶ (a) To represent the rational number $\frac{1}{5}$, divide the line segment between 0 and 1 into 5 equal parts. The *first* point of division to the *right* of 0 corresponds to $\frac{1}{5}$. (*See Figure 5.2(a).*)

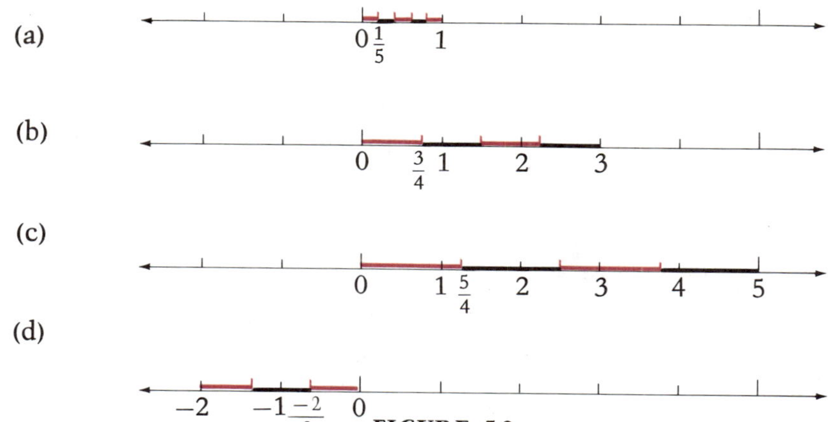

FIGURE 5.2.

(b) To represent the rational number $\frac{\mathbf{3}}{\mathbf{4}}$, divide the line segment between 0 and **3** into **4** equal parts. The *first* point of division to the *right* of 0 corresponds to $\frac{3}{4}$. (*See Figure 5.2(b).*) Note that $\frac{3}{4}$ can also be obtained by dividing the line segment between 0 and 1 into 4 equal parts. Now the *third* point of division to the *right* of 0 corresponds to $\frac{3}{4}$.

(c) To represent the rational number $\frac{\mathbf{5}}{\mathbf{4}}$, or $1\frac{1}{4}$, divide the line segment between 0 and **5** into **4** equal parts. The *first* point of division to the *right* of 0 corresponds to $\frac{5}{4}$. (*See Figure 5.2(c).*)

(d) To represent $\frac{\mathbf{-2}}{\mathbf{3}}$, divide the line segment between **−2** and 0 into **3** equal parts. The *first* point of division to the *left* of 0 corresponds to $\frac{-2}{3}$. (*See Figure 5.2(d).*)

◀

Answers

1. (a) ______ (b) ______

2. ______

PRACTICE EXERCISES FOR TOPIC A

1. (a) What is the numerator of the rational number $\frac{23}{15}$?

(b) What is the denominator?

2. If the line segment between 0 and 2 on the number line were divided into 7 equal parts, which rational number would correspond to the first point of division to the *right* of 0?

3. On the number line in Figure 5.3 locate the point that corresponds to the rational number $\frac{7}{4}$.

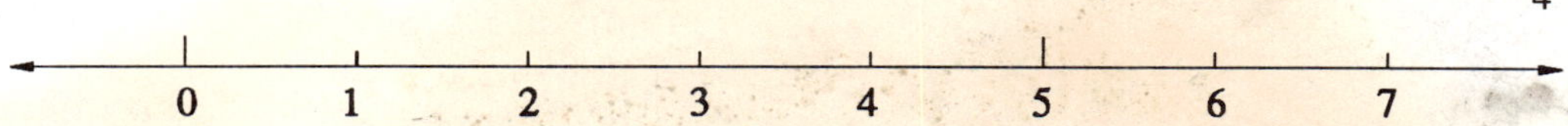

FIGURE 5.3

Now try the exercises for Topic A on page 187.

B. Rational Expressions

Rational numbers were introduced in order to express division of integers. *Rational expressions* will now be introduced to express *division of polynomials.*

DEFINITION

> A **rational expression** is an algebraic expression that can be written in the form $\frac{P}{Q}$, where P and Q are polynomials and $Q \neq 0$ (Q is *not* the zero polynomial). Here P is called the **numerator** and Q the **denominator.**

Example 4 ▶ Determine the numerator and denominator of each rational expression.

(a) $\frac{2x}{x + 2}$ (b) $\frac{t^2 + 2t - 1}{t^3 - t^2 + t + 4}$

Solution:

(a) The numerator is the polynomial $2x$. The denominator is the polynomial $x + 2$.

(b) The numerator is the polynomial $t^2 + 2t - 1$. The denominator is the polynomial $t^3 - t^2 + t + 4$. ◀

Numbers are terms, and therefore (numerical) polynomials. Thus a rational expression can have a number as its numerator or denominator.

Example 5 ▶ Determine the numerator and denominator of each rational expression.

(a) $\frac{5}{x^2 + 1}$ (b) $\frac{x + 4}{2}$ (c) $\frac{-7}{3}$

Solution:

(a) The numerator is the (numerical) polynomial 5. The denominator is the polynomial $x^2 + 1$.

(b) The numerator is the polynomial $x + 4$. The denominator is the polynomial 2.

(c) The numerator is the polynomial -7. The denominator is the polynomial 3. ◀

Recall that an integer, such as 6, can be regarded as the rational number $\frac{6}{1}$, with numerator 6 and denominator 1. Similarly, a polynomial, such as $x + y$, can be considered as the rational expression $\frac{x + y}{1}$, with numerator $x + y$ and denominator 1.

Answers

4. (a) ____

(b) ____

(c) ____

PRACTICE EXERCISES FOR TOPIC B

4. Determine the numerator and denominator of each rational expression.

(a) $\dfrac{x^2 + 3}{4x}$ (b) $\dfrac{1}{2x}$ (c) $x^2 - 2y$

Now try the exercises for Topic B on page 189.

C. Evaluating Rational Expressions

You can evaluate a rational expression, just as you evaluated a polynomial, by substituting numbers for variables. Again *substitute the same number* for *each occurrence* of a variable.

Example 6 ▶ Evaluate $\dfrac{2x}{x + 4}$ three separate times.

(a) when $x = 1$ (b) when $x = 5$ (c) when $x = -1$

Solution:

(a) When x is 1, the numerator is 2(1), or 2, and the denominator is 1 + 4, or 5. The value of the rational expression is therefore $\dfrac{2}{5}$.

(b) $x = 5$:

$$\frac{2(5)}{5 + 4} = \frac{10}{9}$$

(c) $x = -1$:

$$\frac{2(-1)}{-1 + 4} = \frac{-2}{3}$$ ◀

There may be two or more variables present in a rational expression. *For each variable, substitute the same number each time the variable occurs.*

Example 7 ▶ Evaluate $\dfrac{xy - 4}{x^2 + y - 10}$ when $x = 10$ and $y = 1$.

Solution:

Substitute 10 for *each occurrence of* x, and **1** for each occurrence of y:

$$\frac{10 \cdot \mathbf{1} - 4}{10^2 + \mathbf{1} - 10} = \frac{6}{91}$$ ◀

Because division by 0 is undefined, a rational expression $\dfrac{P}{Q}$ is undefined for values of the variable(s) that make $Q = 0$. For example, $\dfrac{x + 1}{x}$ is not defined when $x = 0$, and $\dfrac{1}{x - 4}$ is not defined when $x = 4$.

Answers

5. (a) ______ (b) ______

6. ______

PRACTICE EXERCISES FOR TOPIC C

5. Evaluate $\dfrac{x^2 - 3}{3x + 2}$ (a) when $x = 2$ (b) when $x = 0$.

6. Evaluate $\dfrac{xy + 4x}{x - 2y}$ when $x = 5$ and $y = -1$.

Now try the exercises for Topic C on page 189.

D. Equivalence

A rational number can be expressed in various "equivalent" forms. Figure 5.4 illustrates that $\frac{1}{2} = \frac{2}{4}$. Note that $\frac{2}{4}$ can be obtained from $\frac{1}{2}$ by *multiplying* numerator and denominator each by 2.

$$\frac{1}{2} = \frac{1 \cdot 2}{2 \cdot 2} = \frac{2}{4}$$

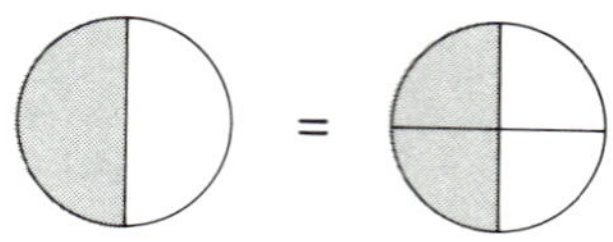

FIGURE 5.4 $\frac{1}{2} = \frac{2}{4}$

And $\frac{2}{4}$ can be "simplified" to $\frac{1}{2}$ by *dividing* numerator and denominator each by 2.

$$\frac{2}{4} = \frac{2 \div 2}{4 \div 2} = \frac{1}{2}$$

DEFINITION

Let N be an integer and let D and k be nonzero integers. Then

$$\frac{N}{D} \text{ and } \frac{Nk}{Dk}$$

are said to be **equivalent forms of the same rational number.** Write

$$\frac{N}{D} = \frac{Nk}{Dk}$$

Similarly, **two rational expressions,**

$$\frac{P}{Q} \text{ and } \frac{P \cdot R}{Q \cdot R}$$

are said to be equivalent, where P is a polynomial and Q and R are nonzero polynomials. Write

$$\frac{P}{Q} = \frac{P \cdot R}{Q \cdot R}$$

Thus an equivalent form of a rational number (or rational expression) is obtained if the numerator and denominator are each multiplied or divided by the same nonzero integer (or polynomial).

Example 8 ▶ (a) $\dfrac{2}{5} = \dfrac{2 \cdot 3}{5 \cdot 3} = \dfrac{6}{15}$

Notice that the "cross products,"

$$\frac{2}{5} \times \frac{6}{15}$$

are equal:

$$2 \cdot 15 = 6 \cdot 5 = 30$$

(b) $\dfrac{1}{x + 2} = \dfrac{1 \cdot 5}{(x + 2) \cdot 5} = \dfrac{5}{5x + 10}$

Again the cross products

$$\frac{1}{x + 2} \times \frac{5}{5x + 10}$$

are equal:

$$1(5x + 10) = 5(x + 2)$$ ◀

In general, let $\dfrac{A}{C}$ and $\dfrac{B}{D}$ be rational numbers (or rational expressions).

If $\dfrac{A}{C} = \dfrac{B}{D}$, then $A \cdot D = B \cdot C$

And if $A \cdot D = B \cdot C$, then $\dfrac{A}{C} = \dfrac{B}{D}$

Geometrically, equivalent *rational numbers* correspond to the same point on the number line. For example,

$$\frac{1}{2} = \frac{1 \cdot 2}{2 \cdot 2} = \frac{2}{4}$$

Thus $\dfrac{1}{2}$ and $\dfrac{2}{4}$ are equivalent rational numbers. To see that they correspond to the same point on the number line, first divide the line segment between 0 and 1 into 2 equal parts. The point of division represents $\dfrac{1}{2}$ (Figure 5.5(a)). Next divide the line segment between 0 and 2 into 4 equal parts, as in Figure 5.5(b). The first point of division to the right of 0 represents $\dfrac{2}{4}$. But this is precisely the point that corresponds to $\dfrac{1}{2}$.

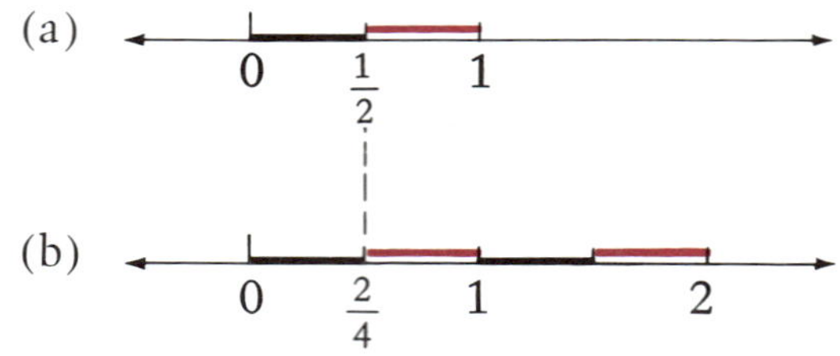

FIGURE 5.5

Equivalent *rational expressions* have the same value whenever the same numbers are substituted for each occurrence of their variables. Thus in Example 8(b) it was shown that $\frac{1}{x+2}$ and $\frac{5}{5x+10}$ are equivalent. In fact, if we replace x by any number other than -2, for which each denominator would equal 0, we obtain the same value. For example, when $x = 1$ the value of $\frac{1}{x+2}$ is given by

$$\frac{1}{1+2} = \frac{1}{3}$$

and the value of $\frac{5}{5x+10}$ by

$$\frac{5}{5(1)+10} = \frac{5}{15} = \frac{1 \cdot 5}{3 \cdot 5} = \frac{1}{3}$$

When $x = -3$, the value of $\frac{1}{x+2}$ is given by

$$\frac{1}{-3+2} = \frac{1}{-1} = -1$$

and the value of $\frac{5}{5x+10}$ by

$$\frac{5}{5(-3)+10} = \frac{5}{-15+10} = \frac{5}{-5} = -1$$

Answers

8. (a) ________

(b) ________

(c) ________

PRACTICE EXERCISES FOR TOPIC D

7. Complete the following statements of equivalence by filling in the numerators or denominators (or both).

(a) $\frac{3}{7} = \frac{3 \cdot 4}{7 \cdot 4} = \frac{\square}{\square}$ (b) $\frac{5}{6} = \frac{\square}{18}$

8. Cross multiply to show whether the given rational numbers or rational expressions are equivalent.

(a) $\frac{3}{5}$ and $\frac{15}{25}$ (b) $\frac{-8}{6}$ and $\frac{12}{-9}$ (c) $\frac{3}{t-3}$ and $\frac{9}{t-9}$

Now try the exercises for Topic D on page 189.

E. Negative Rational Numbers

Let a and b be positive integers, with $b \neq 0$. The following are equivalent ways of expressing the same negative rational number.

$$\frac{-a}{b} \quad (\text{"}-\text{" in the numerator})$$

$$\frac{a}{-b} \quad (\text{"}-\text{" in the denominator})$$

$$-\frac{a}{b} \quad (\text{"}-\text{" before the entire expression})$$

It is customary to place "−" either in the numerator or else before the entire expression so that the resulting expression has a *positive* denominator. Observe that

$$\frac{-a}{b} = \frac{(-a)(-1)}{b(-1)} = \frac{a}{-b}$$

In fact,

$$\frac{-a}{b} = \frac{a}{-b} = -\frac{a}{b}$$

For example,

$$\frac{-1}{2} = \frac{1}{-2} = -\frac{1}{2}$$

(*See Figure 5.6.*)

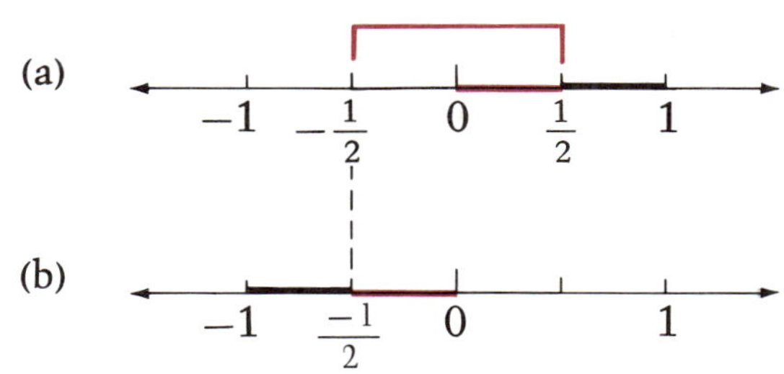

***FIGURE 5.6** (a) $\frac{1}{2}$ is obtained by dividing the line segment from 0 to 1 into 2 equal parts. The point of division represents $\frac{1}{2}$. Then $-\frac{1}{2}$ and $\frac{1}{2}$ lie on opposite sides of, but are the same distance from, the origin. (b) $\frac{-1}{2}$ is obtained by dividing the line segment from −1 to 0 into 2 equal parts. The point of division represents $\frac{-1}{2}$. Note that $-\frac{1}{2} = \frac{-1}{2}$.*

Similarly, for polynomials P and Q, with $Q \neq 0$,

$$\frac{-P}{Q} = \frac{P}{-Q} = -\frac{P}{Q}$$

Answers
9.
10. (a)
(b)
(c)
(d)

PRACTICE EXERCISES FOR TOPIC E

9. Cross multiply to show that $\frac{-3}{5} = \frac{3}{-5}$.

10. For each of the following write another equivalent rational number or rational expression with a positive denominator.

(a) $\frac{9}{-10}$ (b) $\frac{-4}{-5}$ (c) $-\frac{x}{2}$ (d) $\frac{x-4}{-3}$

Now try the exercises for Topic E on page 190.

F. Simplifying

Among the various equivalent forms of a rational number, you usually want to choose the "simplest" one. Thus

$$\frac{2}{3} \text{ is simpler than } \frac{4}{6}$$

Note that the numerator and denominator of $\frac{2}{3}$ have no factors in common other than 1 and -1. However 4 and 6, the numerator and denominator of $\frac{4}{6}$, have the common factor 2.

DEFINITION

> To **simplify a rational number** means to express it in the form $\frac{N}{D}$, where N is an integer, D is a *positive* integer, and N and D have no factors in common other than 1 and -1.

When both numerator and denominator of a rational number are relatively simple, common factors can be divided out, as in Example 9, which follows. In more difficult cases, first find the prime factorizations of the numerator and denominator. Such rational numbers will be discussed in Section 5.2.

► (a) $\frac{45}{-55} = \frac{9 \cdot \cancel{5}}{-11 \cdot \cancel{5}} = \frac{-9}{11}$

(b) $\frac{48}{60} = \frac{4 \cdot \cancel{12}}{5 \cdot \cancel{12}} = \frac{4}{5}$ ◄

The simplification of rational expressions will be discussed in the next few sections.

Answers

11. (a) ______

(b) ______

(c) ______

PRACTICE EXERCISES FOR TOPIC F

11. Simplify:

(a) $\frac{12}{21}$ (b) $\frac{48}{40}$ (c) $\frac{28}{-42}$

Now try the exercises for Topic F on page 191.

G. Rational Coefficients

Rational numbers can occur as the coefficients of polynomials. For example, multiplying a number by $\frac{1}{2}$ is the same as dividing the number by 2. Thus

$$\frac{1}{2}x + \frac{1}{3}, \quad \text{or} \quad \frac{x}{2} + \frac{1}{3}$$

is a polynomial. It is the sum of the term $\frac{1}{2}x$ $\left(\text{or } \frac{x}{2}\right)$ and the numerical term $\frac{1}{3}$.

Example 10 ► Express $\frac{x^2 + 2x + 4}{8}$ as a polynomial with rational coefficients. Simplify these rational coefficients.

Solution:

$$\frac{x^2 + 2x + 4}{8} = \frac{x^2}{8} + \frac{2x}{8} + \frac{4}{8}$$

$$= \frac{1}{8}x^2 + \frac{2}{8}x + \frac{4}{8}$$

$$= \frac{1}{8}x^2 + \frac{1}{4}x + \frac{1}{2}$$ ◄

Answers

12. ______

PRACTICE EXERCISES FOR TOPIC G

12. Express as a polynomial with rational coefficients in simplest form: $\frac{2x^2 + 25x - 30}{10}$

Now try the exercises for Topic G on page 191.

Use this area for work space.

A. In Exercises 1–3 determine (a) the numerator and (b) the denominator of the given rational number. (*See Example 2.*)

1. $\frac{3}{8}$ **2.** $\frac{8}{3}$ **3.** $\frac{-3}{8}$

Use this area for work space.

4. If the line segment between 0 and 5 were divided into 9 equal parts, which rational number would correspond to the first point of division to the *right* of 0?

5. If the line segment between -4 and 0 were divided into 3 equal parts, which rational number would correspond to the first point of division to the *left* of 0?

6. On the number line in Figure 5.7 locate the points that correspond to the following rational numbers. (*See Example 3.*)

(a) $\frac{1}{2}$ (b) $\frac{2}{3}$ (c) $\frac{3}{2}$ (d) $\frac{4}{5}$ (e) $\frac{-1}{4}$ (f) $\frac{-4}{5}$

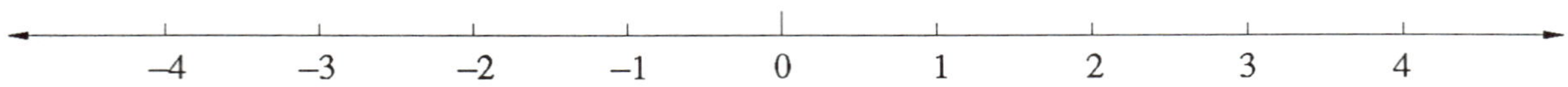

FIGURE 5.7

B. In Exercises 7–9 determine (a) the numerator and (b) the denominator of each rational expression. (*See Examples 4 and 5.*)

7. $\frac{x^2 - x + 1}{x + 3}$ **8.** $4t + 5$ **9.** $\frac{-x}{3}$

10. Determine the rational expression with numerator $2x + 5$ and denominator $3x + 10$.

C. In Exercises 11–23 evaluate each rational expression for the specified number or numbers. (*See Examples 6 and 7.*)

11. $\frac{2x}{x + 1}$ when $x = 10$

12. $\frac{5}{y - 3}$ when $y = 5$

13. $\frac{y - 3}{y + 3}$ when $y = -1$

14. $\frac{1}{x^3 - 2}$ when $x = 2$

15. $\frac{2x - 1}{x + 1}$ when $x = 0$

16. $\frac{t^4 - t^2 + 1}{t + 1}$ when $t = 1$

17. $\frac{u^2}{u + 3}$ when $u = 2$

18. $\frac{z^2 + 1}{z^2 - 1}$ when $z = 4$

19. $\frac{x + 4}{xy}$ when $x = 2$ and $y = 1$

20. $\frac{x + y}{x - y}$ when $x = 6$ and $y = 5$

21. $\frac{x + y + 1}{2x - 3}$ when $x = y = 2$

22. $\frac{xyz}{x + y + z}$ when $x = 2$ and $y = z = 1$

23. $\frac{m^2 - 1}{n^2 + 1}$ when $m = 2$ and $n = 3$

Use this area for work space.

In Exercises 24–26 evaluate each rational expression three separate times (*See Example 6.*)

24. $\dfrac{y+5}{y+2}$ (a) when $y = 2$, (b) when $y = -1$, (c) when $y = 0$

25. $\dfrac{x^9 - 1}{x^6 + 2}$ (a) when $x = 0$, (b) when $x = 1$, (c) when $x = -1$

26. $\dfrac{t^2 + 2t - 3}{5t}$ (a) when $t = 1$, (b) when $t = -1$, (c) when $t = 10$

27. For which value of x is $\dfrac{x+1}{x-2}$ not defined?

28. For which value of y is $\dfrac{y-5}{y+5}$ not defined?

29. The acceleration (or change in velocity) of a body is given by $\dfrac{v-u}{t}$, where u is its initial velocity (in feet per second), v is its final velocity (in feet per second), and t is the time in motion (in seconds). Determine the acceleration if $v = 90$, $u = 45$, and $t = 5$.

30. An object is attracted to earth by a force of magnitude, $\dfrac{1\,000\,000\,000}{h^2}$ kilograms, where h is the distance (in kilometers) of the object from the center of the earth. Determine the magnitude of the force when (a) $h = 1000$. (b) $h = 5000$.

31. An ice-cream plant is able to produce x gallons of chocolate ice cream and y gallons of vanilla ice cream per day, where $y = \dfrac{5000 - 2x}{50 + x}$. Determine the number of gallons of vanilla ice cream produced on a day when 50 gallons of chocolate ice cream are made.

32. (a) The number of gallons of paint required for a room that is l feet long, w feet wide, and h feet high is given by

$$\frac{lw + 2lh + 2wh}{g}$$

where 1 gallon of paint covers g square feet of surface. How many gallons of paint are needed for a room that is 20 feet long, 12 feet wide, and 10 feet high, if 1 gallon of paint covers 200 square feet of surface?

(b) If the paint costs \$12 per gallon, how much does the paint cost for this room?

D. In Exercises 33–38 complete the following statements of equivalence by filling in the numerators or denominators (or both).

33. $\dfrac{2}{5} = \dfrac{2 \cdot 3}{5 \cdot 3} = \dfrac{\square}{\square}$

34. $\dfrac{4}{9} = \dfrac{4(-2)}{9(-2)} = \dfrac{\square}{\square}$

35. $\dfrac{7}{10} = \dfrac{\square}{20}$

36. $\dfrac{-3}{4} = \dfrac{6}{\square}$

37. $\dfrac{3}{x} = \dfrac{3 \cdot 6}{x \cdot 6} = \dfrac{\square}{\square}$

38. $\dfrac{5}{x} = \dfrac{\square}{x^2}$

Use this area for work space.

In Exercises 39–46 cross multiply to show whether the given rational numbers or rational expressions are equivalent.

Sample:	*Solution:* Cross multiply:
$\frac{4}{10}$ and $\frac{2}{5}$	Does $4 \cdot 5 = 2 \cdot 10$?
	Yes. $20 = 20$
	Thus $\frac{4}{10} = \frac{2}{5}$

39. $\frac{4}{16}$ and $\frac{1}{4}$

40. $\frac{-5}{8}$ and $\frac{10}{-16}$

41. $\frac{14}{2}$ and $\frac{1}{7}$

42. $\frac{-36}{48}$ and $\frac{-3}{4}$

43. $\frac{4}{t}$ and $\frac{12}{3t}$

44. $\frac{7}{x}$ and $\frac{7x}{x^3}$

45. $\frac{x - 1}{6x}$ and $\frac{x^3 - x^2}{6x^2}$

46. $\frac{2x}{x - 3}$ and $\frac{2x^2 + 6x}{x^2 - 9}$

E. **47.** Which of the following are equivalent to $-\frac{2}{5}$?

(a) $\frac{-2}{5}$ (b) $\frac{-2}{-5}$ (c) $\frac{2}{-5}$

48. Which of the following are equivalent to $-\frac{5}{-8}$?

(a) $\frac{-5}{8}$ (b) $\frac{5}{8}$ (c) $\frac{-5}{-8}$

49. Cross multiply to show that

$$\frac{-2}{9} = \frac{2}{-9}$$

50. Cross multiply to show that

$$\frac{-3}{x^2} = \frac{3}{-x^2}$$

In Exercises 51–54 write an equivalent rational number or rational expression with a positive denominator.

51. $\frac{7}{-8}$ **52.** $\frac{-2}{-7}$ **53.** $\frac{x + 1}{-3}$ **54.** $\frac{x - 2}{-10}$

Use this area for work space.

F. In Exercises 55–64 simplify each rational number by dividing by factors common to both numerator and denominator. Express the simplified rational number with positive denominator. (*See Example 9.*)

Sample: $\frac{4}{-6}$	*Solution:* Divide numerator and denominator by -2 to obtain $\frac{4}{-6} = \frac{-2}{3}$

55. $\frac{4}{8}$ **56.** $\frac{6}{2}$ **57.** $\frac{-3}{-12}$ **58.** $\frac{10}{4}$ **59.** $\frac{-16}{20}$

60. $\frac{32}{40}$ **61.** $\frac{12}{-18}$ **62.** $\frac{-30}{18}$ **63.** $\frac{-9}{-24}$ **64.** $\frac{15}{-50}$

G. In Exercises 65–68 write each expression as a polynomial with rational coefficients. Simplify these rational coefficients. (*See Example 10.*)

65. $\frac{x^2 + 2x - 4}{5}$

66. $\frac{4x^2 - 6xy + 12}{4}$

67. $\frac{3x^3 - 6x^2 + 8x - 9}{12}$

68. $\frac{5y^4 - 2y^3 + 10y^2 + 12y - 18}{30}$

5.2 Dividing Powers

A. Powers of the Same Base

When simplifying rational numbers or rational expressions, you must often divide powers of a *nonzero* number or variable. In Section 4.1 you have already seen that for $x \neq 0$,

$$\frac{x^5}{x^2} = x^{5-2} = x^3$$

Now you can consider a quotient such as

$$\frac{x^2}{x^5}$$

which represents a rational expression.

Assume a is any nonzero number or variable. Let m and n be positive integers. *To divide a^m by a^n, divide by each factor a, common to numerator and denominator.*

$$\frac{a^m}{a^n} = \frac{\overbrace{a \cdot a \cdot a \ldots a}^{m \text{ factors}}}{\underbrace{a \cdot a \cdot a \ldots a}_{n \text{ factors}}}$$

Three cases occur, depending on the relative sizes of m and n.

1. $\boldsymbol{m > n}$: Every factor a divides out from the denominator. But $(m - n)$ factors a remain in the numerator.

$$\frac{a^m}{a^n} = \frac{\overbrace{a \cdot a \cdot a \ldots a}^{m \text{ factors}}}{\underbrace{a \cdot a \cdot a \ldots a}_{n \text{ factors}}}$$

$$= \frac{\overbrace{\overbrace{\overset{1}{\not a} \cdot \overset{1}{\not a} \cdot \overset{1}{\not a} \ldots \overset{1}{\not a}}^{n \text{ factors}} \cdot \overbrace{a \cdot a \cdot a \ldots a}^{(m-n) \text{ factors}}}^{m \text{ factors}}}{\underbrace{\underset{1}{\not a} \cdot \underset{1}{\not a} \cdot \underset{1}{\not a} \ldots \underset{1}{\not a}}_{n \text{ factors}}}$$

Note that $n + (m - n) = n + m - n = m$.

$$= \frac{\overbrace{a \cdot a \cdot a \ldots a}^{(m-n) \text{ factors}}}{1} = a^{m-n}$$

Thus if $m > n$, then $$\frac{a^m}{a^n} = a^{m-n}$$

2. $\boldsymbol{m = n}$: Every factor a divides out from both numerator and denominator. The resulting fraction equals 1.

$$\frac{a^m}{a^n} = \frac{a^m}{a^m} = \frac{\overbrace{\overset{1}{\not a} \cdot \overset{1}{\not a} \cdot \overset{1}{\not a} \ldots \overset{1}{\not a}}^{m \text{ factors}}}{\underbrace{\underset{1}{\not a} \cdot \underset{1}{\not a} \cdot \underset{1}{\not a} \ldots \underset{1}{\not a}}_{m \text{ factors}}} = 1$$

Thus if $m = n$, then $$\frac{a^m}{a^n} = 1$$

3. $\boldsymbol{m < n}$**:** Every factor a divides out from the numerator. But $(n - m)$ factors a remain in the denominator.

$$\frac{a^m}{a^n} = \frac{\overbrace{a \cdot a \cdot a \ldots a}^{m \text{ factors}}}{\underbrace{a \cdot a \cdot a \ldots a \cdot a \cdot a \cdot a \ldots a}_{n \text{ factors}}}$$

$$= \frac{\overbrace{\overset{1 \cdot 1 \cdot 1 \ldots 1}{\not{a} \cdot \not{a} \cdot \not{a} \ldots \not{a}}}^{m \text{ factors}}}{\underbrace{\underbrace{\underset{1 \cdot 1 \cdot 1 \ldots 1}{\not{a} \cdot \not{a} \cdot \not{a} \ldots \not{a}}}_{m \text{ factors}} \cdot \underbrace{a \cdot a \cdot a \ldots a}_{(n-m) \text{ factors}}}_{n \text{ factors}}}$$

Note that $m + (n - m) = n$

$$= \frac{1}{a^{n-m}}$$

Thus if $m < n$, then $\qquad \dfrac{a^m}{a^n} = \dfrac{1}{a^{n-m}}$

If nothing is said to the contrary, throughout the chapter, assume the denominator of each expression is nonzero.

Example 1 ► (a) $\dfrac{2^6}{2^3} = 2^{6-3} = 2^3$, or $\dfrac{64}{8} = 8$

(b) $\dfrac{(-5)^3}{(-5)^3} = 1$, or $\dfrac{-125}{-125} = 1$

(c) $\dfrac{3^2}{3^4} = \dfrac{1}{3^{4-2}} = \dfrac{1}{3^2}$, or $\dfrac{9}{81} = \dfrac{1}{9}$ ◄

Example 2 ► (a) $\dfrac{y^8}{y^5} = y^{8-5} = y^3$, or

$$\frac{\overset{1 \cdot 1 \cdot 1 \cdot 1 \cdot 1}{\not{y} \cdot \not{y} \cdot \not{y} \cdot \not{y} \cdot \not{y}} \cdot y \cdot y \cdot y}{\underset{1 \cdot 1 \cdot 1 \cdot 1 \cdot 1}{\not{y} \cdot \not{y} \cdot \not{y} \cdot \not{y} \cdot \not{y}}} = y \cdot y \cdot y = y^3$$

(b) $\dfrac{y^4}{y^{10}} = \dfrac{1}{y^{10-4}} = \dfrac{1}{y^6}$, or

$$\frac{\overset{1 \cdot 1 \cdot 1 \cdot 1}{\not{y} \cdot \not{y} \cdot \not{y} \cdot \not{y}}}{\underset{1 \cdot 1 \cdot 1 \cdot 1}{\not{y} \cdot \not{y} \cdot \not{y} \cdot \not{y}} \cdot y \cdot y \cdot y \cdot y \cdot y \cdot y} = \frac{1}{y \cdot y \cdot y \cdot y \cdot y \cdot y} = \frac{1}{y^6}$$ ◄

Example 3 ▶ Divide $\dfrac{15m^8n^{12}p^2}{-5m^{10}n^{12}}$

Solution:

$$\frac{15m^8n^{12}p^2}{-5m^{10}n^{12}} = \frac{-3p^2}{m^{10-8}} = \frac{-3p^2}{m^2}$$ ◀

Answers

1. (a) ____ (b) ____

(c) ____

PRACTICE EXERCISES FOR TOPIC A

1. Simplify:

(a) $\dfrac{k^4}{k^9}$ (b) $\dfrac{10^{12}}{10^8}$ (c) $\dfrac{18x^3y^7}{15x^4y^5}$

Now try the exercises for Topic A on page 195.

B. Prime Factors

When numbers or numerical coefficients are somewhat complicated, it is best to first write out their prime factorizations before dividing.

Example 4 ▶ Simplify $\dfrac{288}{320}$

Solution:

$$288 = 32 \cdot 9 = 2^5 \cdot 3^2$$

$$320 = 64 \cdot 5 = 2^6 \cdot 5$$

$$\frac{288}{320} = \frac{2^5 \cdot 3^2}{2^6 \cdot 5}$$

$$= \frac{3^2}{2 \cdot 5}$$

$$= \frac{9}{10}$$ ◀

Example 5 ▶ Simplify $\dfrac{360x^4yz^2}{-108xy^3z^2}$

Solution:

$$360 = 2^3 \cdot 3^2 \cdot 5$$

$$108 = 2^2 \cdot 3^3$$

$$\frac{360x^4yz^2}{-108xy^3z^2} = \frac{2^3 \cdot 3^2 \cdot 5x^4yz^2}{-2^2 \cdot 3^3xy^3z^2}$$

$$= \frac{-2 \cdot 5x^3}{3y^2}$$

$$= \frac{-10x^3}{3y^2}$$ ◀

Answers

2. (a) ____

(b) ____

PRACTICE EXERCISES FOR TOPIC B

2. Simplify: (a) $\dfrac{225}{135}$ (b) $\dfrac{420xy^2z^5}{700x^2y^3z}$

Now try the exercises for Topic B on page 196.

C. Powers of a Polynomial

Powers of a polynomial can appear in both the numerator and denominator of a rational expression.

Example 6 ▶ Simplify $\dfrac{(x + a)^5}{(x + a)^3}$

Solution:

$$\frac{(x + a)^5}{(x + a)^3} = (x + a)^{5-3} = (x + a)^2$$

The answer can be left in this form. ◀

Example 7 ▶ Simplify $\dfrac{6x(m - n)}{3x(m - n)^4}$

Solution:

$$\frac{6x(m - n)}{3x(m - n)^4} = \frac{2}{(m - n)^{4-1}} = \frac{2}{(m - n)^3}$$ ◀

Answers

3. (a) ____

(b) ____

PRACTICE EXERCISES FOR TOPIC C

3. Simplify: (a) $\dfrac{(x - 3)^2}{(x - 3)^3}$ (b) $\dfrac{18x(x + 2y)^5}{6x^2(x + 2y)^3}$

Now try the exercises for Topic C on page 196.

Use this area for work space.

EXERCISES

A. In Exercises 1–24 simplify each expression. (Multiply out the numbers that remain in each numerator or denominator. *See Examples 1–3.*)

Sample: Simplify: $\dfrac{3^5x^2}{3^2x^4}$	*Solution:* $\dfrac{3^5x^2}{3^2x^4} = \dfrac{3^3}{x^2} = \dfrac{27}{x^2}$

Use this area for work space.

1. $\frac{3}{3^2}$ 2. $\frac{7^4 \cdot 11^2}{7^2 \cdot 11^3}$ 3. $\frac{a^4}{a^4}$ 4. $\frac{b^3}{-b}$ 5. $\frac{y^2}{y^5}$

6. $\frac{m^6}{m^{18}}$ 7. $\frac{ab}{a^2b}$ 8. $\frac{x^4y}{xy^4}$ 9. $\frac{m^8n}{mn^2}$ 10. $\frac{-m^4n^3}{-m^4n^3}$

11. $\frac{a^2bc}{a}$ 12. $\frac{r^2s^2t^3}{rs^2t^4}$ 13. $\frac{x^{10}y^{17}z}{x^8y^{16}z^{16}}$ 14. $\frac{-pqrs}{pr^3s^2}$ 15. $\frac{2ab^2}{4ab}$

16. $\frac{-3xyz}{5xy}$ 17. $\frac{-7abc}{14c^2}$ 18. $\frac{20x^3y^3z}{-10x^2z^2}$ 19. $\frac{9a^2bc}{15a^4bc^2}$ 20. $\frac{36x^{10}yz^8}{12x^7y^2z^6}$

21. $\frac{40m^2n^7}{20m^4n^8}$ 22. $\frac{32a^5cd}{-8abc}$ 23. $\frac{-18m^{10}n^7p^8}{-12mn^8p^7}$ 24. $\frac{-54ax^7}{-27a^3x^7}$

25. Pick out the *error(s)*.

(a) $\frac{x^2}{x^8} = \frac{1}{x^4}$ (b) $\frac{y^5}{y^{15}} = \frac{1}{y^{10}}$ (c) $\frac{a^2b^3}{a^3b^2} = \frac{b}{a}$

(d) $\frac{-a^2}{a^6} = \frac{a^3}{-a^7}$ (e) $\frac{b^5c^4}{bc} = \frac{b^6c^5}{b^2c^2}$

B. In Exercises 26–38 simplify each expression. (Multiply out the numbers that remain in each numerator or denominator. *See Examples 4 and 5.*)

26. $\frac{72}{108}$ 27. $\frac{132}{288}$ 28. $\frac{-6250}{200}$ 29. $\frac{64}{-84}$ 30. $\frac{-144}{450}$

31. $\frac{240}{128}$ 32. $\frac{-52}{-156}$ 33. $\frac{121a^4}{44a^3}$ 34. $\frac{224x^6}{98x^3}$ 35. $\frac{256y^3}{96y^9}$

36. $\frac{560a^8b}{160a^4}$ 37. $\frac{250a^4b^2c}{375ab^2c}$ 38. $\frac{150a^4x^2yz^3}{36axy^4z}$

C. In Exercises 39–58 simplify each expression. (Multiply out the numbers that remain in each numerator or denominator. *See Examples 6 and 7.*)

39. $\frac{(x+1)^2}{x+1}$ 40. $\frac{(x+y)^3}{(x+y)^2}$ 41. $\frac{(a+b)^4}{(a+b)^2}$ 42. $\frac{(a-b)^5}{a-b}$

43. $\frac{(x+y)^2}{(x+y)^2}$ 44. $\frac{m+n}{(m+n)^2}$ 45. $\frac{a-b}{(a-b)^4}$ 46. $\frac{16(c-d)^5}{2(c-d)^8}$

47. $\frac{24(x+y)}{12(x+y)}$ 48. $\frac{-12(x-y)^2}{-4(x-y)}$ 49. $\frac{-36(a+b)^7}{6(a+b)^4}$ 50. $\frac{18(m-n)^5}{24(m-n)^3}$

51. $\frac{(a+b)(x+y)}{a+b}$ 52. $\frac{(x+y)(r+s)^2}{(x+y)(r+s)}$ 53. $\frac{(x+y)^4(x-y)}{(x+y)^2(x-y)^3}$

54. $\frac{4(a+b)^7(a-b)^3}{8(a+b)(a-b)^3}$ 55. $\frac{36(a+b)^{10}(x+y)^7}{144(a+b)^9(x+y)(x-y)}$ 56. $\frac{-84(x-y)(a+b)^7}{98(x-y)^6(a+b)}$

57. $\frac{5^2x^2(a+2)}{5^4x^3(a+2)^2}$ 58. $\frac{108x^2y(x-y)^4}{144xy^2(x-y)}$

5.3 Factoring and Simplifying

A. Monomial Denominators

Consider a rational expression, such as

$$\frac{5a + 5b}{10ab}$$

in which the denominator is a monomial but the numerator contains more than one term. You can often simplify such an expression by first isolating the common factor in the numerator.

Example 1 ▶ Simplify $\dfrac{x^2y^2 + 2x^2}{x^2}$

Solution:
Isolate the common factor in the numerator:

$$x^2y^2 + 2x^2 = x^2(y^2 + 2)$$

Therefore $$\frac{x^2y^2 + 2x^2}{x^2} = \frac{\overset{1}{\cancel{x^2}}(y^2 + 2)}{\underset{1}{\cancel{x^2}}}$$ *Divide numerator and denominator by x^2.*

$$= y^2 + 2$$ ◀

Example 2 ▶ Simplify $\dfrac{4a^2b + 8ab^2}{8a^2b^2}$

Solution:
Isolate the common factor in the numerator:

$$4a^2b + 8ab^2 = 4ab(a + 2b)$$

Therefore $$\frac{4a^2b + 8ab^2}{8a^2b^2} = \frac{\overset{1}{\cancel{4ab}}(a + 2b)}{\underset{2ab}{\cancel{8a^2b^2}}} = \frac{a + 2b}{2ab}$$ ◀

Example 3 ▶ Simplify $\dfrac{5a^2b + 10a^2b^2 + 25a^2b^3}{15ab^2}$

Solution:

$$\frac{5a^2b + 10a^2b^2 + 25a^2b^3}{15ab^2}$$

$$= \frac{\overset{a}{\cancel{5a^2b}}(1 + 2b + 5b^2)}{\underset{3b}{\cancel{15ab^2}}}$$

$$= \frac{a(1 + 2b + 5b^2)}{3b}$$ ◀

Answers

1. (a) ______

(b) ______

PRACTICE EXERCISES FOR TOPIC A

1. Simplify: (a) $\dfrac{9x^2 - 6x^3}{6x^2}$ (b) $\dfrac{8x^6y^2 + 2x^3y^2}{2x^3y}$

Now try the exercises for Topic A on page 200.

B. Monomial Numerators

The same method applies when the numerator is a monomial, but the *denominator* contains more than one term. Now, whenever possible, isolate the common factor in the *denominator*.

Example 4 ▶ Simplify $\dfrac{9xy^2z}{6xy^2 - 9xyz}$

Solution:

$$\frac{9xy^2z}{6xy^2 - 9xyz} = \frac{\overset{3y}{\cancel{9}\cancel{x}\cancel{y^2}}z}{\underset{1}{\cancel{3}\cancel{x}\cancel{y}}(2y - 3z)} = \frac{3yz}{2y - 3z}$$ ◀

Answers

2. ______

PRACTICE EXERCISES FOR TOPIC B

2. Simplify $\dfrac{9n^2}{9n^2 + 6n}$

Now try the exercises for Topic B on page 200.

C. More Complicated Expressions

Suppose both the numerator and denominator of a rational expression contain more than one term. Factoring the numerator or denominator (possibly both), often helps you to simplify the given expression.

Example 5 ▶ Simplify $\dfrac{5x - 5y}{5x + 5y}$

Solution:
Isolate the common factor in both the numerator and denominator:

$$\frac{5x - 5y}{5x + 5y} = \frac{\overset{1}{\cancel{5}}(x - y)}{\underset{1}{\cancel{5}}(x + y)} = \frac{x - y}{x + y}$$ ◀

Note that in Example 5, you **cannot** divide the individual terms. **The following is wrong:**

$$\frac{\overset{1}{\cancel{x}} - \overset{1}{\cancel{y}}}{\underset{1}{\cancel{x}} + \underset{1}{\cancel{y}}} = \frac{0}{2} = 0$$

For example, if $x = 2$ and $y = 1$, then

$$\frac{x - \mathbf{y}}{x + \mathbf{y}} = \frac{2 - \mathbf{1}}{2 + \mathbf{1}} = \frac{1}{3}$$

and *not* 0, as above.

Example 6 ► Simplify $\dfrac{2x + 2y}{x^2 - y^2}$

Solution:
Isolate the common factor in the numerator.

$$2x + 2y = 2(x + y)$$

The denominator is the difference of squares.

$$x^2 - y^2 = (x + y)(x - y)$$

Therefore divide the numerator and denominator by $x + y$:

$$\frac{2x + 2y}{x^2 - y^2} = \frac{2\overset{1}{\cancel{(x + y)}}}{\underset{1}{\cancel{(x + y)}}(x - y)} = \frac{2}{x - y}$$ ◄

Example 7 ► Simplify $\dfrac{a^2 + 3a + 2}{a^2 + 2a + 1}$

Solution:
Factor both the numerator and denominator.

$$a^2 + 3a + 2 = (a + 1)(a + 2)$$
$$a^2 + 2a + 1 = (a + 1)^2$$

Therefore $$\frac{a^2 + 3a + 2}{a^2 + 2a + 1} = \frac{\overset{1}{\cancel{(a + 1)}}(a + 2)}{\underset{a + 1}{\cancel{(a + 1)^2}}} = \frac{a + 2}{a + 1}$$ ◄

Example 8 ► Simplify $\dfrac{25ax^2 - 25ay^2}{5x^2 + 10xy + 5y^2}$

Solution:
Factor both the numerator and denominator.

$$25ax^2 - 25ay^2 = 25a(x^2 - y^2) = 25a(x + y)(x - y)$$

$$5x^2 + 10xy + 5y^2 = 5(x^2 + 2xy + y^2) = 5(x + y)^2$$

Therefore

$$\frac{25ax^2 - 25ay^2}{5x^2 + 10xy + 5y^2} = \frac{\overset{5}{\cancel{25}}a\overset{1}{\cancel{(x + y)}}(x - y)}{\cancel{5}\underset{1(x + y)}{\cancel{(x + y)^2}}} = \frac{5a(x - y)}{x + y}$$ ◄

Example 9 ▶ Simplify $\frac{x - y}{y - x}$

Solution:

Observe that by the Distributive Laws,

$$(-1)(y - x) = (-1)y - (-1)x$$
$$= -y + x$$
$$= x - y$$

Thus $$x - y = (-1)(y - x)$$

and $$\frac{x - y}{y - x} = \frac{(-1)\overset{1}{\cancel{(y - x)}}}{\underset{1}{\cancel{y - x}}} = -1$$ ◀

Answers

3. (a) ______

(b) ______

(c) ______

PRACTICE EXERCISES FOR TOPIC C

3. Simplify:

(a) $\frac{6x^2 + 6x}{3x^3 - 3x^2}$ (b) $\frac{x^2 - 3x}{2x - 6}$ (c) $\frac{12 - 6a}{2a^2 - 8}$

Now try the exercises for Topic C on page 201.

Use this area for work space.

EXERCISES

A. In Exercises 1–18 simplify each rational expression. (*See Examples 1–3.*)

1. $\frac{5m + 10n}{5}$ **2.** $\frac{ax + ay}{a}$ **3.** $\frac{2ax + 2ay}{2a}$

4. $\frac{x^2 + x^2y}{x^2}$ **5.** $\frac{10a - 5b}{20}$ **6.** $\frac{a^2x - a^2y}{a}$

7. $\frac{ab + ac^2}{a^2}$ **8.** $\frac{cx^2 + cy^2}{cx}$ **9.** $\frac{4x^2 + 8xy}{4x}$

10. $\frac{6x^2y + 9xy^2}{6xy}$ **11.** $\frac{18a^2b + 36ab^2}{6ab}$ **12.** $\frac{30abc - 40a^2c^2}{10a^2c}$

13. $\frac{4x^3y^2z - 8x^2y^2z}{8xy^2z}$ **14.** $\frac{-24r^2s - 30rst}{18rs}$ **15.** $\frac{9ax + 9ay - 9az}{3a}$

16. $\frac{6r^2s + 12r^2t - 6r^2u}{36r^2}$ **17.** $\frac{5a^2x - 10ax^2 + 15a^2x^2}{20a^2x^2}$

18. $\frac{20a^2b^3c^2 + 25a^4b^2c - 30a^2b^2c^2}{25a^2b^3c}$

B. In Exercises 19–24 simplify each rational expression. (*See Example 4.*)

19. $\frac{4}{4a - 4b}$ **20.** $\frac{2x}{4x + 6x^2}$

Use this area for work space.

21. $\frac{5xy}{10x^2y + 15xy^2}$

22. $\frac{25ab^2c^3}{50a^3bc^2 - 25ab^2c^6}$

23. $\frac{100a^2xy}{10a^2y - 25a^2x + 50a^2xy}$

24. $\frac{-3a^2b^3c^{10}}{6a^2b^3c + 9ab^4c^4 - 27a^3b^3c^5}$

C. In Exercises 25–64 simplify each rational expression. (*See Examples 5–9.*)

25. $\frac{2a + 2b}{2x + 2y}$

26. $\frac{5a - 5b}{10a + 10b}$

27. $\frac{9x + 9y}{3x - 3y}$

28. $\frac{ax + ay}{ax - ay}$

29. $\frac{2ax + 2ay}{4ax - 4ay}$

30. $\frac{5m + 10n}{10m - 5n}$

31. $\frac{6x + 6y}{5x + 5y}$

32. $\frac{8a - 8b}{5a - 5b}$

33. $\frac{ax + ay}{bx + by}$

34. $\frac{x^2 - y^2}{3x + 3y}$

35. $\frac{5a + 5b}{a^2 - b^2}$

36. $\frac{6x - 6y}{2x^2 - 2y^2}$

37. $\frac{x^2 - 1}{2x + 2}$

38. $\frac{a^2 - 9}{4a + 12}$

39. $\frac{3m + 3n}{5m^2 - 5n^2}$

40. $\frac{5a^2 - 20}{3a + 6}$

41. $\frac{a^2 - 4b^2}{2a - 4b}$

42. $\frac{x^2 - 25y^2}{2x + 10y}$

43. $\frac{4x^2 - 9y^2}{12x + 18y}$

44. $\frac{100a^2 - 64b^2}{5a - 4b}$

45. $\frac{x^2 + 2x + 1}{x^2 + 4x + 3}$

46. $\frac{a^2 + 5a + 6}{a^2 + 4a + 4}$

47. $\frac{y^2 - 9}{y^2 + 4y + 3}$

48. $\frac{z^2 + 7z + 10}{z^2 - 25}$

49. $\frac{u^2 - 36}{u^2 + 7u + 6}$

50. $\frac{r^2 - 1}{2 - 2r^2}$

51. $\frac{m^2 + 7m + 12}{m^2 + 5m + 6}$

52. $\frac{a^2 - 8a + 12}{a^2 - 36}$

53. $\frac{x^2 + 2x - 8}{x^2 - 16}$

54. $\frac{x^2 - y^2}{x^2 + 2xy + y^2}$

55. $\frac{3a^2 - 3b^2}{6a^2 + 18ab + 12b^2}$

56. $\frac{9u^2 - 36}{3u^2 - 9u + 6}$

57. $\frac{5x^2 + 35x + 60}{10x^2 + 10x - 60}$

58. $\frac{a^2u^2 - a^2v^2}{au^2 + 6auv + 5av^2}$

59. $\frac{2x^2 + 11x + 15}{x^2 + 6x + 9}$

60. $\frac{3a^2 - 2ab - b^2}{4a^2 - 4b^2}$

61. $\frac{x - 1}{1 - x}$

62. $\frac{a + 4}{-4 - a}$

63. $\frac{2y - 2z}{z - y}$

64. $\frac{5a^2 - 5b^2}{15b - 15a}$

65. Which of the following *cannot* be simplified?

(a) $\frac{2x + 6}{3x}$ (b) $\frac{4x - 8}{2x}$ (c) $\frac{1 - x^2}{x + 1}$ (d) $\frac{x + 3}{x + 5}$

(e) $\frac{ax + by}{ab}$ (f) $\frac{ax + by}{x + y}$ (g) $\frac{a - b}{a + b}$

66. Pick out the *error(s)*.

(a) $\frac{x + 2}{x - 2} = 0$

(b) $\frac{x - 4}{x + 4} = -1$

(c) $\frac{x^2 + y^2}{x + y} = x + y$

(d) $\frac{x^2 - y^2}{x + y} = x - y$

(e) $\frac{4 - x^2}{2 - x} = x + 2$

(f) $\frac{x^2 + 8x + 15}{x + 5} = x + 3$

5.4 Dividing Polynomials

A. Degree

Consider a *polynomial in a single variable,* such as

$$x^4 + 5x^3 - x^2 + x + 1$$

The **degree of a term** is simply the exponent of the variable of that term, if there is a variable; the **degree of a nonzero numerical term** is 0. (The degree of 0 is not defined.) For the above polynomial,

the degree of x^4 is 4

the degree of $5x^3$ is 3

the degree of $-x^2$ is 2

the degree of x is 1

the degree of 1, the numerical term, is 0

The **degree of a polynomial** is the highest degree of any of its terms. Thus the degree of $x^4 + 5x^3 - x^2 + x + 1$ is 4.

A *polynomial in a single variable* is said to be in **standard form** if its terms are arranged in order of decreasing degree. Thus

$$x^4 + 5x^3 - x^2 + x + 1$$

is in standard form, but

$$5x^3 + x^4 + 1 - x^2 + x$$

is not.

Example 1 ► (a) Express $t^4 - t^3 + 2t^6 + t^8 - 1 + t^2$ in standard form.
(b) What is the degree of this polynomial?

Solution:

(a) Rearrange the terms in order of decreasing degree:

$$t^8 + 2t^6 + t^4 - t^3 + t^2 - 1$$

is now in standard form.

(b) The degree of this polynomial is 8, the highest degree of any of its terms. ◄

Answers	
1. (a)	(b)
(c)	
2. (a)	
(b)	

PRACTICE EXERCISES FOR TOPIC A

1. Consider the polynomial $-4x^5 + 3x + 12$. What is the degree of each term?

(a) $-4x^5$ (b) $3x$ (c) 12

2. (a) Express in standard form:

$$8 + y^3 - 3y^4 - 6y$$

(b) What is the degree of this polynomial?

Now try the exercises for Topic A on *page 208.*

B. The Division Process

There is a method of dividing polynomials that resembles long division.

Recall that to divide 651 by 21, you proceed as follows:

$$\begin{array}{r} 31 \\ 21\overline{)651} \\ \underline{63} \\ 21 \\ \underline{21} \end{array}$$

For polynomials P, Q, and S, where $S \neq 0$,

$$\frac{P}{S} = Q \qquad \text{if} \qquad P = S \cdot Q$$

You can also express division of polynomials by writing

$$\begin{array}{r} Q \\ S\overline{)P} \end{array}$$

Here S is called the **divisor,** P the **dividend,** and Q the **quotient.**

Example 2 ▶ Divide and *check* $x + 2\overline{)2x^2 + 7x + 6}$

Solution:

$$\begin{array}{rrl} & \overbrace{\text{Ⓐ} \rightarrow 2x \quad + 3 \leftarrow \text{Ⓒ}}^{\textit{quotient}} & \\ \textit{divisor} \rightarrow x + 2 & \overline{)2x^2 + 7x + 6} & \leftarrow \textit{dividend} \\ \text{Ⓑ} & \underline{2x^2 + 4x} & \\ & 3x + 6 & \\ \text{Ⓓ} & \underline{3x + 6} & \\ & 0 & \end{array}$$

Ⓐ

$$\frac{2x^2}{x} = 2x$$

Divide $2x^2$, the first term of the dividend, by x, the first term of the divisor, to obtain $2x$, the first term of the quotient.

Ⓑ

$$\begin{array}{rl} & 2x \\ x + 2 & \overline{)\;2x^2 + 7x + 6} \\ & \underline{\mp 2x^2 \mp 4x} \\ & 3x + 6 \leftarrow \textit{1st difference polynomial} \end{array}$$

Now multiply $2x$ by the divisor, $x + 2$, and *subtract* the product, $2x^2 + 4x$, from $2x^2 + 7x$, the sum of the first two terms of the divisor. The difference is $3x$. Bring down 6, the next term of the dividend to obtain the **1st difference polynomial,** $3x + 6$.

Ⓒ

$$\frac{3x}{x} = 3$$

Next, divide $3x$, the first term of the 1st difference polynomial, by x, the first term of the divisor. The second term of the quotient is 3.

Ⓓ

$$\begin{array}{rrl}
\textit{quotient} \rightarrow & & 2x + 3 \\
\textit{divisor} \rightarrow & x + 2 \,\big) & 2x^2 + 7x + 6 \leftarrow \textit{dividend} \\
& & \underline{\mp 2x^2 \mp 4x} \\
& & \quad 3x + 6 \leftarrow \textit{1st difference polynomial} \\
& & \quad \underline{\mp 3x \mp 6} \\
& & \qquad 0 \leftarrow \textit{2nd difference polynomial}
\end{array}$$

Multiply 3 by the divisor, $x + 2$, to obtain $3x + 6$. When you subtract this from the 1st difference polynomial, you obtain 0. There are no more terms to bring down. Thus 0 is the 2nd difference polynomial, and $x + 2$ divides $2x^2 + 7x + 6$ (evenly). The quotient is $2x + 3$.

Check:
You can *check* a division example *by multiplying the quotient by the divisor (in either order).*
The product should be the dividend, as you see below.

$$\begin{array}{rr}
\textit{quotient} \rightarrow & 2x + 3 \\
\textit{divisor} \rightarrow & \underline{\times \; x + 2} \\
& 2x^2 + 3x \qquad \\
& \underline{4x + 6} \\
\textit{dividend} \rightarrow & 2x^2 + 7x + 6
\end{array}$$

◀

Example 3 ▶ Divide and *check* $3x + 1\overline{)6x^2 + 17x + 5}$
Solution:

$$\begin{array}{rrl}
\textit{quotient} \rightarrow & & 2x + 5 \\
\textit{divisor} \rightarrow & 3x + 1 \,\big) & 6x^2 + 17x + 5 \leftarrow \textit{dividend} \\
& & \underline{\mp 6x^2 \mp 2x} \\
& & \quad 15x + 5 \leftarrow \textit{1st difference polynomial} \\
& & \quad \underline{\mp 15x \mp 5} \\
& & \qquad 0 \leftarrow \textit{2nd difference polynomial}
\end{array}$$

The quotient is $2x + 5$.

Check:

$$\begin{array}{rr}
\textit{quotient} \rightarrow & 2x + 5 \\
\textit{divisor} \rightarrow & \underline{\times \; 3x + 1} \\
& 6x^2 + 15x \qquad \\
& \underline{2x + 5} \\
\textit{dividend} \rightarrow & 6x^2 + 17x + 5
\end{array}$$

◀

When you divide polynomials in a single variable, always express them in standard form. Add 0's for missing terms, as in Example 4, which follows.

Continue the process until you have used up all of the terms of the dividend. There may be more than two difference polynomials.

Example 4 ▶ Divide $6 - 7x + x^3$ by $x - 2$.

Solution:

Express the dividend in standard form. Also, add $0x^2$ for the missing term of degree 2:

$$x^3 + 0x^2 - 7x + 6$$

$$\begin{array}{rrrrrl}
 & x^2 & +\,2x & -\,3 & & \leftarrow \text{quotient} \\
x - 2\,\big) & x^3 & +\,0x^2 & -\,7x & +\,6 & \leftarrow \text{divisor } x-2 \\
 & -x^3 & +\,2x^2 & & & \\ \hline
 & & 2x^2 & -\,7x & & \leftarrow \text{1st difference polynomial} \\
 & & -2x^2 & +\,4x & & \\ \hline
 & & & -\,3x & +\,6 & \leftarrow \text{2nd difference polynomial} \\
 & & & +\,3x & -\,6 & \\ \hline
 & & & & 0 & \leftarrow \text{3rd difference polynomial}
\end{array}$$

The quotient is $x^2 + 2x - 3$. ◀

Example 5 ▶ Divide $x + 1\overline{)x^3 + x^2 + x + 1}$

Solution:

$$\begin{array}{rrrrr}
 & x^2 & & +\,1 & \\
x + 1\,\big) & x^3 & +\,x^2 & +\,x & +\,1 \\
 & -x^3 & -\,x^2 & & \\ \hline
 & & & x & +\,1 \\
 & & & -x & -\,1 \\ \hline
 & & & & 0
\end{array}$$

The 1st difference polynomial is 0. Bring down two terms from the dividend because the divisor has two terms.

The quotient is $x^2 + 1$. ◀

PRACTICE EXERCISES FOR TOPIC B

Answers
3. ______
4. ______

3. Divide $3x^2 - 2x - 8$ by $x - 2$.

4. Divide and *check* $2x + 3\overline{)2x^3 + 7x^2 - 9}$

Now try the exercises for Topic B on page 209.

C. Division with a Remainder

Observe that 2 does not divide 7 (evenly). When you divide 7 by 2, there is a "remainder."

$$\frac{7}{2} = 3\frac{1}{2}$$

dividend → 7; divisor → 2; quotient → 3; remainder → 1; divisor → 2

Thus 3 is the quotient and 1, the remainder. Similarly, division of polynomials often results in a remainder. Here is a simple example.

Example 6 ▶ Divide $2x + 1\overline{)4x + 3}$

Solution:

$$\begin{array}{r} 2 \\ 2x + 1\overline{)\ 4x + 3} \\ \underline{\mp 4x \mp 2} \\ 1 \end{array}$$

There are no more terms of the dividend to bring down. Thus 2 is the quotient; 1 is the remainder.

$$\text{dividend} \rightarrow \frac{4x + 3}{2x + 1} = \overset{\text{quotient}}{\overset{\downarrow}{2}} + \frac{1}{2x + 1} \begin{array}{l} \leftarrow \text{remainder} \\ \leftarrow \text{divisor} \end{array}$$

(divisor → $2x + 1$)

Note that the remainder, 1, is a (numerical) polynomial of degree 0, whereas the divisor is a polynomial of degree 1. Thus

$$\underbrace{\text{degree }(1)}_{0} < \underbrace{\text{degree }(2x + 1)}_{1}$$ ◀

Let P and S be polynomials, with $S \neq 0$. Then either S divides P (evenly), as in Examples 2–5, or else

$$\frac{P}{S} = Q + \frac{R}{S}$$

where Q and R are polynomials and degree $R <$ degree S. The polynomial Q is called the **quotient** and R the **remainder.** Thus, *if S does not divide P, add the rational expression* $\frac{R}{S}$ *to the quotient Q.*

When you divide polynomials, there are two possibilities for the *final difference polynomial:*

1. This difference polynomial is 0, as in Examples 2–5. In this case, the remainder is zero.

2. The degree of R, the final difference polynomial, is less than the degree of the divisor, as in Example 6. Then R is the remainder.

Continue the division process until you obtain one of these difference polynomials.

Example 7 ▶ Find the quotient and remainder.

$$5x + 1\overline{)10x^2 + 12x - 2}$$

Solution:

$$
\begin{array}{lrl}
\textit{quotient} \rightarrow & 2x + 2 + \dfrac{-4}{5x+1} & \begin{array}{l}\leftarrow \textit{remainder}\\ \leftarrow \textit{divisor}\end{array} \\
\textit{divisor} \rightarrow 5x+1\,\big) & 10x^2 + 12x - 2 & \leftarrow \textit{dividend} \\
 & \mp 10x^2 \mp 2x & \\
\hline
 & 10x - 2 & \leftarrow \textit{1st difference polynomial} \\
 & \mp 10x \mp 2 & \\
\hline
 & -4 & \leftarrow \textit{2nd difference polynomial (remainder)}
\end{array}
$$

The quotient is $2x + 2$ and the remainder is -4. The result is expressed as

$$2x + 2 + \frac{-4}{5x + 1}$$

◀

To *check* a division example, where there is a remainder:

1. *Multiply the quotient and divisor (in either order).*
2. *Add the remainder to this.*

The resulting polynomial should be the *dividend.*

Check (for Example 7)

$$
\begin{array}{rr}
\textit{quotient} \rightarrow & 2x + 2 \\
\textit{divisor} \rightarrow & \times 5x + 1 \\
\hline
 & 10x^2 + 10x \\
 & 2x + 2 \\
\hline
 & 10x^2 + 12x + 2 \\
\textit{remainder} \rightarrow & + - 4 \\
\hline
\textit{dividend} \rightarrow & 10x^2 + 12x - 2
\end{array}
$$

Example 8 ▶ Find the quotient and remainder. Also *check* the result.

$$3x - 2\,\big)\,\overline{12x^3 - 11x^2 + 5x}$$

Solution:

$$
\begin{array}{lrl}
\textit{quotient} \rightarrow & 4x^2 - x + 1 + \dfrac{2}{3x-2} & \begin{array}{l}\leftarrow \textit{remainder}\\ \leftarrow \textit{divisor}\end{array} \\
\textit{divisor} \rightarrow 3x-2\,\big) & 12x^3 - 11x^2 + 5x & \leftarrow \textit{dividend} \\
 & \mp 12x^3 \pm 8x^2 & \\
\hline
 & - 3x^2 + 5x & \leftarrow \textit{1st difference polynomial} \\
 & \pm 3x^2 \mp 2x & \\
\hline
 & 3x & \leftarrow \textit{2nd difference polynomial} \\
 & \mp 3x \pm 2 & \\
\hline
 & 2 & \leftarrow \textit{3rd difference polynomial}
\end{array}
$$

The quotient is $4x^2 - x + 1$ and the remainder is 2. The result is expressed as

$$4x^2 - x + 1 + \frac{2}{3x - 2}$$

Check:

$$\begin{array}{lr}
\textit{quotient} \rightarrow & 4x^2 - x + 1 \\
\textit{divisor} \rightarrow & \times\ 3x - 2 \\
\hline
 & 12x^3 - 3x^2 + 3x \\
 & - 8x^2 + 2x - 2 \\
\hline
 & 12x^3 - 11x^2 + 5x - 2 \\
\textit{remainder} \rightarrow & + \quad 2 \\
\hline
\textit{dividend} \rightarrow & 12x^3 - 11x^2 + 5x
\end{array}$$

◀

Example 9 ▶ Find the quotient and remainder.

$$x^2 \overline{)x}$$

Solution:
1, the degree of the dividend, x, is less than 2, the degree of the divisor, x^2. Thus the quotient is 0 and the remainder is x.

$$\frac{x}{x^2} = 0 + \frac{x}{x^2} \leftarrow \textit{remainder}$$

◀

Answers

5. ________

6. ________

PRACTICE EXERCISES FOR TOPIC C

5. Find the quotient and remainder. *Check* the result.

$$x + 5\overline{)x^3 + 5x^2 - 2x - 7}$$

6. Express the result of the following division in the form

$$\text{QUOTIENT} + \frac{\text{REMAINDER}}{\text{DIVISOR}}$$

$$x + 3\overline{)4x^3 + 11x^2 - 2x + 1}$$

Now try the exercises for Topic C on page 209.

Use this area for work space.

EXERCISES

A. In Exercises 1–6: (a) Express each polynomial in standard form. (b) What is the degree of each term of the resulting polynomial? (c) What is the degree of the polynomial? (*See Example 1.*)

1. $x^2 + 1 + 2x$

2. $y^3 - 1 + 2y^2 + 4y$

3. $u - u^2 + u^3 - u^4$

4. $x + x^2 + x^3 + x^4$

5. $z^7 - z^4 + z + z^{10}$

6. $x^9 - x^8 + 1 - x^4 - x^{12}$

Use this area for work space.

B. In Exercises 7–30 divide as indicated. *Check* the division in Exercises 7–16. (*See Examples* 2–5.)

7. $x + 1\overline{)x^2 + 5x + 4}$ (*Check.*)

8. $y + 2\overline{)y^2 + 8y + 12}$ (*Check.*)

9. $x + 3\overline{)x^2 + 8x + 15}$ (*Check.*)

10. $x - 1\overline{)x^2 + 5x - 6}$ (*Check.*)

11. $t + 9\overline{)t^2 + 14t + 45}$ (*Check.*)

12. $2x - 1\overline{)6x^2 - 7x + 2}$ (*Check.*)

13. $x + 4\overline{)x^3 + 5x^2 + 5x + 4}$ (*Check.*)

14. $x + 2\overline{)2x^3 + 5x^2 + 3x + 2}$ (*Check.*)

15. $t + 4\overline{)12t^2 + 3t^3 - 4 - t}$ (*Check.*)

16. $y^2 + 2\overline{)y^4 + 3y^2 + 2}$ (*Check.*)

17. $x + 3\overline{)x^2 + 12x + 27}$

18. $5x - 2\overline{)10x^2 + x - 2}$

19. $z^2 - 1\overline{)z^4 - 1}$

20. $t + 3\overline{)t^3 + 27}$

21. Divide $9 + 3a + 3a^2 + a^3$ by $3 + a^2$.

22. Divide $9x^2 + x^3 + 17x + 6$ by $x^2 + 7x + 3$.

23. Divide $15 + 8a^2 + a^4$ by $a^2 + 3$.

24. $x + 5\overline{)x^3 + 6x^2 + 6x + 5}$

25. $a - 5\overline{)2a^3 - 11a^2 + 7a - 10}$

26. $2b + 1\overline{)4b^3 + 4b^2 + 5b + 2}$

27. $x^2 + x - 2\overline{)x^3 + 3x^2 - 4}$

28. $2y^2 + 3y + 1\overline{)2y^3 + 9y^2 + 10y + 3}$

29. $x^2 + x + 1\overline{)x^4 + 3x^3 + 4x^2 + 3x + 1}$

30. $x^2 + x - 1\overline{)2x^4 + 3x^3 + 4x^2 + 4x - 5}$

C. In Exercises 31–36: (a) What is the quotient? (b) What is the remainder? (c) Show that the degree of the remainder is less than the degree of the divisor. (*See Example* 6.)

31. $\dfrac{8x + 3}{2x} = 4 + \dfrac{3}{2x}$

32. $\dfrac{16x + 7}{4x + 2} = 4 + \dfrac{-1}{4x + 2}$

33. $\dfrac{2x^2 + 1}{x} = 2x + \dfrac{1}{x}$

34. $\dfrac{x^2 + 4x + 8}{x + 3} = x + 1 + \dfrac{5}{x + 3}$

35. $\dfrac{2x^3 - 4x + 5}{x^2 - 2} = 2x + \dfrac{5}{x^2 - 2}$

36. $\dfrac{x^4 + 4x^3 + 6x^2 + 8x + 3}{x^2 + x + 2} = x^2 + 3x + 1 + \dfrac{x + 1}{x^2 + x + 2}$

In Exercises 37–46: (a) Find the quotient, (b) find the remainder, and (c) *check* the result. (*See Example* 8.)

37. $x + 1\overline{)3x + 2}$

38. $x + 2\overline{)9x - 4}$

39. $x + 5\overline{)5x + 6}$

40. $1 + x\overline{)1 + 10x}$

41. $x + 4\overline{)x^2 + 7x + 10}$

42. $x - 1\overline{)x^2 - 2x}$

43. $x + 1\overline{)x^3 + 3x^2 + 3x + 5}$

44. $x^2 + x + 1\overline{)2x^3 + 7x^2 + 6x + 10}$

45. $x^2 - x\overline{)3x^3 + 2x^2 + 5x + 2}$

46. $x^2 + 1\overline{)x^4 - x^2 + 3}$

Use this area for work space.

In Exercises 47–60: (a) Find the quotient, (b) find the remainder, and (c) express your result in the form

$$\text{QUOTIENT} + \frac{\text{REMAINDER}}{\text{DIVISOR}}$$

(See Examples 7 and 9.)

47. $x + 3\overline{)x^2 + 8x + 20}$

48. $x - 1\overline{)2x^3 - 5x}$

49. $x - 5\overline{)x^2}$

50. $x - 1\overline{)6x^4 - 5}$

51. $2x + 1\overline{)4x^2 + 6x + 5}$

52. $x^2 + x + 1\overline{)4x^3 + x^2 - 2x + 10}$

53. $3x + 2\overline{)12x^2 + 11x + 11}$

54. $x^2 - 2\overline{)x^4 + x^2}$

55. $x^2 + 1\overline{)x^4 - 1}$

56. $x^2 + x + 1\overline{)x^4 - x^2 + x + 2}$

57. $x + 1\overline{)x^3 - 3x^2 + 2x + 1}$

58. $x^2 + 1\overline{)2x}$

59. $x + 2\overline{)x^3 - 2x^2 + x + 2}$

60. $x^3 - x\overline{)x^2 + x}$

5.5 Multiplying and Dividing Rational Expressions

A. Multiplying Rational Numbers

Suppose you cut a pie into quarters, as in Figure 5.8(a). You then decide to cut each piece in half. Observe that each new portion is $\frac{1}{8}$ of the pie. [See Figure 5.8(b).] Thus

$$\frac{1}{2} \text{ of } \frac{1}{4} \text{ equals } \frac{1}{8}$$

(Here, the word "of" indicates multiplication.)

$$\frac{1}{2} \cdot \frac{1}{4} = \frac{1}{8}$$

Note that

$$\frac{1}{2} \cdot \frac{1}{4} = \frac{1 \cdot 1}{2 \cdot 4} = \frac{1}{8}$$

In Figure 5.8(c) the product $\frac{1}{2} \cdot \frac{1}{4}$ is depicted on the number line.

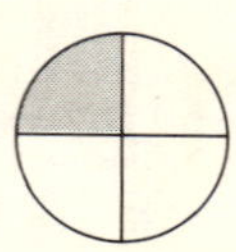

FIGURE 5.8(a). $\frac{1}{4}$

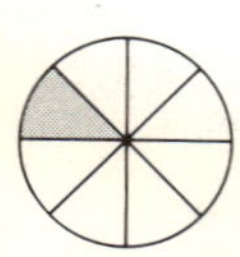

FIGURE 5.8(b). $\frac{1}{2} \cdot \frac{1}{4} = \frac{1}{8}$

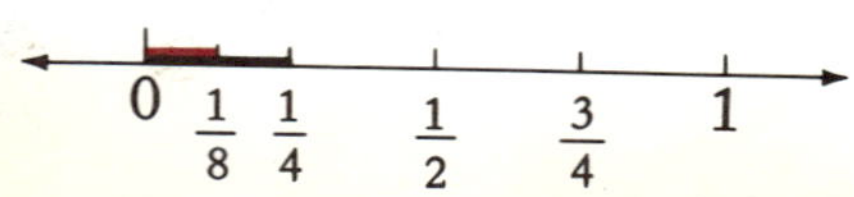

FIGURE 5.8(c). $\frac{1}{2} \cdot \frac{1}{4} = \frac{1}{8}$

To multiply two rational numbers:

1. Multiply their numerators.
2. Multiply their denominators.

$$\frac{a}{b} \cdot \frac{c}{d} = \frac{ac}{bd}, \qquad b \neq 0,\ d \neq 0$$

Example 1 ▶

$$\frac{2}{3} \cdot \frac{1}{5} = \frac{2 \cdot 1}{3 \cdot 5} = \frac{2}{15}$$ ◀

It is best to divide by factors common to the numerators and denominators before multiplying. You will then be working with smaller numbers.

Example 2 ▶ Multiply $\frac{2}{5} \cdot \frac{15}{4}$

Solution:
Divide 15 and 5 by 5. Then, divide 2 and 4 by 2.

$$\frac{\overset{1}{\cancel{2}}}{\underset{1}{\cancel{5}}} \cdot \frac{\overset{3}{\cancel{15}}}{\underset{2}{\cancel{4}}} = \frac{3}{2}$$ ◀

The same procedure applies to **products of three or more rational numbers:**

1. Divide by factors common to numerators and denominators.
2. Multiply the resulting numerators.
3. Multiply the resulting denominators.

Example 3 ▶ Multiply $\frac{-5}{8} \cdot \frac{12}{25} \cdot \frac{10}{-3}$

Solution:

Divide 12 and 8 by 4.
Divide 10 and 5 by 5.
Divide -1 and -1 by -1.

Divide -5 and 25 by 5.
Divide 3 and -3 by 3.
Divide 2 and 2 by 2.

$$\frac{\overset{\overset{1}{\cancel{-1}}}{\cancel{-5}}}{\underset{\underset{1}{\cancel{2}}}{\cancel{8}}} \cdot \frac{\overset{\overset{1}{\cancel{3}}}{\cancel{12}}}{\underset{\underset{1}{\cancel{5}}}{\cancel{25}}} \cdot \frac{\overset{\overset{1}{\cancel{2}}}{\cancel{10}}}{\underset{\underset{1}{\cancel{-1}}}{\cancel{-3}}} = 1$$ ◀

Example 4 ▶ Find $\frac{3}{4}$ of 12.

Solution:
Here the word "of" can be translated as "times." Thus

$$\frac{3}{4} \cdot 12 = \frac{3}{\underset{1}{\cancel{4}}} \cdot \frac{\overset{3}{\cancel{12}}}{1} = 9$$ ◀

Answers

1. (a) ____

(b) ____

(c) ____

PRACTICE EXERCISES FOR TOPIC A

1. Multiply. (Express the product in simplest form.)

(a) $\frac{6}{7}\cdot\frac{5}{12}$ (b) $28\cdot\frac{5}{21}$ (c) $\frac{6}{5}\cdot\frac{8}{9}\cdot\frac{5}{16}$

Now try the exercises for Topic A on page 215.

B. Multiplying Rational Expressions

Example 5 ▶ $$\frac{x^3}{3y}\cdot\frac{a+b}{a-b}=\frac{x^3(a+b)}{3y(a-b)}$$ ◀

Note that the resulting rational expression is left in factored form.

To multiply two or more rational expressions:

1. First divide by factors common to the numerators and denominators.
2. Multiply the resulting numerators.
3. Multiply the resulting denominators.

Leave the resulting expression in factored form.

Example 6 ▶ Multiply $\frac{4x^3}{x+y}\cdot\frac{(x+y)^2}{8x}$

Solution:

$$\frac{\overset{x^2}{\cancel{4x^3}}}{\underset{1}{\cancel{x+y}}}\cdot\frac{\overset{x+y}{\cancel{(x+y)^2}}}{\underset{2}{\cancel{8x}}}=\frac{x^2(x+y)}{2}$$ ◀

Answers

2. (a) ____

(b) ____

3. ____

PRACTICE EXERCISES FOR TOPIC B

2. Multiply: (a) $\frac{2(x+1)}{3ax}\cdot\frac{x+1}{5x}$ (b) $\frac{a^3}{a-2}\cdot\frac{9(a-2)^2}{3a}$

3. Multiply the given polynomial with rational coefficients.

$$\frac{2}{5}a^3b\cdot\frac{10}{9}ab^4\cdot\frac{3}{8}a^2b$$

Now try the exercises for Topic B on page 216.

C. Factoring Numerator or Denominator

You may have to factor a numerator or denominator in order to divide by common factors.

Example 7 ▶

$$\frac{m+n}{m^2-3m+2}\cdot\frac{m-2}{m^2-n^2}=\frac{\cancel{m+n}}{\cancel{(m-2)}(m-1)}\cdot\frac{\cancel{m-2}}{\cancel{(m+n)}(m-n)}$$

$$=\frac{1}{(m-1)(m-n)}$$ ◀

Example 8 ▶

$$\frac{x^2-4}{x^2+5x+4}\cdot\frac{2x+8}{(x+2)^2}\cdot\frac{-1}{16x-32}=\frac{\cancel{(x+2)}\cancel{(x-2)}}{\cancel{(x+4)}(x+1)}\cdot\frac{2\cancel{(x+4)}}{\cancel{(x+2)^2}}\cdot\frac{-1}{\cancel{16}\cancel{(x-2)}}$$

$$=\frac{-1}{8(x+1)(x+2)}$$ ◀

Answers

4. ______

PRACTICE EXERCISES FOR TOPIC C

4. Multiply. (Factor, divide out any factors common to the numerator and denominator, and leave the result in factored form.)

$$\frac{x^2-2x-3}{6x-18}\cdot\frac{3x^3}{x^3-x}$$

Now try the exercises for Topic C on page 216.

D. Dividing Rational Numbers

Observe that

divisor ↓ *quotient* ↓

$$8 \div 4 = 2$$

$$8 \div 2 = 4$$

$$8 \div 1 = 8$$

Each time the divisor is *divided by* 2, the quotient is *multiplied by* 2. According to this pattern:

$$8 \div \frac{1}{2} = 16 \qquad \textit{Note that } 8\cdot\frac{2}{1}=16.$$

$$8 \div \frac{1}{4} = 32 \qquad \textit{Note that } 8\cdot\frac{4}{1}=32.$$

Thus

to divide 8 by $\frac{1}{2}$, invert $\frac{1}{2}$ to obtain $\frac{2}{1}$, and then multiply

to divide 8 by $\frac{1}{4}$, invert $\frac{1}{4}$ to obtain $\frac{4}{1}$, and then multiply

To divide rational numbers or rational expressions,

$$\frac{a}{b} \div \frac{c}{d}$$

invert the divisor and multiply. Thus

$$\frac{a}{b} \div \frac{c}{d} = \frac{a}{b} \cdot \frac{d}{c} = \frac{ad}{bc}$$

Note that when $d = 1$, this becomes

$$\frac{a}{b} \div c = \frac{a}{bc}$$

Example 9 ▶ Find $\frac{1}{2} \div \frac{1}{4}$

Solution:

Invert the *second* rational number $\frac{1}{4}$, and multiply.

$$\frac{1}{2} \div \frac{1}{4} = \frac{1}{\cancel{2}_{\,1}} \cdot \frac{\cancel{4}^{\,2}}{1} = 2$$ ◀

Example 10 ▶

$$\frac{-5}{6} \div \frac{10}{27} = \frac{\cancel{-5}^{\,-1}}{\cancel{6}_{\,2}} \cdot \frac{\cancel{27}^{\,9}}{\cancel{10}_{\,2}} = \frac{-9}{4}$$ ◀

Answers

5. (a) ______

(b) ______

(c) ______

PRACTICE EXERCISES FOR TOPIC D

5. Divide:

(a) $6 \div \frac{2}{3}$ (b) $\frac{-9}{10} \div \frac{6}{5}$ (c) $\frac{2}{5} \div 4$

Now try the exercises for Topic D on page 216.

E. Dividing Rational Expressions

The same procedure applies when you divide rational expressions. Invert the second expression and multiply.

Example 11 ▶ Find $\frac{1}{x} \div \frac{1}{x^2}$

Solution:

Invert the second expression, $\frac{1}{x^2}$, and multiply.

$$\frac{1}{x} \div \frac{1}{x^2} = \frac{1}{\cancel{x}_1} \cdot \frac{\cancel{x^2}^{x}}{1} = x$$ ◀

Example 12 ▶ Find $\dfrac{x^2 + 2xy + y^2}{x^2y^3} \div \dfrac{x^2 - y^2}{x^4y^5}$

Solution:
First factor both numerators.

$$\frac{x^2 + 2xy + y^2}{x^2y^3} \div \frac{x^2 - y^2}{x^4y^5} = \frac{(x + y)^2}{x^2y^3} \div \frac{(x + y)(x - y)}{x^4y^5}$$

$$= \frac{\overset{x + y}{\cancel{(x + y)^2}}}{\underset{1}{\cancel{x^2y^3}}} \cdot \frac{\overset{x^2y^2}{\cancel{x^4y^5}}}{\underset{1}{\cancel{(x + y)}}(x - y)}$$

$$= \frac{x^2y^2(x + y)}{x - y}$$ ◀

Answers

6. (a) ______

(b) ______

PRACTICE EXERCISES FOR TOPIC E

6. Divide. Leave the resulting rational expression in factored form.

(a) $\dfrac{5x}{x + 5} \div \dfrac{x^5}{10}$ (b) $\dfrac{9a - 6}{12a} \div \dfrac{9a^2 - 4}{3a^3 + 2a^2}$

Now try the exercises for Topic E on page 217.

Use this area for work space.

EXERCISES

A. In Exercises 1–12 multiply the given rational numbers. (*See Examples 1–3.*)

1. $\frac{1}{2} \cdot \frac{1}{3}$ **2.** $\frac{2}{3} \cdot \frac{1}{5}$ **3.** $\frac{1}{2} \cdot 3$ **4.** $\frac{-1}{3} \cdot 6$

5. $\frac{5}{8} \cdot \frac{2}{15}$ **6.** $\frac{3}{4} \cdot \frac{8}{9}$ **7.** $\frac{7}{10} \cdot \frac{25}{49}$ **8.** $\frac{5}{8} \cdot \frac{1}{10} \cdot \frac{2}{3}$

9. $\frac{-1}{7} \cdot \frac{-2}{5} \cdot \frac{-14}{3}$ **10.** $\frac{3}{4} \cdot \frac{49}{16} \cdot \frac{64}{21}$ **11.** $\frac{10}{7} \cdot \frac{5}{14} \cdot \frac{49}{25} \cdot \frac{2}{3}$ **12.** $\frac{7}{2} \cdot \frac{5}{8} \cdot \frac{24}{25} \cdot \frac{12}{7}$

13. Find $\frac{1}{2}$ of 48. (*See Example 4.*) **14.** Find $\frac{2}{5}$ of 60. **15.** Find $\frac{3}{4}$ of $\frac{8}{9}$.

16. One-third of the employees in a factory are women. Two-thirds of these women are married. What fraction of the employees in this factory are married women?

17. Ed finds half a pie in the refrigerator and eats three-fourths of this portion of pie. What fraction of the entire pie does he eat?

Use this area for work space.

18. Two-fifths of the students at a college are foreign. One-sixth of the foreign students are Indian. Half of the Indian students are women. What fraction of the students at this college are Indian women?

B. In Exercises 19–28 multiply the given rational expressions. Leave the resulting rational expressions in factored form. (*See Examples 5 and 6.*)

19. $a \cdot \frac{x}{y}$ **20.** $\frac{a^2}{c} \cdot \frac{b}{c}$ **21.** $\frac{2a}{b} \cdot \frac{b^2}{a}$ **22.** $\frac{4xyz}{3ab} \cdot \frac{9ax}{8yz}$

23. $\frac{16x^2y}{5a^3b} \cdot \frac{25a^2b^2}{4xy^2}$ **24.** $\frac{a^2}{b} \cdot \frac{x}{a} \cdot \frac{b^2}{x}$ **25.** $\frac{2x}{3a} \cdot \frac{6a^2}{x^2} \cdot \frac{12a}{x}$

26. $\frac{x+1}{4c^2} \cdot \frac{2c}{(x+1)^3}$ **27.** $\frac{a^4}{(x-y)^2} \cdot \frac{(x-y)^5}{a^3}$

28. $\frac{a+1}{(a-1)^2} \cdot \frac{(a+1)^3}{a-1} \cdot \frac{a-1}{(a+1)^4}$

In Exercises 29–32 multiply the given polynomials.

29. $\frac{2}{3}x^2 \cdot \frac{1}{5}xy$ **30.** $\frac{3}{4}a^2b \cdot \frac{8}{9}ab^3$

31. $\frac{2}{5}x^2yz \cdot \frac{10}{3}xy^4 \cdot \frac{9}{20}x^3yz^2$ **32.** $\frac{1}{2}x\left(\frac{2}{3}x + \frac{4}{9}\right)$

C. In Exercises 33–40 multiply the given rational expressions. Leave the resulting rational expressions in factored form. (*See Examples 7 and 8.*)

33. $\frac{3a+3}{a} \cdot \frac{a^2}{a+1}$ **34.** $\frac{5x+5}{x+2} \cdot \frac{1}{x+1}$

35. $\frac{5a+15b}{x^2y} \cdot \frac{xy^2}{a+3b}$ **36.** $\frac{x^2-1}{x^2+1} \cdot \frac{x^2+1}{x+1}$

37. $\frac{a^2-4}{3a-9} \cdot \frac{2a-6}{a^2+4a+4}$ **38.** $\frac{x^2-16}{2ax} \cdot \frac{4a^2}{3x+12} \cdot \frac{3x}{2x-8}$

39. $\frac{a^2-9}{4x^3} \cdot \frac{12x}{(a+3)^2} \cdot \frac{x^2}{a-3}$ **40.** $\frac{x^2-y^2}{4a^3} \cdot \frac{a}{y-x} \cdot \frac{16a^4}{x+y}$

41. Find the product of $\frac{a}{a-2}$ and $\frac{4a-8}{a^2}$.

42. Find the product of $\frac{x}{x+1}$ and $\frac{x^2-1}{x^3}$.

D. In Exercises 43–52 divide the given rational numbers. (*See Examples 9 and 10.*)

43. $\frac{1}{3} \div \frac{1}{6}$ **44.** $\frac{4}{5} \div \frac{8}{25}$ **45.** $\frac{7}{9} \div \frac{-7}{3}$ **46.** $\frac{2}{7} \div 4$

47. $5 \div \frac{1}{5}$ **48.** $\frac{15}{16} \div \frac{25}{4}$ **49.** $\frac{20}{7} \div \frac{10}{21}$ **50.** $\frac{121}{84} \div \frac{132}{49}$

Use this area for work space.

51. $\frac{-55}{64} \div \frac{-125}{128}$ **52.** $\frac{81}{96} \div \frac{243}{144}$

53. What rational number must be multiplied by $\frac{1}{3}$ to obtain $\frac{5}{6}$?

54. What rational number must be multiplied by $\frac{2}{3}$ to obtain $\frac{3}{4}$?

E. In Exercises 55–68 divide the given rational expressions. Leave the resulting rational expressions in factored form. (*See Examples 11 and 12.*)

55. $\frac{1}{a} \div \frac{1}{b}$ **56.** $\frac{a}{b} \div a$ **57.** $\frac{x}{a^2} \div \frac{x}{a}$ **58.** $\frac{a}{b} \div \frac{b}{a}$ **59.** $\frac{ab}{c} \div \frac{a^2c}{b}$

60. $\frac{abc^3}{d} \div \frac{ab^3}{c^2d}$ **61.** $\frac{x-a}{2} \div \frac{(x-a)^2}{4}$ **62.** $\frac{a+1}{a-1} \div \frac{(a+1)^2}{(a-1)^2}$

63. $\frac{3x-3}{2} \div \frac{x-1}{4}$ **64.** $\frac{5a+5b}{a-b} \div \frac{a^2-b^2}{2}$

65. $\frac{x^2+3x+2}{ab^2x} \div \frac{ax+a}{bx}$ **66.** $\frac{x^2+7x+12}{x^2-16} \div \frac{x^2+6x+9}{2x-8}$

67. $\frac{a^2x^2-a^2y^2}{ax+ay} \div \frac{ax-ay}{a^2}$ **68.** $\frac{x^2-9x+14}{x^2-49} \div \frac{x^2-4}{x+7}$

69. Divide the product of $\frac{ax}{2}$ and $\frac{a^2y}{5}$ by $\frac{axy}{20}$.

70. Divide $\frac{x+y}{3ab}$ by the product of $\frac{x^2-y^2}{9a^2}$ and $\frac{x^2}{6b}$.

71. What rational expression must be multiplied by $\frac{1}{x}$ to obtain $\frac{3}{x^2}$?

72. What rational expression must be multiplied by $\frac{x}{x-a}$ to obtain $\frac{x^2y}{x^2-a^2}$?

73. Pick out the *error(s).*

(a) $\frac{a}{b} \div c = \frac{a}{bc}$ (b) $\frac{x+a}{x-a} \div \frac{x+a}{x-a} = 1$

(c) $\frac{x+1}{x} \div \frac{x+2}{x^2} = \frac{(x+1)(x+2)}{x^3}$ (d) $\frac{y+4}{y-2} \div \frac{y-2}{y+4} = \frac{(y+4)^2}{(y-2)^2}$

(e) $\frac{1}{x^2} \div \frac{1}{x^3} = \frac{1}{x}$

5.6 Adding and Subtracting: Same Denominator

Figure 5.9 illustrates how rational numbers (or rational expressions) with the same denominator are added.

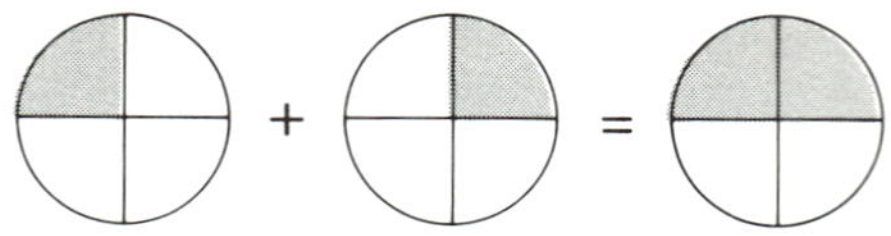

FIGURE 5.9. $\frac{1}{4} + \frac{1}{4} = \frac{1+1}{4} = \frac{2}{4} = \frac{1}{2}$

To add or subtract rational numbers or rational expressions with the *same denominator D:*

1. Add or subtract the numerators.
2. The denominator is *D*.
3. Simplify the resulting expression, if possible.

A. Rational Numbers

Example 1 ► Add $\frac{1}{3} + \frac{1}{3}$

Solution:

$$\frac{1}{3} + \frac{1}{3} = \frac{1+1}{3} = \frac{2}{3}$$ ◄

Example 2 ► Add $\frac{1}{6} + \frac{1}{6}$

Solution:

$$\frac{1}{6} + \frac{1}{6} = \frac{1+1}{6} = \frac{2}{6}$$

Simplify the sum $\frac{2}{6}$ to $\frac{1}{3}$ ◄

Example 3 ► Subtract $\frac{3}{8} - \frac{1}{8}$

Solution:

$$\frac{3}{8} - \frac{1}{8} = \frac{3-1}{8} = \frac{2}{8} = \frac{1}{4}$$ ◄

Example 4 ► Combine $\frac{1}{10} + \frac{7}{10} - \frac{3}{10}$

Solution:

$$\frac{1}{10} + \frac{7}{10} - \frac{3}{10} = \frac{1+7-3}{10} = \frac{5}{10} = \frac{1}{2}$$ ◄

Answers
1. (a)
(b)
(c)

PRACTICE EXERCISES FOR TOPIC A

1. Add or subtract as indicated. Simplify the result.

(a) $\frac{7}{5} + \frac{3}{5}$ (b) $\frac{5}{9} - \frac{2}{9}$ (c) $\frac{11}{12} + \frac{5}{12} - \frac{7}{12}$

Now try the exercises for Topic A on page 220.

B. Rational Expressions

Rational expressions are added in the same way as rational numbers are.

Example 5 ▶ Add $\frac{1}{x} + \frac{2}{x}$

Solution:

$$\frac{1}{x} + \frac{2}{x} = \frac{1 + 2}{x} = \frac{3}{x}$$ ◀

Example 6 ▶ Subtract $\frac{x}{yz} - \frac{1}{yz}$

Solution:

$$\frac{x}{yz} - \frac{1}{yz} = \frac{x - 1}{yz}$$ ◀

Factoring the numerator or denominator of the sum or difference may simplify the result.

Example 7 ▶ Add $\frac{2x}{x + 1} + \frac{2}{x + 1}$

Solution:

$$\frac{2x}{x + 1} + \frac{2}{x + 1} = \frac{2x + 2}{x + 1}$$

$$= \frac{2\overset{1}{\cancel{(x - 1)}}}{\underset{1}{\cancel{x + 1}}}$$

$$= 2$$ ◀

Example 8 ▶ Subtract $\frac{x^2}{x + 1} - \frac{1}{x + 1}$

Solution:

$$\frac{x^2}{x + 1} - \frac{1}{x + 1} = \frac{x^2 - 1}{x + 1}$$

$$= \frac{\overset{1}{\cancel{(x + 1)}}(x - 1)}{\underset{1}{\cancel{x + 1}}}$$

$$= x - 1$$ ◀

Answers

2. (a) ____ (b) ____

(c) ____

(d) ____

PRACTICE EXERCISES FOR TOPIC B

2. Add or subtract as indicated. Simplify the result.

(a) $\dfrac{7}{2a} + \dfrac{3}{2a}$ (b) $\dfrac{3x}{x-3} - \dfrac{3}{x-3}$ (c) $\dfrac{x^2}{x+4} + \dfrac{3x}{x+4} - \dfrac{4}{x+4}$

(d) $\dfrac{5x-2}{5x} + \dfrac{7}{5x}$

Now try the exercises for Topic B on page 221.

C. Rational Coefficients

Example 9 ▶ Add $\frac{2}{5}x + \frac{1}{4}y + \frac{1}{5}x + \frac{3}{4}y$

Solution:

Group together like terms. Then add the like terms by first adding the coefficients.

$$\frac{2}{5}x + \frac{1}{4}y + \frac{1}{5}x + \frac{3}{4}y = \left(\frac{2}{5}x + \frac{1}{5}x\right) + \left(\frac{1}{4}y + \frac{3}{4}y\right)$$

$$= \frac{3}{5}x + \frac{4}{4}y$$

$$= \frac{3}{5}x + y$$ ◀

Answers

3. ____

PRACTICE EXERCISES FOR TOPIC C

3. Combine like terms. $\dfrac{2x}{3} + \dfrac{y}{8} + \dfrac{4x}{3} - \dfrac{5y}{8}$

Now try the exercises for Topic C on page 221.

Use this area for work space.

EXERCISES

A. In Exercises 1–20 add or subtract and simplify. (*See Examples 1–4.*)

1. $\frac{1}{5} + \frac{1}{5}$ **2.** $\frac{2}{7} + \frac{1}{7}$ **3.** $\frac{2}{3} + \frac{1}{3}$ **4.** $\frac{2}{3} - \frac{1}{3}$

5. $\frac{3}{4} + \frac{1}{4}$ **6.** $\frac{3}{5} - \frac{1}{5}$ **7.** $\frac{5}{8} - \frac{1}{8}$ **8.** $\frac{9}{10} - \frac{1}{10}$

9. $\frac{1}{4} + \frac{1}{4} + \frac{1}{4}$ **10.** $\frac{1}{6} + \frac{1}{6} + \frac{3}{6}$ **11.** $\frac{1}{5} + \frac{2}{5} + \frac{2}{5}$

12. $\frac{1}{12} + \frac{5}{12} + \frac{7}{12}$ **13.** $\frac{7}{9} - \frac{4}{9} + \frac{5}{9}$ **14.** $\frac{7}{9} - \left(\frac{4}{9} + \frac{5}{9}\right)$

Use this area for work space.

15. $\frac{-3}{4} + \frac{1}{4} + \frac{5}{4}$ **16.** $\frac{7}{3} - \frac{1}{3} + \frac{2}{3}$

17. $\frac{1}{12} + \frac{1}{12} + \frac{5}{12} + \frac{11}{12}$ **18.** $\frac{7}{10} - \left(\frac{3}{10} + \frac{1}{10} - \frac{9}{10}\right)$

19. $\left(\frac{7}{6} - \frac{5}{6}\right) - \left(\frac{1}{6} + \frac{1}{6}\right)$ **20.** $\frac{11}{20} - \left(\frac{3}{20} - \frac{1}{20} + \frac{7}{20}\right)$

21. A stock goes up $\frac{1}{10}$ of a point on Monday and $\frac{3}{10}$ of a point on Tuesday. It then falls $\frac{7}{10}$ of a point on Wednesday. What is its total gain or loss for these three days?

22. Sam eats $\frac{1}{8}$ of a pie and Al eats $\frac{3}{8}$ of the pie. How much of the pie remains?

B. In Exercises 23–49 add or subtract and simplify. (*See Examples 5–8.*)

23. $\frac{1}{x} + \frac{1}{x}$ **24.** $\frac{3}{y^2} + \frac{1}{y^2}$ **25.** $\frac{4}{y} - \frac{3}{y}$ **26.** $\frac{a}{xy} + \frac{b}{xy}$

27. $\frac{10}{x + 1} + \frac{1}{x + 1}$ **28.** $\frac{7}{x + y} - \frac{5}{x + y}$ **29.** $\frac{6a}{x + 2} + \frac{2a}{x + 2}$

30. $\frac{5x}{x + 3} - \frac{2x}{x + 3}$ **31.** $\frac{6}{a} + \frac{1}{a} + \frac{2}{a}$ **32.** $\frac{5}{x^2} + \frac{3}{x^2} - \frac{1}{x^2}$

33. $\frac{9}{x} + \frac{1}{x} - \frac{4}{x}$ **34.** $\frac{3}{b + 1} + \frac{5}{b + 1} + \frac{2}{b + 1}$ **35.** $\frac{x}{x + 2} + \frac{2}{x + 2}$

36. $\frac{a}{a - 4} - \frac{4}{a - 4}$ **37.** $\frac{3a}{a + 1} + \frac{3}{a + 1}$ **38.** $\frac{6x}{x - 1} - \frac{6}{x - 1}$

39. $\frac{1}{x - 1} - \frac{x}{x - 1}$ **40.** $\frac{a^2}{x^2 - a^2} - \frac{x^2}{x^2 - a^2}$ **41.** $\frac{x}{x^2 - 4} - \frac{2}{x^2 - 4}$

42. $\frac{a}{a^2 - 9} - \frac{3}{a^2 - 9}$ **43.** $\frac{b^2}{b + 1} + \frac{1}{b + 1}$ **44.** $\frac{4m^2}{m + 2} - \frac{16}{m + 2}$

45. $\frac{y^2}{y + 1} + \frac{2y}{y + 1} + \frac{1}{y + 1}$ **46.** $\frac{a^2}{a + 2} + \frac{5a}{a + 2} + \frac{6}{a + 2}$

47. $\frac{3}{2x} + \frac{5 - 4x}{2x}$ **48.** $\frac{5a - 1}{a^2} - \frac{2a - 1}{a^2}$

49. $\frac{2x + 3}{x + 1} + \frac{4x + 3}{x + 1}$

50. Pick out the *error(s).*

(a) $\frac{a + b}{c} = \frac{a}{c} + \frac{b}{c}$ (b) $\frac{a}{b + c} = \frac{a}{b} + \frac{a}{c}$ (c) $\frac{x + 3}{x + 5} = 1 + \frac{3}{5}$

(d) $\frac{y + 2}{y + 4} = 1\frac{1}{2}$ (e) $\frac{z + 2}{2z + 4} = \frac{z}{2z} + \frac{2}{4}$

C. In Exercises 51–60 add or subtract and simplify. (*See Example 9.*)

51. $\frac{1}{3}x + \frac{1}{3}x$ **52.** $\frac{3}{8}a + \frac{1}{8}a$

Use this area for work space.

53. $\frac{3}{10}y + \frac{7}{10}y - \frac{1}{10}y$

54. $\frac{1}{6}x^2 - \frac{5}{6}x^2$

55. $\frac{a}{4} + \frac{3a}{4}$

56. $\frac{ab}{12} - \frac{5ab}{12} + \frac{7ab}{12}$

57. $\frac{1}{9}x + \frac{4}{7}y + \frac{2}{9}x - \frac{3}{7}y$

58. $\frac{1}{20}a + \frac{3}{10}b - \frac{7}{20}a + \frac{9}{10}b$

59. $\frac{x}{6} - \frac{y}{5} + \frac{5x}{6} - \frac{4y}{5}$

60. $\frac{x^2}{100} + \frac{3x}{10} + \frac{2}{5} - \frac{x^2}{100} + \frac{7x}{10} - \frac{4}{5}$

5.7 Least Common Multiples

The key to adding rational numbers or rational expressions with *different* denominators is the notion of the *least common multiple.*

Recall that ab is a *multiple* of both a and b. In order to add or subtract rational numbers or rational expressions with *different denominators,* first convert to equivalent expressions with the *same denominator*. The new denominator will then be a multiple of each of the given denominators. Each denominator is an integer or polynomial. In this section you will learn how to find the smallest multiple of several integers or polynomials.

A. *lcm* of Several Integers

DEFINITION

> The **least common multiple (*lcm*) of two or more integers** is the *smallest* positive integer that is a multiple of each of them.

Write

lcm (2, 5) for the least common multiple of 2 and 5

Example 1 ▶ Find *lcm* (2, 5)

Solution:

$$lcm\ (2, 5) = 10$$

In fact, $10 = 2 \cdot 5$

Thus 10 is a multiple of both 2 and 5. None of the integers

$$1, 2, 3, 4, 5, 6, 7, 8, 9$$

is a multiple of *both* 2 and 5. For example, 4 is a multiple of 2, but not of 5. ◀

Example 2 ▶ Find *lcm* (2, 6)
Solution:
Observe that 6 is a multiple of 2:

$$6 = 3 \cdot 2$$

And 6 is a multiple of 6:

$$6 = 1 \cdot 6$$

By considering the numbers

$$1, 2, 3, 4, 5$$

check that 6 is the *smallest* positive multiple of *both* 2 and 6. Thus

$$lcm\ (2, 6) = 6$$ ◀

Sometimes, the least common multiple can be found almost immediately, as in the above examples. At other times, it is best to consider the prime factors of the integers.

To find the *lcm* of two or more integers:

1. Express them in terms of prime factors.
2. The *lcm* is the product of the *highest* powers of all primes that occur.

Example 3 ▶ Find *lcm* (8, 12)
Solution:
Although this example, too, can possibly be done at sight, it illustrates the prime factoring method.

$$8 = 2^3, \qquad 12 = 2^2 \cdot 3$$

The only primes that occur are 2 and 3. The *highest* power of 2 that occurs is 2^3. The only power of 3 that occurs is 3 itself. Thus

$$lcm\ (8, 12) = 2^3 \cdot 3 = 8 \cdot 3 = 24$$ ◀

Note that the *lcm* is always a *positive* integer. Thus, from Example 3,

$$lcm\ (-8, -12) = 24$$

Example 4 ▶ Find *lcm* (40, 50, 64)
Solution:

$$40 = 2^3 \cdot 5$$

$$50 = 2 \cdot 5^2$$

$$64 = 2^6$$

The highest power of 2 that occurs is 2^6. The highest power of 5 that occurs is 5^2 Thus

$$lcm\ (40, 50, 64) = 2^6 \cdot 5^2$$

$$= 64 \cdot 25$$

$$= 1600$$ ◀

Answers

1. ____________

2. ____________

PRACTICE EXERCISES FOR TOPIC A

1. Find *lcm*(8, 18) **2.** Find *lcm*(14, 54, 63)

Now try the exercises for Topic A on page 225.

B. *lcm* of Several Polynomials

DEFINITION

> The **least common multiple of two or more polynomials** is the product of the *highest* powers of all factors that occur when you factor each of these polynomials.

Example 5 ▶ Find *lcm* $(2x^2y, 4xy^3)$
Solution:
Write

$$4xy^3 = 2^2xy^3$$

The highest power of 2 that occurs is 2^2, or 4. The highest power of x that occurs is x^2. The highest power of y that occurs is y^3. Thus

$$lcm\ (2x^2y, 4xy^3) = 4x^2y^3$$ ◀

Example 6 ▶ Find *lcm* $(x + y, x - y)$
Solution:
$x + y$ and $x - y$ are each in factored form. Thus

$$lcm\ (x + y, x - y) = (x + y)(x - y), \qquad \text{or } x^2 - y^2$$ ◀

Example 7 ▶ Find *lcm* $(x^2 - 1, x^2 + 2x + 1)$
Solution:
Factor each polynomial:

$$x^2 - 1 = (x + 1)(x - 1)$$

$$x^2 + 2x + 1 = (x + 1)^2$$

$$lcm\ (x^2 - 1, x^2 + 2x + 1) = (x + 1)^2(x - 1)$$ ◀

Example 8 ▶ Find *lcm* $(9x + 27, x^2 + 2x - 3, 3x^2 - 6x + 3)$
Solution:
Factor each polynomial:

$$9x + 27 = 9(x + 3) = 3^2(x + 3)$$

$$x^2 + 2x - 3 = (x + 3)(x - 1)$$

$$3x^2 - 6x + 3 = 3(x^2 - 2x + 1) = 3(x - 1)^2$$

$$lcm\ (9x + 27, x^2 + 2x - 3, 3x^2 - 6x + 3) = 3^2(x + 3)(x - 1)^2$$

$$= 9(x + 3)(x - 1)^2$$ ◀

Answers

3. (a) ______

(b) ______

(c) ______

(d) ______

PRACTICE EXERCISES FOR TOPIC B

3. Find the *lcm* of the given polynomials.

(a) $3x, 2y$ (b) $12a^2b, 9ab^3$ (c) $2x + 10, x^2 + 5x$ (d) $6x, 3x + 6, x^2 - 4$

Now try the exercises for Topic B on this page.

Use this area for work space.

EXERCISES

A. In Exercises 1–22 find the *lcm* of the given integers. (*See Examples 1–4.*)

1. 2, 3 **2.** 3, 5 **3.** 2, 4 **4.** 2, 8

5. −3, 6 **6.** −3, −12 **7.** 5, 25 **8.** 10, 50

9. 4, 6 **10.** 6, 9 **11.** 10, 15 **12.** 4, 10

13. 12, 30 **14.** 6, 8 **15.** 12, 27 **16.** 50, 125

17. 2, 3, 5 **18.** 4, 8, 16 **19.** 2, 3, 4 **20.** 6, 9, 16

21. 50, 75, 100 **22.** 48, 64, 72

B. In Exercises 23–50 find the *lcm* of the given polynomials. (*See Examples 5–8.*)

23. a, x **24.** x, y **25.** $5, x$ **26.** x^2, y^2

27. x, x^2 **28.** a, a^3 **29.** $x + 1, (x + 1)^2$ **30.** $a, a + b$

31. a^2x, ax^2 **32.** xy^2, xy^3 **33.** $5xyz, x^2z$ **34.** $4a^2b, 8abc$

35. $4axy, 6a^2bx$ **36.** $12mn^2, 8m^2n$

37. $x + a, x - a$ **38.** $(x + a)^2, (x + a)(x - a)$

39. $x^2 - 4, x + 2$ **40.** $y^2 - 9, (y - 3)^2$

41. $a^2 - 25, a^2 + 10a + 25$ **42.** $b^2 + 3b + 2, b^2 + 5b + 6$

43. $4x - 4, 8x^2 - 16x + 8$ **44.** $z^2 - 100, 7z - 70$

45. $abx - aby, a^2b^2$ **46.** $t^2 + 6t + 8, (t + 2)^3$

47. $abx^2y^3, a^2b^7xy, ax^4yz$ **48.** $(x - 1)^7(x + 1), 4x + 4, (x - 1)^9$

49. $a^2 - 4, 5a - 10, 25a + 50$ **50.** $a^2x^2 - a^2y^2, 4x + 4y, 6x - 6y$

5.8 Adding and Subtracting: Different Denominators

A. Least Common Denominators

In order to add or subtract rational numbers or rational expressions with *different denominators*, first find equivalent expressions with the same denominator. This denominator should be the *lcm* of the original denominators.

DEFINITION

> The **least common denominator (*lcd*) of several rational numbers or rational expressions** is the *lcm* of the individual denominators.

Write

$$lcd\left(\frac{1}{2}, \frac{2}{3}\right) \quad \text{for the least common denominator of } \frac{1}{2} \text{ and } \frac{2}{3}$$

Recall that when the numerator and denominator of a rational number are each multiplied by the same number, an equivalent rational number is obtained.

Example 1 ▶ (a) Find $lcd\left(\frac{1}{2}, \frac{2}{3}\right)$

(b) Write equivalent rational numbers with this *lcd* as denominator.

Solution:

(a) To find the *lcd*, determine the *lcm* of the denominators.

$$lcm\ (2, 3) = 6$$

Therefore

$$lcd\left(\frac{1}{2}, \frac{2}{3}\right) = 6$$

(b) To write $\frac{1}{2}$ with denominator 6, multiply the given denominator 2 by 3. To obtain an *equivalent* rational number, also multiply the numerator by 3. Thus

$$\frac{1}{2} = \frac{1 \cdot 3}{2 \cdot 3} = \frac{3}{6}$$

Similarly, to write $\frac{2}{3}$ with denominator 6, multiply the given denominator 3 by 2. To preserve equivalence, multiply the numerator by 2. Thus

$$\frac{2}{3} = \frac{2 \cdot 2}{3 \cdot 2} = \frac{4}{6}$$ ◀

Example 2 ▶ (a) Find $lcd\left(\frac{3}{40}, \frac{-7}{50}\right)$

(b) Write equivalent rational numbers with this *lcd* as denominator.

Solution:

(a) $lcd\left(\frac{3}{40}, \frac{-7}{50}\right) = lcm\ (40, 50)$

$$40 = 2^3 \cdot 5$$

$$50 = 2 \cdot 5^2$$

$$lcm\ (40, 50) = 2^3 \cdot 5^2 = 8 \cdot 25 = 200$$

Therefore $$lcd\left(\frac{3}{40}, \frac{-7}{50}\right) = 200$$

(b) To write $\frac{3}{40}$ with denominator 200, note that $\frac{200}{40} = 5$. Multiply the numerator and denominator by 5:

$$\frac{3}{40} = \frac{3 \cdot 5}{40 \cdot 5} = \frac{15}{200}$$

To write $\frac{-7}{50}$ with denominator 200, note that $\frac{200}{50} = 4$, or 2^2. Multiply the numerator and denominator by 4:

$$\frac{-7}{50} = \frac{(-7) \cdot 4}{50 \cdot 4} = \frac{-28}{200}$$ ◀

Now consider the same problem for rational expressions.

Example 3 ▶ (a) Find $lcd\left(\frac{1}{x}, \frac{5a}{y^2}\right)$

(b) Write equivalent rational expressions with this *lcd* as denominator.

Solution:

(a) As in the case of rational numbers, find the *lcm* of the denominators x and y^2.

$$lcm\ (x, y^2) = xy^2$$

Therefore $$lcd\left(\frac{1}{x}, \frac{5a}{y^2}\right) = xy^2$$

(b) To find equivalent rational expressions with denominator xy^2, multiply the numerator and denominator of each expression as indicated.

$$\frac{1}{x} = \frac{1 \cdot y^2}{x \cdot y^2} = \frac{y^2}{xy^2}$$

$$\frac{5a}{y^2} = \frac{5a \cdot x}{y^2 \cdot x} = \frac{5ax}{xy^2}$$ ◀

Example 4 ▶ (a) Find $lcd\left(\frac{x}{x^2 - a^2}, \frac{a}{x^2 + 2ax + a^2}\right)$

(b) Write equivalent expressions with this *lcd* as denominator.

Solution:

(a) First factor the given denominators to find their *lcm*.

$$x^2 - a^2 = (x + a)(x - a)$$

$$x^2 + 2ax + a^2 = (x + a)^2$$

$$lcm\ (x^2 - a^2, x^2 + 2ax + a^2) = (x + a)^2(x - a)$$

Thus $$lcd\left(\frac{x}{x^2 - a^2}, \frac{a}{x^2 + 2ax + a^2}\right) = (x + a)^2(x - a)$$

(b) To obtain an expression equivalent to $\frac{x}{x^2 - a^2}$, but with denominator $(x + a)^2(x - a)$, note that the denominator $x^2 - a^2$, or $(x + a)(x - a)$, lacks a factor $x + a$. Thus, multiply the numerator and denominator of

$$\frac{x}{x^2 - a^2} \quad \text{by} \quad x + a$$

Thus

$$\frac{x}{x^2 - a^2} = \frac{x}{(x + a)(x - a)}$$

$$= \frac{x \cdot (x + a)}{(x + a)(x - a) \cdot (x + a)}$$

$$= \frac{x^2 + ax}{(x + a)^2(x - a)}$$

And the denominator $x^2 + 2ax + a^2$, or $(x + a)^2$, lacks the factor $x - a$. Thus multiply the numerator and denominator of

$$\frac{a}{x^2 + 2ax + a^2} \quad \text{by} \quad x - a$$

Therefore

$$\frac{a}{x^2 + 2ax + a^2} = \frac{a}{(x + a)^2}$$

$$= \frac{a \cdot (x - a)}{(x + a)^2 \cdot (x - a)}$$

$$= \frac{ax - a^2}{(x + a)^2(x - a)}$$ ◀

As you will see, *when adding rational expressions, you must multiply out the numerators before adding.*

Answers

1. (a) ____
(b) ____
2. (a) ____
(b) ____
____
3. (a) ____
(b) ____
____

PRACTICE EXERCISES FOR TOPIC A

For each of the following,

(a) find the *lcd* of the given expressions, and

(b) write equivalent expressions with this *lcd* as denominator.

1. $\frac{4}{9}$ and $\frac{7}{15}$ **2.** $\frac{3b}{5a^2}$ and $\frac{2}{15ab}$ **3.** $\frac{3}{10x^2}$ and $\frac{7}{2x^2 - 4x}$

Now try the exercises for Topic A on page 233.

B. Combining Rational Numbers

Figure 5.10 illustrates how rational numbers (or rational expressions) with *different denominators* are added.

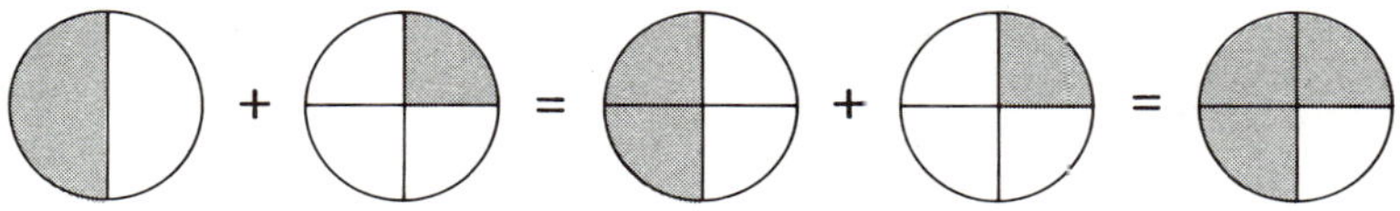

FIGURE 5.10. $\frac{1}{2} + \frac{1}{4} = \frac{2}{4} + \frac{1}{4} = \frac{3}{4}$

To add or subtract rational numbers or rational expressions with *different denominators:*

1. First find their *lcd*.
2. Find equivalent expressions with this *lcd* as denominator.
3. Add or subtract the numerators.
4. Simplify the resulting expression, if possible.

Steps (1) and (2) are what you found in Examples 1–4. Steps (3) and (4) are the rules for adding expressions with the same denominator (Section 5.6).

Example 5 ▶ Add $\frac{3}{4} + \frac{5}{8}$

Solution:

$$lcd\left(\frac{3}{4}, \frac{5}{8}\right) = lcm\ (4, 8) = 8$$

Find a rational number equivalent to $\frac{3}{4}$, but with denominator 8.

$$\frac{3}{4} = \frac{3 \cdot 2}{4 \cdot 2} = \frac{6}{8}$$

$\left(\frac{5}{8}\right.$ already has denominator 8.$\left.\right)$ Now add expressions with the *same* denominator.

$$\frac{3}{4} + \frac{5}{8} = \frac{6}{8} + \frac{5}{8} = \frac{6+5}{8} = \frac{11}{8}$$ ◀

Example 6 ▶ Subtract $\frac{9}{20} - \frac{7}{12}$

Solution:

$$20 = 2^2 \cdot 5$$

$$12 = 2^2 \cdot 3$$

$$lcd\left(\frac{9}{20}, \frac{7}{12}\right) = lcm\ (20, 12)$$

$$= 2^2 \cdot 5 \cdot 3$$

$$= 4 \cdot 15$$

$$= 60$$

Find equivalent rational numbers with denominator 60.

$$\frac{9}{20} = \frac{9 \cdot 3}{20 \cdot 3} = \frac{27}{60}$$

$$\frac{7}{12} = \frac{7 \cdot 5}{12 \cdot 5} = \frac{35}{60}$$

Now add rational numbers with the *same* denominator.

$$\frac{9}{20} - \frac{7}{12} = \frac{27}{60} - \frac{35}{60}$$
$$= \frac{27 - 35}{60}$$
$$= \frac{-8}{60}$$
$$= \frac{-2}{15}$$ ◀

Note that when adding or subtracting rational numbers, you do *not* work with the simplest form. Thus, in Example 6, you expressed $\frac{9}{20}$ as $\frac{27}{60}$ and $\frac{7}{12}$ as $\frac{35}{60}$. You did this in order to obtain rational numbers with the *same* denominator.

Example 7 ▶ Find the value of $\frac{5}{24} + \frac{1}{18} - \frac{7}{36}$

Solution:

$$24 = 2^3 \cdot 3$$
$$18 = 2 \cdot 3^2$$
$$36 = 2^2 \cdot 3^2$$

$$lcd\left(\frac{5}{24}, \frac{1}{18}, \frac{7}{36}\right) = lcm\ (24, 18, 36)$$
$$= 2^3 \cdot 3^2$$
$$= 8 \cdot 9$$
$$= 72$$

$$\frac{5}{24} = \frac{5 \cdot 3}{24 \cdot 3} = \frac{15}{72}$$
$$\frac{1}{18} = \frac{1 \cdot 4}{18 \cdot 4} = \frac{4}{72}$$
$$\frac{7}{36} = \frac{7 \cdot 2}{36 \cdot 2} = \frac{14}{72}$$

$$\frac{5}{24} + \frac{1}{18} - \frac{7}{36} = \frac{15}{72} + \frac{4}{72} - \frac{14}{72}$$
$$= \frac{15 + 4 - 14}{72}$$
$$= \frac{5}{72}$$ ◀

Answers
4. (a)
(b)

PRACTICE EXERCISES FOR TOPIC B

4. Add or subtract, as indicated:

(a) $\frac{7}{10} + \frac{1}{6}$ (b) $\frac{8}{9} + \frac{7}{12} - \frac{3}{4}$

Now try the exercises for Topic B on page 234.

C. Mixed Numbers

Example 8 ▶ Express (a) $3\frac{1}{4}$ and (b) $-3\frac{1}{4}$ as rational numbers.

Solution:

(a)
$$\begin{aligned} 3\frac{1}{4} &= 3 + \frac{1}{4} \\ &= \frac{3 \cdot 4}{1 \cdot 4} + \frac{1}{4} \\ &= \frac{12}{4} + \frac{1}{4} \\ &= \frac{13}{4} \end{aligned}$$

(b) Note that $-3\frac{1}{4} = -\left(3 + \frac{1}{4}\right) = -3 - \frac{1}{4} = \frac{-12}{4} - \frac{1}{4} = \frac{-13}{4}$. ◀

Answers
5. (a)
(b)

PRACTICE EXERCISES FOR TOPIC C

5. Express the following mixed numbers as rational numbers.

(a) $4\frac{2}{5}$ (b) $-2\frac{7}{8}$

Now try the exercises for Topic C on page 234.

D. Combining Rational Expressions

The method for adding rational numbers carries over to the addition of rational expressions.

Example 9 ▶ Add $\frac{1}{ax^2} + \frac{2}{a^2x}$

Solution:

$$\text{lcd}\left(\frac{1}{ax^2}, \frac{2}{a^2x}\right) = \text{lcm}\,(ax^2, a^2x) = a^2x^2$$

Find equivalent expressions with denominator a^2x^2.

$$\frac{1}{ax^2} = \frac{1 \cdot a}{ax^2 \cdot a} = \frac{a}{a^2x^2}$$

$$\frac{2}{a^2x} = \frac{2 \cdot x}{a^2x \cdot x} = \frac{2x}{a^2x^2}$$

Now add expressions with the *same* denominator.

$$\frac{1}{ax^2} + \frac{2}{a^2x} = \frac{a}{a^2x^2} + \frac{2x}{a^2x^2} = \frac{a + 2x}{a^2x^2}$$ ◀

Example 10 ▶ Subtract $\frac{2}{x - 4} - \frac{1}{x + 4}$

Solution:

$$lcd\left(\frac{2}{x - 4}, \frac{1}{x + 4}\right) = lcm\ (x - 4, x + 4)$$

$$= (x - 4)(x + 4), \qquad \text{or } x^2 - 16$$

Thus

$$\frac{2}{x - 4} = \frac{2 \cdot (x + 4)}{(x - 4) \cdot (x + 4)} = \frac{2x + 8}{x^2 - 16}$$

$$\frac{1}{x + 4} = \frac{1 \cdot (x - 4)}{(x + 4) \cdot (x - 4)} = \frac{x - 4}{x^2 - 16}$$

And

$$\frac{2}{x - 4} - \frac{1}{x + 4} = \frac{2x + 8}{x^2 - 16} - \frac{x - 4}{x^2 - 16}$$

Note that $-(x - 4) = -x + 4$.

$$= \frac{2x + 8 - x + 4}{x^2 - 16}$$

$$= \frac{x + 12}{x^2 - 16}$$ ◀

Example 11 ▶ Add $\frac{1}{a^2 - ax} + \frac{1}{x^2 - ax}$

Solution:

$$a^2 - ax = a(a - x) = -a(x - a)$$

$$x^2 - ax = x(x - a)$$

The *lcd* is $ax(x - a)$. Thus $-a(x - a)$ must be multiplied by $-x$ in order to obtain $ax(x - a)$.

$$\frac{1}{a^2 - ax} + \frac{1}{x^2 - ax} = \frac{1 \cdot (-x)}{-a(x - a) \cdot (-x)} + \frac{1 \cdot a}{x(x - a) \cdot a}$$

$$= \frac{-x + a}{ax(x - a)}$$

$$= \frac{-\overset{1}{\cancel{(x - a)}}}{ax\underset{1}{\cancel{(x - a)}}}$$

$$= \frac{-1}{ax}$$ ◀

Example 12 ▶ Add $\frac{y+1}{y+2} + \frac{2y}{y-2}$

Solution:

The *lcd* is $(y+2)(y-2)$. Thus

$$\frac{y+1}{y+2} + \frac{2y}{y-2} = \frac{(y+1)\cdot(y-2)}{(y+2)\cdot(y-2)} + \frac{2y\cdot(y+2)}{(y-2)\cdot(y+2)}$$

$$= \frac{(y+1)(y-2) + 2y(y+2)}{(y+2)(y-2)}$$

$$= \frac{y^2 - y - 2 + 2y^2 + 4y}{(y+2)(y-2)}$$

$$= \frac{3y^2 + 3y - 2}{(y+2)(y-2)}$$ ◀

Answers

6. (a) ______

(b) ______

(c) ______

PRACTICE EXERCISES FOR TOPIC D

6. Add or subtract, as indicated.

(a) $\frac{5y}{4x^2} + \frac{3}{10xy}$ (b) $\frac{7}{x} + \frac{x}{x-1}$ (c) $\frac{10}{a^2+a} - \frac{2a}{a^2-1}$

Now try the exercises for Topic D on page 234.

Use this area for work space.

EXERCISES

A. In Exercises 1–18: (a) Find the *lcd,* and (b) write equivalent expressions with this *lcd* as denominator. (*See Examples 1–4.*)

1. $\frac{1}{3}$ and 1

2. $\frac{2}{3}$ and $\frac{-1}{5}$

3. $\frac{1}{2}$ and $\frac{5}{6}$

4. $\frac{-3}{4}$ and $\frac{-1}{12}$

5. $\frac{5}{4}$ and $\frac{7}{10}$

6. $\frac{3}{8}$ and $\frac{5}{12}$

7. $\frac{7}{20}$ and $\frac{3}{28}$

8. $\frac{5}{36}$ and $\frac{1}{27}$

9. $\frac{1}{8}$, $\frac{-1}{12}$, and $\frac{2}{15}$

10. $\frac{1}{x}$ and $\frac{1}{y}$

11. $\frac{1}{a^2}$ and $\frac{a}{b}$

12. $\frac{a}{x^2y}$ and $\frac{b}{xy^2z}$

13. $\frac{x-a}{x^2y^3z^2}$ and $\frac{a}{xy^4z}$

14. $\frac{2}{x(x-a)}$ and $\frac{4}{x(x+a)}$

15. $\frac{1}{x^2-1}$ and $\frac{x}{(x-1)^2}$

16. $\frac{3}{x^2+6x+9}$ and $\frac{-1}{x^2+4x+3}$

17. $\frac{1}{ax-2a}$, $\frac{a}{x^2-4}$, $\frac{-x}{ax+2a}$

18. $\frac{a}{x^2-25}$, $\frac{b}{(x+5)^2}$, $\frac{c}{ax+5a}$

19. A 1-yard-long pole is immersed in a pond so that $\frac{1}{4}$ of the pole is in bottom sand and $\frac{2}{9}$ is in water. What is the length of pole above water?

Use this area for work space.

20. Bob bikes for $\frac{2}{3}$ of a mile and jogs for $\frac{5}{6}$ of a mile during a 2-mile trip. If he walks the rest of the way, how far does he walk?

B. In Exercises 21–36 add or subtract, as indicated. (*See Examples 5–7.*)

21. $\frac{1}{2} + \frac{1}{3}$ **22.** $\frac{1}{4} - \frac{1}{5}$ **23.** $\frac{1}{2} + \frac{1}{4}$

24. $\frac{1}{8} - \frac{1}{4}$ **25.** $\frac{1}{2} + \frac{5}{6}$ **26.** $\frac{3}{4} - \frac{1}{12}$

27. $\frac{2}{9} + 1$ **28.** $\frac{7}{8} + \frac{1}{12}$ **29.** $\frac{7}{10} + \frac{1}{25}$

30. $\frac{5}{24} + \frac{7}{30}$ **31.** $\frac{3}{64} - \frac{1}{48}$ **32.** $\left(\frac{1}{96} + \frac{5}{9}\right)$

33. $\frac{4}{5}\left(\frac{13}{40} - \frac{3}{100}\right)$ **34.** $\frac{5}{8} - \frac{1}{4} + \frac{1}{16}$ **35.** $\frac{1}{10} + \frac{7}{5} - \frac{2}{25}$

36. $\frac{10}{21} - \left(\frac{3}{7} + \frac{2}{9}\right)$

C. In Exercises 37–40 express each mixed number as a rational number. (*See Example 8.*)

37. $1\frac{1}{2}$ **38.** $-2\frac{1}{5}$ **39.** $6\frac{5}{8}$ **40.** $99\frac{9}{10}$

D. In Exercises 41–68 add or subtract, as indicated. (*See Examples 9–12.*)

41. $\frac{1}{x} + \frac{1}{y}$ **42.** $2 - \frac{1}{b}$ **43.** $\frac{a}{x} + \frac{a}{x^2}$

44. $\frac{x}{2} + \frac{y}{3}$ **45.** $\frac{2}{xyz} + xyz$ **46.** $\frac{a}{s^2t} - \frac{b}{st^3}$

47. $\frac{1}{x + a} + \frac{2}{(x + a)^2}$ **48.** $\frac{1}{at + a} - \frac{a}{t + 1}$

49. $\frac{3}{x + 5} + \frac{1}{x^2 + 10x + 25}$ **50.** $\frac{x}{x^2 - 4} - \frac{1}{x^2 - 2x}$

51. $\frac{a}{x^2 + 5x + 6} - \frac{a}{x^2 - 4}$ **52.** $\frac{1}{a^2 - 9a} + \frac{9}{a^2 - 81}$

53. $\frac{1}{x - 1} + \frac{1}{x - 2} + \frac{1}{x - 3}$

54. $\frac{5}{a^2 - 25} - \left(\frac{1}{a^2 + 10a + 25} - \frac{1}{a^2 - 10a + 25}\right)$

55. $\frac{1}{a^2}\left(\frac{x}{a^2 - 4a} + \frac{1}{a^2 - 16}\right)$ **56.** $\frac{1}{x^2 + 6x + 9} - \left(\frac{1}{x + 3} + \frac{1}{x^2 - 9}\right)$

57. $\frac{4x}{x - 1} + \frac{4}{1 - x}$ **58.** $\frac{a}{a - 3} - \frac{3}{3 - a}$

59. $\frac{1}{x + 1} + \frac{1}{x + 2} + \frac{1}{x^2 + 3x + 2}$ **60.** $\frac{x}{x + 2} + \frac{6}{x - 2} - \frac{8}{x^2 - 4}$

Use this area for work space.

61. $\frac{a+1}{a^2} - \frac{1}{a+1} + \frac{1}{a^3+a^2}$

62. $\frac{1}{a+4} - \frac{1}{a-5} + \frac{9}{a^2-a-20}$

63. $\frac{x+1}{x+4} + \frac{x}{x+2}$

64. $\frac{y-2}{y+3} - \frac{y+1}{y-3}$

65. $\frac{a+4}{a-1} + \frac{a+1}{a^2}$

66. $\frac{b-1}{b+1} + \frac{b-1}{b} - \frac{b+2}{b^2}$

67. $\frac{2}{3}x + \frac{4}{5}x$

68. $\frac{1}{2}a + \frac{3}{5}b - \frac{3}{4}a + \frac{9}{10}b$

69. Pick out the *error(s)*.

(a) $\frac{1}{x} + \frac{1}{a} = \frac{1}{x+a}$

(b) $\frac{1}{a} + \frac{1}{a^2} = \frac{1}{a^3}$

(c) $\frac{1}{a-1} + \frac{1}{1-a} = 0$

(d) $\frac{3}{a} + \frac{5}{2a} = \frac{11}{2a}$

(e) $\frac{x}{y} - \frac{y}{x} = \frac{x-y}{xy}$

5.9 Complex Expressions

A. What Are Complex Expressions?

DEFINITION

> A **complex expression** is a rational number (or rational expression) that contains other rational numbers (or rational expressions) in its numerator or denominator. There may be rational numbers (or rational expressions) in both the numerator and denominator.

$$\frac{\frac{1}{4}}{3}, \quad \frac{-2}{\frac{2}{3}+1}, \quad \text{and} \quad \frac{\frac{x}{x+1}}{\frac{x-2}{x+2}}$$

are complex expressions. You can simplify complex expressions by first expressing them in terms of division.

For example, the numerator of $\frac{\frac{1}{4}}{3}$ is $\frac{1}{4}$; the denominator is 3. Thus

$$\frac{\frac{1}{4}}{3} = \frac{1}{4} \div 3$$

Similarly, the numerator of $\dfrac{-2}{\frac{2}{3}+1}$ is -2; the denominator is $\frac{2}{3}+1$. Thus

$$\frac{-2}{\frac{2}{3}+1} = (-2) \div \left(\frac{2}{3}+1\right)$$

And the numerator of $\dfrac{\frac{x}{x+1}}{\frac{x-2}{x+2}}$ is $\dfrac{x}{x+1}$; the denominator is $\dfrac{x-2}{x+2}$. Therefore

$$\frac{\frac{x}{x+1}}{\frac{x-2}{x+2}} = \frac{x}{x+1} \div \frac{x-2}{x+2}$$

Answers

1. (a) ______ (b) ______

2. (a) ______

(b) ______

3. (a) ______

(b) ______

PRACTICE EXERCISES FOR TOPIC A

In Practice Exercises 1 and 2, (a) what is the numerator and (b) what is the denominator of each complex expression?

1. $\dfrac{-a}{a+\frac{1}{2}}$ **2.** $\dfrac{\frac{a+4}{a^2-1}}{\frac{a}{a^2-1}}$

3. Express each of the following complex expressions in terms of division.

(a) $\dfrac{\frac{x}{2}}{5}$ (b) $\dfrac{\frac{a}{a-2}}{\frac{a^2}{a+2}}$

Now try the exercises for Topic A on page 239.

B. Rational Numbers

Bear in mind that complex expressions call for division.

Example 1 ▶ Simplify $\dfrac{\frac{1}{2}}{5}$

Solution:

$$\frac{\frac{1}{2}}{5} = \frac{1}{2} \div 5$$

$$= \frac{1}{2} \cdot \frac{1}{5}$$

$$= \frac{1}{10}$$ ◀

Example 2 ▶ Simplify $\dfrac{\frac{2}{3}}{\frac{4}{9}}$

Solution 1:

$$\frac{\frac{2}{3}}{\frac{4}{9}} = \frac{2}{3} \div \frac{4}{9}$$

$$= \frac{\overset{1}{\cancel{2}}}{\underset{1}{\cancel{3}}} \cdot \frac{\overset{3}{\cancel{9}}}{\underset{2}{\cancel{4}}}$$

$$= \frac{3}{2}$$

Solution 2:

The *lcd* of $\frac{2}{3}$ and $\frac{4}{9}$ is 9. Multiply the numerator and denominator of the complex expression by 9.

$$\frac{\frac{2}{3}}{\frac{4}{9}} = \frac{\frac{2}{\cancel{3}} \cdot \overset{3}{\cancel{9}}}{\frac{4}{\cancel{9}} \cdot \cancel{9}} = \frac{6}{4} = \frac{3}{2}$$ ◀

Example 3 ▶ Simplify $\dfrac{1 + \frac{1}{2}}{\frac{1}{4}}$

Solution:

$$\frac{1 + \frac{1}{2}}{\frac{1}{4}} = \left(1 + \frac{1}{2}\right) \div \frac{1}{4}$$

$$= \frac{2 + 1}{\underset{1}{\cancel{2}}} \cdot \frac{\overset{2}{\cancel{4}}}{1}$$

$$= 3 \cdot 2$$

$$= 6$$ ◀

PRACTICE EXERCISES FOR TOPIC B

Answers

4. (a) ______

(b) ______

4. Simplify: (a) $\dfrac{10}{\frac{4}{5}}$ (b) $\dfrac{4\frac{1}{5}}{\frac{7}{10}}$

Now try the exercises for Topic B on page 239.

C. Rational Expressions

You also simplify a complex rational expression by first rewriting it in terms of division.

Example 4 ▶ Simplify $\dfrac{\frac{1}{x}}{\frac{1}{y}}$

Solution:

$$\frac{\frac{1}{x}}{\frac{1}{y}} = \frac{1}{x} \div \frac{1}{y} = \frac{1}{x} \cdot \frac{y}{1} = \frac{y}{x}$$ ◀

Example 5 ▶ Simplify $\dfrac{\frac{1}{x-4}}{\frac{x}{x^2-16}}$

Solution:

$$\frac{\frac{1}{x-4}}{\frac{x}{x^2-16}} = \frac{1}{x-4} \div \frac{x}{x^2-16}$$

$$= \frac{1}{\cancel{x-4}_{\,1}} \cdot \frac{\overset{1}{\frac{(x+4)\cancel{(x-4)}}{\cancel{x^2-16}}}}{x}$$

First factor $x^2 - 16$.

$$= \frac{x+4}{x}$$ ◀

Example 6 ▶ Simplify $\dfrac{\frac{2}{a-2} - \frac{1}{a+2}}{\frac{a}{a^2-4}}$

Solution:

$$\frac{\frac{2}{a-2} - \frac{1}{a+2}}{\frac{a}{a^2-4}} = \left(\frac{2}{a-2} - \frac{1}{a+2}\right) \div \frac{a}{a^2-4}$$

$$= \frac{2(a+2) - 1(a-2)}{\cancel{(a-2)}\cancel{(a+2)}} \cdot \frac{\frac{\cancel{(a-2)}\cancel{(a+2)}}{\cancel{a^2-4}}}{a}$$

$$= \frac{2a + 4 - a + 2}{a}$$

$$= \frac{a+6}{a}$$ ◀

Answers

5. (a) ______

(b) ______

PRACTICE EXERCISES FOR TOPIC C

5. Simplify: (a) $\dfrac{\dfrac{x}{3}}{\dfrac{x+2}{6x}}$ (b) $\dfrac{a^2+8a}{\dfrac{2}{a}+\dfrac{1}{4}}$

Now try the exercises for Topic C on this page.

Use this area for work space.

EXERCISES

A. In Exercises 1–3 (a) what is the numerator and (b) what is the denominator of each complex expression? (*See Practice Exercises 1 and 2.*)

1. $\dfrac{\dfrac{3}{4}}{7}$ **2.** $\dfrac{5}{\dfrac{x-1}{4}}$ **3.** $\dfrac{\dfrac{x^2}{x+y}}{\dfrac{x-1}{(x+y)^2}}$

In Exercises 4–6 express each complex expression in terms of division. (*See Practice Exercise 3.*)

4. $\dfrac{\dfrac{5}{x+2}}{x^2+1}$ **5.** $\dfrac{9-\dfrac{x}{4}}{\dfrac{x+1}{x+4}}$ **6.** $\dfrac{3\frac{1}{2}}{5-\dfrac{2}{x-2}}$

B. In Exercises 7–26 simplify. (*See Examples 1–3.*)

7. $\dfrac{\dfrac{1}{2}}{2}$ **8.** $\dfrac{5}{\dfrac{1}{3}}$ **9.** $\dfrac{\dfrac{3}{4}}{-3}$ **10.** $\dfrac{-7}{\dfrac{1}{3}}$ **11.** $\dfrac{\dfrac{1}{2}}{\dfrac{1}{6}}$ **12.** $\dfrac{\dfrac{2}{5}}{\dfrac{1}{10}}$

13. $\dfrac{\dfrac{-1}{8}}{\dfrac{3}{4}}$ **14.** $\dfrac{\dfrac{-2}{9}}{\dfrac{-1}{3}}$ **15.** $\dfrac{\dfrac{4}{3}}{\dfrac{16}{9}}$ **16.** $\dfrac{\dfrac{2}{7}}{\dfrac{1}{14}}$ **17.** $\dfrac{1+\dfrac{1}{2}}{\dfrac{1}{2}}$ **18.** $\dfrac{2\frac{1}{3}}{\dfrac{5}{3}}$

19. $\dfrac{1-\dfrac{1}{4}}{\dfrac{1}{8}}$ **20.** $\dfrac{6-\dfrac{3}{4}}{\dfrac{-1}{2}}$ **21.** $\dfrac{\dfrac{1}{2}+\dfrac{1}{3}}{\dfrac{1}{6}}$ **22.** $\dfrac{1-\dfrac{1}{2}}{1+\dfrac{1}{4}}$ **23.** $\dfrac{\dfrac{3}{4}+\dfrac{1}{12}}{\dfrac{5}{6}+\dfrac{1}{8}}$

24. $\dfrac{\dfrac{1}{9}-\dfrac{1}{6}}{\dfrac{2}{3}+\dfrac{1}{9}}$ **25.** $\dfrac{\dfrac{3}{5}-\dfrac{1}{10}}{\dfrac{2}{25}-\dfrac{1}{50}}$ **26.** $\dfrac{\dfrac{1}{12}-\dfrac{1}{144}}{\dfrac{1}{3}-\dfrac{2}{9}}$

C. In Exercises 27–48 simplify. (*See Examples 4–6.*)

27. $\dfrac{\dfrac{a}{x}}{y}$ **28.** $\dfrac{a}{\dfrac{x}{y}}$ **29.** $\dfrac{\dfrac{-1}{x}}{\dfrac{y}{2}}$ **30.** $\dfrac{\dfrac{2}{x}}{\dfrac{4}{y}}$ **31.** $\dfrac{\dfrac{a}{b^2}}{\dfrac{b}{a^2}}$

Use this area for work space.

32. $\dfrac{\frac{xy}{z}}{\frac{y}{x^2}}$ **33.** $\dfrac{\frac{-ab}{c^2}}{\frac{a}{c^2}}$ **34.** $\dfrac{\frac{r^2s}{t^2}}{\frac{t}{r^2s^3}}$ **35.** $\dfrac{\frac{6x}{y^2}}{\frac{12y}{x}}$ **36.** $\dfrac{\frac{8a^2}{b}}{\frac{12a}{b^2}}$

37. $\dfrac{\frac{20a^2b}{cd}}{\frac{15ad}{bc}}$ **38.** $\dfrac{\frac{1}{x+1}}{\frac{x}{x-1}}$ **39.** $\dfrac{\frac{x^2-1}{x^2-4}}{\frac{x+1}{x+2}}$ **40.** $\dfrac{\frac{a^2-3a}{a+3}}{\frac{a^2}{a^2-9}}$ **41.** $\dfrac{\frac{1}{a}+1}{a^2}$

42. $\dfrac{1-\frac{1}{x}}{\frac{1}{x^2}}$ **43.** $\dfrac{\frac{1}{x+1}-\frac{1}{x-1}}{\frac{2}{x^2-1}}$ **44.** $\dfrac{\frac{x}{x-a}+\frac{a}{x+a}}{\frac{ax}{x^2-a^2}}$

45. $\dfrac{\frac{1}{a^2-b^2}+\frac{1}{a+b}}{\frac{a}{b^2-a^2}}$ **46.** $\dfrac{\frac{a^3-a^2}{a^2-1}}{\frac{a^3}{a^2+2a+1}}$ **47.** $\dfrac{\frac{1}{4}x^3}{\frac{1}{2}x^2}$

48. $\dfrac{\frac{3}{5}x^2y^5}{\frac{-9}{10}x^4y}$

49. Pick out the *error(s)*.

(a) $\dfrac{\frac{a}{b}}{c} = \dfrac{a}{\frac{b}{c}}$

(b) $\dfrac{\frac{a}{b}}{\frac{-a}{-b}} = -1$

(c) $\dfrac{\frac{a}{b}}{\frac{b}{a}} = 1$

(d) $\dfrac{\frac{a}{b}+2}{\frac{a}{b}+1} = 2$

(e) $\dfrac{\frac{a}{a^2-b^2}}{\frac{b}{b^2-a^2}} = -\dfrac{a}{b}$

What Have You Learned in Chapter 5?

You have learned that rational numbers express division of integers and that rational expressions express division of polynomials.

You can evaluate rational expressions by substituting numbers for variables.

You can simplify rational expressions by dividing by powers of numbers and variables, as well as by applying factoring techniques.

You have learned a method of dividing polynomials that resembles long division.

You can add, subtract, multiply, and divide rational numbers and rational expressions.

And you can simplify complex expressions, which are rational numbers (or rational expressions) that contain rational numbers (or rational expressions) in their numerators or denominators.

REVIEW

Let's Review Chapter 5

5.1 Rational Numbers and Rational Expressions

1. Cross multiply to show that $\frac{3}{5}$ and $\frac{-9}{-15}$ are equivalent.

In Exercises 2–4 simplify each rational number by dividing by factors common to both the numerator and denominator. Express the simplified rational number with a positive denominator.

2. $\frac{6}{9}$ **3.** $\frac{15}{-18}$ **4.** $\frac{-32}{-40}$

5. Evaluate $\frac{3x + 2}{x - 9}$ when $x = 10$.

6. Evaluate $\frac{xy}{x + y + 3}$ when $x = 2$ and $y = 3$.

7. Evaluate $\frac{x^2 + 1}{x^2 - 2}$ three separate times: (a) when $x = 1$, (b) when $x = -1$, (c) when $x = 2$

8. Write $\frac{3x^2 - 2x + 4}{12}$ as a polynomial with rational coefficients. Simplify these rational coefficients.

In Exercises 9–20 simplify each rational expression. (Multiply out the numbers that remain in numerator or denominator.)

5.2 Dividing Powers

9. $\frac{a^4b}{a^2b^5}$ **10.** $\frac{150a}{-135a}$

REVIEW

11. $\dfrac{-48a^4bc^7}{-36abc^9}$

12. $\dfrac{54(x+y)^3(x-y)}{81(x+y)(x-y)^2}$

5.3 *Factoring and Simplifying*

13. $\dfrac{4x-4y}{2}$

14. $\dfrac{ax^2+bx^2}{x^3}$

15. $\dfrac{3a^2bc+6abc^2}{9abc}$

16. $\dfrac{20xy^2z}{10xy^2z-50xy^3z}$

17. $\dfrac{5a+5b}{25a+50b}$

18. $\dfrac{m^2-9}{m^2+6m+9}$

19. $\dfrac{s^3-s^2}{s^4-s^3}$

20. $\dfrac{2x^2-2y^2}{x^2-2xy+y^2}$

5.4 *Dividing Polynomials*

21. (a) Express $-x^4+7x^6+x^2+x^2+5x^5$ in standard form.
(b) What is the degree of this polynomial?

22. Divide and *check*

$$y+2\overline{\smash{)}y^2+7y+10}$$

23. Divide

$$x+4\overline{\smash{)}x^3+3x^2-3x+4}$$

24. Divide a^4+4a^2-5 by a^2+5.

25. It can be shown that

$$\frac{5x+7}{x+1}=5+\frac{2}{x+1}$$

(a) What is the quotient? (b) What is the remainder? (c) Show that the degree of the remainder is less than the degree of the divisor.

26. (a) Find the quotient, (b) find the remainder, and (c) *check* the result.

$$x+2\overline{\smash{)}4x^2-7x+9}$$

In Exercises 27 and 28 (a) find the quotient, (b) find the remainder, and (c) express your result in the form

$$\text{QUOTIENT}+\frac{\text{REMAINDER}}{\text{DIVISOR}}$$

27. $a+7\overline{\smash{)}a^2+4a-32}$

28. $x+4\overline{\smash{)}x^3-x^2+1}$

5.5 *Multiplying and Dividing Rational Expressions*

In Exercises 29–34 multiply or divide the given rational numbers or rational expressions.

29. $\frac{-2}{3} \cdot \frac{-1}{9} \cdot \frac{6}{5}$

30. $\frac{a^2}{bc} \cdot \frac{b^3c}{a}$

31. $\frac{(x+3)^2}{4yz} \cdot \frac{8y^2}{x^2-9}$

32. $\frac{12}{25} \div \frac{3}{50}$

33. $\frac{4x^2y^2}{3ab} \div \frac{12xy}{9a^2b}$

34. $\frac{x^2-y^2}{x+2} \div \frac{x+y}{x^2+3x+2}$

5.6 *Adding and Subtracting: Same Denominator*

In Exercises 35–38 add or subtract, and simplify.

35. $\frac{1}{6} + \frac{1}{6}$

36. $\frac{5}{8} - \frac{3}{8}$

37. $\frac{2x}{x+3} + \frac{6}{x+3}$

38. $\frac{x}{x^2-9} - \frac{3}{x^2-9}$

5.7 *Least Common Multiples*

In Exercises 39–42 find the *lcm* of the given integers or polynomials.

39. 10 and 15

40. 6, 20, and 27

41. $4x^2$ and x^3y

42. $x^2 - 4$ and $x^2 - 3x + 2$

5.8 *Adding and Subtracting: Different Denominators*

43. (a) Find $lcd\left(\frac{1}{3}, \frac{2}{5}\right)$.

(b) Write equivalent rational numbers with this *lcd* as denominator.

In Exercises 44–46 add or subtract, as indicated.

44. $\frac{3}{4} + \frac{1}{2} - \frac{5}{8}$

45. $\frac{3}{x} - \frac{1}{x^2} + \frac{1}{x^3}$

46. $\frac{x}{x+a} + \frac{a}{x^2-a^2}$

5.9 *Complex Expressions*

In Exercises 47–50 simplify each expression.

47. $\dfrac{\frac{3}{8}}{\frac{1}{4}}$

48. $\dfrac{1-\frac{1}{3}}{\frac{5}{6}}$

49. $\dfrac{\frac{a}{b}}{\frac{c}{b^2}}$

50. $\dfrac{\frac{x-a}{x^2-1}}{\frac{x^2-a^2}{x+1}}$

REVIEW

REVIEW

And These from Chapters 1-4

51. Multiply $5x^2y^3(x^4y - 3x^2yz + 2x^4z^4)$

52. Divide $\dfrac{4a^5b^4c^2 - 12a^2b^3c^4 + 8a^4b^2c^3}{-4a^2b^2c^2}$

53. Factor $3x^2 + 11xy + 6y^2$

54. Factor $4a^2x^2 - 36a^2$

55. Find the prime factorization of 1080.

56. Combine and simplify:

$$-4(x^2 + 2x - 5) + \frac{8x^2 - 12x + 20}{4}$$

57. Multiply the sum of $4x - 5$ and $3x + 2$ by the product of $2x + 1$ and 3.

58. Evaluate $5x - 3y^2 + 2xy$ when $x = 2$ and $y = -1$.

59. Find the area of a square in which each side is of length 15 centimeters.

60. Let x represent an integer. Express

one less than 5 times the integer

in terms of x.

61. Solve $10x - 3 = 4x + 9$

62. Solve $5 - (3y - 4) = 8y - 2$

63. Find the value of $\left(\dfrac{14 - 5}{3} + 2\right)^2 - (3 + 4)^2$

64. Find *gcd* (96, 108, 144)

Practice Exam 1 on Chapter 5

In Questions 1–3 simplify each rational expression.

1. $\dfrac{36x^2(x + 1)^4}{42x^3(x + 1)}$ **2.** $\dfrac{5x^4 - 10x^2}{15x^3}$ **3.** $\dfrac{x^2 - 9}{x^2 + 5x + 6}$

4. Evaluate $\dfrac{5x + 3y}{x - 2y}$ when $x = 2$ and $y = -2$.

5. Evaluate $\dfrac{x^2 - 2x + 4}{x^4 + 1}$ three separate times: (a) when $x = 1$, (b) when $x = -1$, (c) when $x = 2$

6. (a) Find the quotient, (b) find the remainder, and (c) check the result:

$$x + 2\overline{)x^3 + 7x^2 + 2x}$$

In Questions 7–10 combine and simplify.

7. $\dfrac{3}{4} \cdot \dfrac{-8}{9} \cdot \dfrac{12}{-7}$ **8.** $\dfrac{x^2}{x + 1} \div \dfrac{x}{x^2 - 1}$ **9.** $\dfrac{a}{x} + \dfrac{1}{2x} - \dfrac{1}{x^2}$

10. $\dfrac{1}{x+4} + \dfrac{2}{x-4} + \dfrac{8}{x^2-16}$

11. Simplify: $\dfrac{\dfrac{a^2}{b^2-a^2}}{\dfrac{a^3}{a+b}}$

REVIEW

Practice Exam 2 on Chapter 5

1. Evaluate $\dfrac{x-9}{3x^2-x}$ when $x = -4$.

2. Evaluate $\dfrac{x^3-2y}{3xy+4}$ when $x = -2$ and $y = 5$.

In Questions 3–5 simplify each rational expression.

3. $\dfrac{30x^4(x-2)^2}{18x(x-2)^3}$

4. $\dfrac{12x^3}{3x^2+12x}$

5. $\dfrac{8x^3+16x^2}{x^5-4x^3}$

6. (a) Find the quotient, (b) find the remainder, and (c) *check* the result:

$$x+5\overline{)x^3-15x+40}$$

In Questions 7–11 combine and simplify.

7. $\dfrac{x^2-4x+3}{x^3} \cdot \dfrac{5x}{5x-15}$

8. $\dfrac{a^4-3a^3}{10b} \div \dfrac{a^2}{25b^5}$

9. $\dfrac{2}{a-2} - \dfrac{1}{a}$

10. $\dfrac{x+1}{x^2-9} + \dfrac{x}{2x+6}$

11. $\dfrac{5}{xy} + \dfrac{y-1}{x^2} - \dfrac{2}{y}$

12. Simplify: $\dfrac{3\frac{1}{3}}{\frac{3}{8}-\frac{1}{6}}$

13. Simplify: $\dfrac{\dfrac{x+1}{x^2}}{x-\dfrac{1}{x}}$

6

More Equations and Inequalities

6.1 Proportions and Equations with Rational Expressions

Recall that an equation is a statement of equality. You solve an equation by transforming it into simpler and simpler equivalent equations. You want to bring variables to one side, numerical terms to the other side. You can add the same expression to, or subtract the same expression from, both sides of an equation. You can also multiply or divide both sides of an equation by the same nonzero number or, with some restrictions, by the same algebraic expression.

A. Equations with Rational Numbers

Rational numbers often appear on one or both sides of an equation.

Example 1 ► Solve and *check* $x + \frac{1}{2} = 2$

Solution:

$$x + \frac{1}{2} = 2 \qquad \text{Subtract } \frac{1}{2} \text{ from both sides.}$$

$$x + \frac{1}{2} - \frac{1}{2} = 2 - \frac{1}{2} \qquad 2 - \frac{1}{2} = \frac{4-1}{2} = \frac{3}{2}$$

$$x = \frac{3}{2}$$

Check:

$$\frac{3}{2} + \frac{1}{2} \stackrel{?}{=} 2$$

$$\frac{3+1}{2} \stackrel{?}{=} 2$$

$$\frac{4}{2} \stackrel{?}{=} 2$$

$$2 \stackrel{\checkmark}{=} 2$$

◀

Example 2 ▶ Solve $\frac{3}{4}x = 12$

Solution:

$$\frac{3}{4}x = 12 \qquad \text{Divide both sides by } \frac{3}{4}\text{, that is, multiply by } \frac{4}{3}.$$

$$\frac{\overset{1}{\cancel{4}}}{\underset{1}{\cancel{3}}} \cdot \frac{\overset{1}{\cancel{3}}}{\underset{1}{\cancel{4}}}x = \frac{4}{\underset{1}{\cancel{3}}} \cdot \overset{4}{\cancel{12}}$$

$$x = 16$$

◀

An equation with integral coefficients can have a rational root.

Example 3 ▶ Solve $10x = 7$

Solution:

$$10x = 7 \qquad \text{Divide both sides by 10.}$$

$$\frac{10x}{10} = \frac{7}{10}$$

$$x = \frac{7}{10}$$

◀

PRACTICE EXERCISES FOR TOPIC A

Answers

1. ______

1. Solve $t - \frac{3}{4} = \frac{1}{2}$

2. ______________

3. ______________

2. Solve $\frac{7}{10}x = 35$

3. Solve $4x - 1 = \frac{2}{3}$

Now try the exercises for Topic A on page 251.

B. Solving Proportions

In a *proportion*, quantities in fractional form are equated. Here is a more precise definition.

DEFINITION

> A **proportion** is an equation of the form
>
> $$\frac{a}{c} = \frac{b}{d}$$
>
> and is often read, "*a* is to *c* as *b* is to *d*."

Example 4 ▶ Find the value of c in the following proportion.

$$\frac{8}{c} = \frac{2}{3}$$

Solution:

$$\frac{8}{c} = \frac{2}{3}$$ Cross multiply. $\frac{8}{c} \times \frac{2}{3}$

$$8 \cdot 3 = 2c$$

$$24 = 2c$$ Divide both sides by 2.

$$12 = c$$ ◀

Proportions occur naturally in everyday situations.

Example 5 ▶ The width and length of an 8×10 photograph are enlarged proportionally. If the length of the enlargement is 15 inches, what is its width? (*See Fig. 6.1.*)

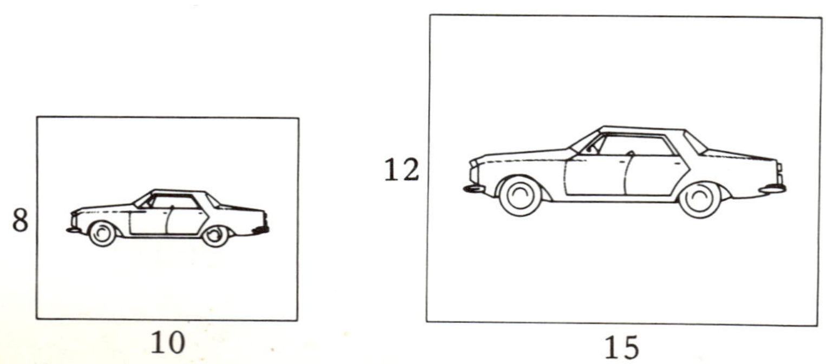

FIGURE 6.1

Solution:
The width of the original photograph is 8 inches; the length is 10 inches. Let w be the width of the enlargement (in inches). Then

8 is to 10 as w is to 15

Set up the proportion:

$$\frac{8}{10} = \frac{w}{15} \qquad \text{Cross multiply.}$$

$$8 \cdot 15 = w \cdot 10$$

$$120 = 10w$$

$$12 = w$$

The width of the enlargement is 12 inches. ◄

PRACTICE EXERCISES FOR TOPIC B

Answers

4. ____________

5. ____________

4. Find the value of k in the following proportion.

$$\frac{20}{36} = \frac{k}{27}$$

5. Bob's Mustang used exactly 3 gallons of gas to travel 80 miles. At this rate, how many gallons of gas will it take to travel 150 miles?

Now try the exercises for Topic B on page 251.

C. Multiplying by the *lcd*

An equation, such as

$$\frac{x}{2} = \frac{x+6}{5}$$

has the form of a proportion. To solve this type of equation, *cross multiply.*

Example 6 ► Solve $\frac{x}{2} = \frac{x+6}{5}$

Solution:

$$\frac{x}{2} = \frac{x+6}{5} \qquad \text{Cross multiply.}$$

$$5x = (x+6)2$$

$$5x = 2x + 12$$

$$3x = 12$$

$$x = 4$$

◄

Note that when you cross multiplied, you actually multiplied both sides by $2 \cdot 5$. In fact, $2 \cdot 5$ is the *lcd* of $\frac{x}{2}$ and $\frac{x+6}{5}$. Equations that involve rational numbers or rational expressions can often be simplified by multiplying both sides by the *lcd* of these expressions. (In Exercise 23, the cross-multiplication method and the *lcd* method are different, but lead to the same solution.)

Example 7 ▶ Solve $\frac{4}{y+3} = \frac{1}{y-3}$

Solution:

$$\frac{4}{y+3} = \frac{1}{y-3}$$ Multiply both sides by $(y+3)(y-3)$, the *lcd* (or cross multiply).

$$4(y-3) = 1 \cdot (y+3)$$
$$4y - 12 = y + 3$$
$$3y = 15$$
$$y = 5$$ ◀

In Example 8 the equation does *not* have the form of a proportion, so that cross multiplication does not directly apply. Instead, multiply by the *lcd* to clear the equation of "fractions."

Example 8 ▶ Solve $\frac{6}{x} + \frac{4}{x+1} = \frac{9}{x}$

Solution:

$$lcd\left(\frac{6}{x}, \frac{4}{x+1}, \frac{9}{x}\right) = x(x+1)$$

$$\frac{6}{x} + \frac{4}{x+1} = \frac{9}{x}$$
$$\left(\frac{6}{x} + \frac{4}{x+1}\right) \cdot x(x+1) = \frac{9}{x} \cdot x(x+1)$$
$$6(x+1) + 4x = 9(x+1)$$
$$6x + 6 + 4x = 9x + 9$$
$$10x + 6 = 9x + 9$$
$$10x + 6 - 9x - 6 = 9x + 9 - 9x - 6$$
$$x = 3$$ ◀

PRACTICE EXERCISES FOR TOPIC C

Answers
6.
7.
8.

6. Solve $\frac{x+9}{4} = \frac{2x-2}{3}$

7. Solve $\frac{5}{y} = \frac{8}{y-6}$

8. Solve $\frac{5}{3z} + \frac{2}{3} = \frac{3}{z}$

Now try the exercises for Topic C on page 251.

Use this area for work space.

EXERCISES

A. In Exercises 1–6 solve each equation. (*See Examples 1–3.*)

1. $x + \frac{1}{2} = 5$ **2.** $x - \frac{3}{4} = \frac{5}{8}$ **3.** $9x = 3$

4. $\frac{-1}{2}x = \frac{5}{6}$ **5.** $\frac{-2}{9}y = 18$ **6.** $\frac{3y}{4} - \frac{1}{2} = \frac{1}{3}$

In Exercises 7–10 solve and *check.* (*See Example 1.*)

7. $z - \frac{1}{4} = \frac{1}{2}$ **8.** $-3t = 8$ **9.** $1 + 5u = 3$ **10.** $\frac{4 - 3x}{7} = \frac{1}{2}$

B. In Exercises 11–16 find the value of *a, b, c,* or *d* in each proportion. (*See Example 4.*)

11. $\frac{a}{9} = \frac{2}{3}$ **12.** $\frac{1}{c} = \frac{4}{20}$ **13.** $\frac{12}{30} = \frac{b}{5}$

14. $\frac{7}{11} = \frac{-21}{d}$ **15.** $\frac{-15}{9} = \frac{10}{d}$ **16.** $\frac{108}{144} = \frac{3b}{48}$

In Exercises 17–22 set up a proportion and solve. (*See Example 5.*)

17. A salesman receives a commission of \$36 on a \$2000 sale. At this rate, how much does he receive on a \$4500 sale?

18. A hostess makes 40 cups of coffee for 28 guests. How many cups should she make for 42 guests?

19. A typist makes 4 errors on 10 pages. At this rate, how many errors will he make on 55 pages?

20. The price of grapefruit is three for 80¢. How much does a dozen grapefruit cost?

21. A batter gets 2 hits for every 7 times at bat. At this rate, how many hits will he have for 420 times at bat?

22. A basketball player makes 4 out of 5 free throws. At this rate, how many free throws must she attempt in order to make 16 free throws?

C. In Exercises 23–48 solve each equation. (*See Examples 6–8.*)

23. $\frac{x}{4} = \frac{x - 2}{2}$ **24.** $\frac{x + 2}{5} = \frac{x}{4}$ **25.** $\frac{x + 5}{2} = \frac{1 - x}{4}$

26. $\frac{x}{3} - 1 = \frac{x}{4}$ **27.** $\frac{x}{5} + \frac{x}{2} = 7$ **28.** $\frac{x}{3} - \frac{x}{9} = 2$

29. $\frac{x + 1}{2} + \frac{x - 1}{4} = 4$ **30.** $\frac{x}{3} + \frac{x + 2}{4} = 4$ **31.** $\frac{2x}{3} - 1 = \frac{x + 1}{2}$

Use this area for work space.

32. $\frac{x+5}{2} - \frac{x}{7} = 0$ **33.** $\frac{y+3}{5} = \frac{2y+3}{7}$ **34.** $\frac{z-2}{3} = \frac{4-z}{4}$

35. $\frac{u}{u-1} = 2$ **36.** $\frac{v}{v+6} = -1$ **37.** $\frac{x}{x+10} = 0$

38. $\frac{y+1}{y-2} = 2$ **39.** $\frac{z-1}{z+1} = 2$ **40.** $\frac{2}{u-3} = \frac{5}{u}$

41. $\frac{-1}{u+2} = \frac{2}{u+5}$ **42.** $\frac{3}{u-4} = \frac{2}{u-1}$ **43.** $\frac{1}{z} - \frac{1}{2} = \frac{1}{3}$

44. $\frac{7}{t} - \frac{3}{2} = \frac{5}{2t}$ **45.** $\frac{4}{x-2} - \frac{1}{x} = \frac{5}{x-2}$ **46.** $\frac{18}{v+2} - \frac{5}{v+1} = \frac{10}{v+1}$

47. $\frac{9}{y^2-1} - \frac{3}{y+1} = \frac{-4}{y-1}$ **48.** $\frac{2}{t+2} - \frac{10}{t^2-4} = \frac{1}{2-t}$

In Exercises 49–52 solve and *check*.

49. $\frac{y+3}{y-2} = 2$ **50.** $\frac{x}{4} + \frac{x}{2} = 9$

51. $\frac{2}{u+2} = \frac{5}{1-u}$ **52.** $\frac{2}{u} + \frac{2}{u+3} = \frac{6}{u+3}$

6.2 Decimals and Equations

Equations with decimals play an important role in applications. You have already encountered problems involving dollars and cents, which make use of decimal notation. Before studying more general problems, here is a brief review of the most important facts about decimals.

A. Decimal Notation

A rational number, such as

$$\frac{7}{10}, \quad \text{or} \quad \frac{59}{100}, \quad \text{or} \quad \frac{361}{1000}$$

whose denominator is a *power of* 10, can be expressed in **decimal notation** as follows:

decimal point

$$\frac{7}{10} = .7$$

tenths

Read this decimal as "seven tenths" or as "point seven."

$$\frac{59}{100} = .59$$

Read this decimal as "fifty-nine hundredths" or as "point five, nine."

tenths
hundredths

$$\frac{5}{10} + \frac{9}{100} = \frac{50}{100} + \frac{9}{100} = \frac{50 + 9}{100} = \frac{59}{100}$$

$$\frac{361}{1000} = .361$$

Read this decimal as "three hundred sixty-one thousandths" or as "point three, six, one."

tenths
hundredths
thousandths

$$\frac{3}{10} + \frac{6}{100} + \frac{1}{1000} = \frac{300}{1000} + \frac{60}{1000} + \frac{1}{1000} = \frac{300 + 60 + 1}{1000} = \frac{361}{1000}$$

Digits (0, 1, 2, 3, 4, 5, 6, 7, 8, 9) to the *right* of the decimal point are called **decimal digits.** Thus in the case of .361, there are three decimal digits: 3 is the tenths digit, 6 is the hundredths digit, and 1 is the thousandths digit.

Note that $\frac{3}{10} = .3$ whereas $\frac{3}{100} = \frac{\mathbf{0}3}{100} = .\mathbf{0}3$

tenths
hundredths

Also $\frac{57}{1000} = \frac{\mathbf{0}57}{1000} = .\mathbf{0}57$

tenths
hundredths
thousandths

and $\frac{7}{1000} = \frac{\mathbf{00}7}{1000} = .\mathbf{00}7$

tenths
hundredths
thousandths

Example 1 ► Express each rational number as a decimal.

(a) $\frac{1}{10}$ (b) $\frac{1}{100}$ (c) $\frac{1}{1000}$ (d) $\frac{11}{1000}$

Solution:

(a) $\frac{1}{10} = .1$ (b) $\frac{1}{100} = .01$ (c) $\frac{1}{1000} = .001$ (d) $\frac{11}{1000} = .011$ ◄

An integer is expressed as a decimal simply by placing a decimal point after the rightmost digit. Thus

for 475, write 475. ⟵ *decimal point*

A mixed number, such as $41\frac{7}{10}$, can be expressed as a decimal by first writing the *integral part*, 41, *to the left of the decimal point. The fractional part*, $\frac{7}{10}$, *is then expressed to the right of the decimal point.* Thus

$$\frac{7}{10} = .7$$

and

$$41\frac{7}{10} = 41 + \frac{7}{10}$$
$$= 41 + .7 \quad \text{Omit the "+".}$$
$$= 41.7$$

Read: "forty-one and seven tenths" or "forty-one *point* seven."

Example 2 ▶ Express each decimal as a rational number.
(a) .25 (b) .009 (c) 1.17
Solution:

(a)
$$.25 = \frac{25}{100}$$

This can be simplified by dividing the numerator and denominator by 25. Thus

$$.25 = \frac{25}{100} = \frac{1}{4}$$

(b)
$$.009 = \frac{9}{1000}$$

(c)
$$1.17 = 1 + \frac{17}{100}$$
$$= 1\frac{17}{100} \quad \textit{(as a mixed number)}$$
$$= \frac{117}{100} \quad \textit{(as a rational number)}$$ ◀

Observe that $\frac{3\mathbf{0}}{10\mathbf{0}} = \frac{3}{10}$

Thus in terms of decimals $.3\mathbf{0} = .3$

Similarly $.21\mathbf{0} = .21$

and $7.\mathbf{00} = 7.\mathbf{0} = 7.$

In general, you can add 0's to the right of the last decimal digit without changing the value. In other words, the new decimal is equivalent to the given one.

Answers	
1. (a)	(b)
(c)	(d)
2. (a)	(b)
(c)	(d)

PRACTICE EXERCISES FOR TOPIC A

1. Express each rational number or mixed number as a decimal.

(a) $\frac{6}{10}$ (b) $\frac{9}{100}$ (c) $\frac{27}{1000}$ (d) $\frac{59}{10\,000}$

2. Express each decimal as a rational number.

(a) .08 (b) .607 (c) .90 (d) 4.023

Now try the exercises for Topic A on page 260.

B. Infinite Repeating Decimals*

A rational number such as $\frac{3}{4}$ can be represented as a decimal by *dividing the numerator by the denominator:*

$$4\overline{)3.00} \quad .75$$

← *Line up the decimal points.*
← *Add as many 0's as necessary to the right of the decimal point.*

Thus $$\frac{3}{4} = .75$$

The rational number $\frac{1}{3}$ can be expressed as an **infinite repeating decimal.** Divide the numerator by the denominator:

$$3\overline{)1.000\,000} \quad .333\,333$$

Thus $\frac{1}{3} = .333\,333\ldots$ *Here the digit 3 repeats, as indicated by the 3 dots.*

The rational number $\frac{2}{11}$ can be expressed as an infinite repeating decimal in which *two digits,* 1 and 8, *repeat:*

$$11\overline{)2.00\,00\,00} \quad .18\,18\,18$$

Thus $\frac{2}{11} = .18\,18\,18\ldots$ *Again 3 dots indicate this repetition.*

Every rational number can be written as either a (regular) decimal or as an infinite repeating decimal.

Example 3 ► Express as decimals.

(a) $\frac{7}{8}$ (b) $\frac{2}{9}$ (c) $\frac{1}{6}$

*This topic (including Exercises 23–26 on page 260) will not be used in any subsequent discussion.

Solution:
(a) Divide the numerator, 7, by the denominator, 8:

$$8\overline{)7.000}\quad .875$$

Thus $$\frac{7}{8} = .875$$

(b) $$9\overline{)2.000\ 000}\quad .222\ 222$$

Thus $$\frac{2}{9} = .222\ 222\ldots$$

(c) $$6\overline{)1.0\ 000\ 000}\quad .1\ 666\ 666$$

Here just the 6 repeats. Thus

$$\frac{1}{6} = .1\ 666\ 666\ldots$$ ◀

Answers
3. (a)
(b)
(c)

PRACTICE EXERCISES FOR TOPIC B

3. Express as decimals.

(a) $\frac{3}{25}$ (b) $\frac{9}{11}$ (c) $\frac{7}{16}$

Now try the exercises for Topic B on page 260.

C. Adding and Subtracting Decimals

In order to solve equations involving decimals, you must be familiar with the arithmetic operations on decimals. To add or subtract, line up the decimal points and combine in columns.

Example 4 ▶ Add
.409
.36
.5812
.764

Solution:

$$\begin{array}{r} .4090 \\ .3600 \\ .5812 \\ .7640 \\ \hline 2.1142 \end{array}$$ ◀

Example 5 ▶

$$\begin{array}{r} 4.53 \\ -2.27 \\ \hline 2.26 \end{array}$$ ◀

Answers

4. (a) ____________

(b) ____________

(c) ____________

5. ____________

PRACTICE EXERCISES FOR TOPIC C

4. Perform the indicated operations.

(a) Add:
$$\begin{array}{r} 1.09 \\ 9.8 \\ \underline{.56} \end{array}$$

(b) $2.71 + 13.5 + .064$

(c) Subtract: $8.4 - 1.07$

5. Combine terms: $.4x + .8x - .2x$

Now try the exercises for Topic C on page 261.

D. Multiplying and Dividing Decimals

$$.3 \times 10 = \frac{3}{10} \times \frac{10}{1} = 3. \qquad 3.$$

To multiply a decimal by 10, move the decimal point one digit to the *right*. To multiply by 10^n, move the decimal point n digits to the *right*. If necessary, add zeros to the right of the right-most digit, as in part (d) of Example 6.

Example 6 ▶ Multiply

(a) $.174 \times 10$ (b) $.174 \times 100$ (c) $.174 \times 1000$ (d) $.174 \times 10\,000$

Solution:

(a) $.174 \times 10 = 1.74$ ⟶ 1.74

(b) $100 = 10^2$. Thus $.174 \times 100 = 17.4$ ⟶ 17.4

(c) $1000 = 10^3$. Thus $.174 \times 1000 = 174$ ⟶ 174.

(d) $10\,000 = 10^4$. Thus $.174 \times 10\,000 = 174\mathbf{0}$ ⟶ 174**0**. ◀

When one or both factors of a product is a decimal, count the total number of decimal digits in the factors to place the decimal point in the product. Thus

$$(.3)(.11) = \frac{3}{10} \times \frac{11}{100} = \frac{33}{1000} = .\mathbf{0}33$$

1 decimal digit ***2 decimal digits*** $\mathbf{10^1}$ $\mathbf{10^2}$ $\mathbf{10^3}$ ***3 (or 1 + 2) decimal digits***

Note that you add a 0 immediately to the right of the decimal point in order to obtain three decimal digits.

Example 7 ▶ Multiply

(a) $4 \times .25$ (b) $(.04)(.025)$

Solution:

(a)

$$\begin{array}{rl} .25 & \leftarrow \textbf{2 decimal digits} \\ \times\ 4 & \leftarrow \textbf{+0 decimal digits} \\ \hline 1.00 & \leftarrow \textbf{2 decimal digits} \end{array}$$

Thus $.25 \times 4 = 1$

(b)

$$\begin{array}{rl} .025 & \leftarrow \textbf{3 decimal digits} \\ \times .04 & \leftarrow \textbf{+2 decimal digits} \\ \hline .\mathbf{00}100 & \leftarrow \textbf{5 decimal digits} \end{array}$$

Add two 0's immediately to the right of the decimal point to make five decimal digits in the product. Note that the 0's to the right of the 1 can be dropped because

$$.001\ 00 = .001$$

Thus $(.025)(.04) = .001$ ◀

Next observe that

$$.3 \div 10 = \frac{3}{10} \times \frac{1}{10} = \frac{3}{100} = .\mathbf{0}3 \qquad .\mathbf{0}3$$

To divide a decimal by 10, move the decimal point one digit to the *left.* To divide by 10^n, move the decimal point n digits to the *left.* If necessary, add 0's immediately to the right of the decimal point to obtain the correct number of decimal digits, as in parts (b) and (c) of Example 8.

Example 8 ▶

(a) $3.92 \div 10 = .392$ $\quad .3\ 92$

(b) $3.92 \div 100 = .\mathbf{0}392$ $\quad .03\ 92$

(c) $3.92 \div 1000 = .\mathbf{00}392$ $\quad .003\ 92$ ◀

If the divisor has n decimal digits, multiply both dividend and divisor by 10^n, so that the *divisor* will then be an integer. In Example 9 the divisor has 2 decimal digits, so that both the dividend and divisor are multiplied by 10^2, or 100.

Example 9 ▶

$$\begin{array}{r} \textbf{3 decimal digits} \rightarrow \\ \textbf{2 decimal digits} \rightarrow \end{array} \frac{.008}{.02} = \frac{.008 \times 100}{.02 \times 100} = \frac{00.8}{2}$$

$$= .4 \leftarrow \textbf{1 or (3 − 2) decimal digit}$$ ◀

PRACTICE EXERCISES FOR TOPIC D

Answers	
6. (a)	(b)
7. (a)	(b)
8.	

6. Multiply: (a) 6.5×100 (b) $9.1 \times .003$

7. Divide: (a) $.21 \div 1000$ (b) $8.4 \div .007$

8. Compute, as indicated.

$$\frac{.9 + .5 + .7}{.07}$$

9. (a) ______ (b) ______

9. Find each value: (a) $(.2)^3$ (b) $(.01)^2$

Now try the exercises for Topic D on page 261.

E. Basic Equations

Here are some equations involving decimals.

Example 10 ▶ Solve $y + .05 = .7$
Solution:

$$y + .05 = .70 \qquad \textit{Subtract .05 from both sides.}$$

$$y + .05 - .05 = .70 - .05$$

$$y = .65$$ ◀

Example 11 ▶ Solve $\frac{x}{.5} = 2$
Solution:

$$\frac{x}{.5} = 2 \qquad \textit{Multiply both sides by .5.}$$

$$\frac{x}{.5}(.5) = 2(.5)$$

$$x = 1.0, \quad \text{or } 1$$ ◀

Answers

10. ______

11. ______

PRACTICE EXERCISES FOR TOPIC E

10. Solve $\frac{x}{8.2} = 3$

11. Solve $4y - 5 = .08$

Now try the exercises for Topic E on page 261.

F. Eliminating Decimals

Frequently, it is best to solve an equation by first eliminating decimals, as in Examples 12 and 13.

Example 12 ▶ Solve $.3t - .7 = .2$
Solution:
Multiply both sides by 10 to eliminate decimals.

$$(.3t - .7) \times 10 = .2 \times 10 \qquad \textit{On the left side use the Distributive Laws.}$$

$$3t - 7 = 2$$

$$3t = 9$$

$$t = 3$$ ◀

Example 13 ▶ Solve $.4x + .05 = .15x + .3$

Solution:

Multiply both sides by 100 to eliminate decimals.

$$(.4x + .05) \times 100 = (.15x + .3)100$$ *On both sides use the Distributive Laws.*

$$40x + 5 = 15x + 30$$

$$40x + 5 - 15x - 5 = 15x + 30 - 15x - 5$$

$$25x = 25$$

$$x = 1$$ ◀

Answers

12. (a) ______ (b) ______

PRACTICE EXERCISES FOR TOPIC F

12. Solve by eliminating decimals:

(a) $.9x + 12 = 22.8$ (b) $.3y - 2.5 = .12y + 6.5$

Now try the exercises for Topic F on page 262.

Use this area for work space.

EXERCISES

A. In Exercises 1–8 express as a decimal. (*See Example 1.*)

1. $\frac{13}{100}$ **2.** $\frac{-107}{1000}$ **3.** $\frac{9}{100}$ **4.** $\frac{9}{1000}$

5. $\frac{9}{10\,000}$ **6.** $\frac{77}{100\,000}$ **7.** 12 **8.** $12\frac{99}{100}$

In Exercises 9–16 express each decimal as a rational number. Simplify. (*See Example 2.*)

9. .29 **10.** .371 **11.** .4 **12.** .25

13. .36 **14.** .05 **15.** .004 **16.** −4.2

17. Which of the following decimals are equivalent to .1?
(a) .01 (b) .10 (c) .100 (d) .0100

18. Which of the following decimals are equivalent to 3.2?
(a) 3.02 (b) 3.20 (c) 30.2 (d) 3.200

B. In Exercises 19–26 express as a decimal. (*See Example 3.*)

19. $\frac{1}{5}$ **20.** $\frac{-3}{20}$ **21.** $\frac{2}{25}$ **22.** $\frac{7}{400}$

23. $\frac{2}{3}$ **24.** $\frac{5}{6}$ **25.** $\frac{1}{7}$ **26.** $\frac{11}{15}$

Use this area for work space.

C. In Exercises 27–30 add or subtract as indicated. (*See Examples 4 and 5.*)

27.
$$\begin{array}{r} .9284 \\ .2713 \\ .0495 \\ +.3182 \\ \hline \end{array}$$

28.
$$\begin{array}{r} 24.309 \\ .5412 \\ 1.003 \\ +12.0609 \\ \hline \end{array}$$

29. $4.038 - 2.129$

30. $27.625 - 19.9854$

In Exercises 31 and 32 combine terms, as indicated. (*See Practice Exercise 5.*)

31. $.6x + .3x + .1x$

32. $.36x^2 + .05x^2 - .21x^2$

D. In Exercises 33–47 compute, as indicated. (*See Examples 6–9 and Practice Exercise 9.*)

33. $.238 \times 1000$

34. $.0095 \times 100$

35. $(.5)(.6)$

36. $(.004)(.25)$

37. $(1.2)(.05)(.001)$

38. $49.4 \div 100$

39. $3865 \div 1000$

40. $.412 \div 100$

41. $\frac{.24}{.6}$

42. $\frac{.24}{.012}$

43. $\frac{.0016}{.004}$

44. $(.3)^3$

45. $(.2)(.6) + (.4)^2$

46. $\frac{.6 + .2 + .4}{.3}$

47. $\frac{2.4 - 1.2 + 3.6}{.012}$

48. Pick out the *error(s).*

(a) $(.3)(.2) = .6$

(b) $(.5) \times (.2) = 1.0$

(c) $\frac{.8}{.2} = .4$

(d) $(.01)^3 = .0001$

(e) $\frac{1.69}{1.3} = 1.3$

In Exercises 49 and 50 combine terms, as indicated.

Sample: $(.3xy)(.2x^2y)$	*Solution:* $(.3xy)(.2x^2y) = (.3)(.2)x \cdot x^2 \cdot y \cdot y = .06x^3y^2$

49. $(.2xy)(.5x^2)(.1xy^3)$

50. $\frac{.28x^2y}{.04xy}$

E. In Exercises 51–60 solve each equation. (*See Examples 10 and 11.*)

51. $x - .3 = .7$

52. $z + .4 = 2.4$

53. $y - .03 = 4.97$

54. $t + .48 = .9$

55. $x - (.2 + .03) = 0$

56. $\frac{x}{.4} = 5$

Use this area for work space.

57. $\frac{z}{.03} = 100$ **58.** $.5x = 60$ **59.** $\frac{z}{.8} = 30$ **60.** $\frac{y}{1.5} = 2$

In Exercises 61–64 solve and *check*.

61. $x - .02 = 9.98$ **62.** $x + .42 = .8$ **63.** $\frac{x}{.06} = 200$ **64.** $.01x = 50$

65. When .5 is added to a number, the sum is 5.5. What is the number?

66. When .5 is subtracted from a number, the difference is 5.5. What is the number?

67. What number must be multiplied by .5 to obtain 25?

68. What number must be divided by .5 to obtain 25?

F. In Exercises 69–78 solve each equation by first eliminating decimals. (*See Examples 12 and 13.*)

69. $.07y = 14$ **70.** $.001x = 5$ **71.** $2.75x = 5.5$ **72.** $.2(x - .1) = 3$

73. $.4x + .05 = 9$ **74.** $.4y + .3 = .1 - .5y$ **75.** $.05(t + 1) = .25$

76. $t + (1 - .2t) = 2$ **77.** $\frac{x - .03}{1} = 4.1x$ **78.** $\frac{y + 1}{.08} = \frac{y - 1}{.4}$

6.3 Interest Problems

Before solving financial problems concerning interest, commissions, markups, and discounts, you must be familiar with some basic facts about percentage.

A. Percent

Percent means hundredths. For example,

$$17\% \text{ means } .17, \text{ or } \frac{17}{100}$$

$$9\% \text{ means } .09, \text{ or } \frac{9}{100}$$

$$50\% \text{ means } .50, \text{ or } \frac{50}{100}, \text{ which equals } \frac{1}{2}$$

$$100\% \text{ means } 1.00, \text{ or } 1$$

and

$$150\% \text{ means } 1.50, \text{ or } \frac{150}{100}, \text{ which equals } \frac{3}{2} \text{ or } 1\frac{1}{2}$$

Note that

$$\frac{1}{2}\% \text{ means } \frac{1}{2} \textit{ of } 1\%$$

Here the word "of" indicates multiplication. Thus

$$\frac{1}{2}\% = .5 \times .01 = .005$$

Also

$$8\frac{1}{2}\% = 8\% + \frac{1}{2}\%$$

$$= \begin{array}{r} .080 \\ +.005 \\ \hline \end{array}$$

$$= .085$$

To convert a percent to a decimal:*

1. If there is no decimal point, insert one to the left of the percent symbol.
2. Remove the percent symbol.
3. Move the decimal point two digits to the left.

Example 1 ▶ Express each percent as a decimal.

(a) 120%

(b) $7\frac{1}{2}\%$

Solution:

(a) 120% = 120.% = 1.20 = 1.2

(b) Because $\frac{1}{2} = .5$, it follows that

$$7\frac{1}{2}\% = 7.5\% = .075$$

◀

To express a decimal as a percent:

1. If there are less than two decimal digits, add 0's at the right.
2. Move the decimal point two digits to the right.
3. Insert the percent symbol at the right.

Example 2 ▶ Express each decimal as a percent.

(a) 2.7 (b) .008

Solution:

(a) 2.7 = 2.70 = 270.% = 270%

(b) .008 = 00.8% = .8% (*read:* eight-tenths of one percent) ◀

Example 3 ▶ Find $7\frac{1}{2}\%$ of 2000.

Solution:

$$7\tfrac{1}{2}\% = .075$$

Multiply

$$\begin{array}{r} .075 \\ \times \quad 2000 \\ \hline 150.000 \end{array}$$

Thus $7\frac{1}{2}\%$ of 2000 is 150. ◀

*Conversions between rational numbers and percents are treated in Section 6.3 of the accompanying Student Workbook.

Answers	
1. (a)	(b)
(c)	(d)
2. (a)	(b)
(c)	
3.	

PRACTICE EXERCISES FOR TOPIC A

1. Express each percent as a decimal.

(a) 4% (b) 90% (c) 167% (d) $3\frac{1}{2}\%$

2. Express each decimal as a percent.

(a) .43 (b) 1.9 (c) .005

3. Find 9% of 450.

Now try the exercises for Topic A on page 268.

B. Annual Interest

When money is invested, it earns *interest* over a period of time. To simplify matters, *the basic period of time will at first be one year.* The amount of money invested, called the **principal,** times the **annual,** or **yearly, interest rate** equals the **interest earned in one year.** Let

R be the interest rate, P be the principal, and I be the interest earned

Then $R \cdot P = I$

Thus \$100 (principal) invested for one year at a rate of 8% earns

$$.08 \times 100 \text{ dollars (interest)}$$

or \$8 in one year. Also \$300 invested for one year at 7% earns

$$.07 \times 300 \text{ dollars}$$

or \$21 in one year.

Example 4 ► Jose deposits \$850 in a bank that pays an annual interest rate of 6%. How much interest does he earn in one year?
Solution:

R ·	P =	I
.06	850	I

$$I = .06 \times 850 = 51$$

He earns \$51 interest in one year. ◄

Example 5 ► How much must Alice invest at $8\frac{1}{2}\%$ in order to earn \$340 interest in one year? *Check your result.*

Solution:

$$I = 340$$

$$R = 8\frac{1}{2}\% = 085$$

R	· P	= I
.085	P	340

$$.085P = 340$$

$$\frac{.085P}{.085} = \frac{340 \times 1000}{.085 \times 1000} = \frac{340\ 000}{85}$$

$$P = 4000$$

Alice must invest \$4000 at $8\frac{1}{2}\%$ to earn \$340 interest in one year.

Check:

$$.085 \times 4000 \stackrel{?}{=} 340$$

$$340 \stackrel{\checkmark}{=} 340$$

◀

Answers

4. ______

5. ______

6. ______

PRACTICE EXERCISES FOR TOPIC B

4. Kevin invested \$3000 in a money-market fund that paid a $9\frac{1}{2}\%$ annual interest. How much interest did he earn in one year?

5. Kate invested her money in a savings account that paid a 7% annual interest. If her earnings for one year were \$343, how much did she invest?

6. In one year Hank earned \$30 on an investment of \$500. What was the annual interest rate?

Now try the exercises for Topic B on page 268.

C. Interest for Part of a Year

When money is invested for part of a year, the above interest formula must be modified. Let

R be the annual interest rate

P be the principal

T be the time (expressed as a "fraction" of a year)

I be the interest earned

Then $R \cdot P \cdot T = I$

$$\underbrace{R \cdot P}_{\text{annual interest}} \cdot \underbrace{T}_{\text{fraction of a year}} = I$$

Example 6 ▶ How much interest is earned on \$3000 invested at an annual rate of 6% for $\frac{1}{4}$ of a year?

Solution:

R	P	T	I
.6	3000	$\frac{1}{4}$	I

$$I = .6 \times 3000 \times \frac{1}{4} = \frac{180}{4} = 45$$

Thus \$45 interest is earned for $\frac{1}{4}$ of a year. ◀

Answers

7. ____

8. ____

PRACTICE EXERCISES FOR TOPIC C

7. How much interest is earned on \$1120 invested for $\frac{3}{4}$ of a year at an annual interest rate of 8%?

8. How much interest is earned on \$840 invested for 5 months at an annual interest rate of $5\frac{1}{2}$%?

Now try the exercises for Topic C on page 269.

D. Compound Interest

In practice, interest is generally *compounded*. **Compound interest** is interest paid on previously earned interest, as well as on the principal. Compounding results in greater interest, as you will see in Example 7. There, the basic period for the interest rate is 1 year. In this case, the interest is said to be **compounded annually.**

Example 7 ▶ Jerry deposits \$500 in a bank that pays an annual interest rate of 6%. How much interest does he earn if he leaves all his money, including his interest, in for two years?

Solution:

There are two different amounts involved—the original principal (for the *first* year) as well as the principal at the end of 1 year (for the *second* year). Use *subscripts* on P to distinguish between these principals. Thus let P_1 be the principal of \$500 for the *first* year.

There are also different amounts of interest earned for each year. Let I_1 be the interest earned for the *first* year. The interest *rate* for each year is 6%, or .06.

R	P_1	I_1
.06	500	I_1

$$I_1 = .06 \times 500 = 30$$

Jerry earns \$30 interest for the first year. He leaves this in the bank and, therefore, has \$530 at the end of 1 year. Thus his principal, P_2, for the *second* year is \$530. Let I_2 be the interest earned during the *second* year. Again the interest rate is .06.

R ·	P_2 =	I_2
.06	530	I_2

$$I_2 = .06 \times 530 = 31.80$$

He earns $31.80 interest for the second year. For 2 years he earns

$$\begin{array}{rl} \$30.00 & \text{------ first year's interest} \\ +\ 31.80 & \text{------ second year's interest (compounded)} \\ \hline \$61.80 & \text{------ total interest for 2 years} \end{array}$$

If interest were paid *only on principal,* he would have earned

$$\begin{array}{rl} \$30.00 & \text{------ first year's interest} \\ +\ 30.00 & \text{------ second year's interest (not compounded)} \\ \hline \$60.00 & \text{------ total interest for 2 years} \end{array}$$

Thus he earns $1.80 more because of compounding the interest. ◀

When the basic period for the interest rate is $\frac{1}{4}$ of a year, the interest is said to be **compounded quarterly.** When the basic period is $\frac{1}{2}$ of a year, the interest is **compounded semiannually.**

Example 8 ▶ Juanita invests $1000 at 8% compounded *quarterly.* How much interest does she earn in *half* a year?

Solution:

$$\frac{1}{2} = \frac{2}{4}$$

Thus half a year is the same as 2 quarters (of a year). Let P_1 be the principal and I_1 the interest for the first quarter. The interest rate for each quarter is 8% or .08.

R ·	P_1 ·	T =	I_1
.08	1000	$\frac{1}{4}$	I_1

$$I_1 = .08 \times 1000 \times \frac{1}{4} = \frac{80}{4} = 20$$

Juanita earns $20 interest for the first quarter. She leaves this in the bank and has $1020 at the end of the first quarter. This is her principal, P_2, for the second quarter. Let I_2 be the interest earned for the second quarter.

R ·	P_2 ·	T =	I_2
.08	1020	$\frac{1}{4}$	I_2

$$I_2 = .08 \times 1020 \times \frac{1}{4} = \frac{81.60}{4} = 20.40$$

She earns \$20.40 for the second quarter. Thus, for half a year she earns \$40.40.

$$\begin{array}{rl} \$20.00 & \text{-----} \ I_1 \\ +\ 20.40 & \text{-----} \ I_2 \\ \hline \$40.40 & \end{array}$$

◀

Answers

9. ______

PRACTICE EXERCISES FOR TOPIC D

9. How much interest is earned on an investment of \$400 deposited for $1\frac{1}{2}$ years at a rate of 6% compounded semiannually?

Now try the exercises for Topic D on page 269.

Use this area for work space.

EXERCISES

A. In Exercises 1–8 express each percent as a decimal. (*See Example 1.*)

1. 17% **2.** 8% **3.** 42% **4.** $9\frac{1}{2}\%$

5. $10\frac{1}{2}\%$ **6.** 200% **7.** 250% **8.** $7\frac{1}{4}\%$

In Exercises 9–14 express each decimal as a percent. (*See Example 2.*)

9. .52 **10.** .3 **11.** .004 **12.** 2 **13.** 1.1 **14.** .0011

15. Find 10% of 6000. (*See Example 3.*)

16. Find 5% of 7500. **17.** Find 8% of 12 000. **18.** Find $5\frac{1}{2}\%$ of 200.

19. Find $6\frac{1}{2}\%$ of 5000. **20.** Find $8\frac{1}{2}\%$ of 6600.

B. **21.** How much interest is earned in 1 year on a principal of \$1200 if the annual interest rate is 7%? (*See Examples 4 and 5.*)

22. How much interest is earned in 1 year on a principal of \$400 if the annual interest rate is $7\frac{1}{2}\%$?

23. What is the annual interest rate if \$45 interest is paid in a year on a principal of \$750? *Check your result.* (*See Practice Exercise 6.*)

24. What is the annual interest rate if \$144 is paid in a year on a principal of \$1800?

25. How much money must be invested for a year at an annual interest rate of 9% in order to earn \$54 interest? *Check your result.*

26. How much money must be invested for a year at an annual interest rate of $8\frac{1}{2}\%$ in order to earn \$102 interest?

27. Which earns more interest: \$900 invested at 8% or \$1000 at $7\frac{1}{2}\%$?

Use this area for work space.

28. How much money must be invested at 8% in order to earn the same as $800 at 12%? *Check your result.*

C.

29. How much interest is earned on $18 000 invested at an annual rate of 5% for half a year? (*See Example 6.*)

30. How much interest is earned on $4800 invested at an annual rate of 8% for $\frac{3}{4}$ of a year?

31. For what part of a year must $1200 be invested at an annual rate of 8% in order to earn $24 interest?

32. At what annual rate must $8000 be invested in order to earn $400 in half a year?

33. Which earns more interest, $5000 at an annual rate of 7% for half a year or $3000 at an annual rate of 8% for $\frac{3}{4}$ of a year?

D.

34. A sum of $250 is left as security on an apartment. The landlord must pay an interest rate of 5% annually. If all the money, including the interest, is left for 2 years, how much interest is earned? (*See Example 7.*)

35. Hillel invests $10 000 at 6% compounded quarterly. How much interest does he earn in half a year? (*See Example 8.*)

36. Maria invests $1000 at 6% compounded semiannually. How much interest does she earn in a year?

37. Roberto deposits $3200 in a bank that pays an annual interest rate of 8% compounded semiannually. How much interest will he earn if he leaves all his money, including the interest, in for 2 years? (*Round to the nearest cent after each interest period.**)

38. Teresa invests $2000 at 5% compounded semiannually. How much interest does she earn in a year and a half?

6.4 Commissions, Markups, and Discounts

A. Commissions

Many salespeople earn commissions based on what they sell. A **commission** is a payment determined by the **sale price.** The **rate of commission** is a *percentage* of the sale price. Let

$$R = \text{rate of commission}$$

$$S = \text{sale price}$$

$$C = \text{commission}$$

Then

$$R \cdot S = C$$

*That is, drop all decimal digits beyond the second, but if the third decimal digit is 5 or more, increase the second decimal digit by 1.

For example, if the sale price of a suit is \$100 and the rate of commission is 6%, the commission is given by:

R ·	S =	C
.06	100	6

Thus there is a \$6 commission on this sale.

Example 1 ▶ A car salesman earns a \$128 commission upon selling a car for \$3200. What is his rate of commission?

Solution:

R ·	S =	C
R	3200	128

$$3200R = 128$$

$$R = \frac{128}{3200}$$

Here the rate of commission, R, is expressed as a rational number. To obtain a percent, convert R to decimal form. To divide, place a decimal point after the 8:

$$3200\overline{)128.00} = .04$$

The rate of commission, R, is 4%. ◀

Example 2 ▶ An encyclopedia saleswoman works on an 8% rate of commission. If a set of encyclopedias sells for \$250, how many sets must she sell to earn \$300 in commissions? *Check your result.*

Solution:

Let x be the number of sets she must sell. For each set the sale price is \$250. For x sets the total sale is \$250x.

R ·	S =	C
.08	250x	300

Solve

$$(.08)(250x) = 300$$

$$8(250x) = 30\,000$$

$$2000x = 30\,000$$

$$x = 15$$

She must sell 15 sets to earn \$300 in commissions.

Check:

R ·	S =	C
.08	(250)(15) └3750┘	300

$$(.08)(250)(15) \stackrel{?}{=} 300$$

$$(.08)(3750) \stackrel{?}{=} 300$$

$$300 \stackrel{\checkmark}{=} 300$$

◀

Answers

1. ______

2. ______

PRACTICE EXERCISES FOR TOPIC A

1. A real-estate agent earns a $3\frac{1}{2}$% commission on the sale of an $80 000 house. What is her commission?

2. A salesman earns a 9% commission on the sale of vacuum cleaners that sell for $200 each. How many must he sell in order to make $450 in commissions?

Now try the exercises for Topic A on page 274.

B. Markups

It costs a craftsman $10 to make a pair of sandals. He sells the sandals for $40. The **cost** of the item is $10 and its **sale price** is $40. The **markup** (on cost) is

$$\$40 - \$10, \text{ or } \$30$$

In general, $\quad \text{markup} = \text{sale price} - \text{cost}$

The **rate of markup** (on cost) is defined by the formula

$$\text{rate of markup} = \frac{\text{markup}}{\text{cost}}$$

In the illustration on the preceding page, the rate of markup is

$$\frac{30}{10}, \text{ which equals } \frac{3}{1}, \text{ or } 300\%$$

Let

$$R = \text{rate of markup}$$

$$M = \text{markup}$$

$$C = \text{cost}$$

Then

$$R = \frac{M}{C}$$

or

$$R \cdot C = M$$

Example 3 ▶ A wholesaler buys baby carriages for $60, marks them up 25%, and sells them to a retail store. The store then marks the carriages up an additional 20%.
(a) At what price does the wholesaler sell the carriages?
(b) At what price does the retail store sell them?
Solution:
(a) Here the cost is $60 and the rate of markup is 25%.

$$R \cdot C = M$$

.25	60	M

$$(.25)(60) = M$$

$$15 = M$$

The markup is \$15.

$$\text{cost} + \text{markup} = \text{sale price}$$

Fill in the left side of the equation, and add to obtain the right side.

$$60 + 15 = 75$$

Thus the wholesaler sells the carriages for \$75.

(b) Now \$75 is the *cost* to the retailer. The new markup is 20%.

$$R \cdot C = M$$

.20	75	M

$$(.20)75 = M$$

$$15 = M$$

The markup (by the retailer) is also \$15.

$$\text{cost} + \text{markup} = \text{sale price}$$

$$75 + 15 = 90$$

The retail store sells the carriages for \$90. ◀

Answers

3. ____________

4. ____________

PRACTICE EXERCISES FOR TOPIC B

3. A retail store marks up its merchandise 85% from the wholesale cost. If the store acquires clocks from the wholesaler at \$20 apiece, what is the retail price of a clock?

4. An antique dealer buys an oak table for \$60 and sells it for \$210. What is the rate of markup?

Now try the exercises for Topic B on page 274.

C. Discounts

During a sale, a store will sell merchandise at a discount. For example, a radio, which normally sells for \$100, is marked down to \$60. Here \$100 is the **marked price,** \$60 is the **sale price,** and \$40 is the **discount** (or **reduction**).

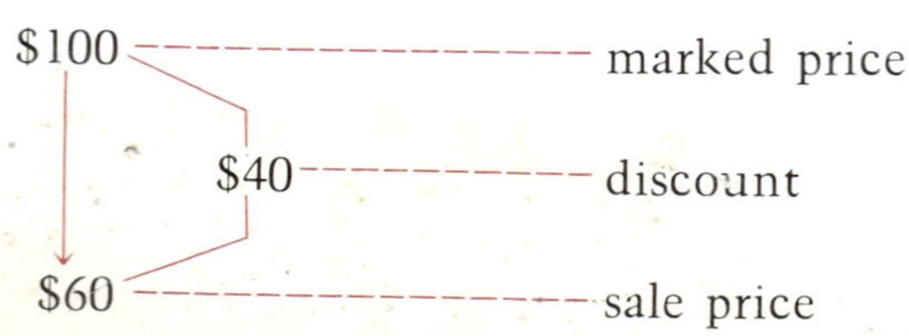

Thus $\text{discount} = \text{marked price} - \text{sale price}$

The **rate of discount** is defined by

$$\text{rate of discount} = \frac{\text{discount}}{\text{marked price}}$$

The rate of discount for the radio is therefore $\frac{40}{100}$, or 40%. In general,

$$\text{rate of discount} \cdot \text{marked price} = \text{discount}$$

In symbols,

$$R \cdot M = D$$

Example 4 ► All dresses in a boutique are reduced (discounted) by 30%.
(a) What is the discount for a dress marked $60?
(b) What is the sale price for this dress?
Solution:

(a)

$R \cdot M = D$

R	M	D
.30	60	D

$$(.30)(60) = D$$

$$18.00 = D$$

The discount is $18.

(b) Let S be the sale price.

$$\text{discount} = \text{marked price} - \text{sale price}$$

$$D = M - S$$

$$S = M - D$$

$$S = 60 - 18$$

$$S = 42$$

The sale price is $42. ◄

PRACTICE EXERCISES FOR TOPIC C

Answers

5. ________

6. ________

7. ________

5. A hotel offers senior citizens a 15% discount on room charges. If the standard room rate is $50, what is the charge for a senior citizen?

6. At a "30% OFF" sale Nancy saved $12 off the original price of a skirt. What was the original price?

7. A department store marks a $20 shirt down to $16. What is the rate of discount?

Now try the exercises for Topic C on page 274.

Use this area for work space.

EXERCISES

A.

1. How much commission is earned on a total sales of $1750 at a 7.5% rate of commission? *(See Examples 1 and 2.)*

2. How much commission is earned on a total sales of $2200 at a 9% rate of commission?

3. An automobile salesman earns a commission of $400 by selling a $5000 automobile. What is his rate of commission? *Check your result.*

4. A cosmetics saleswoman earns a commission of $45 by selling $900 worth of perfume. What is her rate of commission?

5. Suppose a salesman works on a 12% rate of commission. What must his sales be to earn commissions of $300?

6. A book saleswoman works on a 14% rate of commission. If each book sells for $10.50, how many books must she sell to earn $294 in commissions? *Check your result.*

7. A real estate agent sells a house for $42 000, for which she receives a commission from the seller.
(a) If the rate of commission is 6%, how much commission does she receive?
(b) How much money does the seller receive?

8. A shoe salesman receives a weekly salary of $150 and a 6% rate of commission on his sales. One week he sells $950 worth of shoes. How much does the salesman earn that week?

9. A saleswoman receives a 4% commission on the first $1000 of sales each week, 5% on the next $2000, and 6% on all sales thereafter. What is her commission on $8000 of sales for a week?

B.

10. A vendor buys ice cream pops for 20¢ apiece and marks them up at the rate of 25%. For how much does the vendor sell an ice cream pop? *Check your result. (See Example 3.)*

11. A wholesaler buys gloves for $8 a pair and marks them up 50%. He sells them to a retailer who then marks them up 20%.
(a) At what price does the wholesaler sell the gloves?
(b) At what price does the retailer sell them?

12. Recently, New York City allowed landlords of rent-controlled apartments to increase the rent by $7\frac{1}{2}$% each year. If an apartment was originally rented for $400 per month, what was the monthly rent two years later?

13. A markup of $35 on a typewriter is at the rate of 25%. What is the sale price?

14. A tie sells for $7. Its rate of markup is 75%. What is the markup?

15. A wholesaler buys sweaters for $50 a dozen and sells them to a store for $80 a dozen. What is the rate of markup? *(See Practice Exercise 4.)*

16. A ticket agency sells $12.50 seats for $20. What is the rate of markup?

C.

17. A television set is marked at $400 and is on sale with 20% off.
(a) What is the discount?
(b) What is the sale price? *(See Example 4.)*

18. A $60 pocketbook is reduced by 25%. What is its sale price?

19. A book that normally sells for $15 is on sale for $9. What is the rate of discount? *(See Practice Exercise 7.)*

Use this area for work space.

20. A coat marked at \$120 is on sale with a \$30 discount.
(a) What is the rate of discount?
(b) What is the sale price?

21. Two sweaters are on sale, each with the same rate of discount. The gray turtleneck is marked at \$20 and is on sale for \$14. The black cardigan is marked at \$22. What is its sale price? *Check your result.*

22. The price of a used car is reduced by 15%. The discount amounts to \$450. What was its marked price?

23. A book is on sale for \$10. The rate of discount is 20%.
(a) What was the marked price?
(b) What is the discount?

24. A kitchen table is on sale for \$150. The rate of discount is 25%.
(a) What was the marked price?
(b) What is the discount?

25. A \$200 table is discounted by 20% during a sale. After the sale, what must be the rate of markup to bring the *discounted price* back up to \$200?

26. A typewriter is on sale for \$200. The rate of discount is 20%. (a) What is the marked price? (b) What is the discount?

6.5 Literal Equations

A. What Are Literal Equations?

Scientific formulas may involve *several variables*. It is often necessary to solve for one variable in terms of the other variables. These formulas are expressed in terms of *literal equations.*

DEFINITION

> A **literal equation** is one that contains at least two variables and that must be solved for one variable in terms of the others.

Example 1 ►

$$x + y = 5$$

is a literal equation. You can solve for x in terms of y by subtracting y from both sides.

Thus

$$x = 5 - y$$

Sometimes you want to have information *directly* about y. You can solve for y in terms of x by subtracting x from both sides of the given equation:

$$y = 5 - x$$ ◄

To solve a literal equation, bring terms containing *the variable for which you are solving* to one side. Bring *all other terms* to the other side.

Example 2 ▶ Solve $5x - 3a + 1 = 0$ for x. *Check your result.*
Solution:

$$5x - 3a + 1 = 0$$

Leave 5x on the left side. Bring all other terms to the right side. Thus add $3a - 1$ to both sides.

$$5x - 3a + 1 + 3a - 1 = 0 + 3a - 1$$

$$5x = 3a - 1$$

$$\frac{5x}{5} = \frac{3a - 1}{5}$$

$$x = \frac{3a - 1}{5}$$

Check:

Substitute $\frac{3a - 1}{5}$ for x in the given equation:

$$5\left(\frac{3a - 1}{5}\right) - 3a + 1 \stackrel{?}{=} 0$$

$$3a - 1 - 3a + 1 \stackrel{?}{=} 0$$

$$0 \stackrel{\checkmark}{=} 0$$ ◀

Example 3 ▶ Solve $4x - 3y = 12$ (a) for x, (b) for y.
Solution:

(a)

$$4x - 3y = 12$$

Add 3y to both sides.

$$4x = 12 + 3y$$

$$\frac{4x}{4} = \frac{12 + 3y}{4}$$

$$x = \frac{12 + 3y}{4}$$

(b)

$$4x - 3y = 12$$

Subtract 4x from both sides.

$$-3y = 12 - 4x$$

$$\frac{-3y}{-3} = \frac{12 - 4x}{-3}$$

$$y = \frac{4x - 12}{3}$$

Note that $12 - 4x = -(4x - 12)$. ◀

Example 4 ▶ Solve $2x - 5y + 3z = 20$ for z.
Solution:

$$2x - 5y + 3z = 20$$

Bring the terms containing x and y to the right side. Thus add $-2x + 5y$ to both sides.

$$3z = 20 - 2x + 5y$$

$$z = \frac{20 - 2x + 5y}{3}$$ ◀

Answers

1. ____________

2. (a) ____________

(b) ____________

PRACTICE EXERCISES FOR TOPIC A

1. Solve for x in terms of y: (*Check your result.*)

$$\frac{x}{5} + 6 = 2y$$

2. Solve the equation $x - 4y = 9 - 2x$

(a) for x and (b) for y

Now try the exercises for Topic A on page 279.

B. Solving by Factoring

Example 5 ► Solve $xy + x = 1$ for x.

Solution:

Here both terms containing x are on the left side. In order to bring y to the right side, first factor the polynomial on the left.

$$xy + x = x(y + 1)$$

Thus the given equation becomes

$$x(y + 1) = 1 \qquad \textit{Divide both sides by } y + 1.$$

$$x = \frac{1}{y + 1}$$ ◄

In Example 5, when you divide both sides by $y + 1$, you must check that $y + 1 \neq 0$ (because division by 0 is undefined), and thus that $y \neq -1$. But if $y = -1$, then from the given equation,

$$x(-1) + x = 1$$

But $$x(-1) + x = -x + x = 0$$

or $$0 = 1$$

This cannot be. Thus $y + 1 \neq 0$.

Answers

3. ____________

PRACTICE EXERCISES FOR TOPIC B

3. Solve $8k + 1 = r(6 - k)$ for k.

Now try the exercises for Topic B on page 280.

C. Applications

Example 6 ► The formula

$$C = \frac{5}{9}(F - 32)$$

relates the Celsius and Fahrenheit temperature scales. Here C stands for degrees Celsius and F for degrees Fahrenheit. (*See Figure 6.2.*)

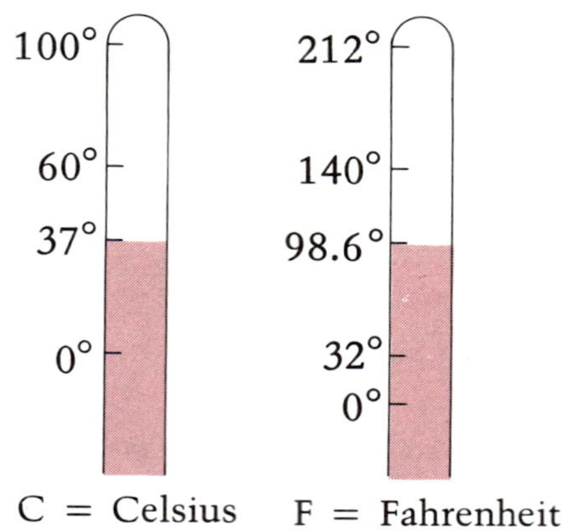

FIGURE 6.2. $C = \frac{5}{9}(F - 32)$

(a) Find C when F is 212, the boiling point of water.
(b) Solve for F in terms of C.
(c) Find F when C = 37.

Solution:

(a) Substitute 212 for F in the given formula.

$$C = \frac{5}{9}(212 - 32)$$

$$C = \frac{5}{\cancel{9}_{1}} \cdot \cancel{180}^{20}$$

$$C = 100$$

(b)

$$C = \frac{5}{9}(F - 32)$$ *Divide both sides by* $\frac{5}{9}$ $\left(\text{or } \textit{multiply by } \frac{9}{5}\right)$.

$$\frac{9}{5}C = \frac{9}{5} \cdot \frac{5}{9}(F - 32)$$

$$\frac{9}{5}C = F - 32$$

$$\frac{9}{5}C + 32 = F$$

(c) Substitute 37 for C in the formula obtained in part (b).

$$\frac{9}{5} \cdot 37 + 32 = F$$

$$\frac{333}{5} + 32 = F$$

$$66.6 + 32 = F$$

$$98.6 = F$$

Thus 37° C (37 degrees Celsius) corresponds to normal body temperature, 98.6° F (98.6 degrees Fahrenheit). ◀

Answers

4. ____

PRACTICE EXERCISES FOR TOPIC C

4. The formula for the area of a trapezoid is $A = \dfrac{h(B + b)}{2}$, where h is the height and the two bases are of lengths B and b. (*See Figure 6.3.*) Solve this equation for B.

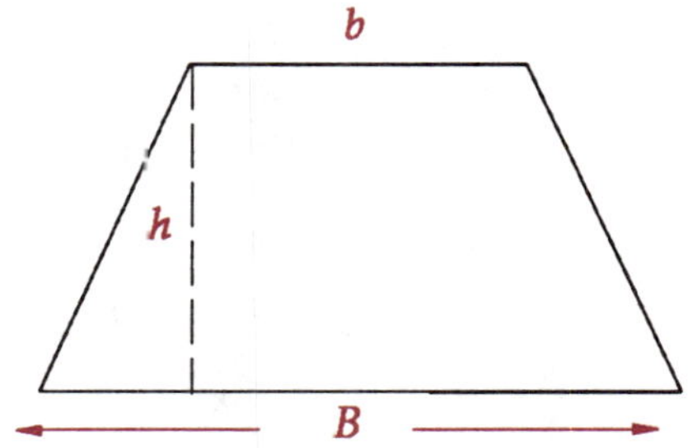

FIGURE 6.3

Now try the exercises for Topic C on page 280.

Use this area for work space.

EXERCISES

A. In Exercises 1–12 solve for x in terms of the other variable(s). (*See Example 2.*)

1. $5x = y$

2. $6x = 3y$

3. $x + 2 = y$

4. $x - 3 = 2y$

5. $x + y + 4 = 0$

6. $3x - 2y + 1 = 0$

7. $2x - 5t = 7x + 2t$

8. $x + 2t - 1 = 5x + 2$

9. $x + y + z = 1$

10. $3x - y + 2z = 0$

11. $5x + y - z = 2x + y - 3z$

12. $x - 1 + y = z + 3(x - 1)$

In Exercises 13–18 solve for the indicated variable. (*See Example 4.*)

13. $2x + 3y = 5$ (for y)

14. $5a - 4b = 10$ (for a)

15. $2u - 3v = 6$ (for v)

16. $6m - \dfrac{n}{2} = 5$ (for n)

17. $2a + b - 3c = 12$ (for a)

18. $4 - 3s + 2t = 5$ (for s)

In Exercises 19–26 solve for the indicated variables (*See Example 3.*)

19. $5x - y = 20$ (a) for x, (b) for y

20. $3a - 4b = 7$ (a) for a, (b) for b

21. $5(c - 2b) = 1 + c$ (a) for b, (b) for c

22. $s - 3t = 2s + 3t$ (a) for s, (b) for t

23. $2y - z + 1 = 10z - y$ (a) for y, (b) for z

24. $u - 3v + 2 = 2u - v$ (a) for u, (b) for v

25. $\dfrac{a}{b} = 1$ (a) for a, (b) for b

26. $\dfrac{2x}{y} = -3$ (a) for x, (b) for y

Use this area for work space.

In Exercises 27–32 (a) solve for the indicated variable, and (b) *check your result.* (*See Example 2.*)

27. $2x - y = 10$ (for x)

28. $5a - 2b = 4$ (for b)

29. $\frac{3}{u} = \frac{v}{2}$ (for v)

30. $\frac{12}{x} = \frac{y}{5}$ (for y)

31. $x - y + 2z = 8$ (for y)

32. $\frac{a}{2} + \frac{b}{3} - \frac{c}{4} = 12$ (for c)

B. In Exercises 33–40 solve for the indicated variable. (*See Example 5.*)

33. $4x - xy = 10$ (for x)

34. $10x + 2xy = 1$ (for x)

35. $ax + bx = 5$ (for x)

36. $7x - 2y + 3xy = 0$ (for x)

37. $ab - bc = 1$ (for b)

38. $10xy - 3yz = 1 + 5y$ (for y)

39. $xy - y = 3$ (for y)

40. $ab + 2a = 1$ (for a)

In Exercises 41–44 solve for the indicated variables.

41. $xy + 4y = 1$ (a) for x, (b) for y

42. $3m - mn = 2n$ (a) for m, (b) for n

43. $5x - xy + 2z = 10$ (a) for x, (b) for y

44. $uv - 2v + 6w = 7uw$ (a) for v, (b) for w

C. **45.** The formula

$$A = lw$$

expresses the area, A, of a rectangle in terms of the length, l, and width, w. Solve for l. (*See Example 6.*)

46. (a) The formula

$$P = 2l + 2w$$

expresses the perimeter, P, of a rectangle in terms of the length, l, and width, w. Solve for l.

(b) Find l when $P = 60$ feet and $w = 10$ feet. (*See Figure 6.4.*)

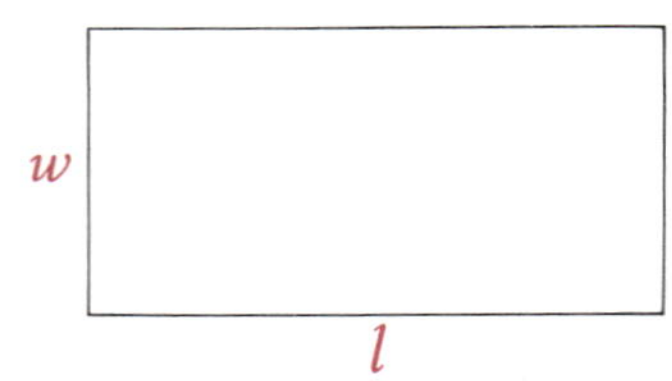

FIGURE 6.4. $P = 2l + 2w$

47. (a) The formula

$$A = \frac{bh}{2}$$

expresses the area, A, of a triangle in terms of b, the length of the base, and h, the length of the altitude. Solve for h.

(b) Find h when $A = 100$ square inches and $b = 4$ inches.

48. (a) The formula

$$V = lwh$$

expresses the volume, V, of a rectangular box in terms of the length, l, the width, w, and the height, h. Solve for h. (*See Figure 6.5.*)

Use this area for work space.

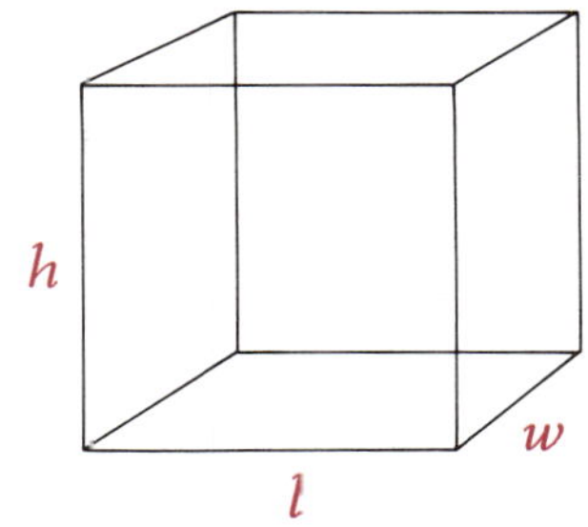

FIGURE 6.5. $V = lwh$

(b) Find h when $V = 2000$ cubic inches, $l = 20$ inches, and $w = 5$ inches.

49. (a) The formula

$$d = rt$$

expresses the distance, d, that can be traveled at a constant rate, r, over a period of time, t. Solve for r.

(b) Find r when $d = 300$ miles and $t = 5$ hours.

50. (a) The formula

$$R = \frac{kl}{t^2}$$

expresses the resistance, R, of an electrical wire in terms of the length, l, of the wire, the thickness, t, of the wire, and a numerical factor, k. Solve for l.

(b) Suppose the resistance is 40 ohms when the wire is .1 inch thick and when $k = \frac{1}{60\,000}$. Find the length of the wire.

51. In the interest formula

$$R \cdot P \cdot T = I$$

(a) Solve for P.

(b) Find P when $R = 8\%$, $T = \frac{3}{4}$, and $I = \$600$.

6.6 Intervals

Let a, b, and c be any numbers.* Recall that a is less than b,

$$a < b$$

or equivalently, b is greater than a,

$$b > a$$

if a lies to the left of b on the number line.

A. Less than or Equal to

It is convenient to write

$$a \leq b$$

*Although only rational numbers have been introduced thus far, the concepts discussed in the remainder of the chapter apply to real numbers, in general.

if either $a < b$ or if $a = b$. Thus $5 \leq 6$ because $5 < 6$. But also $6 \leq 6$ because $6 = 6$. In general, $a \leq b$ indicates that a **is less than or equal to** b. Similarly, write

$$a \geq b$$

if either $a > b$ or else $a = b$. Thus $a \geq b$ indicates that a **is greater than or equal to** b.

Example 1 ► (a) $4 \leq 8$ because $4 < 8$
(b) $7 \geq 5$ because $7 > 5$
(c) $-2 \leq -1$ because $-2 < -1$
(d) $9 \leq 9$ because $9 = 9$. Similarly $9 \geq 9$ because $9 = 9$ ◄

Observe that for every number a,

$$a \leq a \qquad \text{and} \qquad a \geq a$$

are both true.

The following **Laws of Transitivity** hold. Let a, b, and c be any numbers.

$$\text{If } a < b \text{ and } b < c, \text{ then } a < c$$
$$\text{And if } a \leq b \text{ and } b \leq c, \text{ then } a \leq c$$

PRACTICE EXERCISES FOR TOPIC A

Answers

1. (a) ______
(b) ______
(c) ______
3. ______

1. Answer "true" or "false."

(a) $-12 \leq 0$ (b) $-8 \geq 8$ (c) $20 \geq 20$

2. Fill in "≤" or "≥". If both symbols apply, write "both."

(a) 3 ☐ 0 (b) −6 ☐ −3 (c) −7 ☐ −7

3. If $x \leq y$ and $y \leq 0$, what conclusion can you draw about x?

Now try the exercises for Topic A on page 285.

B. Open Intervals, Closed Intervals, Rays

Instead of considering the entire number line, there are times when you are only concerned with the points "between a and b." Let

$$a < x < b \qquad \text{stand for both} \qquad a < x \quad \text{and} \quad x < b$$

and let

$$a \leq x \leq b \qquad \text{stand for both} \qquad a \leq x \quad \text{and} \quad x \leq b$$

For example,

$$5 < x < 7 \qquad \text{if } 5 < x \text{ as well as } x < 7$$

Clearly 6 is one number that satisfies this condition because $5 < 6$ and $6 < 7$ are both true. Thus $5 < 6 < 7$

DEFINITION

> Let a and b be any numbers such that $a < b$. The **open interval** (a, b) consists of all numbers* x such that $a < x < b$. The **closed interval** $[a, b]$ consists of all numbers x such that $a \leq x \leq b$. Here a is called the **left endpoint** and b the **right endpoint** of each of these types of intervals.

Observe that neither endpoint a nor b is in the open interval (a, b). *Both* endpoints a and b are in the closed interval $[a, b]$.

Geometrically, the open interval (a, b) consists of all points of the number line that lie *both* to the right of a *and* to the left of b. (*See Figure 6.6(a).*) The closed interval $[a, b]$ consists of all points in (a, b) as well as the two endpoints a and b. (*See Figure 6.6(b).*)

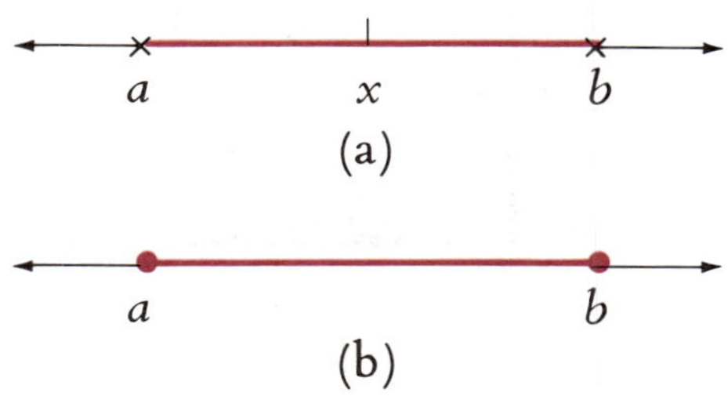

FIGURE 6.6. (a) The open interval (a, b) consists of all x that lie both to the right of a and to the left of b. The points a and b are excluded. (b) The closed interval [a, b] consists of the numbers in (a, b) plus the endpoints a and b.

Example 2 ▶ (a) The open interval $(0, 1)$ consists of all numbers x such that $0 < x < 1$. Thus $\frac{1}{2}$, $\frac{2}{3}$, and $\frac{9}{10}$ are in this open interval. For example, $0 < \frac{1}{2} < 1$. Neither 0 nor 1 is in this open interval. Also negative numbers, such as -2 and -1, are less than 0 and therefore are not in this interval. Furthermore 2, 3, and 4 are each greater than 1, and are therefore not in this interval.

(b) The closed interval $[0, 1]$ consists of all numbers in the open interval $(0, 1)$ as well as the two endpoints 0 and 1. Thus $\frac{1}{10}$ and $\frac{3}{5}$ are in this closed interval; $\frac{-1}{10}$ and 10 are not in it. ◀

Example 3 ▶ (a) The open interval $(-6, -1)$ consists of all numbers x satisfying $-6 < x < -1$. Thus -5, -2, and -1.1 are in this open interval; -7, -6, -1, and 0 are not in it.

(b) The closed interval $[-6, -1]$ consists of all numbers in the open interval $(-6, -1)$ as well as the two endpoints, -6 and -1. ◀

---

*Irrational numbers as well as rational numbers (Section 8.2)

Example 4 ▶ Which of the following numbers are in the interval $(-3, 2)$?

(a) -3 (b) -2 (c) $\frac{3}{2}$ (d) 2 (e) 2.01

Solution:
(b) and (c). Note that $(-3, 2)$ is an open interval. Thus the two endpoints, -3 and 2, are excluded. Also $\frac{3}{2} = 1\frac{1}{2} < 2$. Thus $-3 < \frac{3}{2} < 2$, so that $\frac{3}{2}$ is in the interval. ◀

Sometimes it is necessary to speak about all $x > 4$ or about all $x \leq 1$.

DEFINITION

Let a and b be numbers.

The **open right ray** (a, ∞) consists of all numbers x such that $a < x$.

The **closed right ray** $[a, \infty)$ consists of all numbers x such that $a \leq x$.

The **open left ray** $(-\infty, b)$ consists of all numbers x such that $x < b$.

The **closed left ray** $(-\infty, b]$ consists of all numbers x such that $x \leq b$.

Geometrically, the *open right ray* consists of all x to the *right* of a; the *closed right ray* consists of these points together with the *left* endpoint a. The *open left ray* $(-\infty, b)$ consists of all x to the *left* of b; the *closed left ray* consists of these points together with the *right* endpoint b. (*See Figure 6.7.*) [The symbol "∞" (infinity) is used to suggest that these rays extend indefinitely in one direction.]

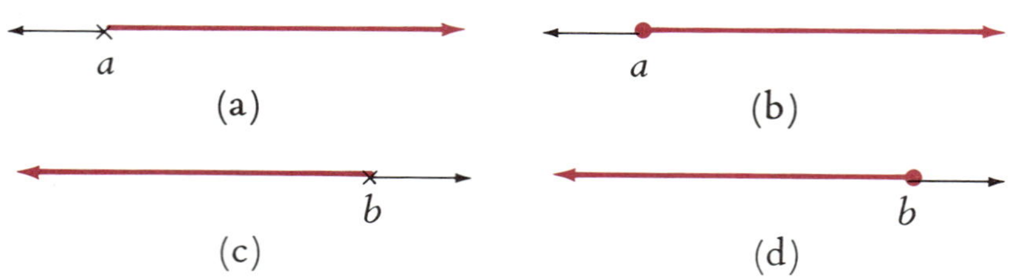

FIGURE 6.7. (a) The open right ray (a, ∞). The point a is excluded. (b) The closed right ray $[a, \infty)$. The point a is included. (c) The open left ray $(-\infty, b)$. The point b is excluded. (d) The closed left ray $(-\infty, b]$. The point b is included.

Example 5 ▶ (a) The open right ray $(4, \infty)$ consists of all numbers greater than 4. Thus 5 and 10 are in this ray, but 3 and 4 are not.
(b) The closed right ray $[4, \infty)$ consists of all numbers in $(4, \infty)$ as well as the left endpoint 4.
(c) The open left ray $(-\infty, 1)$ consists of all numbers less than 1. Thus every negative number is in the ray. Among the nonnegative numbers in this ray are 0 and $\frac{1}{2}$.
(d) The closed left ray $(-\infty, 1]$ consists of all numbers in $(-\infty, 1)$ together with the right endpoint, 1. ◀

Answers

4. (a) ____
(b) ____
(c) ____
5. (a) ____
(b) ____
(c) ____
(d) ____
6. (a) ____
(b) ____
(c) ____
(d) ____
7. (a) ____
(b) ____

PRACTICE EXERCISES FOR TOPIC B

4. Indicate whether each of the following is an open interval or a closed interval.

(a) $[0, 9]$ (b) $(-10, 3)$

(c) the set of numbers that satisfy the condition $3 \leq x \leq 100$.

5. Answer "true" or "false."

(a) -9 is in $[-9, 5]$. (b) $.002$ is in $(0, 1)$. (c) 5 is in $(-8, 5)$.

(d) $-4 < 6 < 6$

6. Describe each of the following as an open left ray, a closed left ray, an open right ray, or a closed right ray.

(a) $(10, \infty)$ (b) $(-\infty, -3]$

(c) the set of numbers satisfying the condition $x \geq 2$.

(d) the set of all negative numbers.

7. Answer "true" or "false."

(a) 0 is in $(-1, \infty)$. (b) -2 is in $(-\infty, -2]$.

Now try the exercises for Topic B on page 286.

Use this area for work space.

EXERCISES

A. In Exercises 1–8 answer "true" or "false." (*See Example 1 and Practice Exercise 1.*)

1. $1 \leq 2$ **2.** $5 \geq 3$ **3.** $-3 \leq -4$ **4.** $-10 \geq -8$

5. $7 \leq 7$ **6.** $8 \geq 8$ **7.** $0 \leq 0$ **8.** $-2 \geq -2$

In Exercises 9–18 fill in "≤" or "≥". If both symbols apply, write "both." (*See Practice Exercise 2.*)

9. $4 \ \square\ 10$ **10.** $9 \ \square\ 6$ **11.** $3 \ \square\ -4$

12. $1 \ \square\ 1$ **13.** $-7 \ \square\ -2$ **14.** $-12 \ \square\ -8$

15. $-6 \ \square\ -6$ **16.** $\frac{1}{2} \ \square\ \frac{1}{4}$ **17.** $\frac{-1}{2} \ \square\ \frac{-1}{4}$

18. $3.1 \ \square\ 3.01$

In Exercises 19–22 what conclusion can you draw about x? (*See Practice Exercise 3.*)

19. $x < y$ and $y < 4$

20. $x > y$ and $y > 0$

21. $x \leq y$ and $y \leq -1$

22. $x \geq y$ and $y \geq 10$

Use this area for work space.

B. In Exercises 23–36 describe each set of numbers as one of the following.

(a) an open interval
(b) a closed interval
(c) an open right ray
(d) a closed right ray
(e) an open left ray
(f) a closed left ray

(See Examples 2, 3, and 5.)

23. $(5, 9)$
24. $[2, 8]$
25. $(-4, -2)$
26. $[6, \infty)$
27. $(-\infty, 2]$
28. $(-\infty, -4)$
29. $[0, \infty)$
30. $(-3, \infty)$
31. $(-\infty, -3)$
32. $[-100, 100]$
33. $(1, \infty)$
34. $(-\infty, -100]$
35. $(-\infty, 100]$
36. $[-100, \infty)$

In Exercises 37–50 answer "true" or "false." *(See Example 4.)*

37. 5 is in $(1, 6)$.
38. -2 is in $(-4, 0)$.
39. 2.1 is in $[2, 3]$.
40. 1 is in $(1, 2)$.
41. 0 is in $[-1, 1]$.
42. $\frac{1}{2}$ is in $(0, 3)$.
43. $\frac{-1}{2}$ is in $(-1, 0)$.
44. -4 is in $(-\infty, -1)$.
45. 4 is in $(4, \infty)$.
46. 3 is in $[-2, \infty)$.
47. -7 is in $(-\infty, -6)$.
48. $1 \leq 4 \leq 5$
49. $-5 \leq -4 \leq -1$
50. $-1 \leq -2 \leq -3$

In Exercises 51–60 describe in one of the forms

(a, b), $[a, b]$, (a, ∞), $[a, \infty)$, $(-\infty, b)$, $(-\infty, b]$

the set of numbers satisfying each of the following.

Sample $4 \leq x \leq 8$	*Solution:* $[4, 8]$

51. $0 < x < 2$
52. $-4 \leq x \leq -3$
53. $x > 5$
54. $x < -5$
55. $-1 \leq x \leq 1$
56. $x > -3$
57. $x < 0$
58. $x \leq -2$
59. Both $x > 1$ and $x < 2$
60. Both $x \geq 2$ and $x \leq 6$

61. Pick out the *error(s)*.

(a) -4 is in $(-\infty, -4)$.
(b) 2 is in $[2, \infty)$.
(c) $(0, \infty)$ consists of all the positive numbers.
(d) 0 is in both $(-\infty, 0]$ and $[0, \infty)$.
(e) Every number in $(2, 3)$ is also in $(1, 4)$.

6.7 Arithmetic of Inequalities

There are a number of important properties that you will be using when you "solve inequalities." As you read this material, recall how you used the Addition and Multiplication Properties when you solved equations (pages 92–94).

Throughout this section, unless otherwise specified, assume a, b, c, and d are any numbers.

A. Addition and Inequalities

First note that

$$5 < 10$$

and that

$$5 + 2 < 10 + 2$$

or

$$7 < 12$$

Property A If $a < b$, then $a + c < b + c$ and $a - c < b - c$

Property **A** asserts that *if you add or subtract the same number on both sides of an inequality, the sense (or direction) of the inequality is preserved.*

In Figure 6.8 note that a lies to the left of b. Also $a + c$ lies the same number of units to the left of $b + c$. Thus $a + c < b + c$.

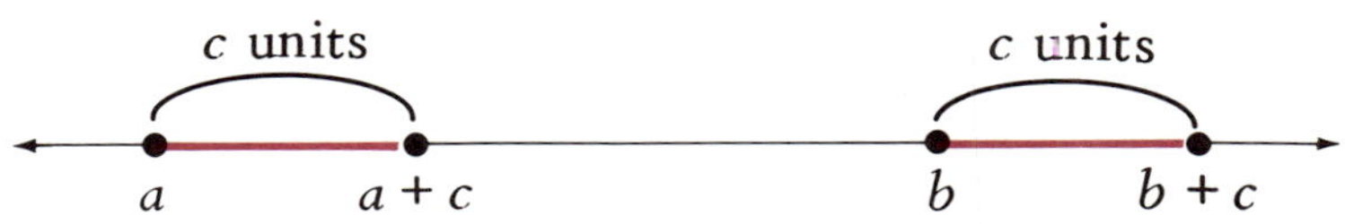

***FIGURE 6.8.** If $a < b$, then $a + c < b + c$. (The diagram is for the case of $c > 0$ and $a + c < b$.)*

Example 1 ► (a)

$$3 < 6$$

Thus

$$3 + 1 < 6 + 1$$

or

$$4 < 7$$

(b)

$$4 < 9$$

Thus

$$4 - 3 < 9 - 3$$

or

$$1 < 6$$ ◄

Property B If $a < b$ and $c < d$, then $a + c < b + d$

Thus *inequalities may be added.* In fact, because $c < d$,

$$a + c < \mathbf{b} + c < \mathbf{b} - d$$

By transitivity,

$$a + c < b + d$$

Example 2 ► $10 < 20$ and $1 < 2$

Thus

$$10 + 1 < 20 + 2$$

or

$$11 < 22$$ ◄

Example 3 ▶ Let a be any number. Then
(a) $a + 3 < a + 4$ because $3 < 4$
(b) $a - 3 < a - 2$ because $-3 < -2$ ◀

PRACTICE EXERCISES FOR TOPIC A

1. Let n be any number. Fill in "<" or ">".

(a) $n \square n + 5$ (b) $n - 3 \square n$ (c) $n + 4 \square n + 1$

(d) $1 + n \square -2 + n$

Now try the exercises for Topic A on page 292.

B. Multiplication and Inequalities

Observe that $3 < 4$

and that $3(2) < 4(2)$

or $6 < 8$ (*See Figure 6.9(a).*)

whereas $3(-2) > 4(-2)$

or $-6 > -8$ (*See Figure 6.9(b).*)

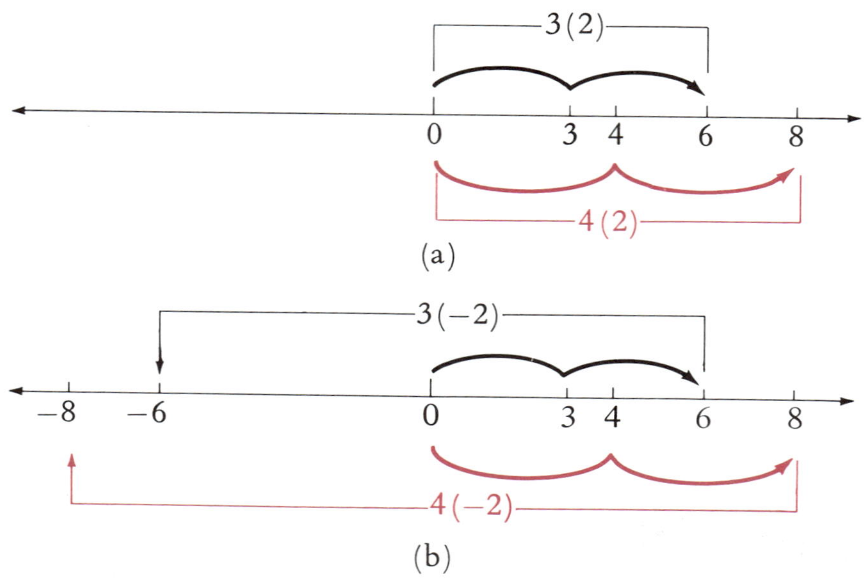

FIGURE 6.9. ***(a)* 3(2) < 4(2)** ***(b)* 3(−2) > 4(−2)**

When you multiplied both sides of the given inequality $3 < 4$ by the *positive* number 2, the inequality was preserved. But when you multiplied both sides of the given inequality by the *negative* number -2, the inequality was reversed.

Recall that c is *positive* if $c > 0$; c is *negative* if $c < 0$.

Property C_1 If $a < b$ and $c > 0$, then $ac < bc$

Property C_2 If $a < b$ and $c < 0$, then $ac > bc$

Thus *multiplication by a* **positive** *number* **preserves** *the sense of an inequality. Multiplication by a* **negative** *number* **reverses** *the sense of an inequality.*

Example 4 ▶ $2 < 3$

(a) $2(4) < 3(4)$

or $8 < 12$

(*See Figure 6.10(a).*)

(b) $2(-4) > 3(-4)$

or $-8 > -12$

(*See Figure 6.10(b).*) ◀

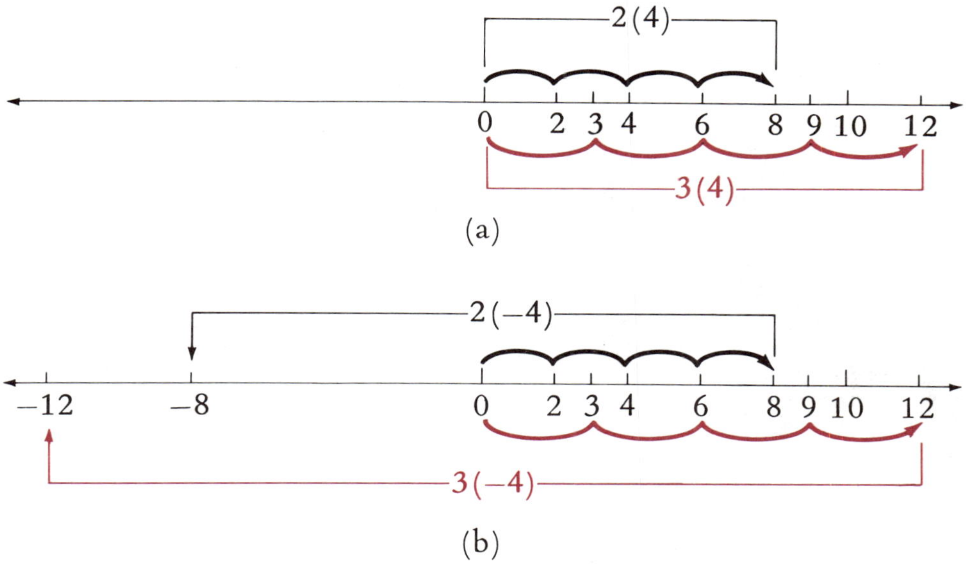

FIGURE 6.10. **(*a*) 2(4) < 3(4) (*b*) 2(−4) > 3(−4)**

Example 5 ▶ $-5 < -3$

(a) $(-5)(2) < -3(2)$

or $-10 < -6$

(*See Figure 6.11(a)* on page 290.)

(b) $(-5)(-2) > (-3)(-2)$

or $10 > 6$

(*See Figure 6.11(b).*) ◀

Division by c amounts to multiplication by $\frac{1}{c}$. Note that if c is positive, then $\frac{1}{c}$ is positive; and if c is negative, then $\frac{1}{c}$ is negative. Thus the following properties also hold.

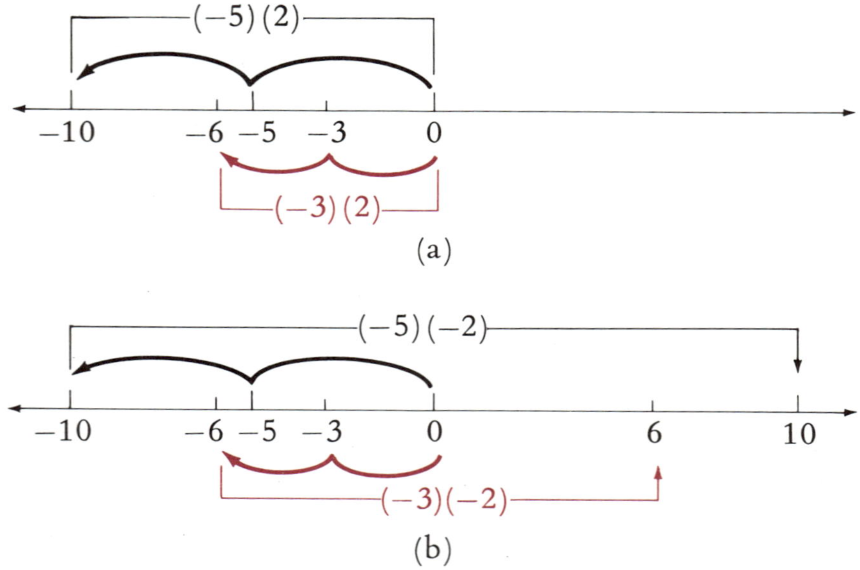

FIGURE 6.11. **(*a*) $(-5)(2) < (-3)(2)$ (*b*) $(-5)(-2) > (-3)(-2)$**

Property C_3 If $a < b$ and $c > 0$, then $\frac{a}{c} < \frac{b}{c}$

Property C_4 If $a < b$ and $c < 0$, then $\frac{a}{c} > \frac{b}{c}$

Example 6 ▶ Let a be any positive number. Then

(a) $4a < 5a$ because $4 < 5$

(b) $\frac{a}{5} < \frac{a}{4}$

(c) $-5a < -4a$ ◀

Here are some further properties that are often used.

Assume that a and b are nonzero numbers and that $a < b$. Then

Property D_1 $\frac{1}{b} < \frac{1}{a}$

If $c > 0$, then $c \cdot \frac{1}{b} < c \cdot \frac{1}{a}$ so that

Property D_2 $\frac{c}{b} < \frac{c}{a}$

If $c < 0$, then $c \cdot \frac{1}{a} < c \cdot \frac{1}{b}$ so that

Property D_3 $\frac{c}{a} < \frac{c}{b}$

Example 7 ▶ (a)

$$2 < 10$$

Thus $\frac{1}{10} < \frac{1}{2}$, by Property D_1.

Also, $3 > 0$ so that

$$\frac{3}{10} < \frac{3}{2}, \text{ by Property } D_2$$

And $-3 < 0$ so that

$$\frac{-3}{2} < \frac{-3}{10}, \text{ by Property } D_3$$

(b)

$$-10 < -2$$

Thus $\frac{1}{-2} < \frac{1}{-10}$, by Property D_1

that is, $\frac{-1}{2} < \frac{-1}{10}$

Also, $3 > 0$ so that

$$\frac{-3}{2} < \frac{-3}{10}, \text{ by Property } D_2$$

And $-3 < 0$ so that

$$\frac{-3}{-10} < \frac{-3}{-2}, \text{ by Property } D_3$$

or

$$\frac{3}{10} < \frac{3}{2}$$ ◀

All of these properties can be restated in terms of $\leq$ and $\geq$. For example:

Property A′ If $a \leq b$, then $a + c \leq b + c$ and $a - c \leq b - c$

Finally, observe that

$$\text{if } a < b, \text{ then } 0 < b - a$$

For by Property **A,**

$$\text{if } a < b, \text{ then } \underbrace{a - a}_{0} < b - a, \text{ that is, } 0 < b - a$$

It is also true that

$$\text{if } 0 < b - a, \text{ then } a < b$$

PRACTICE EXERCISES FOR TOPIC B

2. Let n be any number. Fill in "<" or ">".

(a) $10n$ ☐ $10(n - 1)$ (b) $-5(n + 4)$ ☐ $-5(n + 3)$

3. Fill in "<" or ">".

(a) If $k > 0$, then $\frac{2}{k}$ ☐ $\frac{-3}{k}$ (b) If $k < 0$, then $-5k$ ☐ $2k$

Now try the exercises for Topic B on page 292.

Use this area for work space.

EXERCISES

A. Let a, c, x, and y be any numbers.

In Exercises 1–8 fill in "$<$" or "$>$". (*See Examples 1–3.*)

1. $2 + c$ ☐ $4 + c$

2. $-3 + c$ ☐ $-1 + c$

3. $a + 6$ ☐ $a + 4$

4. $a - 8$ ☐ $a - 3$

5. $-2 + c$ ☐ c

6. If $x < y$, then $y - x$ ☐ 0

7. If $x < 3$, then $x + 2$ ☐ 5

8. If $x < y$, then $x - y$ ☐ 0

In Exercises 9–12 fill in "$\leq$" or "$\geq$".

9. $4 + c$ ☐ $9 + c$

10. $-3 + c$ ☐ $-7 + c$

11. $-b$ ☐ $2 - b$

12. If $x \leq y$, then $y - x$ ☐ 0

13. Suppose the temperature in downtown Los Angeles one day ranges from 75° (Fahrenheit) to 92°, and that the temperature at a local beach is always 8° cooler. Express the temperature range that day at the beach in terms of an inequality.

14. A man has at most \$75 to spend on spirits. He would like to buy wine at \$5 per bottle, gin at \$8 per bottle, and scotch at \$12 per bottle. Let x, y, and z be, respectively, the number of bottles of wine, gin, and scotch he can buy. Find an inequality (involving a sum) that expresses the number of bottles of each that he can buy.

15. The sum of the measures of the angles of any triangle is 180°. Suppose one angle of a triangle measures more than 90° but less than 180°. Let x and y be the number of degrees of the other two angles. Express the restrictions on $x + y$ in terms of inequalities.

B. In Exercises 16–36 fill in "$<$" or "$>$". (*See Examples 4–7 and Practice Exercises 2 and 3.*)

16. $10(7)$ ☐ $20(7)$

17. $10(-3)$ ☐ $20(-3)$

18. $\frac{1}{4}$ ☐ $\frac{1}{3}$

19. $\frac{1}{-5}$ ☐ $\frac{1}{-2}$

20. $(-7)(-2)$ ☐ $(-4)(-2)$

21. If $c > 0$, then $2c$ ☐ $5c$

22. If $c > 0$, then $-3c$ ☐ $-4c$

23. If $c < 0$, then $3c$ ☐ $4c$

24. If $c < 0$, then $-6c$ ☐ $-5c$

25. If $x < y$ and $z < 0$, then xz ☐ yz

26. If $b < -3$, then $\frac{b}{3}$ ☐ -1

27. If $x < 4$, then $-3x$ ☐ -12

28. If $x < 10$, then $\frac{-x}{10}$ ☐ -1

29. $5x$ ☐ $5(x + 1)$

30. $5x$ ☐ $5(x - 2)$

31. $5(x + 6)$ ☐ $5(x + 3)$

32. $-5x$ ☐ $-5(x + 1)$

33. $-5x$ ☐ $-5(x - 1)$

34. $\frac{x}{5}$ ☐ $\frac{x + 1}{5}$

35. If $x > 0$, then $\frac{5}{x}$ ☐ $\frac{5}{x + 1}$

36. If $x > 0$, then $\frac{-5}{x}$ ☐ $\frac{-5}{x + 1}$

Use this area for work space.

In Exercises 37–44 fill in "≤" or "≥".

37. If $c < 0$, then $5c$ ☐ $8c$

38. If $c < 0$, then $-5c$ ☐ $-9c$

39. $\frac{1}{2}$ ☐ $\frac{1}{4}$

40. $\frac{1}{8}$ ☐ $\frac{1}{5}$

41. If $x \leq 10$, then $4x$ ☐ 40

42. If $x \leq -2$, then $3x$ ☐ -6

43. If $x \leq -5$, then $-5x$ ☐ 25

44. If $x \leq 8$, then $\frac{-x}{8}$ ☐ -1

45. Pick out the *error(s)*. Assume that a, b, and c are any numbers.

(a) If $a < b$ and $c \neq 0$, then $ac^2 < bc^2$

(b) If $a > b$ and $c < 0$, then $ac < bc$

(c) If $a \leq b$ and $c > 0$, then $\frac{a}{c} \leq \frac{b}{c}$

(d) If $a \leq b$ and $c < 0$, then $ac - bc \leq 0$

(e) If $a \leq b$ and $c \leq 0$, then $a - c \geq b - c$

6.8 Solving Inequalities

Recall that an equation is a statement of equality, and has the form:

☐ = ☐

DEFINITION

An **inequality** is a statement of one of the forms:*

☐ < ☐

☐ ≤ ☐

☐ > ☐

☐ ≥ ☐

Example 1 ▶ Each of the following is an inequality:

(a) $1 < 2$ (b) $x + 2 \leq 6$ (c) $3x - 2 \geq 2x + 1$ ◀

A solution (or root) of an equation (in a single variable) is a number that when substituted in the equation yields a true statement. For each of the equations thus far considered there was only one solution.

*Strictly speaking, a statement of the form ☐ ≠ ☐ is also an inequality, but one that we will not be concerned with.

DEFINITION

> A **solution of an inequality** (in a single variable) is a number that when substituted in the inequality yields a true statement. The **solution set of an inequality** is the set (or collection) of all solutions.

For example, for the inequality $x > 0$, the solution set consists of all the positive numbers. Thus the solution set is the open ray $(0, \infty)$.

For the inequalities you will consider here, the solution set will be either an interval or a ray.

Recall that equivalent equations are equations with exactly the same solutions (or roots). You solve an equation by successively transforming it into simpler equivalent equations. This is also the way you solve an inequality.

DEFINITION

> **Equivalent inequalities** are inequalities with the same solution sets.

The arithmetic properties you studied in the preceding section enable you to transform an inequality into simpler equivalent inequalities. Thus adding or subtracting the same number, or multiplying or dividing by the same *positive* number on both sides of an inequality *preserves* the sense of the inequality. On the other hand, multiplying or dividing both sides of an inequality by the same *negative* number *reverses* the sense of the inequality. Here is how you use these properties to simplify an inequality.

A. Addition Principle (for Inequalities)

The **Addition Principle** asserts that you can add the same quantity to both sides of an inequality, and preserve the sense of the inequality.

Let ***R*** be a rational expression (possibly a polynomial or even a constant). Then

$$\square < \square$$

and

$$\square + \boldsymbol{R} < \square + \boldsymbol{R}$$

are equivalent inequalities.

Example 2 ► Solve $x - 3 < 5$

Solution:

Use the Addition Principle as you would for the *equation* $x - 3 = 5$. Thus add 3 to both sides of the given *inequality:*

$$x - 3 + 3 < 5 + 3$$
$$x < 8$$

The solution set is the open left ray $(-\infty, 8)$. ◄

Because $a - b = a + (-b)$, you can also subtract by means of the Addition Principle.

Example 3 ▶ Solve $x + 4 > 7$

Solution:

Subtract 4 from (or add -4 to) both sides.

$$x + 4 - 4 > 7 - 4$$

$$x > 3$$

The solution set is the open right ray $(3, \infty)$. ◀

Answers

1. ______

2. ______

PRACTICE EXERCISES FOR TOPIC A

1. Solve $x - 10 > -3$

2. Solve $2y < y + 6$

Now try the exercises for Topic A on page 298.

B. Multiplication Principle (for Inequalities)

The Multiplication Principle asserts that *you can multiply both sides of an inequality by the same positive number, and preserve the sense of the inequality.* But *if you multiply both sides by the same negative number, the sense of inequality is reversed.*

(a) Let $\boldsymbol{p} > 0$. Then

$$\square < \square$$

and

$$\square \cdot \boldsymbol{p} < \square \cdot \boldsymbol{p}$$

are equivalent inequalities.

(b) Let $\boldsymbol{n} < 0$. Then

$$\square < \square$$

and

$$\square \cdot \boldsymbol{n} > \square \cdot \boldsymbol{n}$$

are equivalent inequalities.

Example 4 ▶ Solve $\frac{x}{2} < 3$

Solution:

Multiply both sides of the inequality by the positive number 2. This preserves the sense of inequality.

$$\frac{x}{2} < 3$$

$$\frac{x}{2} \cdot 2 < 3 \cdot 2$$

$$x < 6$$

The solution set is the open left ray $(-\infty, 6)$. ◀

Example 5 ▶ Solve $-x < 2$

Solution:

Multiply both sides of the inequality by -1. This reverses the sense of inequality.

$$-x < 2$$

$$(-x)(-1) > 2(-1)$$

or

$$x > -2$$

The solution set is the open right ray $(-2, \infty)$. ◀

Because $\dfrac{a}{b} = a \cdot \dfrac{1}{b}$, $b \neq 0$ and because b and $\dfrac{1}{b}$ are either both positive or else both negative, the Multiplication Principle can be applied to divide both sides of an inequality by the same nonzero number.

Example 6 ▶ Solve $10y > 100$

Solution:

Divide both sides by the *positive* number 10 $\left(\text{or multiply by } \dfrac{1}{10}\right)$. This preserves the sense of inequality.

$$10y > 100$$

$$\frac{10y}{10} > \frac{100}{10}$$

$$y > 10$$

The solution set is the open right ray $(10, \infty)$. ◀

The Addition and Multiplication Principles also apply to inequalities involving $\leq$ or $\geq$.

Example 7 ▶ Solve $2x + 7 \leq 16 - x$

Solution:

Bring variables to one side, numerical terms to the other.

$$2x + 7 + x - 7 \leq 16 - x + x - 7$$

$$3x \leq 9$$

$$\frac{3x}{3} \leq \frac{9}{3}$$

$$x \leq 3$$

The solution set is the closed left ray $(-\infty, 3]$. ◀

Example 8 ▶ Solve $\dfrac{x}{4} \geq \dfrac{2}{3}$

Solution:

$$\frac{x}{4} \geq \frac{2}{3}$$ *Multiply both sides by 12, the lcd.*

$$\frac{x}{4} \cdot 12 \geq \frac{2}{3} \cdot 12$$

$$3x \geq 8$$

$$x \geq \frac{8}{3}$$

The solution set is the closed right ray $\left[\frac{8}{3}, \infty\right)$. ◀

Answers

3. ______
4. ______
5. ______
6. ______
7. ______

PRACTICE EXERCISES FOR TOPIC B

3. Solve $\frac{x}{20} > -1$

4. Solve $-5x < 30$

5. Solve $\frac{2}{3}x > 8$

6. Solve $x + 1 \leq 4(x + 4)$

7. Solve $18 - \frac{x}{10} \geq 0$

Now try the exercises for Topic B on page 299.

C. Simultaneous Inequalities

Sometimes you must solve two inequalities simultaneously.

Example 9 ▶ Solve $5 < x + 3 < 10$

Solution:

You are given two inequalities that hold for x:

$$5 < x + 3 \quad \text{and} \quad x + 3 < 10$$

You can work with both inequalities at the same time by subtracting 3 from each expression.

$$5 - 3 < x + 3 - 3 < 10 - 3$$

$$2 < x < 7$$

The solution set is the open interval (2, 7). ◀

Example 10 ▶ Solve $-3 \le 2x - 1 \le 3$

Solution:

$$-3 \le 2x - 1 \le 3 \qquad \textit{Add 1 to each expression.}$$

$$-3 + 1 \le 2x - 1 + 1 \le 3 + 1$$

$$-2 \le 2x \le 4 \qquad \textit{Divide each expression by 2.}$$

$$\frac{-2}{2} \le \frac{2x}{2} \le \frac{4}{2}$$

$$-1 \le x \le 2$$

The solution set is the closed interval $[-1, 2]$. ◀

Example 11 ▶ Solve $3 < -t < 7$

Solution:

Multiply each expression by -1. This *reverses* the senses of both inequalities.

$$3(-1) > (-t)(-1) > 7(-1)$$

$$-3 > t > -7$$

or, as this is usually written,

$$-7 < t < -3$$

The solution set is the open interval $(-7, -3)$. ◀

Answers

8. ______

9. ______

10. ______

PRACTICE EXERCISES FOR TOPIC C

8. Solve $0 \le 3x - 3 \le 18$

9. Solve $-10 < 6 - 2x < 7$

10. A car travels for 2 hours at a rate varying from 50 to 65 miles per hour. Describe the distance traveled in terms of an interval.

Now try the exercises for Topic C on page 299.

Use this area for work space.

EXERCISES

A. In Exercises 1–8 solve the indicated inequalities. (*See Examples 2 and 3.*)

1. $x - 2 < 4$ **2.** $x - 4 < 0$ **3.** $x + 6 < 9$

4. $y + 1 > -1$ **5.** $t + 3 < 12$ **6.** $2 - z < 4$

7. $u + 4 > 9$ **8.** $t - \frac{1}{2} > \frac{1}{2}$

Use this area for work space.

B. In Exercises 9–30 solve the indicated inequalities. (*See Examples 4–8.*)

9. $4x < 20$

10. $5x \le -15$

11. $\frac{y}{3} > 2$

12. $\frac{z}{4} \le 0$

13. $-t > 4$

14. $-2x < 18$

15. $-5y \ge -25$

16. $\frac{x}{3} \le \frac{1}{9}$

17. $\frac{-3x}{4} > 2$

18. $x + 3 < 4x - 30$

19. $5y + 2 > 3y + 8$

20. $10 - 4z \le 6z - 10$

21. $3x - 1 \ge x + 8$

22. $4(x + 1) - 2(x + 5) > 0$

23. $\frac{x}{2} < \frac{3}{4}$

24. $\frac{2x}{5} > \frac{3}{10}$

25. $\frac{3 - y}{3} < \frac{y}{4}$

26. $\frac{u}{4} + \frac{1}{2} \le \frac{3}{8}$

27. $4(u - 3) - 3(u + 2) \le 0$

28. $\frac{1 - 2z}{10} \le \frac{z}{5}$

29. $\frac{t}{3} - \frac{1}{6} \le \frac{3t}{4}$

30. $\frac{5x - 1}{2} + \frac{1}{5} > \frac{x + 1}{10}$

C. In Exercises 31–46 solve the given simultaneous inequalities. (*See Examples 9–11.*)

31. $0 < x - 5 < 10$

32. $-3 \le x + 4 \le 3$

33. $6 \le 3x \le 18$

34. $-12 \le 2y \le -4$

35. $-\frac{1}{2} \le \frac{x}{4} \le \frac{1}{2}$

36. $-1 \le 2y + 3 \le 11$

37. $0 \le \frac{x}{4} \le 16$

38. $-16 \le 5x - 1 \le 14$

39. $-4 < \frac{x + 4}{2} < 4$

40. $-3 \le -x \le 2$

41. $-6 < -x < 6$

42. $0 < -3x < 9$

43. $-3 \le 1 - 2y \le 9$

44. $-2 < 3 - 5z < 4$

45. $\frac{1}{2} < \frac{3 - 2t}{4} < 1$

46. $\frac{1}{2} \le \frac{3}{4} - 2x \le 1$

47. A turnpike has a minimum speed of 40 miles per hour and a maximum speed of 60 miles per hour. A truck travels on this turnpike for 4 hours within the legal range. Describe the distance traveled in terms of an interval. (*See Practice Exercise 10.*)

48. During the preliminary phase of labor, the expectant mother must take 6 to 9 deep breaths per minute. What is the range of breaths she should take during a 40-second period?

49. One May morning the temperature in Baltimore ranges between 68° and 77° Fahrenheit. Celsius (C) is related to Fahrenheit (F) by the formula

$$C = \frac{5}{9}(F - 32)$$

What is the temperature range on the Celsius scale in Baltimore that morning?

50. In Paris one January morning the temperature ranges between 5° and 10° Celsius. What is the range on the Fahrenheit scale? (*See Exercise 49.*)

What Have You Learned in Chapter 6?

You have learned how to solve equations that involve rational numbers, rational expressions, or decimals.

You know that a proportion is an equation of the form $\frac{a}{c} = \frac{b}{d}$. You can solve a proportion by cross multiplying, or by multiplying by the *lcd* of the expressions.

You have solved applied problems that concern interest, commissions, markups, and discounts.

You know that a literal equation is one that contains at least two variables and that must be solved for one variable in terms of the others.

You have learned that $a \leq b$ indicates that a is less than or equal to b. You know that the open interval (a, b) consists of all numbers x such that $a < x < b$, and that the closed interval $[a, b]$ consists of all numbers in (a, b) as well as the two endpoints, a and b. You have also considered various types of rays, such as, for example, the open right ray (a, ∞), which consists of all numbers x such that $a < x$.

You know that if you add or subtract the same number or if you multiply or divide by the same *positive* number on both sides of an inequality, the sense of inequality is preserved. But if you multiply or divide by the same *negative* number on both sides of an inequality, the sense of inequality is reversed.

You have also learned how to solve various inequalities. A solution of an inequality (in a single variable) is a number that when substituted in the inequality yields a true statement. The solution set of an inequality is the collection of all solutions. You solve an inequality by successively transforming it into simpler equivalent inequalities, that is, inequalities with the same solution sets.

Let's Review Chapter 6

6.1 *Proportions and Equations with Rational Expressions*

1. Find the value of c in the proportion $\frac{6}{c} = \frac{-2}{9}$.

In Exercises 2–4 solve each equation.

2. $\frac{x}{7} + \frac{x}{2} = 9$ **3.** $\frac{1}{y} + \frac{1}{2} = \frac{3}{4}$ **4.** $\frac{3}{x-2} + \frac{5}{x+2} = \frac{6}{x-2}$

5. A hostess prepares 55 hamburgers for 20 guests. How many hamburgers should she prepare for 28 guests?

6.2 *Decimals and Equations*

6. (a) Express $\frac{7}{20}$ as a decimal. (b) Express .04 as a rational number. Simplify.

7. Compute (a) $24.2 \times .002$ (b) $\frac{1 - .001}{.03}$

In Exercises 8–10 solve each equation.

8. $\frac{x}{.04} = 3$ **9.** $3.2y = .064$ **10.** $.005(t + 2) = 4$

REVIEW

6.3 Interest Problems

11. How much interest is earned in 1 year on a principle of $900 if the annual interest rate is $6\frac{1}{2}\%$?

12. Frank invests $500 more at 8% than at 9%. His total yearly interest from these two investments is $210. How much does he invest at each rate?

13. Joan invests $4000 at 10% compounded quarterly. How much does she earn in half a year?

6.4 Commissions, Markups, and Discounts

14. A salesman works on a $12\frac{1}{2}\%$ rate of commission. What must his sales be in order to earn a commission of $250?

15. A bookstore buys an art book for $12 and marks it up 40%. At what price does the bookstore sell the art book?

16. An encyclopedia is on sale for $120. The rate of discount is 25%.
(a) What is the marked price? (b) What is the discount?

17. A coat normally sells for $200.
(a) If the price is increased to $250, what is the rate of markup?
(b) If the price is reduced from $200 to $150, what is the rate of discount?

6.5 Literal Equations

18. Solve for x: $2x - y = 3$

19. Solve for y: $3 - 2x + 7y - 9 = 0$

20. Solve $\frac{3a}{b} = -2$ (a) for a, (b) for b.

21. (a) Solve $7u - 3v = 0$ for v. (b) *Check your result.*

22. Solve $ax + by - cx = 1$ for x.

6.6 Intervals

23. Fill in "$\leq$" or "$\geq$". If both symbols apply, write "both."
(a) $12 \square 6$ (b) $-4 \square -3$ (c) $7 \square 7$

24. Answer "true" or "false."
(a) 3 is in $(0, 3)$. (b) -5 is in $[-6, 0)$. (c) 4 is in $[4, \infty)$.

25. Describe in one of the forms

$$(a, b), \quad [a, b], \quad (a, \infty), \quad [a, \infty), \quad (-\infty, b), \quad (-\infty, b]$$

the set of numbers satisfying each of the following.
(a) $4 < x < 7$ (b) $-2 \leq x \leq -1$ (c) $x < 2$

6.7 Arithmetic of Inequalities

26. Fill in "<" or ">".
(a) $3 + c \ \square \ 5 + c$
(b) If $c > 0$, then $6c \ \square \ 2c$
(c) If $c < 0$, then $-4c \ \square \ -3c$

27. Fill in "≤" or "≥".
(a) $10 - c \ \square \ 7 - c$
(b) If $a \leq b$, then $b - a \ \square \ 0$
(c) If $x \leq 5$, then $-6x \ \square \ -30$

6.8 Solving Inequalities

In Exercises 28–31 solve the indicated inequalities.

28. $2 - x < 5$

29. $5(x + 1) - 2(x + 2) \geq 0$

30. $\frac{t}{2} - \frac{3}{4} \leq \frac{5t}{8}$

31. $3 \leq 2y - 1 \leq 9$

32. A bakery finds that it sells at least 50 and at most 100 loaves of bread every day. What is the range of loaves it sells during a 7-day week?

And These from Chapters 1-5

33. Find the value of

$$20 \div 4 - 7 \cdot 2 + 3 \cdot 3$$

34. Evaluate $2t^4 - 3t^2 - 5t + 1$ when $t = -1$.

35. Evaluate $4xy - y^2 + 3$ when $x = 2$ and $y = 10$.

36. Add $3x - 1$ to the product of $5x - 2$ and $x - 3$.

37. Multiply the sum of $4y^2 - 1$ and $3y + 1$ by $y - 2$.

38. Divide $\dfrac{25a^4 - 10a^2 + 15a}{5a}$

39. Solve the equation

$$5x - 2 - 3x = 1 - 7x + 7$$

40. Solve the equation

$$1 - (2 - t) = 3(4 - 2t)$$

41. Two cars leave a toll booth at the same time traveling in the same direction along a straight road. If one car averages 45 miles per hour and the other car averages 62 miles per hour, how far apart are they after 4 hours?

42. Factor $9x^2y^2 - 16x^2$

43. Factor $3x^2 - 11x + 10$

44. Factor $5a^4 - 45a^2b^2$

45. Solve the equation

$$2(x - 3) + 5(2x - 1) = 10x - 1$$

46. Multiply $\frac{4}{(x-1)^2} \cdot \frac{x^2-1}{8x} \cdot \frac{2}{x+1}$

47. Divide $\frac{3x^2y}{x^2-16} \div \frac{12x^6y}{x^2+5x+4}$

48. Add $\frac{4}{y+2} + \frac{3y}{2y+4} + \frac{y}{y^2-4}$

Practice Exam 1 on Chapter 6

1. Solve $\frac{x+3}{4} = \frac{2x}{5}$

2. The price of melons is four for a dollar. How many melons can be bought for $2.25?

3. Solve $\frac{3}{t-3} + \frac{6}{t} = \frac{21}{2t}$

4. Solve $.01x - 1 = .4$

5. How much money must be invested for a year at an annual interest rate of $9\frac{1}{2}\%$ in order to earn $285 interest?

6. A woman sells her house for $50 000. If she must pay the real estate agent 6% of the sales price, how much does the woman actually receive?

7. Solve $3x + 5y = 10$ for y.

8. Solve $3ab + 2a = 5$
   (a) for a
   (b) for b.

9. Answer "true" or "false."
   (a) $4 \le 4$
   (b) 3 is in $(-\infty, 3)$.
   (c) -2 is in $[-3, 0]$.

10. Fill in "<" or ">".
   (a) $6 - c$ ☐ $-6 - c$
   (b) If $x > y$, then $x - y$ ☐ 0
   (c) If $c < 0$, then $2c$ ☐ $3c$

11. Solve the inequality $2y - 1 \le 8 - y$

12. Solve the inequality $4 < \frac{x+1}{2} < 6$

Practice Exam 2 on Chapter 6

1. Solve $\frac{2}{3}x - 3 = \frac{x+3}{9}$

2. Solve $\frac{1}{x-5} = \frac{7}{2x}$

3. If 8 ounces of milk contain 140 calories, how many calories are contained in 10 ounces of milk?

REVIEW

4. Express $\frac{9}{75}$ (a) as a decimal and (b) as a percent.

5. Solve and *check* $.04(x + 3) = 2.2$

6. If \$8400 is invested in an account that pays 7% interest annually, how much interest is earned in 3 months?

7. A furniture dealer sold a sofa originally priced at \$900 at a 40% discount. What was the sale price of the sofa?

8. Solve $3(a - 5) = 2b + 1$ for a.

9. Solve $10x - 3xy = 4$

   (a) for x

   (b) for y.

10. Answer "true" or "false."

    (a) $-7 \geq -7$

    (b) $-4 \leq -5$

    (c) -100 is in $(-\infty, 100)$.

    (d) 4 is in $(-4, 4)$.

11. Fill in "<" or ">".

    (a) $-4 + c \ \square \ 7 + c$

    (b) If $k > 0$, then $\frac{-2}{k} \ \square \ \frac{1}{k}$.

    (c) If $k < 0$, then $10k \ \square \ -3k$.

12. Solve the inequality

$$15 - 3x \geq 1 - x$$

13. Solve $6 \leq \frac{x}{3} + 7 \leq 9$

7

Lines and Linear Systems

7.1 Rectangular Coordinates

A. Coordinate Axes

Recall that numbers correspond to points on a horizontal line, the *number line*. Now you will see how *pairs of numbers*, in a definite order, correspond to points on a plane.

Draw a vertical line through the origin, 0, on the number line. (*See Figure 7.1.*) Because the number line is horizontal, the two lines are perpendicular. The horizontal line will now be called the **x-axis,** and the vertical line will be called the **y-axis.** The two lines intersect at the origin, which will represent 0 on the y-axis (as well as on the x-axis). For convenience, choose the same distance unit on the y-axis as on the x-axis. *On the y-axis, positive numbers are plotted upward from 0, and negative numbers downward.* (*See Figure 7.2.*)

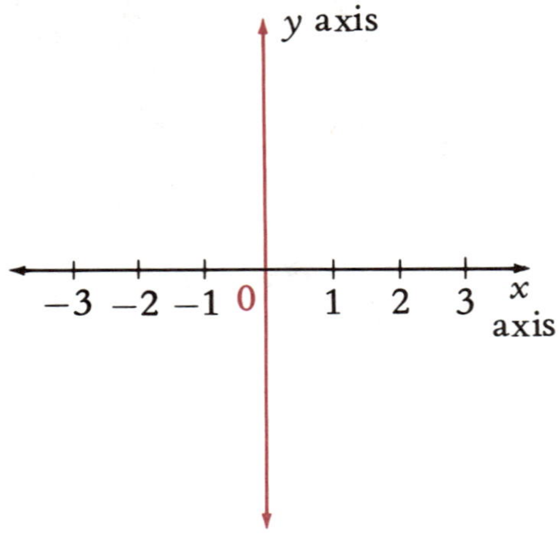

FIGURE 7.1

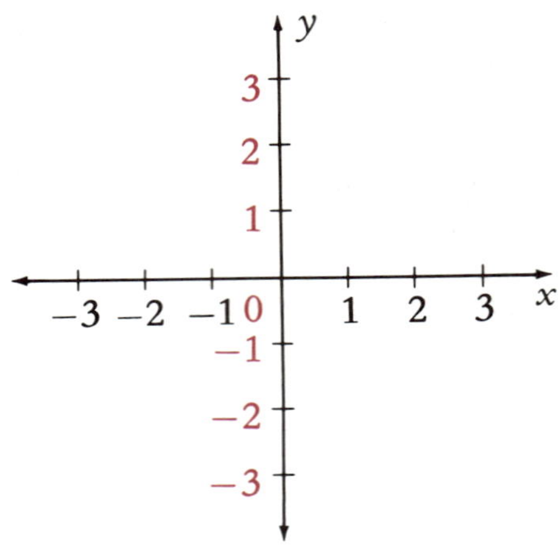

FIGURE 7.2. **On the y-axis, positive numbers are plotted upward from 0, and negative numbers downward.**

Example 1 ▶ Plot the following points on the y-axis.

(a) 3 (b) $\frac{1}{2}$ (c) -2 (d) $\frac{-5}{2}$

Solution:
See Figure 7.3. ◀

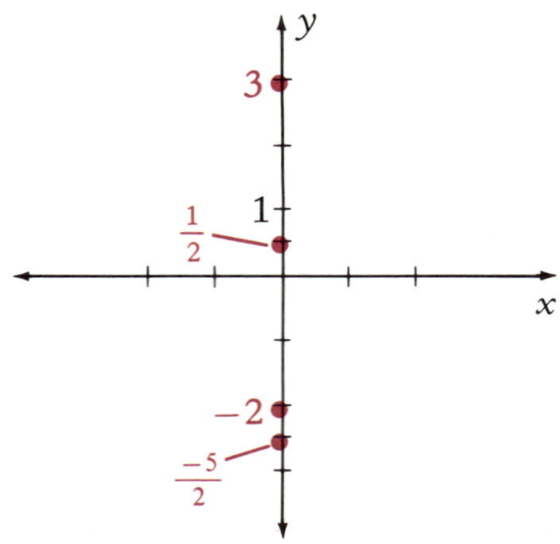

FIGURE 7.3

PRACTICE EXERCISES FOR TOPIC A

1. Plot the following points on the y-axis.

(a) 1.5 (b) −.5 (c) −3

Now try the exercises for Topic A on page 309.

B. Coordinates

Every number corresponds to a point on the x-axis and to a point on the y-axis. If you consider two numbers, a and b, then a corresponds to a point on the x-axis and

b to a point on the y-axis. Locate these points and draw perpendiculars to the axes through each of them, as in Figure 7.4(a). The intersection of these perpendiculars is the point P of the "x, y-plane" associated with a and b in this order. To indicate the order of a and b, write

$$(a, b)$$

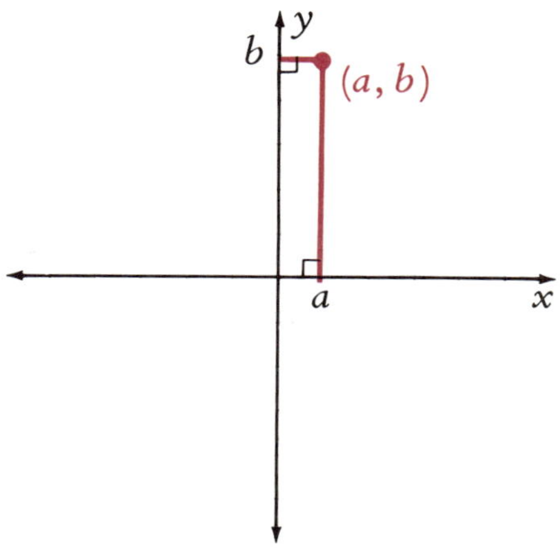

FIGURE 7.4(a)

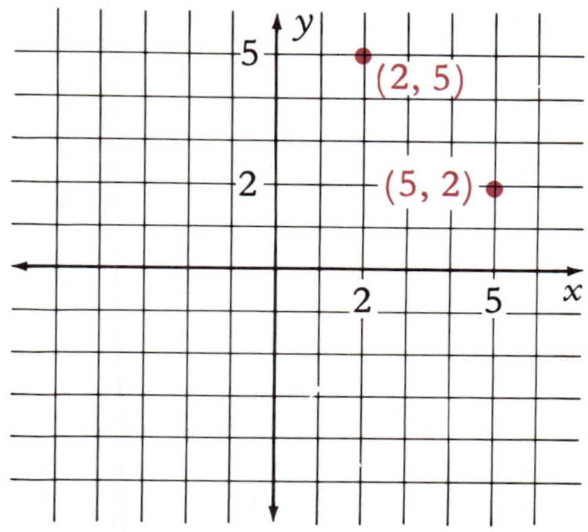

FIGURE 7.4(b). **(2, 5) ≠ (5, 2)**

Call (a, b) an **ordered pair.** *The ordered pair* (a, b) *indicates two numbers* a *and* b *in the order written.* The number a is called the **x-coordinate,** or **abscissa,** of (a, b) and b the **y-coordinate,** or **ordinate,** of (a, b).

Thus 2 is the x-coordinate of (2,5), and 5 is the y-coordinate. On the other hand, if you change the order of the coordinates, then 5 is the x-coordinate and 2, the y-coordinate of (5,2). Observe that

$$(2, 5) \neq (5, 2)$$

(*See Figure* 7.4(*b*).)

Every ordered pair of numbers, (a, b), corresponds to exactly one point P on the x,y-plane, as indicated. Moreover, every point P on the x,y-plane also corresponds to exactly one ordered pair of numbers, (a, b). Together, a and b are called the **rectangular coordinates** of P or, simply, the **coordinates** of P. You will *identify* a point P of the plane with its ordered pair of coordinates (a, b), and write

$$P = (a, b)$$

The x-axis together with the y-axis are called the **coordinate axes.**

Example 2 ► For each point in Figure 7.5, find
(a) its x-coordinate,
(b) its y-coordinate.
(c) Identify the point with its ordered pair of coordinates.

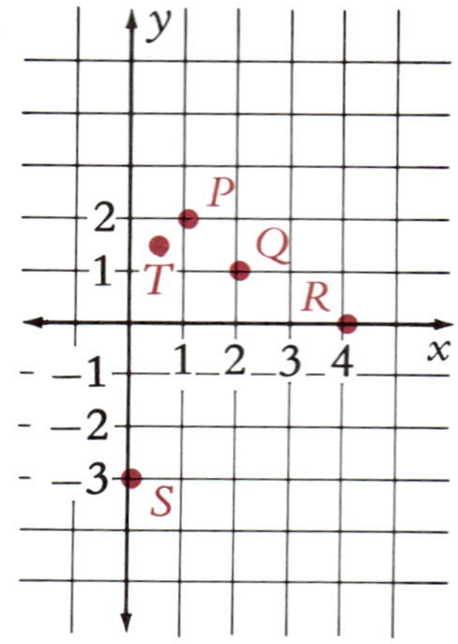

FIGURE 7.5

Solution:

P:	(a) 1,	(b) 2,	(c) $P = (1, 2)$
Q:	(a) 2,	(b) 1,	(c) $Q = (2, 1)$
R:	(a) 4,	(b) 0,	(c) $R = (4, 0)$
S:	(a) 0,	(b) −3,	(c) $S = (0, -3)$
T:	(a) $\frac{1}{2}$,	(b) $\frac{3}{2}$,	(c) $T = \left(\frac{1}{2}, \frac{3}{2}\right)$

◀

Observe that points on the *x-axis* have *y-coordinate* 0 and points on the *y-axis* have *x-coordinate* 0. The origin, which is on both coordinate axes, has both of its coordinates equal to 0. (*See Figure 7.6.*)

$$\text{Origin} = (0, 0)$$

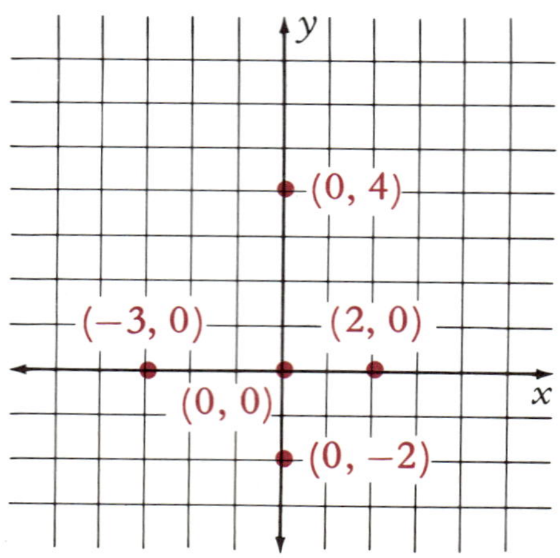

FIGURE 7.6. Points on the x-axis have y-coordinate 0. Points on the y-axis have x-coordinate 0.

Example 3 ▶ Plot the following ordered pairs on a rectangular coordinate system.

$$(4, 1),\ (1, 4),\ (-2, -2),\ (0, -1),\ (.5, -.5)$$

Solution:
See Figure 7.7. ◀

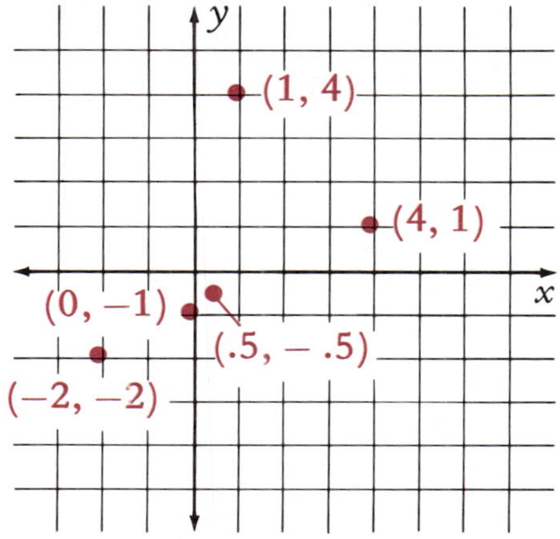

FIGURE 7.7

Answers

2. (a) ____

(b) ____

(c) ____

PRACTICE EXERCISES FOR TOPIC B

2. For each point in Figure 7.8, find

(a) its x-coordinate, (b) its y-coordinate.

(c) Identify the point with its ordered pair of coordinates.

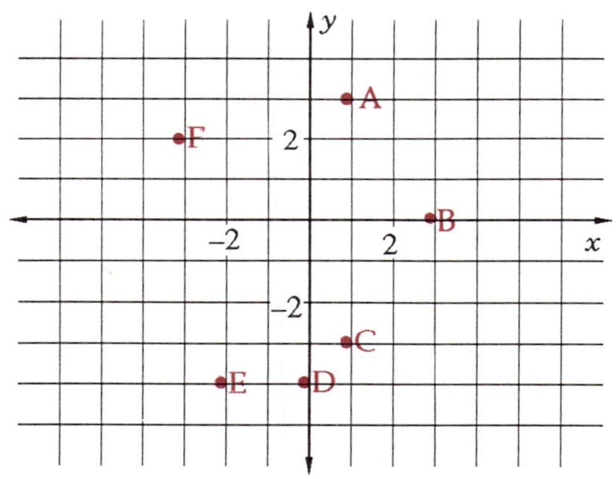

FIGURE 7.8

3. Plot the following ordered pairs on a rectangular coordinate system.

$(-2, 0)$, $(0, 1)$, $(2, -3)$, $(-1, 3.5)$

Now try the exercises for Topic B on page 310.

Use this area for work space.

EXERCISES

A. **1.** Draw a rectangular coordinate system. Plot the following points on the y-axis. (*See Example 1.*)

(a) 4, (b) -4, (c) $\frac{-1}{2}$, (d) $\frac{3}{2}$

Use this area for work space.

2. In Figure 7.9 indicate which numbers are represented by the following points on the y-axis:
(a) P, (b) Q, (c) R, (d) S, (e) T

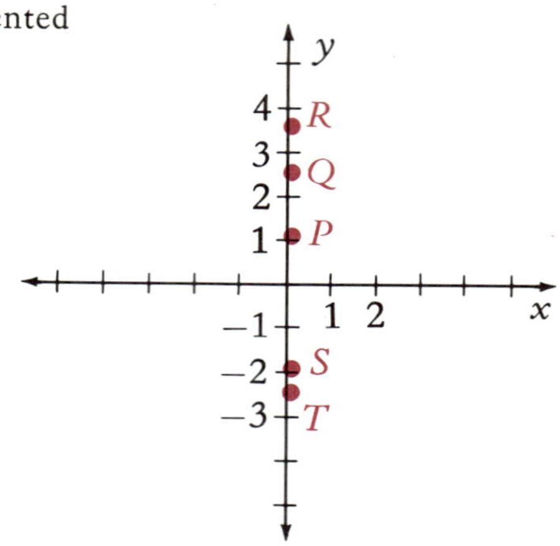

FIGURE 7.9

B. **3.** For each point in Figure 7.10, find
(a) its x-coordinate,
(b) its y-coordinate.
(c) Identify the point with its ordered pair of coordinates. (*See Example 2.*)

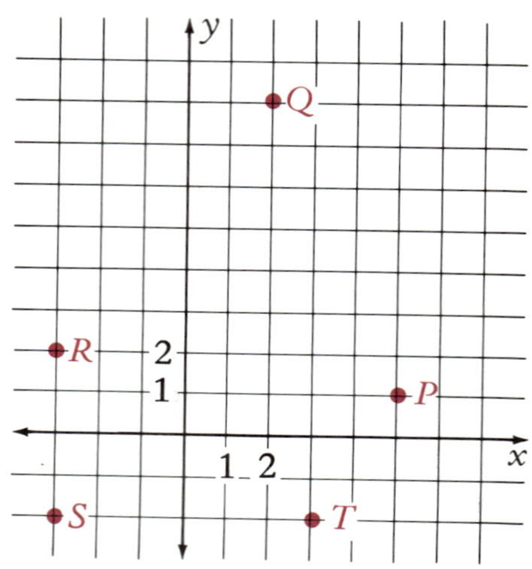

FIGURE 7.10

4. For each point in Figure 7.11, find
(a) its x-coordinate,
(b) its y-coordinate.
(c) Identify the point with its ordered pair of coordinates.

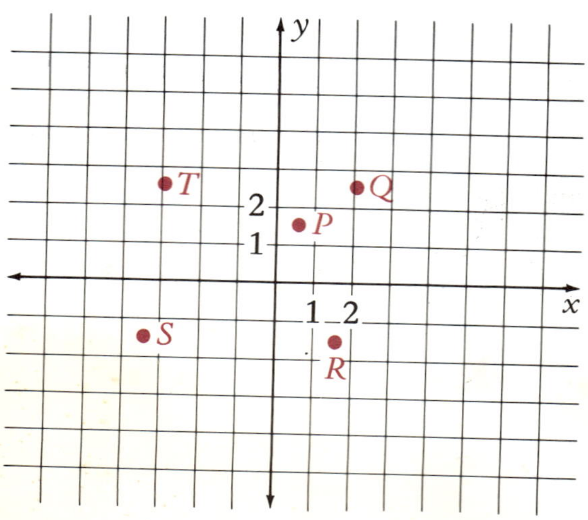

FIGURE 7.11

Use this area for work space.

Plot the following ordered pairs on a rectangular coordinate system. (You may wish to draw a coordinate system for the odd-numbered exercises first, and then a separate coordinate system for the even-numbered exercises.) (*See Example* 3.)

5. (5, 0) **6.** (0, 3) **7.** (−2, 0) **8.** (0, −3)

9. $\left(\frac{1}{2}, 0\right)$ **10.** $\left(0, \frac{-1}{2}\right)$ **11.** $\left(\frac{7}{2}, 0\right)$ **12.** $\left(0, \frac{-3}{2}\right)$

13. (2, 2) **14.** (−1, −1) **15.** (4, 2) **16.** (2, 4)

17. (−4, 2) **18.** (−2, 4) **19.** (−2, −4) **20.** (−4, −2)

21. $\left(6, \frac{1}{2}\right)$ **22.** $\left(\frac{-1}{2}, 4\right)$ **23.** (2, −5) **24.** (−6, −6)

25. (6, −1) **26.** (1, −10) **27.** (.5, 1.5) **28.** (−.5, −2.5)

7.2 Graphs of Equations

A. Equations in Two Variables

In this chapter you will consider equations in two variables, x and y, such as

$$y = x + 5$$

$$y = 2x$$

$$y = 3x - 1$$

and

$$5x - 2y = 10$$

Consider the equation

$$y = x + 5$$

Note that when you replace x by 1 and y by 6, a true statement

$$6 = 1 + 5$$

results. In any equation in x and y, you may want to consider what happens when you replace the variable x by a *number* x_1 (*read:* ex-sub-one) and the variable y by another *number* y_1. Note that here the *letters with subscripts* x_1 and y_1 stand for specific *values* of the variables x and y.

DEFINITION

> An ordered pair (x_1, y_1) is a **solution** of an equation in the variables x and y if a true statement results when x_1 replaces x and y_1 replaces y.

Thus (0, 3) is a solution of the equation

$$y = 2x + 3$$

because when you replace x by 0 and y by 3, a true statement,

$$3 = 2 \cdot 0 + 3$$

results. Note that (1, 5) is also a solution because when you replace x by 1 and y by 5, a true statement,

$$\mathbf{5} = 2 \cdot 1 + 3$$

results. But (2, 6) is *not* a solution because when you replace x by 2 and y by 6, the resulting statement

$$\mathbf{6} = 2 \cdot 2 + 3$$

is *false*.

Answers

1. ______

2. (a) ______

(b) ______

(c) ______

PRACTICE EXERCISES FOR TOPIC A

1. Which of the following ordered pairs are solutions of the equation $y = 3x - 4$?

(a) (2, 2) (b) (1, 1) (c) (6, 2) (d) (0, −4) (e) (3, 5)

2. Find the missing coordinates to make each of the following ordered pairs satisfy the equation $x + 2y = 6$.

(a) (0,) (b) (, 1) (c) (, 0)

Now try the exercises for Topic A on page 318.

B. The Graphing Procedure

DEFINITION

> The **graph** of an equation is its pictorial representation on a rectangular coordinate system. The graph consists of all points (x, y) that are solutions of the equation.

Example 1 ► Graph the equation $y = x + 2$

Solution:

Choose values of x and find the corresponding values of y. Several such ordered pairs are indicated in Table 7.1. Plot the corresponding points, as in Figure 7.12. Note that a straight line appears to go through these points. ◄

TABLE 7.1

x	$y = x + 2$
0	2
1	3
2	4
3	5
4	6

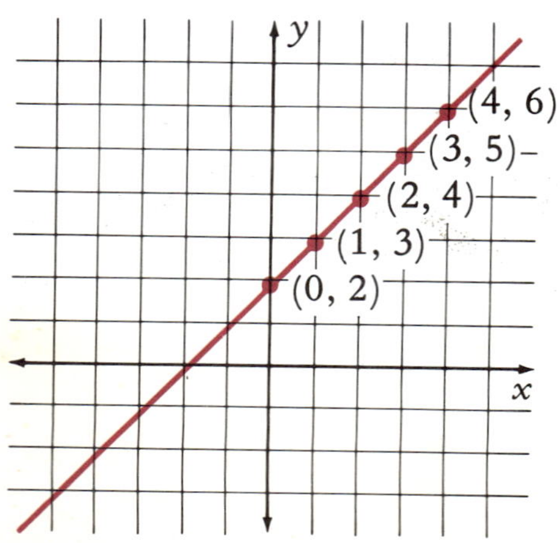

FIGURE 7.12. The graph of $y = x + 2$

The graph of an equation such as

$$y = x + 2$$
$$y = 2x$$
$$y = 3x - 5$$
$$y = -3x$$
$$y = 6$$
$$x = 6$$

or

$$2x + 3y = 9$$

is a line. In such an equation, x or y or both of these variables may appear and *only to the first power.* Thus

$$y = x^2$$

is *not* the equation of a line because x appears to the *second* power.

A line is determined by two points. A line extends indefinitely in both directions.

To graph a line:

1. Find two ordered pairs that are solutions of the given equation.
2. Plot the corresponding points.
3. Draw the line connecting these points.

Example 2 ▶ Graph the equation $y = x - 1$

Solution:

x and y each appear only to the first power. Therefore this is the equation of a line. Choose two values of x—for simplicity, 0 and 1. Then find the corresponding values of y in order to obtain two points on the line. See Table 7.2. The graph is drawn in Figure 7.13. ◀

TABLE 7.2

x	$y = x - 1$
0	-1
1	0

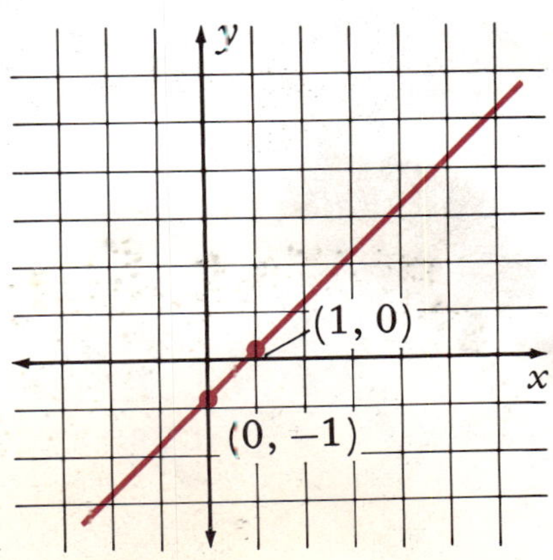

FIGURE 7.13. The graph of $y = x - 1$

Example 3 ▶ Graph the equation $y = 4x$
Solution:
This is the equation of a line. Corresponding values of x and y are given in Table 7.3. The graph is drawn in Figure 7.14. ◀

TABLE 7.3

x	$y = 4x$
0	0
1	4

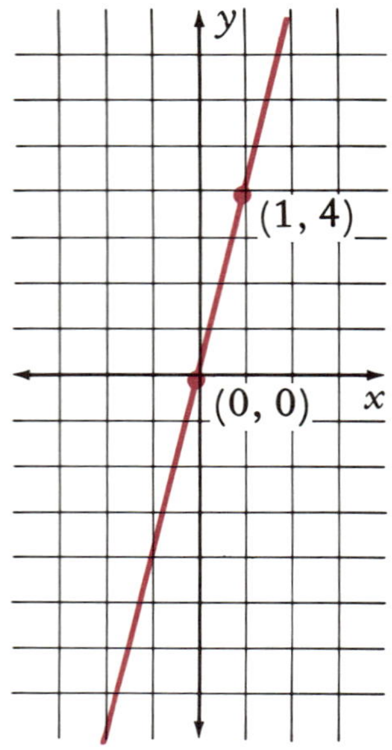

FIGURE 7.14. The graph of y = 4x

Example 4 ▶ Graph the equation $y = -2x$
Solution:
See Table 7.4 and Figure 7.15. ◀

TABLE 7.4

x	$y = -2x$
0	0
1	−2

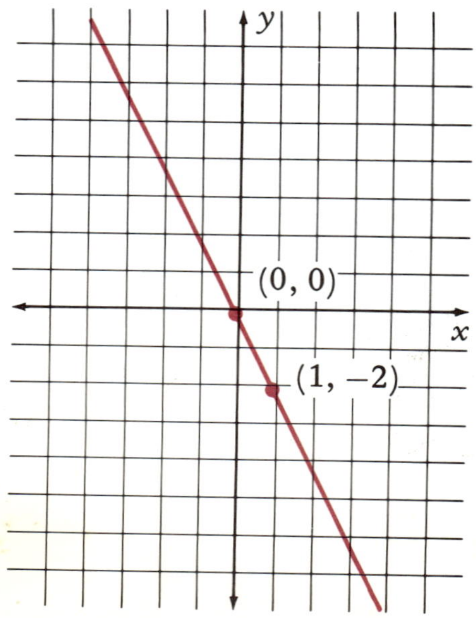

FIGURE 7.15. The graph of y = −2x

Example 5 ▶ Graph the equation $y = 2x + 1$
Solution:
See Table 7.5 and Figure 7.16. ◀

TABLE 7.5

x	$y = 2x + 1$
0	$2(0) + 1 = 0 + 1 = 1$
1	$2(1) + 1 = 2 + 1 = 3$

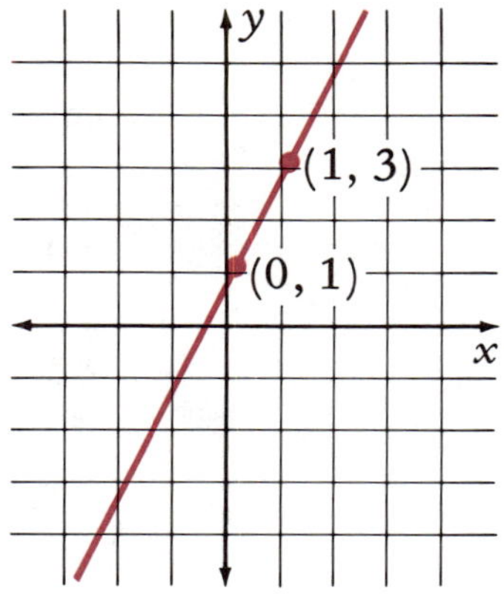

***FIGURE 7.16.* The graph of y = 2x + 1**

Example 6 ▶ Graph the equation $y = 3x - 4$
Solution:
See Table 7.6 and Figure 7.17. ◀

TABLE 7.6

x	$y = 3x - 4$
0	$3(0) - 4 = 0 - 4 = -4$
1	$3(1) - 4 = 3 - 4 = -1$

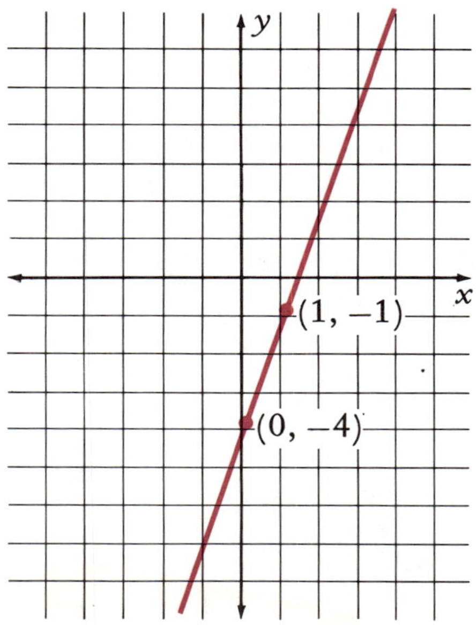

***FIGURE 7.17.* The graph of y = 3x − 4**

Example 7 ▶ Graph the equation $2x + 3y = 6$
Solution:
Here both x and y appear on the left side. It is convenient to first let $x = 0$ in the given equation and then let $y = 0$.

$x = 0$:
$$2(0) + 3y = 6$$
$$3y = 6$$
$$y = 2$$

$y = 0$:

$$\begin{aligned} 2x + 3(0) &= 6 \\ 2x &= 6 \\ x &= 3 \end{aligned}$$

These corresponding values of x and y are given in Table 7.7. The graph of the equation is drawn in Figure 7.18. ◀

TABLE 7.7

x	y
0	2
3	0

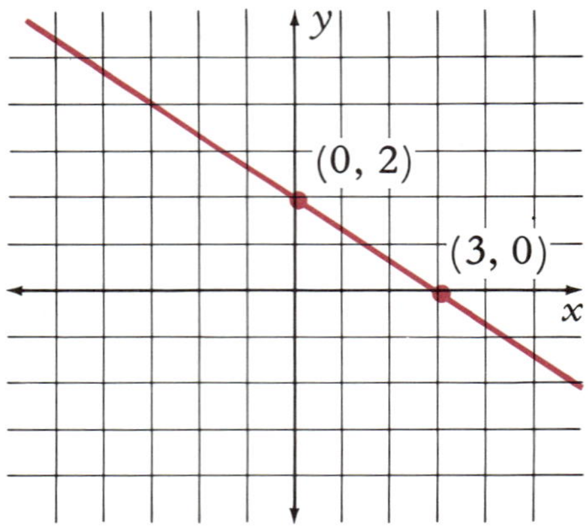

FIGURE 7.18. The graph of the equation $2x + 3y = 6$

PRACTICE EXERCISES FOR TOPIC B

3. Fill in Table 7.8 with the appropriate y-values and graph the equation $y = 3 - x$.

TABLE 7.8

x	$y = 3 - x$
−2	
0	
2	

4. Fill in the missing values in Table 7.9 and graph the equation $2x - y = 4$.

TABLE 7.9

x	y
0	
	0

Now try the exercises for Topic B on page 319.

C. Horizontal Lines and Vertical Lines

Example 8 ▶ Graph the equation $y = 3$

Solution:

Because x does not appear in the given equation, an ordered pair (x_1, y_1) is a solution of this equation only if $y_1 = 3$. Thus

$$\underset{\substack{\uparrow\\ y_1}}{(0, 3)} \text{ is a solution}$$

because its y-coordinate is 3. Also

$$\underset{\substack{\uparrow\\ y_1}}{(2, 3)} \text{ is a solution}$$

because its y-coordinate is 3. Consequently (0, 3) and (2, 3) are points on the line with equation

$$y = 3$$

Draw the line through (0, 3) and (2, 3) and note that it is *horizontal.* (*See Figure 7.19.*) Other points on this horizontal line are (5, 3), (−1, 3), and (−6, 3). In each case the y-coordinate is 3. ◀

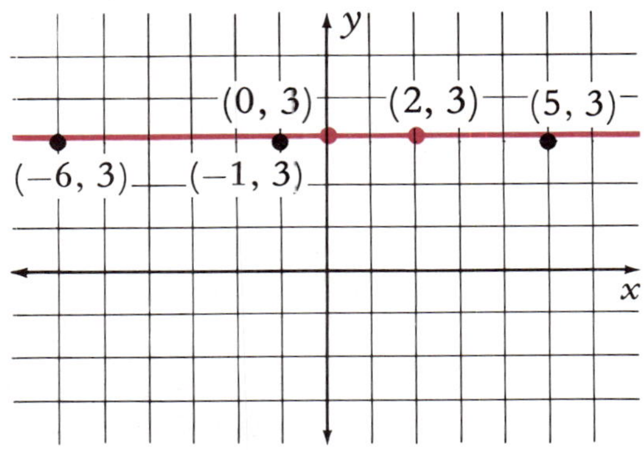

FIGURE 7.19. The graph of y = 3

Example 9 ▶ Graph the equation $x = -2$

Solution:

Here y does not appear in the given equation. Thus an ordered pair (x_1, y_1) is a solution of this equation only if $x_1 = -2$. Therefore

$$\underset{\substack{\uparrow\\ x_1}}{(-2, 0)} \text{ is a solution}$$

because its x-coordinate is −2. Also

$$\underset{\substack{\uparrow\\ x_1}}{(-2, 4)} \text{ is a solution}$$

because its x-coordinate is −2. Thus (−2, 0) and (−2, 4) are points on the line with equation

$$x = -2$$

Draw the line through $(-2, 0)$ and $(-2, 4)$ and observe that it is a *vertical* line. (*See Figure 7.20.*) Other points on this vertical line are $(-2, 1)$, $(-2, -2)$, and $(-2, -6)$. In each case the x-coordinate is -2. ◀

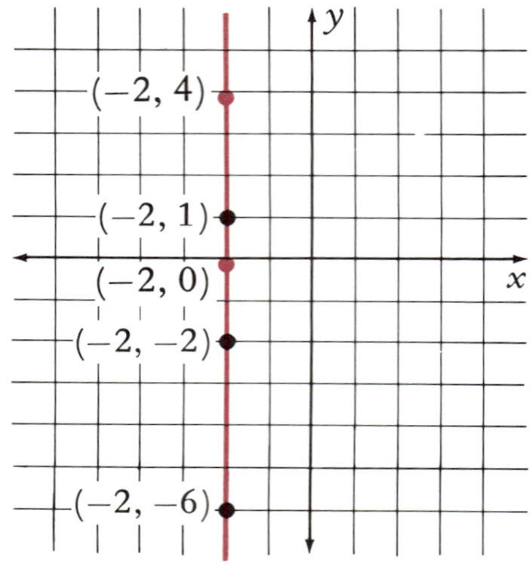

FIGURE 7.20. ***The graph of*** $x = -2$

PRACTICE EXERCISES FOR TOPIC C

5. Graph the equation $y = -2$.

6. Graph the equation $x - 3 = 0$.

Now try the exercises for Topic C on page 320.

Use this area for work space.

EXERCISES

A.

1. Which of the following ordered pairs are solutions of the equation $y = 4x + 1$? (*See Practice Exercise 1.*)

(a) $(4, 1)$ (b) $(1, 5)$ (c) $(0, 1)$

(d) $\left(\frac{1}{4}, 2\right)$ (e) $(-1, -4)$

2. Which of the following ordered pairs are solutions of the equation $y = 10x - 5$?

(a) $(0, -5)$ (b) $(1,5)$ (c) $(10, -5)$

(d) $(.1, -4)$ (e) $\left(\frac{1}{2}, 0\right)$

3. Which of the following ordered pairs are solutions of the equation $y = 5 - 2x$?

(a) $\left(\frac{2}{5}, 0\right)$ (b) $(1, 3)$ (c) $(0, 5)$

(d) $\left(\frac{5}{2}, 0\right)$ (e) $(-2, 9)$

4. Find the missing coordinates to make each of the following ordered pairs satisfy the equation $y = 3x - 8$. (*See Practice Exercise 2.*)

Use this area for work space.

(a) (0,) (b) (1,) (c) (, 7)

(d) (, −11)

5. Find the missing coordinates to make each of the following ordered pairs satisfy the equation $x + 3y = 5$.

(a) (0,) (b) (, 0) (c) (, 1)

(d) (, 2)

6. Find the missing coordinates to make each of the following ordered pairs satisfy the equation $x - y = 3$.

(a) (3,) (b) (7,) (c) (, 1)

(d) (, −1)

B.

7. Fill in Table 7.10 with the appropriate y-values and graph the equation $y = 7 - 2x$. (*See Practice Exercise 3.*)

TABLE 7.10

x	$y = 7 - 2x$
0	
1	
2	

TABLE 7.11

x	y
0	
	0

8. Fill in the missing values in Table 7.11 and graph the equation $3x + 2y = 12$. (*See Practice Exercise 4.*)

In Exercises 9–30 graph each equation. (*See Examples 1–7.*)

9. $y = x + 1$ **10.** $y = x + 4$ **11.** $y = x - 2$

12. $y = x - 5$ **13.** $y = x$ **14.** $y = 2x$

15. $y = -x$ **16.** $y = \frac{-3x}{2}$ **17.** $y = 2x + 3$

18. $y = 2x - 3$ **19.** $y = 3x + 1$ **20.** $y = -2x + 3$

21. $y = 3x - 1$ **22.** $y = 4x - 6$ **23.** $y = 1 - x$

24. $y = 4 - x$ **25.** $y = 6 - 2x$ **26.** $y = 1 - \frac{x}{2}$

27. $x + y = 4$ **28.** $x - y = 3$ **29.** $2x + y = 5$

30. $3x + 2y = 8$

31. A piece worker earns \$2 per item (and proportionally for uncompleted items). (a) Graph her earnings. Let the x-values represent hours worked and let the y-values represent dollars earned. (b) Use the graph to determine how much she earns after finishing 4 items. (c) Use the graph to determine how many items she must complete to earn \$15.

Use this area for work space.

32. A baby-sitter earns \$4 per hour plus \$2 for carfare. (a) Graph her total pay. Let the x-values represent hours worked and let the y-values represent the total she is paid in dollars. (b) Use the graph to determine the total she is paid for 3 hours. (c) Use the graph to determine how many hours she must work to be paid a total of \$30.

C. In Exercises 33–36 graph each equation. (*See Examples 8 and 9.*)

33. $y = 4$ **34.** $x = 4$ **35.** $y = -4$ **36.** $x = -4$

37. Match each graph in Figure 7.21 with one of the equations in the following list.

(i) $x = 2$

(ii) $y = 2$

(iii) $x = -2$

(iv) $y = -2$

(v) $y = 2x$

(vi) $y = -2x$

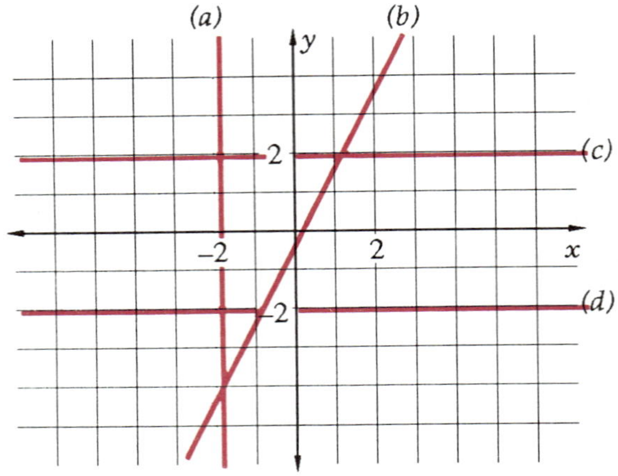

FIGURE 7.21

7.3 Slope

A. Rise and Run

Now you will study lines from an algebraic point of view.

The first matter of importance is the change in the y-coordinates along the line as the x-coordinates change. Do the y-coordinates increase or decrease as the x-coordinates increase? Thus let x_1 and x_2 be the x-coordinates of two different points on a line, and let y_1 and y_2 be the corresponding y-coordinates. Also let

$$P_1 = (x_1, y_1)$$

and

$$P_2 = (x_2, y_2)$$

represent these two points. (*See Figure 7.22.*)

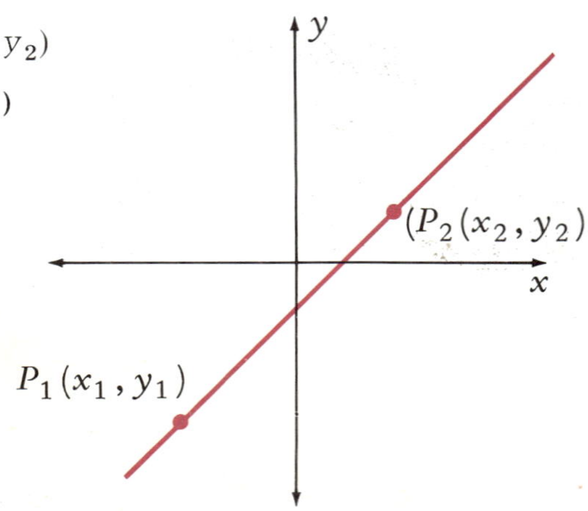

FIGURE 7.22

DEFINITION

SLOPE. The **rise** or the **change in y along the line from P_1 to P_2** is defined to be

$$y_2 - y_1$$

The **run** or the **change in x along the line from P_1 to P_2** is defined to be

$$x_2 - x_1$$

The **slope of the line** is defined to be

$$\frac{y_2 - y_1}{x_2 - x_1}$$

provided $x_2 \neq x_1$. (*See Figure 7.23.*)

Thus, if $x_2 \neq x_1$ (so that $x_2 - x_1 \neq 0$),

$$\text{slope of the line} = \frac{\text{rise}}{\text{run}} = \frac{y_2 - y_1}{x_2 - x_1}$$

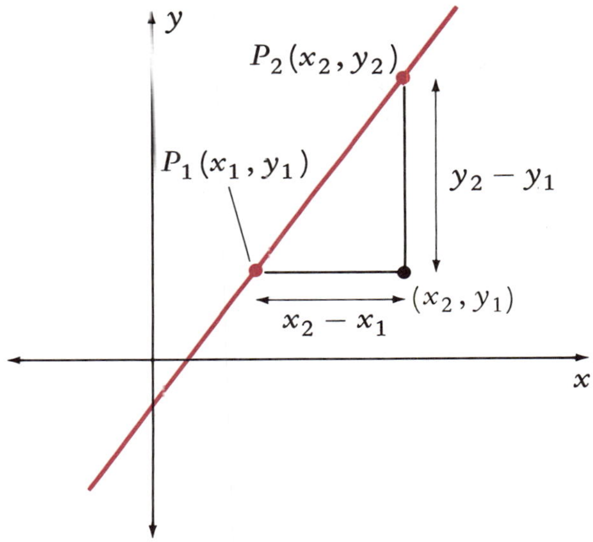

FIGURE 7.23. ***rise*** $= y_2 - y_1$, ***run*** $= x_2 - x_1$, ***slope*** $= \dfrac{y_2 - y_1}{x_2 - x_1}$

Example 1 ▶ Suppose $P_1 = (4, 2)$ and $P_2 = (6, 3)$ are two points on a line. (*See Figure 7.24.*)
(a) Find the rise from P_1 to P_2.
(b) Find the run from P_1 to P_2.
(c) Find the slope of the line.

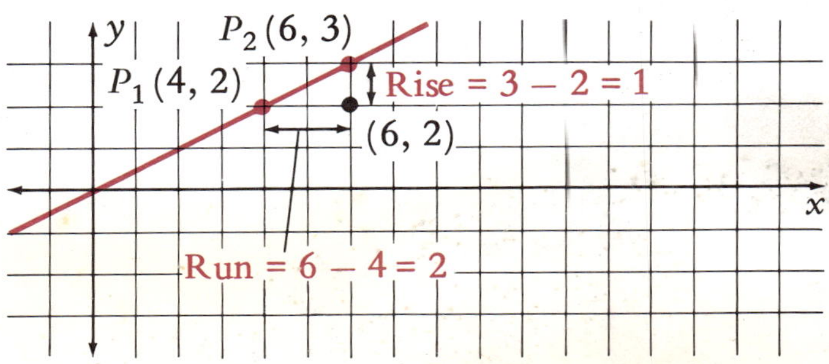

FIGURE 7.24

Solution:
Here $x_1 = 4$, $y_1 = 2$, $x_2 = 6$, and $y_2 = 3$
(a) rise $= y_2 - y_1 = 3 - 2 = 1$
(b) run $= x_2 - x_1 = 6 - 4 = 2$
(c) slope $= \dfrac{\text{rise}}{\text{run}} = \dfrac{1}{2}$ ◀

Example 2 ▶ Suppose $P_1 = (5, 3)$ and $P_2 = (1, 5)$ are two points on a line. (*See Figure 7.25.*)
(a) Find the rise from P_1 to P_2.
(b) Find the run from P_1 to P_2.
(c) Find the slope of the line.

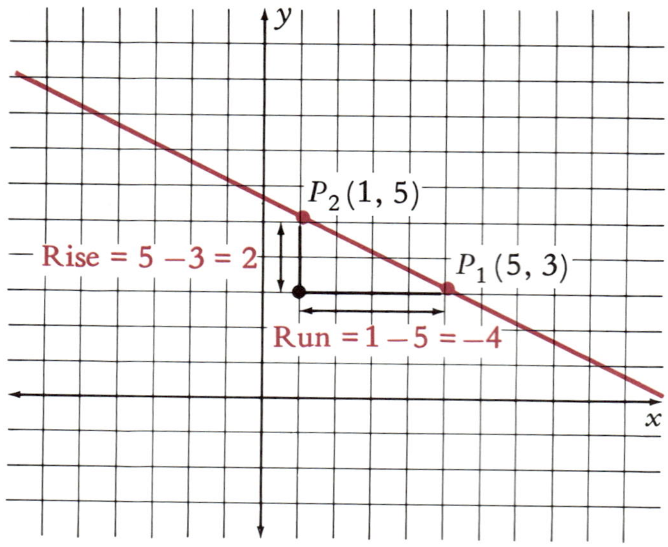

FIGURE 7.25

Solution:
Here $x_1 = 5$, $y_1 = 3$, $x_2 = 1$, and $y_2 = 5$
(a) rise $= y_2 - y_1 = 5 - 3 = 2$
(b) run $= x_2 - x_1 = 1 - 5 = -4$
(c) slope $= \dfrac{\text{rise}}{\text{run}} = \dfrac{2}{-4} = \dfrac{-1}{2}$ ◀

Note that in Example 1 the slope was positive; in Example 2 the slope was negative. *When the rise and run are both of the same sign, the slope* $\left(= \dfrac{\text{rise}}{\text{run}}\right)$ *is positive. In this case the line slopes upward to the right.* (*See Figure 7.26(a).*) *When the rise and run are of different signs, the slope is negative. In this case the line slopes downward to the right (which is the same as upward to the left).* (*See Figure 7.26(b).*)

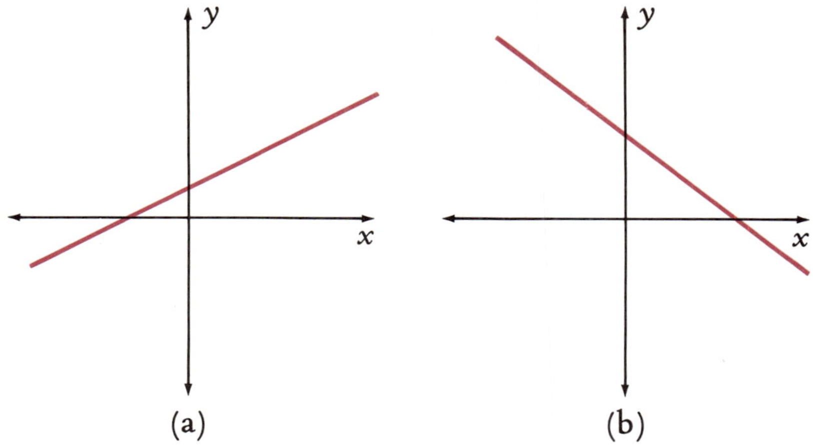

***FIGURE 7.26(a).* This line slopes upward to the right. Its slope is positive. (b). This line slopes downward to the right. Its slope is negative.**

The slope of a line does not depend upon which points P_1 and P_2 are considered. It can be shown that no matter which points of the line are chosen, the slope of the line is the same. Also, in choosing two points, it makes no difference which is called P_1 and which P_2. In fact,

$$\frac{y_1 - y_2}{x_1 - x_2} = \frac{-(y_2 - y_1)}{-(x_2 - x_1)} = \frac{y_2 - y_1}{x_2 - x_1}$$

Example 3 ▶ Find the slope of the line through (1, 3) and (0, −2)
(a) by letting $P_1 = (1, 3)$ and $P_2 = (0, -2)$,
(b) by letting $P_1 = (0, -2)$ and $P_2 = (1, 3)$.
(c) Suppose you are given that (2, 8) is also on the line. Find the slope by letting $P_1 = (1, 3)$ and $P_3 = (2, 8)$.
(d) Does this line slope upward or downward to the right?

Solution:
(a) (*See Figure 7.27(a).*) Here $x_1 = 1$, $y_1 = 3$, $x_2 = 0$, and $y_2 = -2$

$$\text{slope} = \frac{y_2 - y_1}{x_2 - x_1} = \frac{-2 - 3}{0 - 1} = \frac{-5}{-1} = 5$$

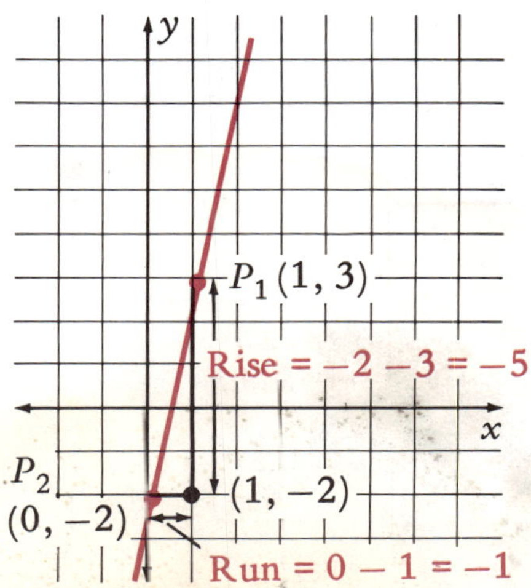

***FIGURE 7.27(a)* slope $= \frac{-5}{-1} = 5$.**

(b) (*See Figure 7.27(b).*) Here $x_1 = 0$, $y_1 = -2$, $x_2 = 1$, and $y_2 = 3$

$$\text{slope} = \frac{y_2 - y_1}{x_2 - x_1} = \frac{3 - (-2)}{1 - 0} = \frac{5}{1} = 5$$

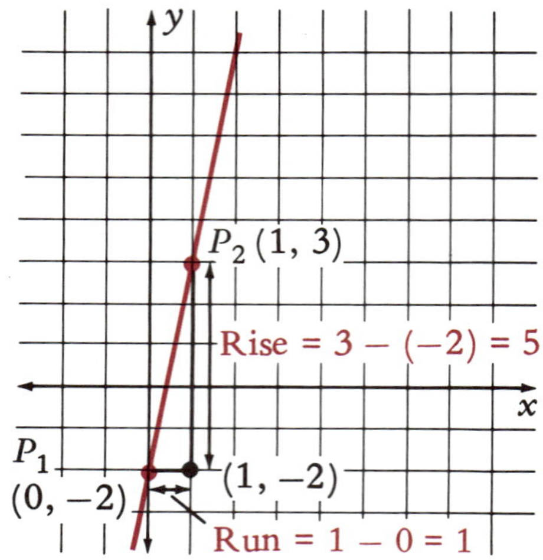

FIGURE 7.27(b) **slope** $= \frac{5}{1} = 5.$

(c) (*See Figure 7.27(c).*) Let $P_3 = (x_3, y_3)$. Here $x_1 = 1$, $y_1 = 3$, $x_3 = 2$, and $y_3 = 8$

$$\text{slope} = \frac{y_3 - y_1}{x_3 - x_1} = \frac{8 - 3}{2 - 1} = \frac{5}{1} = 5$$

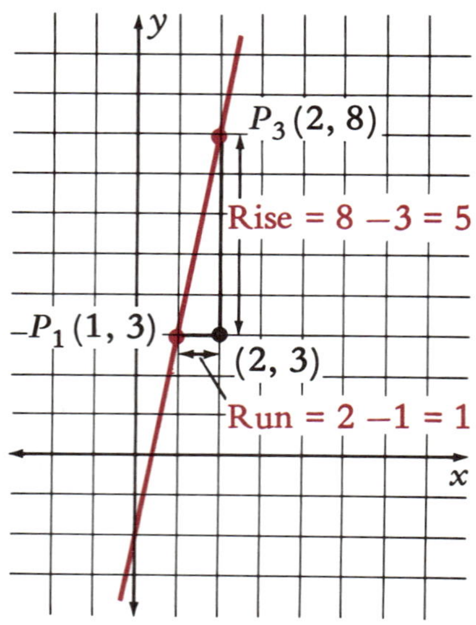

FIGURE 7.27(c) **slope** $= \frac{8 - 3}{2 - 1} = 5.$

(d) The line slopes upward to the right. (*See Figure 7.27(a), (b), or (c).*) ◀

Answers

1. (a) ______
(b) ______
(c) ______
(d) ______
2. (a) ______
(b) ______

PRACTICE EXERCISES FOR TOPIC A

1. Suppose $P_1(1, 2)$ and $P_2(-2, 4)$ are two points on a line. (a) Find the rise from P_1 to P_2. (b) Find the run from P_1 to P_2. (c) Find the slope of the line. (d) Does this line slope upward or downward to the right?

2. Let $P_1 = (-2, -1)$, $P_2 = (1, 0)$, and $P_3 = (4, 1)$. Suppose P_1, P_2, and P_3 are three points on the same line. Find the slope of this line (a) using P_1 and P_2, and (b) using P_1 and P_3.

Now try the exercises for Topic A on page 327.

B. Locating Points on a Line

Example 4 ▶ Draw the line through (3, 5) with slope 2.

Solution:

Let $P_1 = (3, 5)$. To find a second point on the line, increase x by 1 unit. The new x-coordinate is thus 3 + 1, or 4. In order to find the y-coordinate of this point, consider the change in y. Because

$$\text{slope} = 2 \qquad \text{and} \qquad \text{slope} = \frac{\text{change in } y}{\text{change in } x}$$

it follows that

$$2 = \frac{\text{change in } y}{1}$$

The change in y is 2, and the new y-coordinate is 5 + 2, or 7.

Let $P_2 = (4, 7)$. Now draw the line through P_1 and P_2, as in Figure 7.28. ◀

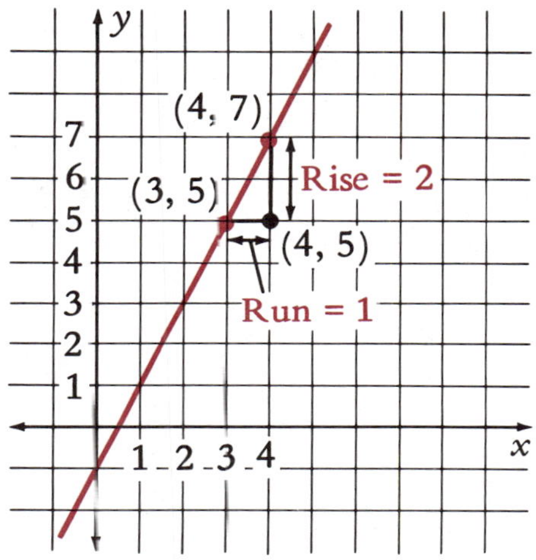

FIGURE 7.28

PRACTICE EXERCISES FOR TOPIC B

3. Draw the line through (3, 1) with slope $\frac{3}{4}$.

Now try the exercises for Topic B on page 327.

C. Horizontal Lines and Vertical Lines

Every point on a *horizontal* line has the same y-coordinate. For example, if one point has y-coordinate 4, then every point on the line has y-coordinate 4. Therefore $y = 4$ for every value of x. (*See Figure 7.29.*) Thus the rise (or the change in y) from P_1 to P_2 is always 0. Also

$$\text{slope} = \frac{\text{rise}}{\text{run}} = \frac{0}{\text{run}} = 0$$

The slope of a horizontal line is 0.

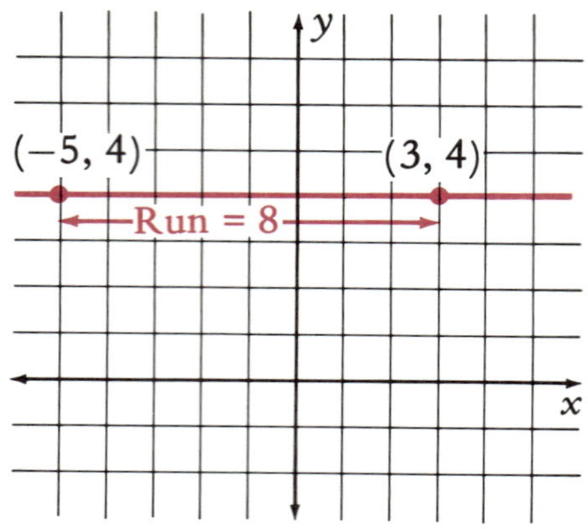

FIGURE 7.29. **slope** $= \frac{\text{rise}}{\text{run}} = \frac{0}{8} = 0$. ***The slope of a horizontal line is 0.***

Every two points on a *vertical* line have the same x-coordinate. Thus if some point on the line has x-coordinate 5, then every point has x-coordinate 5. Consequently, $x = 5$ for every value of y. (*See Figure 7.30.*) Therefore the run from P_1 to P_2 is always 0. The slope is defined as $\frac{\text{rise}}{\text{run}}$. Because the run is 0 and division by 0 is not defined, the slope is not defined. Thus *the slope of a vertical line is undefined.*

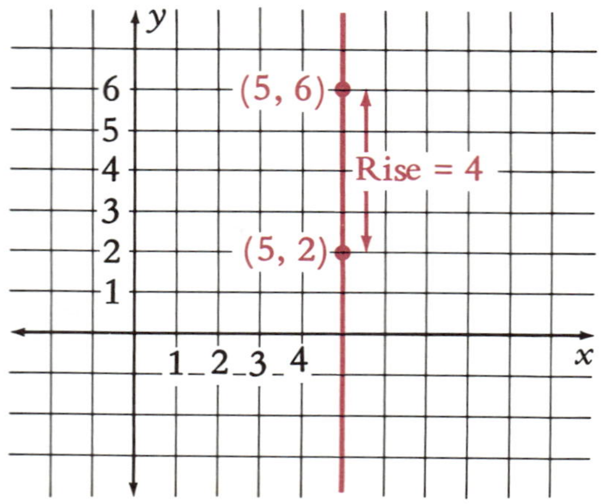

FIGURE 7.30. The run is 0. The slope of a vertical line is undefined. Note that $\frac{4}{0}$ ***is undefined.***

When two different points on a line have the same y-coordinate, the line is horizontal. When two different points have the same x-coordinate, the line is vertical.

A **slanting line** is a line that is neither horizontal nor vertical. The slope of a slanting line is a nonzero number.

Answers

4. (a) ______

(b) ______

5. (a) ______

(b) ______

6. (a) ______

(b) ______

PRACTICE EXERCISES FOR TOPIC C

In Exercises 4–6 consider the line through P_1 and P_2.

(a) Find the slope of this line or say that the slope is undefined.

(b) Indicate whether this line (*i*) is horizontal, (*ii*) is vertical, (*iii*) slopes upward to the right, or (*iv*) slopes downward to the right.

4. $P_1 = (6, -8)$, $P_2 = (8, 6)$

5. $P_1 = (9, 5)$, $P_2 = (9, -1)$

6. $P_1 = (5, -4)$, $P_2 = (0, -4)$

Now try the exercises for Topic C on page 328.

Use this area for work space.

EXERCISES

A. In Exercises 1–20 suppose P_1 and P_2 are two points on a line. (a) Find the rise from P_1 to P_2, (b) find the run from P_1 to P_2, (c) find the slope of the line, and (d) does this line slope upward to the right, downward to the right, or neither? (*See Examples 1–3.*)

1. $P_1 = (1, 1)$, $P_2 = (2, 2)$

2. $P_1 = (1, 2)$, $P_2 = (2, 5)$

3. $P_1 = (3, 3)$, $P_2 = (4, 5)$

4. $P_1 = (0, 0)$, $P_2 = (4, 1)$

5. $P_1 = (-1, -1)$, $P_2 = (1, 0)$

6. $P_1 = (-2, -1)$, $P_2 = (0, 0)$

7. $P_1 = (8, 3)$, $P_2 = (9, 1)$

8. $P_1 = (2, 4)$, $P_2 = (-2, 5)$

9. $P_1 = (3, 2)$, $P_2 = (0, 1)$

10. $P_1 = (1, 0)$, $P_2 = (0, 2)$

11. $P_1 = (4, 0)$, $P_2 = (0, 2)$

12. $P_1 = (-3, 0)$, $P_2 = (0, 3)$

13. $P_1 = (-7, -4)$, $P_2 = (-3, -3)$

14. $P_1 = (-3, -2)$, $P_2 = (-1, -1)$

15. $P_1 = \left(\frac{1}{2}, \frac{1}{2}\right)$, $P_2 = (1, 2)$

16. $P_1 = \left(\frac{1}{4}, \frac{1}{2}\right)$, $P_2 = \left(\frac{1}{2}, \frac{1}{4}\right)$

17. $P_1 = (2, 3)$, $P_2 = (4, 3)$

18. $P_1 = (-1, -1)$, $P_2 = (1, -1)$

19. $P_1 = (-7, 1)$, $P_2 = (7, -1)$

20. $P_1 = \left(3, \frac{1}{2}\right)$, $P_2 = \left(\frac{1}{2}, \frac{-1}{2}\right)$

21. Suppose P_1, P_2, and P_3 are three points on a line, where $P_1 = (0, 2)$, $P_2 = (2, 5)$, and $P_3 = (6, 11)$. Find the slope of this line (a) using P_1 and P_2, and (b) using P_1 and P_3. (*See Example 3.*)

22. Suppose P_1, P_2, and P_3 are three points on a line, where $P_1 = (4, 3)$, $P_2 = (6, -1)$, and $P_3 = (7, -3)$. Find the slope of this line (a) using P_1 and P_2, and (b) using P_2 and P_3.

B. In Exercises 23–32 draw the line through the point P with the given slope. (*See Example 4.*)

Sample: Draw the line through (1, 4) with slope 3.	*Solution:* See Figure 7.31.

Use this area for work space.

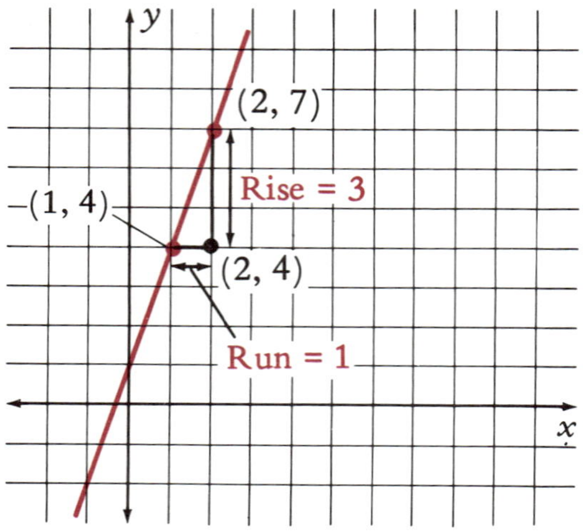

FIGURE 7.31

23. $P = (1, 1)$, slope $= 1$

24. $P = (0, 2)$, slope $= 2$

25. $P = (3, 1)$, slope $= -1$

26. $P = (5, -2)$, slope $= -2$

27. $P = (3, 3)$, slope $= 0$

28. $P = (1, -1)$, slope $= \frac{1}{2}$

29. $P = (4, 0)$, slope $= \frac{-1}{2}$

30. $P = (-1, -2)$, slope $= \frac{1}{4}$

31. $P = (1, 5)$, slope $= 5$

32. $P = (2, 4)$, slope undefined

C. In Exercises 33–40 consider the line through P_1 and P_2. (a) Find the slope of this line or indicate that the slope is undefined. (b) Indicate whether this line (*i*) is horizontal, (*ii*) is vertical, (*iii*) slopes upward to the right, or (*iv*) slopes downward to the right.

Sample: $P_1 = (4, 2)$, $P_2 = (4, 8)$	*Solution:* (*ii*) [The line through (4, 2) and (4, 8) is vertical because two points on the line have the same x-coordinate.]

33. $P_1 = (1, 2)$, $P_2 = (2, 6)$

34. $P_1 = (2, 4)$, $P_2 = (2, 3)$

35. $P_1 = (3, 8)$, $P_2 = (5, 8)$

36. $P_1 = (0, 6)$, $P_2 = (-6, 0)$

37. $P_1 = (4, 0)$, $P_2 = (-4, 0)$

38. $P_1 = (1, 10)$, $P_2 = (1, -10)$

39. $P_1 = \left(\frac{1}{2}, \frac{1}{4}\right)$, $P_2 = \left(\frac{1}{4}, \frac{1}{2}\right)$

40. $P_1 = \left(\frac{1}{2}, \frac{1}{3}\right)$, $P_2 = \left(\frac{1}{4}, \frac{1}{3}\right)$

7.4 Equation of a Line

Here we will learn how to find an equation of a line. This will enable us to utilize algebraic tools for working with a notion that is geometric in nature. In Subsections **A** and **B** the lines considered will be slanting lines; in Subsection **C** horizontal lines and vertical lines will be studied.

A. Slope-Intercept Form

The graph of an equation of the form

$$y = mx + b$$

is a line because only the first powers of x and y occur. To see the significance of m and b, first consider the line with equation

$$y = 2x + 5$$

$$\text{When } x = 0,\ y = 2 \cdot 0 + 5 = 5$$

$$\text{When } x = 1,\ y = 2 \cdot 1 + 5 = 7$$

Thus the points $P_1 = (0, 5)$ and $P_2 = (1, 7)$ are on the line. The slope of the line is given by

$$\frac{\text{rise}}{\text{run}} = \frac{7 - 5}{1 - 0} = \frac{2}{1} = 2$$

Therefore the line with equation

$$y = 2x + 5 \qquad \textit{has slope } 2$$

Similarly, it can be shown that the line with equation

$$y = \frac{x}{2} - 3 \quad \text{or} \quad y = \frac{1}{2}x - 3 \qquad \textit{has slope } \frac{1}{2}$$

and that the line with equation

$$y = -4x + 5 \qquad \textit{has slope } -4$$

In general, *the line with equation*

$$y = mx + b \qquad \textit{has slope } m$$

Let us return to the line given by

$$y = 2x + 5$$

Replace x by 0 in this equation, and obtain

$$y = 2 \cdot 0 + 5$$

or

$$y = \mathbf{5}$$

Thus the point (0, **5**) is on the line because its coordinates satisfy the equation

$$y = 2x + \mathbf{5}$$

Points with x-coordinate 0 are on the y-axis. Therefore (0, **5**) is the point at which the line crosses the y-axis. The y-coordinate, **5**, of this intersection point is known as the "y-intercept" of the line. (*See Figure 7.32.*)

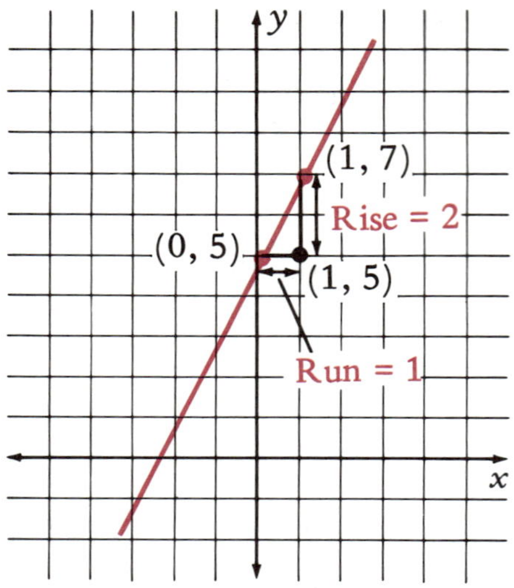

FIGURE 7.32. The line with equation $y = 2x + 5$ has slope 2 and y-intercept 5.

In general, a **y-intercept** of a line is the *y*-coordinate of an intersection with the *y*-axis. To find the *y*-intercept of the line given by

$$y = mx + b$$

replace x by 0 in this equation, and obtain

$$y = m \cdot 0 + b$$

or

$$y = b$$

Thus *the line with equation*

$$y = mx + b$$

has y-intercept b.

To sum up, for the line with equation

$$y = mx + \boldsymbol{b}$$

m = slope and $\boldsymbol{b}$ = *y*-intercept

And the equation

$$y = mx + \boldsymbol{b}$$

is said to be in **slope-intercept form.**

Example 1 ▶ Find (a) the slope and (b) the *y*-intercept of the line with equation

$$y = 4x - 1$$

Solution:
This equation is of the form

$$y = mx + b$$

(a) $m = 4$. Thus the slope is 4.
(b) $b = -1$. Thus the y-intercept is -1. (*See Figure 7.33.*) ◀

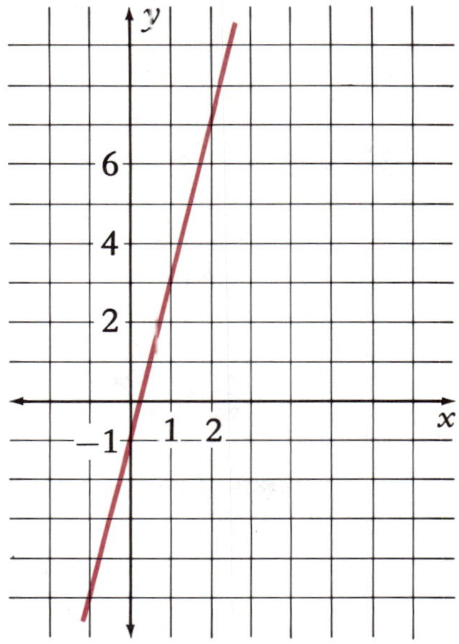

FIGURE 7.33

When you know the equation of a line in the form

$$y = mx + b$$

you already know one point on the line—namely $(0, b)$. This is because b is the y-intercept. To graph the line, find another point. If you let

$$x = 1, \text{ then } y = m \cdot 1 + b = m + b$$

Thus in Example 1, $m + b = 4 - 1 = 3$, and the point $(1, 3)$ is on the line.

Example 2 ▶ Find (a) the slope and (b) the y-intercept of the line with equation

$$y = \frac{3x - 8}{4}$$

(c) Graph the line.
Solution:
Rewrite the equation of the line in the form

$$y = \frac{3}{4}x - \frac{8}{4}$$

or

$$y = \frac{3}{4}x - 2$$

(a) $m = \text{slope} = \dfrac{3}{4}$

(b) $b = y\text{-intercept} = -2$

(c) When $x = 0$, $y = b = -2$. When $x = 1$, $y = \dfrac{3}{4} - 2 = \dfrac{3 - 8}{4} = \dfrac{-5}{4}$

(*See Figure 7.34.*) ◀

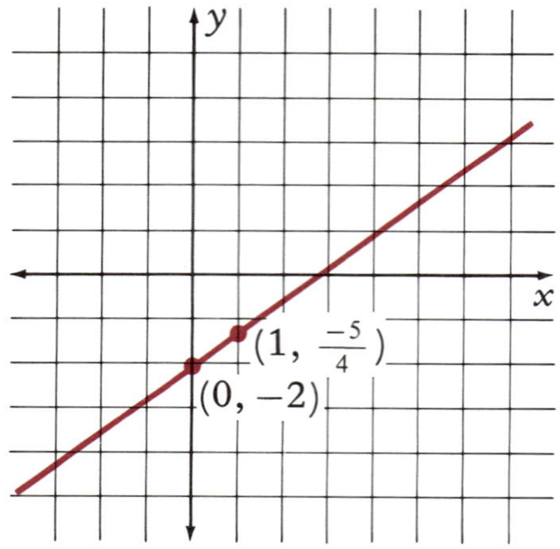

FIGURE 7.34

Example 3 ▶ (a) Find the equation of the line with slope 4 and y-intercept 3.
(b) Graph this line.
Solution:
(a) $m = 4$ and $b = 3$. Thus the equation of the line is

$$y = 4x + 3$$

(b) When $x = 0$, $y = 3$. When $x = 1$, $y = 4 + 3 = 7$. *(See Figure 7.35.)* ◀

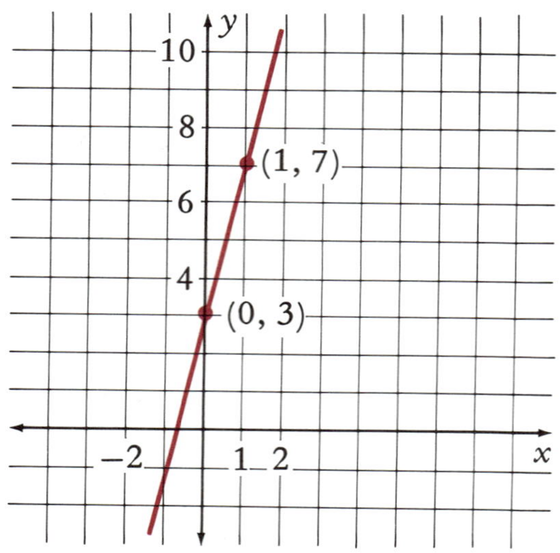

FIGURE 7.35

Example 4 ▶ (a) Find the equation of the line with slope $\frac{-1}{2}$ and y-intercept 0.

(b) Graph this line.
Solution:

(a) $m = \frac{-1}{2}$ and $b = 0$. The equation of this line is

$$y = \frac{-1}{2}x$$

or

$$y = \frac{-x}{2}$$

(b) When $x = 0$, $y = 0$. When $x = 1$, $y = \frac{-1}{2}$. *(See Figure 7.36.)* ◀

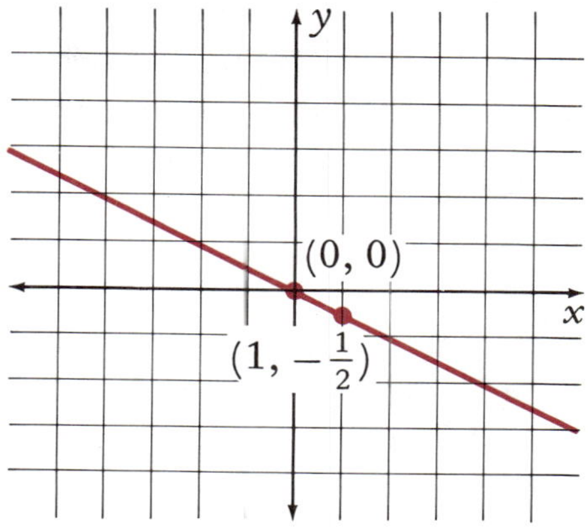

FIGURE 7.36

In general, a slanting line passes through the origin if its y-intercept is 0. Its equation is then of the form

$$y = mx, \text{ where } m \neq 0$$

For the slanting line through the origin in Example 4, $m = \frac{-1}{2}$.

Answers

1. (a) ______

(b) ______

2. ______

PRACTICE EXERCISES FOR TOPIC A

1. Find (a) the slope and (b) the y-intercept of the line with equation $y = \frac{x - 10}{2}$. (c) Graph the line.

2. Find the equation of the line with slope 4 and y-intercept -2.

Now try the exercises for Topic A on page 338.

B. Intercept Form

An **x-intercept** of a line is the x-coordinate of an intersection with the x-axis.

Suppose you know that a is the x-intercept and that b is the y-intercept of a slanting line, where $a \neq 0$ and $b \neq 0$. (This condition indicates that the line does *not* pass through the origin.) Then the equation of the line can be written in the form

$$\frac{x}{a} + \frac{y}{b} = 1$$

This is known as the **intercept form** of the equation of a line. For example, if the x-intercept of a line is 2 and the y-intercept is 3, then the line has the equation

$$\frac{x}{2} + \frac{y}{3} = 1$$

(*See Figure 7.37.*)

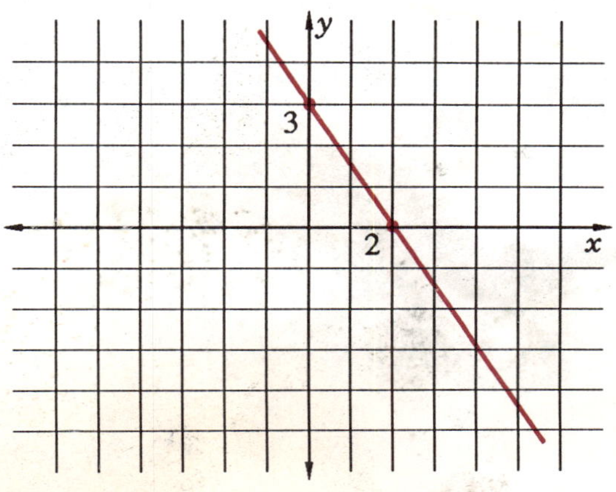

FIGURE 7.37

Suppose you are given the graph of a slanting line that does not pass through the origin. It is often easiest to find the equation of this line in intercept form. See where the line crosses the x-axis to find the x-intercept, a, and see where the line crosses the y-axis to find the y-intercept, b. Write the equation in intercept form

$$\frac{x}{a} + \frac{y}{b} = 1$$

Sometimes it is preferable to clear the equation of fractions. To do so, multiply both sides of the equation by ab, which is the *lcd* on the left side. Thus, for the equation

$$\frac{x}{2} + \frac{y}{3} = 1$$

which is in intercept form, multiply both sides by 6, the *lcd* of $\frac{x}{2}$ and $\frac{y}{3}$.

$$\left(\frac{x}{2} + \frac{y}{3}\right) \cdot 6 = 1 \cdot 6$$

or

$$3x + 2y = 6$$

Example 5 ► Find an equation of the line in Figure 7.38.
(a) Express the equation in intercept form.
(b) Clear the equation of fractions.

Solution:

(a) Clearly, the x-intercept is 6 and the y-intercept is 3. Thus $a = 6$ and $b = 3$. The intercept form of the equation of the line is

$$\frac{x}{6} + \frac{y}{3} = 1$$

(b) Alternatively, multiply both sides by 6 to obtain

$$\left(\frac{x}{6} + \frac{y}{3}\right) \cdot 6 = 1 \cdot 6$$

or

$$x + 2y = 6$$ ◄

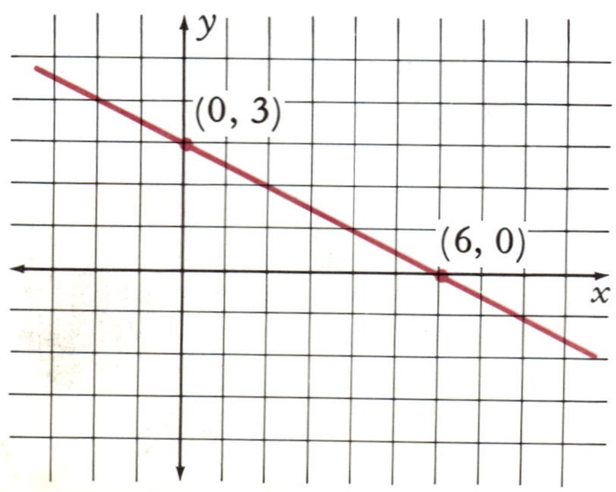

FIGURE 7.38

Example 6 ► Find an equation for the line in Figure 7.39.
(a) Express the equation in intercept form.
(b) Clear the equation of fractions.

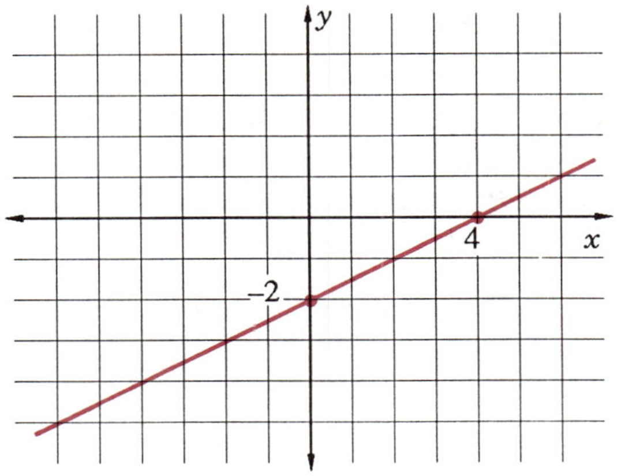

FIGURE 7.39

Solution:

(a) The x-intercept is 4 and the y-intercept is -2. Thus the intercept form of the line is

$$\frac{x}{4} + \frac{y}{-2} = 1$$

(b) Multiply both sides by 4, the *lcd* on the left side, to obtain

$$x - 2y = 4$$

◀

Answers

3. (a) ______

(b) ______

4. (a) ______

(b) ______

5. ______

PRACTICE EXERCISES FOR TOPIC B

3. (a) Find an equation in intercept form for the line in Figure 7.40. (b) Clear the equation of fractions.

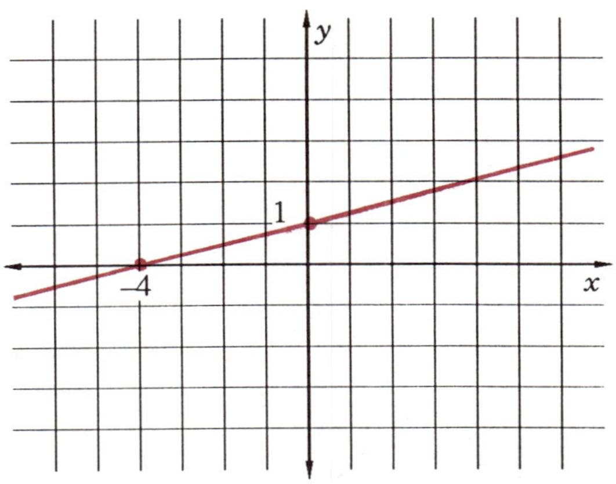

FIGURE 7.40

4. (a) Find the equation in intercept form for the line with x-intercept 5 and y-intercept 2. (b) Clear the equation of fractions. (c) Graph this line.

5. Find the equation of the line that passes through the origin and (6, 3).

Now try the exercises for Topic B on page 339.

C. Horizontal Lines and Vertical Lines

Horizontal lines have slope 0. Thus $m = 0$, and, instead of

$$y = mx + b$$

the equation of a horizontal line is simply

$$y = b$$

where b is the y-intercept.

Example 7 ► (a) Find the equation of the horizontal line with y-intercept 5.
(b) Graph this line.
Solution:
(a) $b = 5$. Thus the equation of the line is

$$y = 5$$

(b) When $x = 0$, $y = 5$. When $x = 1$, $y = 5$. (*See Figure 7.41.*) ◄

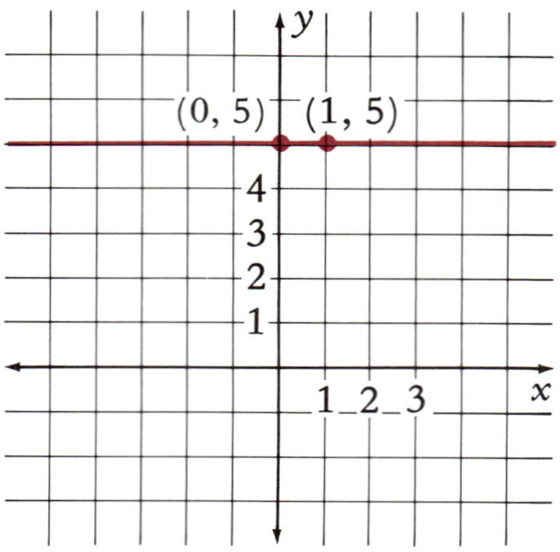

FIGURE 7.41

Actually, if you know *any* point (x_1, y_1) on a *horizontal* line, you can find its equation. For every point on a horizontal line has the same y-coordinate, y_1. Therefore the equation of the line is

$$y = y_1$$

Example 8 ► Find the equation of the horizontal line that passes through $(2, -2)$.
Solution:
Here $y_1 = -2$, the y-coordinate of $(2, -2)$. Thus the equation of the line is

$$y = -2$$

(*See Figure 7.42.*) ◄

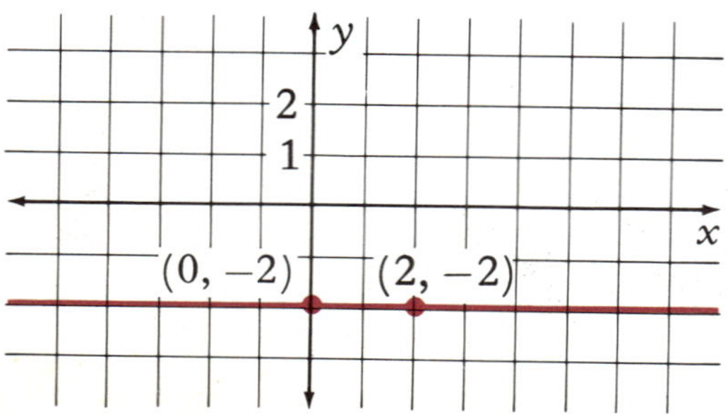

FIGURE 7.42

The x-axis is the horizontal line with y-intercept 0. Thus the equation of the x-axis is

$$y = 0$$

Note that *every* point on the x-axis is an x-intercept. (*See Figure 7.43(a).*)

A horizontal line other than the x-axis does not pass through the x-axis. Thus its x-intercept is undefined. (*See Figure 7.43(b).*)

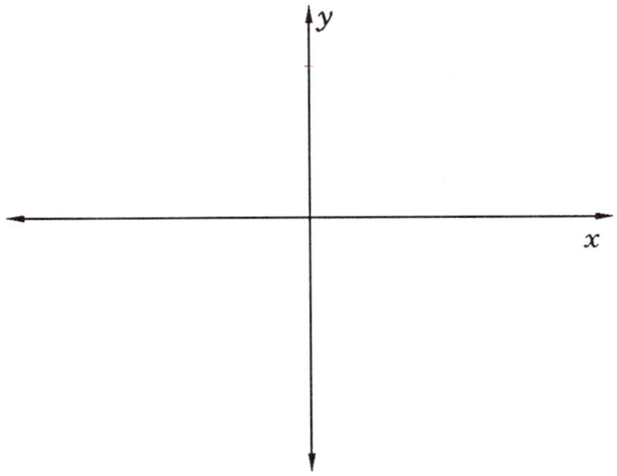

FIGURE 7.43(a). The x-axis has y-intercept 0 and has the equation y = 0. Every point on the x-axis is an x-intercept.

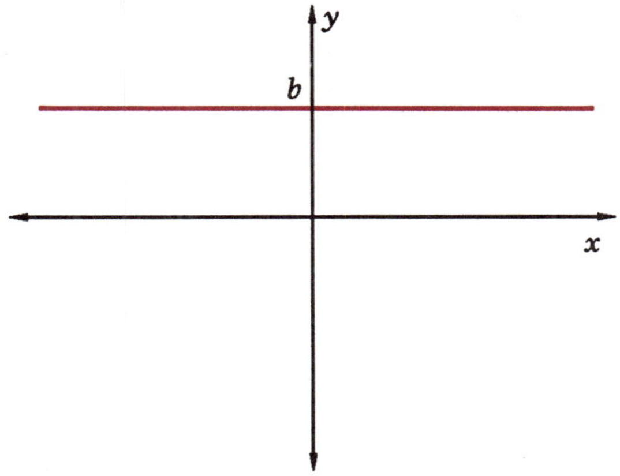

FIGURE 7.43(b). This horizontal line has y-intercept b and has the equation y = b. Its x-intercept is undefined.

For a vertical line the slope is undefined, and the equation

$$y = mx + b$$

no longer applies. Instead, observe that every point on a vertical line has the same x-coordinate. Thus all you need to know is a single point, (x_1, y_1), on a vertical line. The equation of the vertical line is then

$$x = x_1$$

Example 9 ▶ Find the equation of the vertical line that passes through $(4, -5)$.

Solution:

The x-coordinate of the given point is 4. Thus the equation of the vertical line is

$$x = 4 \qquad \text{(See Figure 7.44.)}$$ ◀

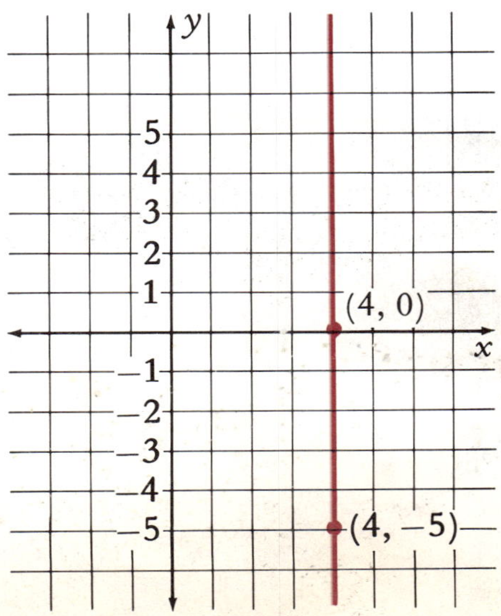

FIGURE 7.44

The y-axis is the *vertical* line that passes through (0, 0) and that has x-intercept 0. Thus the equation of the y-axis is

$$x = 0$$

Note that every point on the y-axis is a y-intercept. (*See Figure 7.45(a).*)

A vertical line other than the y-axis does not pass through the y-axis. Thus its y-intercept is undefined. (*See Figure 7.45(b).*)

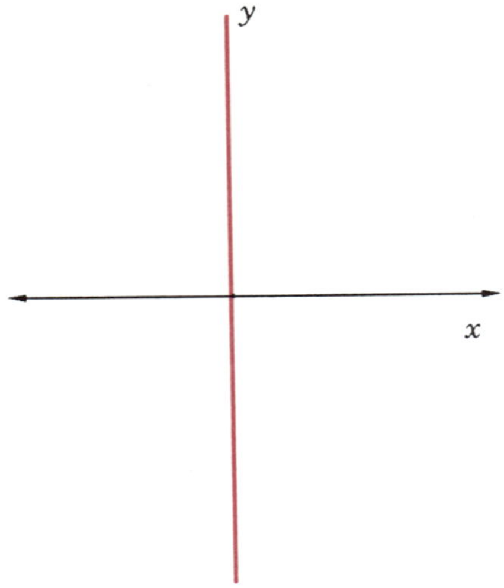

FIGURE 7.45(a). The y-axis has x-intercept 0 and has the equation x = 0. Every point on the y-axis is a y-intercept.

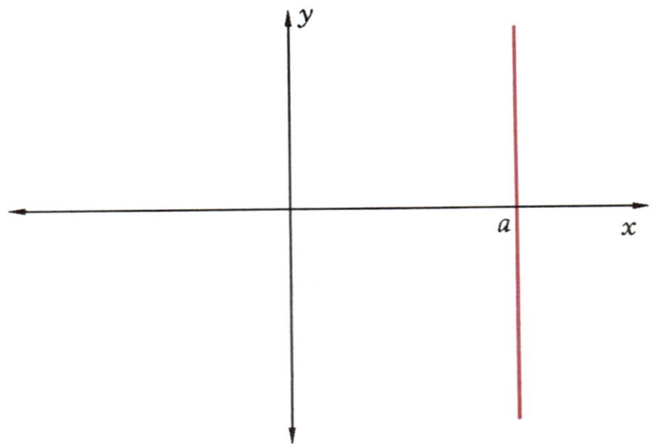

FIGURE 7.45(b). This vertical line has x-intercept a and has the equation x = a. Its y-intercept is undefined.

Answers

6. (a) ________

(b) ________

(c) ________

PRACTICE EXERCISES FOR TOPIC C

6. (a) Write the equation of the horizontal line with y-intercept 7.

(b) Write the equation of the vertical line that passes through $(-2, 3)$.

(c) At what point do these two lines intersect?

Now try the exercises for Topic C on page 340.

Use this area for work space.

EXERCISES

A. In Exercises 1–18 find (a) the slope, and (b) the y-intercept of the line with the given equation. (*See Examples 1 and 2.*)

1. $y = 2x + 4$ **2.** $y = 3x - 2$ **3.** $y = \frac{x}{2} + 5$

4. $y = -x - 1$ **5.** $y = 2 - x$ **6.** $y = \frac{3x + 1}{4}$

7. $y = \frac{x + 2}{2}$ **8.** $y = \frac{4 - x}{2}$ **9.** $y = 3(x + 1)$

10. $y = -2(x + 2)$ **11.** $y = -(2 - 3x)$ **12.** $x + y = 1$

13. $2y = x + 8$ **14.** $3y = x - 6$ **15.** $\frac{y}{2} = x + 5$

Use this area for work space.

16. $-y = x + \frac{1}{2}$ **17.** $y = 5$ **18.** $y + 5 = 0$

In Exercises 19–26 find the equation of the line with the given slope and y-intercept. *(See Examples 3 and 4.)*

19. slope 4, y-intercept 1

20. slope 2, y-intercept 2

21. slope -1, y-intercept 5

22. slope 10, y-intercept 10

23. slope $\frac{1}{2}$, y-intercept $\frac{1}{4}$

24. slope $\frac{-1}{2}$, y-intercept -3

25. slope -2, y-intercept $\frac{1}{2}$

26. slope 0, y-intercept 3

In Exercises 27–32 find (a) the slope, and (b) the y-intercept of the line with the given equation. (c) Graph this line.

27. $y = 2x + 1$ **28.** $y = -x + 4$ **29.** $y = -2x + 1$

30. $y = 3x$ **31.** $y = \frac{x + 1}{2}$ **32.** $y = \frac{-3 - x}{2}$

B. In Exercises 33–38 (a) find the equation in intercept form for each line drawn in Figures 7.46–7.51. (b) Then clear the equation of fractions. *(See Examples 5 and 6.)*

33.

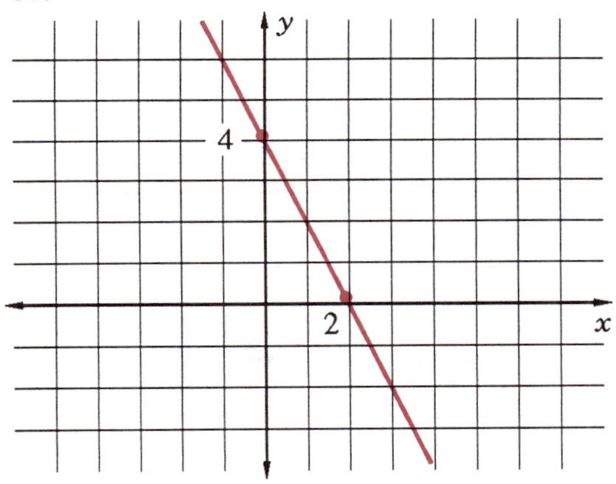

FIGURE 7.46

34.

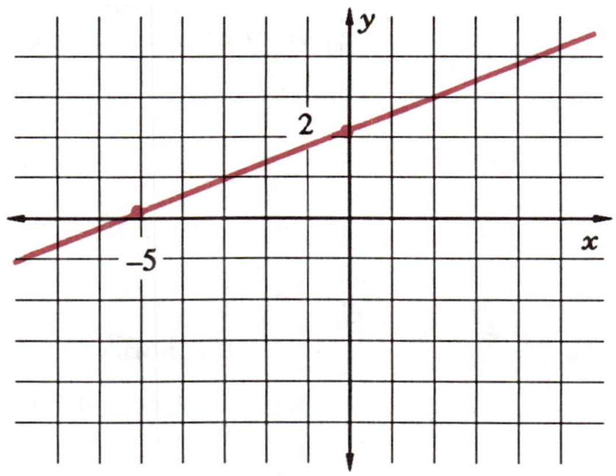

FIGURE 7.47

35.

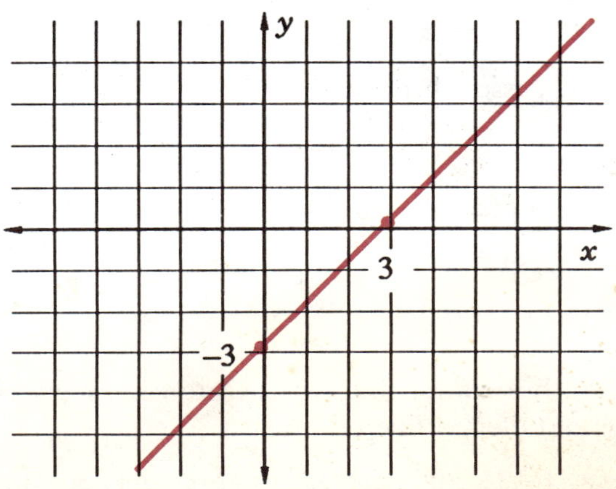

FIGURE 7.48

36.

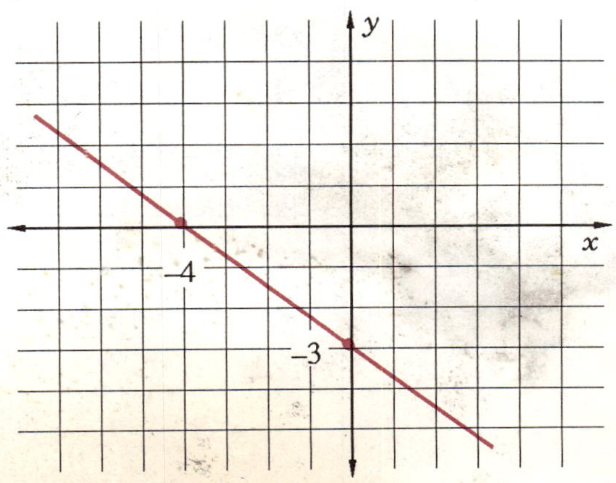

FIGURE 7.49

Use this area for work space.

37.

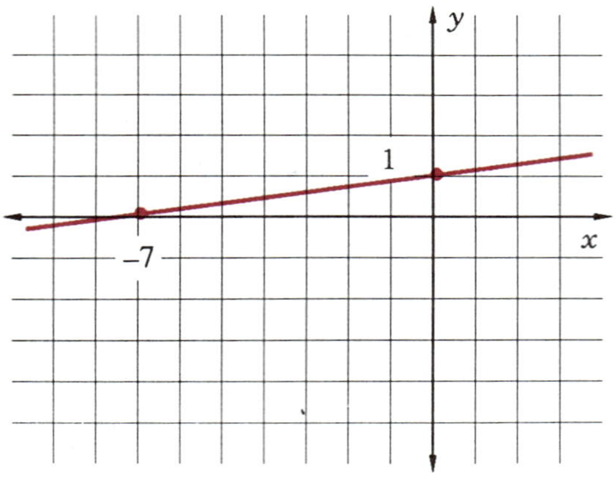

FIGURE 7.50

38.

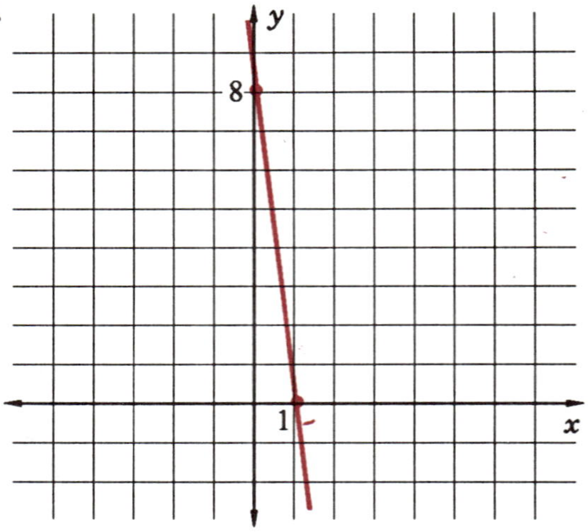

FIGURE 7.51

In Exercises 39–42 (a) find the equation in intercept form for the line with the given x- and y-intercepts. (b) Clear the equation of fractions. (c) Graph this line. (*See Practice Exercise 4.*)

39. x-intercept 3, y-intercept 3.

40. x-intercept 1, y-intercept 2

41. x-intercept 2, y-intercept 1

42. x-intercept 2, y-intercept -1

In Exercises 43–46 find the equation of the line that passes through the origin and the following additional point.

43. $(4, 2)$

44. $(2, 4)$

45. $(-8, 3)$

46. $(5, -1)$

C. In Exercises 47–52 find the equation of the indicated line. (*See Examples 7–9.*)

47. the horizontal line with y-intercept 3

48. the horizontal line that passes through $(1, -2)$

49. the horizontal line that passes through $(-1, 2)$

50. the vertical line with x-intercept 1

51. the vertical line that passes through $(2, -3)$

52. the vertical line that passes through $(-2, 3)$

In Exercises 53–58 (a) find the equation of the indicated line, and (b) graph this line.

53. the horizontal line with y-intercept -3

54. the horizontal line that passes through $(4, -1)$

55. the vertical line that passes through $(4, -1)$

56. the vertical line with x-intercept -10

57. the horizontal line with y-intercept 0

58. the vertical line that passes through $(0, 1)$

59. At what point does the horizontal line with y-intercept 4 intersect the vertical line with x-intercept 7? (*See Practice Exercise 6.*)

60. At what point does the horizontal line with y-intercept -2 intersect the y-axis?

Use this area for work space.

61. Pick out the *error(s)*. Assume m and b are any numbers.

(a) Every line has both an x-intercept and a y-intercept.

(b) The equation of the y-axis is $x = 0$.

(c) The y-intercept of a line with equation $y = mx + b$ is $(0, b)$.

(d) A line with equation $y = mx + b$ cannot be vertical.

(e) If a line has x-intercept a and y-intercept b, its equation is $\frac{x}{a} + \frac{y}{b} = 1$.

7.5 Intersection of Lines

Two *different* lines (on the x,y-plane) either intersect at a single point (Figure 7.52(a)) or else are parallel (Figure 7.52(b)).

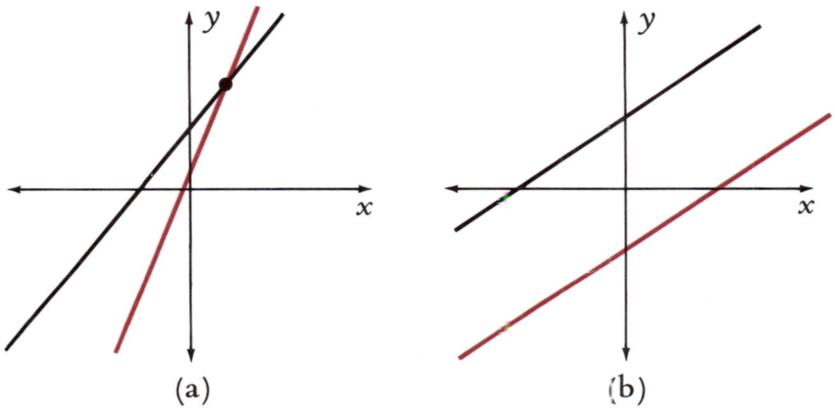

FIGURE 7.52(a). Two different lines (on the x,y-plane) that are not parallel intersect at a single point. (b). Parallel lines are lines (on the same plane) that do not intersect.

A. Intersecting Lines

First consider the case of intersecting lines. For the present you will find the intersection point by graphing both lines on the same coordinate system. In Section 7.6 you will solve this problem algebraically.

Example 1 ▶ Find the intersection of the lines L_1, given by

$$y = x + 1$$

and L_2, given by

$$y = 2x - 1$$

Solution:
Consider Table 7.12.

TABLE 7.12

L_1:

x	y
0	1
1	2

L_2:

x	y
0	−1
1	1

Graph both lines on the same coordinate system. As you see in Figure 7.53, the intersection is the point (2, 3).

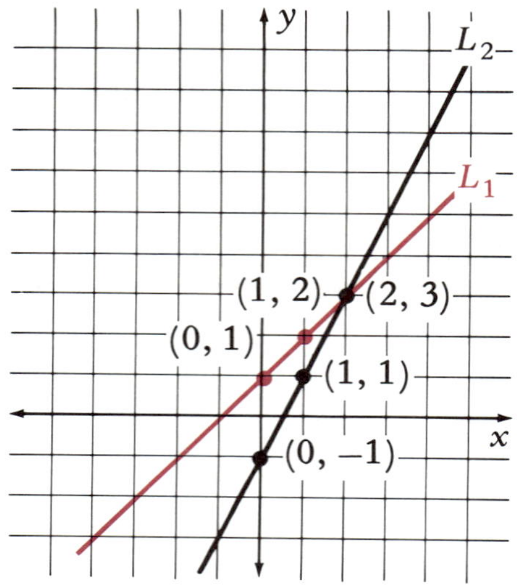

FIGURE 7.53

Check:
You can check your graphical solution by replacing x by 2 and y by 3 in each of the given equations:

$\mathbf{y} = x + 1$	$\mathbf{y} = 2x - 1$
$\mathbf{3} \stackrel{?}{=} 2 + 1$	$\mathbf{3} \stackrel{?}{=} 2 \cdot 2 - 1$
$3 \stackrel{\checkmark}{=} 3$	$3 \stackrel{\checkmark}{=} 3$

◀

In Example 1 the check showed that you found the *exact* point by graphing. However, in general, the intersection point can only be *approximated* by graphing. It is often difficult to approximate closely an intersection point, such as $\left(\frac{3}{5}, \frac{4}{7}\right)$, where the coordinates are not integers.

Example 2 ▶ Find the intersection of the lines L_1, given by

$$y - 2 = 3(x - 5)$$

and L_2, given by

$$2y = x - 6$$

Solution:
Add 2 to both sides of the first equation:

$$y - 2 = 3(x - 5)$$
$$y = 3(x - 5) + 2$$

or

$$y = 3x - 13$$

Divide both sides of the second equation by 2:

$$2y = x - 6$$
$$y = \frac{x - 6}{2}$$

Consider Table 7.13.

TABLE 7.13

L_1: x	$y = 3x - 13$
5	2
6	5

L_2: x	$y = \frac{x - 6}{2}$
0	-3
2	-2

As you see in Figure 7.54, the intersection is the point $(4, -1)$. ◀

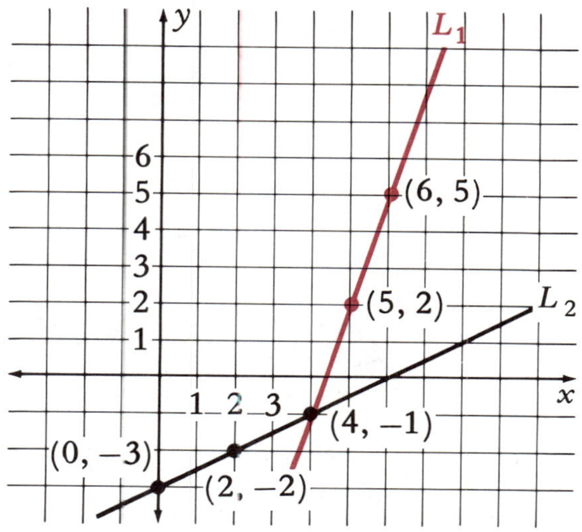

FIGURE 7.54

Example 3 ▶ Find the intersection of the lines L_1, given by

$$y = -4x$$

and L_2, given by

$$y + 3 = 2x$$

Solution:

Subtract 3 from both sides of the second equation:

$$y + 3 = 2x$$

$$y = 2x - 3$$

Consider Table 7.14.

TABLE 7.14

L_1: x	y
0	0
1	-4

L_2: x	y
0	-3
1	-1

As you see in Figure 7.55, the intersection is the point $\left(\frac{1}{2}, -2\right)$. ◀

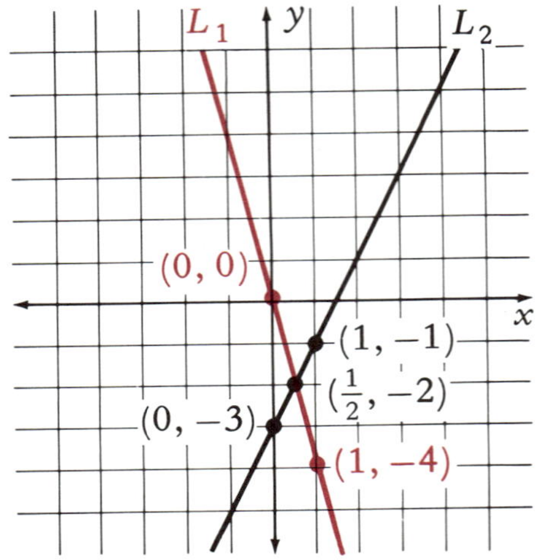

FIGURE 7.55

PRACTICE EXERCISES FOR TOPIC A

1. Find the intersection of the given lines graphically.

L_1: $2y = 4 - x$

L_2: $y = -3(x + 1)$

Check your solution.

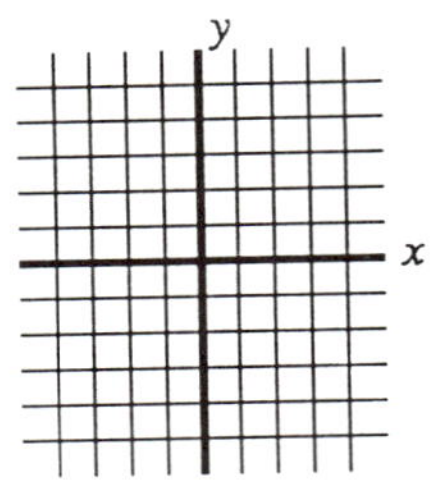

Now try the exercises for Topic A on page 346.

B. Parallel Lines

DEFINITION

Parallel lines are lines (in the same plane) that do not intersect.

Example 4 ► Show that the lines L_1, given by

$$y = 5 - 2x$$

and L_2, given by

$$y = \frac{1 - 4x}{2}$$

are parallel.

Solution:

Consider Table 7.15.

TABLE 7.15

L_1:

x	y
0	5
1	3

L_2:

x	y
0	$\frac{1}{2}$
1	$\frac{-3}{2}$

For any value of x, the y-coordinates of the corresponding points on L_1 and L_2 are $4\frac{1}{2}$ units apart. For example, for each x, let y_1 be the y-coordinate of the point on L_1, and let y_2 be the y-coordinate of the corresponding point on L_2. *(See Figure 7.56.)*

$$\text{When } x = 0,\ y_1 - y_2 = 5 - \frac{1}{2} = 4\frac{1}{2}$$

$$\text{When } x = 1,\ y_1 - y_2 = 3 - \left(\frac{-3}{2}\right) = 4\frac{1}{2}$$

◀

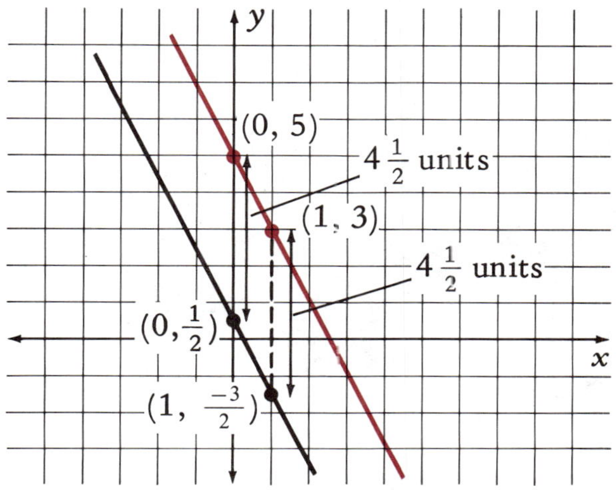

FIGURE 7.56

In Example 4 observe that both lines have slope −2. *Two different* (nonvertical) *lines are parallel if they have the same slope, and intersect if they have different slopes.* Thus the lines given by the equation

$$y = 2x + 3 \quad \text{and} \quad y = 2x - 1$$

are parallel because each of these lines has slope 2. But the lines given by

$$y = 3x \quad \text{and} \quad y = x + 2$$

intersect because the first line has slope 3, whereas the second has slope 1.

Answers

3. __________

PRACTICE EXERCISES FOR TOPIC B

2. Show, graphically, that L_1 and L_2 are parallel.

L_1: $2x + y = 6$

L_2: $2y + 2 = -4x$

3. Without graphing, indicate which of these lines are parallel.

L_1: $\frac{y}{3} = x$ L_2: $y = \frac{x}{3}$ L_3: $y = 3x + 2$ L_4: $\frac{x + y}{3} = 3$

Now try the exercises for Topic B on page 346.

Use this area for work space.

EXERCISES

A. In Exercises 1–12 find the intersection of the given lines graphically. (*See Examples 1–3.*) In Exercises 1–4 *check your solution, as in Example 1.*

1. L_1: $y = x$
L_2: $y = 4 - x$
(*Check.*)

2. L_1: $y = 2x$
L_2: $y - 3 = \frac{x}{2}$
(*Check.*)

3. L_1: $y - 4 = x$
L_2: $y - 2 = 2(x + 1)$
(*Check.*)

4. L_1: $y + 2 = \frac{x + 4}{2}$
L_2: $y = 2x$
(*Check.*)

5. L_1: $2y = x + 3$
L_2: $y = 2x$

6. L_1: $y = 10x - 7$
L_2: $y + 5 = 2(x - 1)$

7. L_1: $y - 4 = 3x$
L_2: $y = x$

8. L_1: $y = 2x + 5$
L_2: $y = x + 1$

9. L_1: $y = \frac{x}{4}$
L_2: $y + 2 = 2(x - 6)$

10. L_1: $3y + 1 = 2x$
L_2: $y + 1 = x$

11. L_1: $y + 3 = 2(x - 5)$
L_2: $2y = \frac{-x}{3}$

12. L_1: $y + 8 = 3(x - 1)$
L_2: $3y = x - 1$

B. In Exercises 13–20 show, graphically, that the indicated lines are parallel. (*See Example 4.*)

13. L_1: $y = x + 3$
L_2: $2y = 2x - 4$

14. L_1: $2y = x - 1$
L_2: $y = \frac{x + 1}{2}$

15. L_1: $y = 5x - 3$
L_2: $2y = 5(2x - 1)$

16. L_1: $y = 3x$
L_2: $\frac{y}{3} = x + 4$

17. L_1: $y - 4 = 2(x - 1)$
L_2: $y = 2x + 1$

18. L_1: $y - 1 = 4(x - 2)$
L_2: $\frac{y}{2} = 2(x - 1)$

19. L_1: $y + 3 = 3(x + 2)$
L_2: $1 - y = 3(2 - x)$

20. L_1: $x + 3y = 4$
L_2: $2 + y = \frac{1 - x}{3}$

In Exercises 21–28 graph the indicated lines to determine whether or not they intersect. If they do, find the intersection point.

21. L_1: $y = x - 1$
L_2: $2y = x + 1$

22. L_1: $y = 2x - 1$
L_2: $2y = 2x + 1$

23. L_1: $y = x - 7$
L_2: $y = x + 3$

24. L_1: $2y = x + 5$
L_2: $4y = 2(x - 1)$

25. L_1: $y = x + 6$
L_2: $y - 3 = 2x + 2$

26. L_1: $y = 1 + x$
L_2: $y = 1 - x$

27. L_1: $\frac{x}{2} + \frac{y}{4} = 0$
L_2: $y = 1 - 2x$

28. L_1: $y = 4$
L_2: $x = 5$

Use this area for work space.

29. Without graphing, indicate which two of these lines are parallel. (*See Practice Exercise 3.*)

L_1: $y = x$ L_2: $y = 2x$

L_3: $y = 3x + 1$ L_4: $y = x + 1$

30. Without graphing, indicate which two of these lines are parallel.

L_1: $x + y = 0$ L_2: $y = 1 - x$

L_3: $y = x + 1$ L_4: $y = 0$

31. Without graphing, indicate which two of these lines are parallel.

L_1: $y = 2x + 3$ L_2: $2y = x + 3$

L_3: $\frac{y}{2} = x + 3$ L_4: $\frac{x}{2} + \frac{y}{2} = 3$

32. Without graphing, indicate which two of these lines are parallel.

L_1: $y = 4(x + 1)$ L_2: $\frac{y}{4} = \frac{x}{4} + 1$

L_3: $y = x + 1$ L_4: $y = \frac{x}{4} + 1$

33. Pick out the *error(s).*

(a) If two different lines intersect, they are not parallel.

(b) If two different lines are parallel, they do not intersect.

(c) Any two different horizontal lines are parallel.

(d) A vertical line always intersects a horizontal line.

(e) Any two different slanting lines intersect.

7.6 Systems of Linear Equations

A. Systems and Their Solutions

The graphical method that you used to find the intersection of two lines tends to be inaccurate, particularly when the coordinates of the intersection point are not integers. There are also algebraic methods of determining this intersection.

An equation of a line is called a **linear equation.** Thus

$$y = 2x + 3$$

is a linear equation.

Two linear equations, taken together, form a **system of linear equations (in two variables).** For example, the two equations

$$y = 2x + 3$$

$$y + 1 = 3(x + 1)$$

together form such a system.

DEFINITION

An ordered pair (x_1, y_1) is a **solution of a system of linear equations** if (x_1, y_1) is a solution of both equations of the system.

The ordered pair (1, 5) is a solution of

$$\mathbf{y} = 2x + 3$$

because $\mathbf{5} = 2(1) + 3$

Note that (1, 5) is also a solution of

$$\mathbf{y} + 1 = 3(x + 1)$$

because $\mathbf{5} + 1 = 3(1 + 1)$

$$[\text{or } 6 = 6]$$

Thus (1, 5) is a solution of the *system:*

$$y = 2x + 3$$
$$y + 1 = 3(x + 1)$$

Geometrically, a solution of a system of linear equations represents the intersection of the lines whose equations comprise the system. (*See Figure 7.57.*)

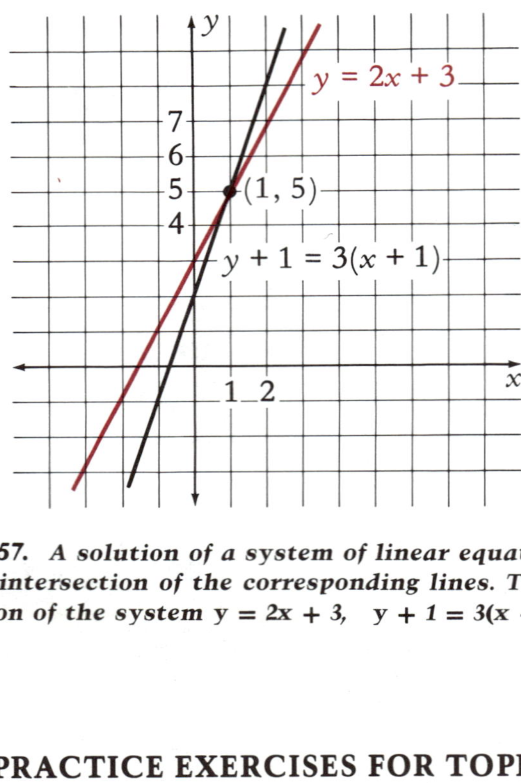

FIGURE 7.57. ***A solution of a system of linear equations represents the intersection of the corresponding lines. Thus (1, 5) is the solution of the system y = 2x + 3, y + 1 = 3(x + 1).***

Answers

1. ______

PRACTICE EXERCISES FOR TOPIC A

1. *Check* whether (10, 1) is a solution of the system of linear equations

$$x - 4y = 6$$
$$2y = x - 8$$

Now try the exercises for Topic A on page 356.

B. Substituting

You can solve a system of equations by substituting for one of the variables, as in Example 1.

Example 1 ▶ Solve the system:

$$y = 2x + 3$$
$$y + 2 = 4x + 1$$

Check the solution.

Solution:

Substitute

$$\mathbf{2x + 3} \text{ for } \mathbf{y}$$

in the second equation.

$$\mathbf{2x + 3} + 2 = 4x + 1$$
$$2x + 5 = 4x + 1$$
$$4 = 2x$$
$$2 = x$$

Now replace x by 2 in the first equation, and find y:

$$y = 2 \cdot 2 + 3$$
$$y = 7$$

The solution of the system is (2, 7). (*See Figure 7.58.*)

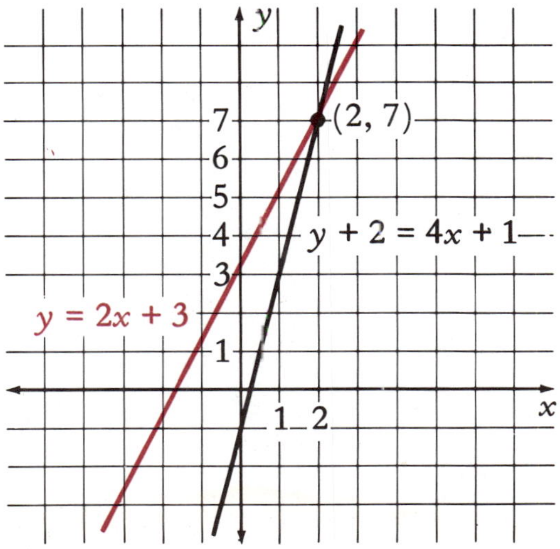

FIGURE 7.58

Check:

You can check the solution by replacing x by 2 and y by 7 in each of the given equations, $y = 2x + 3$ and $y + 2 = 4x + 1$.

$7 \stackrel{?}{=} 2 \cdot 2 + 3$	$7 + 2 \stackrel{?}{=} 4 \cdot 2 + 1$
$7 \stackrel{\checkmark}{=} 7$	$9 \stackrel{\checkmark}{=} 9$

◀

In the **substitution method:**

1. Solve one equation for one of the variables, say y, in terms of the other, x.
2. Replace y by this expression in the other equation.

3. Solve this other equation for x.
4. To find y, replace x by this value in the equation for y in Step **1.**

Example 2 ▶ Solve the system:

$$y + 3 = 4(x - 1)$$
$$y - 2 = -5x$$

Solution:

1. From the second equation:

$$y = 2 - 5x$$

2. Replace y by the expression $2 - 5x$ in the first equation:

$$\mathbf{2 - 5x} + 3 = 4(x - 1)$$

3. Solve this equation for x:

$$5 - 5x = 4x - 4$$
$$9 = 9x$$
$$1 = x$$

4. Replace x by 1 in the equation $y = 2 - 5x$ in Step 1:

$$y = 2 - 5 \cdot 1$$
$$y = -3$$

The solution is $(1, -3)$. (*See Figure 7.59.*) ◀

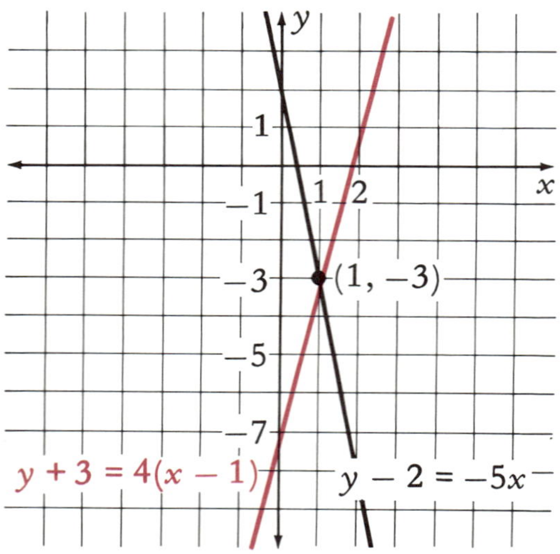

FIGURE 7.59

A linear equation in the form $y = mx + b$ can also be written in the form

$$Ax + By = C$$

where A, B, and C are numbers, and where A and B are not both 0. For example,

$$y = 2x + 5$$

becomes

$$-2x + y = 5$$

Here $A = -2$, $B = 1$, and $C = 5$. Also, the linear equation

$$y - 3 = \frac{1 - x}{2}$$

which is not in either form, becomes

$$2y - 6 = 1 - x$$

and then

$$x + 2y = 7$$

Here $A = 1$, $B = 2$, and $C = 7$.

Often the equations of a system are given in the form

$$Ax + By = C$$
$$ax + by = c$$

Example 3 ▶ Solve the system:

$$x + 2y = 10$$
$$2x - 3y = 6$$

Solution:

1. Because x appears in the first equation with coefficient 1, solve this equation for x in terms of y:

$$x = 10 - 2y$$

2. Replace x by the expression $10 - 2y$ in the second equation:

$$2(10 - 2y) - 3y = 6$$

3. Solve this for y:

$$20 - 4y - 3y = 6$$
$$20 - 7y = 6 \qquad \textit{Add } 7y - 6 \textit{ to both sides.}$$
$$14 = 7y$$
$$2 = y$$

4. Replace y by 2 in the equation $x = 10 - 2y$ in Step 1:

$$x = 10 - 2 \cdot \mathbf{2}$$
$$x = 6$$

The solution is (6, 2). (*See Figure 7.60.*) ◀

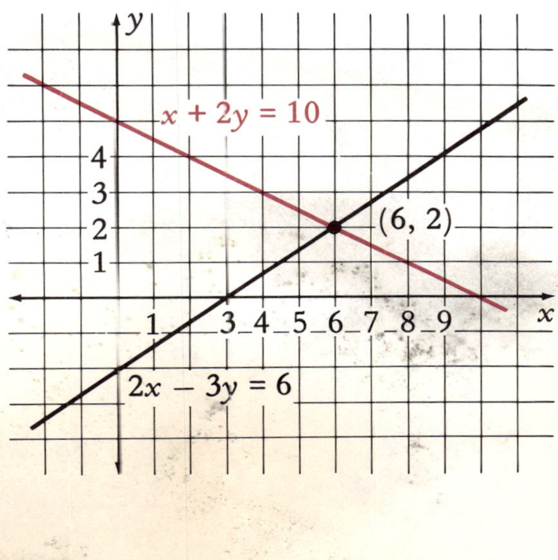

FIGURE 7.60

Answers

2. ______________

3. ______________

PRACTICE EXERCISES FOR TOPIC B

2. Solve by substituting:

$$y = \frac{x - 5}{2}$$
$$x + 2y = 3$$

3. Solve by substituting and *check*:

$$x - 3y = 6$$
$$3y - 2x = 3$$

Now try the exercises for Topic B on page 356.

C. Adding

You can also solve a system of equations by first adding the corresponding sides of the equations to eliminate one of the variables.

Example 4 ▶ Solve the system:

$$2x + 3y = 5$$
$$4x - 3y = 1$$

Solution:
Add the corresponding sides of the equations to eliminate y:

$$\begin{aligned} 2x + 3y &= 5 \\ 4x - 3y &= 1 \\ \hline 6x \quad &= 6 \\ x \quad &= 1 \end{aligned}$$

Now replace x by 1 in either equation. (For example, use the first equation.)

$$2 \cdot 1 + 3y = 5$$
$$3y = 3$$
$$y = 1$$

The solution is (1, 1). (*See Figure 7.61.*) ◀

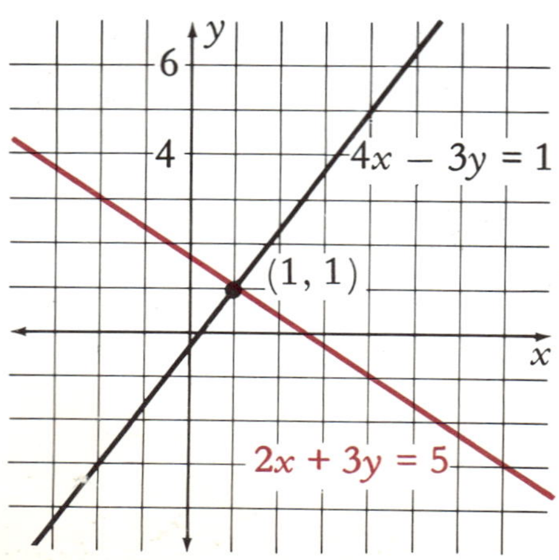

FIGURE 7.61

Example 5 ▶ Solve the system:

$$2x + y = 10$$
$$x + y = 6$$

Solution:
Subtract each side of the second equation from the corresponding side of the first equation to eliminate y:

$$\begin{array}{r} 2x + y = 10 \\ \underline{\mp x \mp y = \mp 6} \\ x \qquad = 4 \end{array}$$

Replace x by 4 in the second equation.

$$4 + y = 6$$
$$y = 2$$

The solution is (4, 2). (*See Figure 7.62.*) ◀

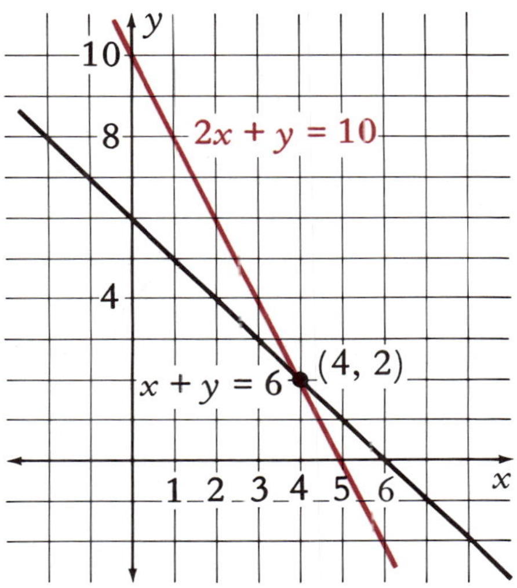

FIGURE 7.62

In order to use this method effectively, you may have to change one or both of the given equations into an equivalent equation.

Example 6 ▶ Solve the system:

$$3x + 4y = 2$$
$$7x + 2y = 12$$

Solution:
Note that $4y$ occurs in the first equation and $2y$ in the second. Multiply both sides of the second equation by -2:

$$-14x - 4y = -24$$

Now add the corresponding sides of this equation and of the given first equation to obtain:

$$\begin{array}{rcl} 3x + 4y &=& 2 \\ -14x - 4y &=& -24 \\ \hline -11x &=& -22 \\ x &=& 2 \end{array}$$

Divide both sides by −11.

Replace x by 2 in the first equation.

$$3(2) + 4y = 2$$
$$6 + 4y = 2$$
$$4y = -4$$
$$y = -1$$

The solution is (2, −1). (*See Figure 7.63.*) ◀

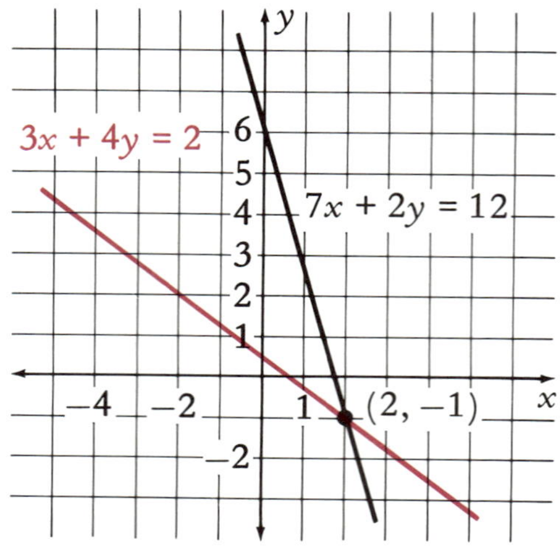

FIGURE 7.63

These algebraic methods apply to systems in variables other than x and y.

In Example 7 *both* equations are first transformed, so that when corresponding sides of the resulting equations are added, a variable is eliminated.

Example 7 ▶ Solve the system:

$$3s + 2t = 10$$
$$2s + 3t = 5$$

Solution:
Multiply both sides of the first equation by 2, and both sides of the second equation by −3. You will then eliminate s.

$$\begin{aligned} 2(3s + 2t) &= 2 \cdot 10 \quad \text{or} \quad & 6s + 4t &= 20 \\ -3(2s + 3t) &= -3 \cdot 5 \quad \text{or} \quad & -6s - 9t &= -15 \\ & & \hline -5t &= 5 \\ & & t &= -1 \end{aligned}$$

Replace t by −1 in the (given) first equation:

$$3s + 2(\mathbf{-1}) = 10$$
$$3s = 12$$
$$s = 4$$

The solution is (4, −1).

You can graph this system by letting the horizontal axis be the s-axis and the vertical axis be the t-axis, as in Figure 7.64. ◀

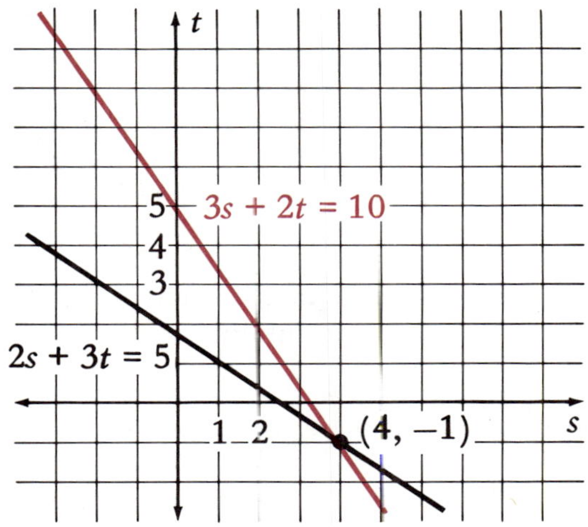

FIGURE 7.64

Answers

4. ______

5. ______

6. ______

PRACTICE EXERCISES FOR TOPIC C

Solve by the "adding" method. In Practice Exercise 6 *check your solution.*

4. $x - 2y = 3$
$3x + 2y = 5$

5. $5x - 4y = -5$
$4x - y = 7$

6. $-7x + 10y = 30$
$3x + 4y = 12$
(*Check.*)

Now try the exercises for Topic C on page 357.

D. Applications

Here is a problem that is solved by setting up and solving a system of two linear equations in two unknowns.

Example 8 ▶ A copy center has two machines. The faster machine in 2 minutes and the slower one in 3 minutes together produce 310 copies. The faster machine in 5 minutes and the slower one in 6 minutes together produce 700 copies. How many copies per minute does each machine produce?

Solution:

Let x be the number of copies the faster machine produces per minute. Then in 2 minutes it produces 2x copies.

Let y be the number of copies the slower machine produces per minute. In 3 minutes it produces $3y$ copies.

Faster in 2 min. and slower in 3 min. produce 310 copies.

$$\underbrace{\text{Faster in 2 min.}}_{2x} + \underbrace{\text{slower in 3 min.}}_{3y} = 310$$

Faster in 5 min. and slower in 6 min. produce 700 copies.

$$\underbrace{\text{Faster in 5 min.}}_{5x} + \underbrace{\text{slower in 6 min.}}_{6y} = 700$$

You obtain the system:

$$2x + 3y = 310 \qquad \textit{Multiply both sides by } -2.$$

$$\begin{aligned} 5x + 6y &= 700 \\ -4x - 6y &= -620 \\ \hline x \quad &= 80 \end{aligned} \qquad \textit{Add.}$$

Replace x by 80 in the first equation.

$$\begin{aligned} 2(80) + 3y &= 310 \\ 160 + 3y &= 310 \\ 3y &= 150 \\ y &= 50 \end{aligned}$$

The faster machine produces 80 copies per minute; the slower machine produces 50 copies per minute. ◀

Answers

7. ______

PRACTICE EXERCISES FOR TOPIC D

7. Ann and Bob both have part-time jobs. Ann's hourly wage is \$.25 more than Bob's. On Saturday Ann worked 4 hours, Bob worked 5 hours, and together they earned a total of \$46. Find the hourly wage of each.

Now try the exercises for Topic D on page 357.

Use this area for work space.

EXERCISES

A.

1. Check whether (2, 3) is a solution of the system

$$\begin{aligned} y &= x + 1 \\ 3y &= 4x + 1 \end{aligned}$$

(*See Practice Exercise 1.*)

2. Check whether (4, −1) is a solution of the system

$$\begin{aligned} y &= 3 - x \\ x - y &= 3 \end{aligned}$$

3. Check whether (−2, −4) is a solution of the system

$$\begin{aligned} 3x - 2y &= 2 \\ 5x + y &= -6 \end{aligned}$$

4. Check whether (−1, 8) is a solution of the system

$$\begin{aligned} 5(x + 1) + 2y &= 16 \\ 3(x - 1) - y &= -2 \end{aligned}$$

B.

In Exercises 5–14 solve by substituting. In Exercises 5–8 check your solution. (*See Examples 1–3.*)

5. $y = x + 1$
$y = 3x - 5$
(*Check.*)

6. $y = x - 2$
$y = 2x - 8$
(*Check.*)

7. $y = 2x - 3$
$y = 3x - 7$
(*Check.*)

8. $y = 4x - 1$
$y = 2x + 3$
(*Check.*)

9. $y = 4 - 3x$
$y - 3 = 2(x - 2)$

10. $y - 8 = 5x$
$y + 1 = 4(x + 2)$

11. $v + 2 = 4(u - 3)$
$v - 4 = \dfrac{u - 1}{2}$

12. $t + 5 = s$
$t + 3 = 4(s - 8)$

Use this area for work space.

13. $b - 2 = -3(a - 1)$
$b = -2a$

14. $2x + 3y = 19$
$x - 2y = -1$

C. In Exercises 15–26 solve by the "adding" method. In Exercises 15–18 check your solution. (*See Examples 4–7.*)

15. $x + y = 4$
$x - y = 2$
(*Check.*)

16. $x - y = 3$
$-x - y = 1$
(*Check.*)

17. $a + 2b = -2$
$3a - 2b = 6$
(*Check.*)

18. $2x + y = 7$
$3x + y = 12$
(*Check.*)

19. $c + 5d = 2$
$5c + 5d = 6$

20. $x + 3y = -1$
$5x + 3y = 7$

21. $r + 2s = 8$
$4r - s = 5$

22. $3m - n = 2$
$2m - 3n = -1$

23. $6x - 5y = 7$
$2x + 3y = 7$

24. $2x + 3y = 9$
$3x + 2y = 11$

25. $4x - 3y = 7$
$3x + 5y = 27$

26. $4x + 5y = 4$
$7x + 7y = 14$

In Exercises 27–38 solve by either substituting or adding.

27. $x + 2y = 6$
$3x - y = 4$

28. $u + v = 8$
$5v = 3u$

29. $3y = 4x$
$y - 4 = x + 2$

30. $x + y = 3$
$2x - 3y = 6$

31. $a + 2b = 5$
$4a - 6b = 6$

32. $6x - 3y = 3$
$3x + y = 9$

33. $3y = 2x$
$y - 4 = 2x$

34. $2x + 5y = 9$
$4x + 3y = 11$

35. $x - 3y = 2$
$4x - 10y = 10$

36. $2s + 3t = 10$
$3s + 2t = 10$

37. $2x - 9y = 2$
$5x - 13y = 24$

38. $5x + 9y = 8$
$7x + 16y = 1$

D. **39.** Five times a number is one more than seven times another number. Twice the first number is one less than three times the second number. Find these two numbers. (*See Example 8.*)

40. A shipping clerk finds that 3 cartons of paper clips and 2 cartons of pens together weigh 46 pounds. Also 6 cartons of paper clips together with 12 cartons of pens weigh 180 pounds. How much does a carton of paper clips weigh?

41. A bookstore shelves books vertically. A 2-foot shelf holds 12 copies of an atlas and 3 copies of a dictionary. A 3-foot shelf holds 4 copies of the atlas and 15 copies of the dictionary. How thick is each book?

42. With your \$5 you can buy 3 hot dogs and an order of french fries. Or you can save 20¢ by ordering 2 of each item. How much does a hot dog cost?

Use this area for work space.

43. One shopper buys 5 pounds of potatoes and 6 pounds of onions together for \$7.32. Another shopper buys 7 pounds of potatoes and 3 pounds of onions for \$7.02. Find the price per pound of each item.

44. The postage for a package comes to \$5.15. A woman uses a total of twenty-two 25-cent and 20-cent stamps on this package. How many of each stamp does she use?

7.7 Direct Variation

A. What Is Direct Variation?

A linear equation of the form

$$y = kx$$

describes many important relationships in science and everyday affairs. For example, if a car is traveling at the constant rate of 60 miles per hour, the distance, y, traveled in x hours is given by the linear equation

$$y = 60x$$

Thus, in half an hour, the car travels $60 \cdot \frac{1}{2}$ or 30 miles; in 3 hours, the car travels $60 \cdot 3$ or 180 miles.

DEFINITION

> Let x and y be variables. Then y **varies directly as** x if
>
> $$y = kx$$
>
> for some nonzero number k. This number k is called the **constant of variation.**

Example 1 ▶ Suppose

$$y = 3x$$

Then y varies directly as x. The constant of variation is 3. Some corresponding values of x and y are given in Table 7.16. The relationship between x and y is described by a line through the origin with slope 3. (*See Figure 7.65.*) ◀

TABLE 7.16

x	y
1	3
2	6
4	12
-1	-3
-3	-9

When y varies directly as x, then

$$y = kx, \quad \text{where } k \neq 0$$

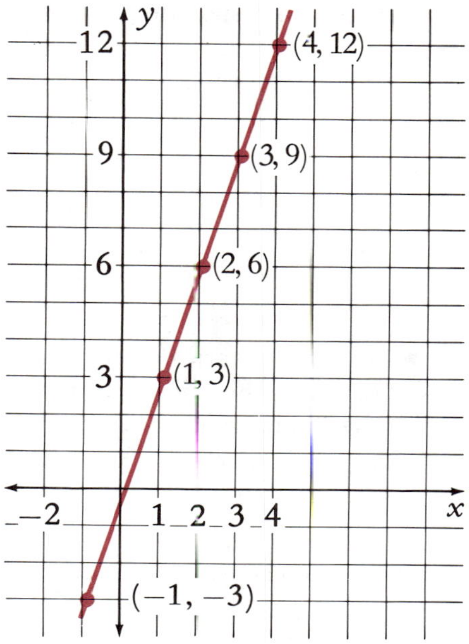

FIGURE 7.65

It follows that x *also varies directly as* y because if you divide both sides of the equation by k, you obtain

$$\frac{1}{k}y = x$$

The constant of variation is now $\frac{1}{k}$. Also, if y varies directly as x, then

$$\frac{y}{x} = k, \quad \text{where } x \neq 0$$

Thus *the quotient,* $\frac{y}{x}$, *remains fixed.* Let x_1 and x_2 be values of x and let y_1 and y_2 be the corresponding values of y. Then

$$\frac{y_1}{x_1} = \frac{y_2}{x_2} = k$$

Thus *corresponding values of y and x are proportional. If k > 0, when one of these variables, x or y, increases the other variable also increases. If k < 0, when x or y increases, the other variable decreases.*

Example 2 ▶ Suppose

$$y = 5x$$

The constant of variation is 5. Then you can also write

$$x = \frac{1}{5}y$$

Here the constant of variation is $\frac{1}{5}$. Also

$$\frac{y}{x} = 5$$

Some corresponding values of x and y are given in Table 7.17. ◀

TABLE 7.17

x	y
1	5
2	10
3	15
4	20

Example 3 ▶ Suppose y varies directly as x, and $y = 12$ when $x = 3$.
(a) Find the constant of variation.
(b) Find y when $x = 5$.
(c) Find x when $y = 8$.

Solution:

(a)
$$y = kx$$

Replace y by 12 and x by 3.

$$\mathbf{12} = k \cdot 3$$
$$4 = k$$

Thus the constant of variation is 4, and $y = 4x$.

(b) Replace x by 5.

$$y = 4 \cdot 5 = 20$$

(c) Replace y by 8.

$$\mathbf{8} = 4x$$
$$2 = x$$ ◀

PRACTICE EXERCISES FOR TOPIC A

Answers
1. (a) ______
(b) ______
(c) ______
2. (a) ______
(b) ______
(c) ______

1. Suppose y varies directly as x, and $y = 2$ when $x = \frac{1}{4}$.

(a) Find the constant of variation.

(b) Find y when $x = 7$.

(c) Find x when $y = 12$.

2. Suppose y varies directly as x, and $y = 4$ when $x = 10$.

(a) Find the constant of variation.

(b) Find y when $x = 45$.

(c) Find x when $y = 9$.

Now try the exercises for Topic A on page 361.

B. Applications

Example 4 ▶ Suppose the amount of bread consumed in a town varies directly as the population. If 70 000 pounds per week are consumed in a town of 20 000 inhabitants, how many pounds per week are consumed in a town of 15 000 inhabitants?

Solution:
Let x be the population of a town, and let y be the amount of bread consumed per week in that town. Then

$$y = kx$$

Replace y by 70 000 and x by 20 000 to find k, the constant of variation.

$$\mathbf{70\,000} = k \cdot 20\,000$$

$$\frac{7}{2} = k$$

Thus the constant of variation is $\frac{7}{2}$ and

$$y = \frac{7}{2}x$$

Now replace x by 15 000 to find the corresponding value of y.

$$y = \frac{7}{\cancel{2}_{1}} \cdot \cancel{15\,000}^{7\,500} = 52\,500$$

Thus 52 500 pounds of bread are consumed each week by the 15 000 inhabitants. ◀

Answers

3. ____________

PRACTICE EXERCISES FOR TOPIC B

3. The social-security tax withheld from a worker's paycheck varies directly with gross earnings. If a $12 tax is withheld when a worker earns $200, how much is withheld when he or she earns $350?

Now try the exercises for Topic B on page 362.

Use this area for work space.

EXERCISES

A. In Exercises 1–8 suppose y varies directly as x. Find the constant of variation. (*See Example 3(a).*)

1. $y = 4$ when $x = 2$

2. $y = 10$ when $x = 2$

3. $y = 9$ when $x = 3$

4. $y = 36$ when $x = 6$

5. $y = 4$ when $x = -2$

6. $y = 10$ when $x = 4$

7. $y = \frac{1}{2}$ when $x = \frac{1}{4}$

8. $y = \frac{1}{2}$ when $x = \frac{1}{3}$

In Exercises 9–25 assume y varies directly as x. (*See Example 3.*)

9. Suppose $y = 6$ when $x = 3$. Find y when $x = 5$.

10. Suppose $y = 12$ when $x = 2$. Find y when $x = 6$.

11. Suppose $y = 15$ when $x = 3$. Find y when $x = 2$.

12. Suppose $y = 20$ when $x = 4$. Find y when $x = -2$.

13. Suppose $y = 100$ when $x = 10$. Find x when $y = 400$.

Use this area for work space.

14. Suppose $y = 49$ when $x = 7$. Find x when $y = 7$.

15. Suppose $y = -12$ when $x = 36$. Find x when $y = 48$.

16. Suppose $y = -64$ when $x = -16$. Find x when $y = 16$.

17. Suppose $y = 6$ when $x = 9$. Find y when $x = 3$.

18. Suppose $y = -8$ when $x = -4$. Find y when $x = 16$.

19. Suppose $y = \frac{1}{2}$ when $x = \frac{1}{4}$. Find y when $x = 3$.

20. Suppose $y = \frac{3}{5}$ when $x = \frac{4}{15}$. Find y when $x = \frac{1}{5}$.

21. Suppose $y = -18$ when $x = 12$. Find x when $y = \frac{1}{2}$.

22. Suppose $y = 250$ when $x = 100$. Find x when $y = 12.5$.

23. Suppose $y = .3$ when $x = .15$. Find x when $y = .7$.

24. Suppose $y = 9$ when $x = 4$. Find x when $y = 6$.

25. Suppose $y = .01$ when $x = .1$. Find x when $y = .001$.

26. Pick out the *error(s)*.

(a) If y varies directly as x, then x also varies directly as y.

(b) If y varies directly as x, then as x increases y must also increase.

(c) If y varies directly as x, then the quotient $\frac{y}{x}$, for $x \neq 0$, is constant.

(d) The equation $xy = 1$ represents direct variation.

(e) The equation $\frac{x}{y} = 1$, $y \neq 0$, represents direct variation.

B.

27. The cost of a slice of cheese varies directly as its weight. If a 12-ounce slice costs \$3.60, how much does a 14-ounce slice cost? (*See Example 4.*)

28. The tension in a spring varies directly as the distance it is stretched. If the tension is 36 pounds when the distance stretched is 9 inches, what is the tension when the distance stretched is 1 foot?

29. At a constant rate the distance traveled varies directly as time. A car travels 135 miles in 3 hours. At this rate (a) how far does it travel in 5 hours? (b) How long does it take to complete a 300-mile trip?

30. Assume the amount of work a gardener does varies directly as time. If she mows $\frac{2}{3}$ of the lawn in an hour and a half, how long does it take her to mow the entire lawn?

31. The number of gene mutations caused by exposure to x-ray dosage varies directly as the size of the x-ray dosage. Suppose the x-ray dosage is tripled. By what factor is the number of gene mutations increased?

32. According to Hooke's law, the distance a vertical spring stretches varies directly with the weight placed on it. If a 60-kilogram weight stretches a vertical spring 8 centimeters, how large a weight must be used to stretch the spring 20 centimeters?

What Have You Learned in Chapter 7?

REVIEW

You can locate points on a rectangular coordinate system.

You know that the graph of an equation is its pictorial representation on a rectangular coordinate system.

You can graph a line.

You have learned how to find the slope and *y*-intercept of a line. You can find the equation of a line with a given slope and *y*-intercept. You can also find the equation of a line from its graph.

You can find the intersection of two lines graphically. Also you can solve the corresponding system of linear equations algebraically.

And you have learned that y varies directly as x if $y = kx$, for some nonzero number k.

Let's Review Chapter 7

7.1 *Rectangular Coordinates*

1. For each point in Figure 7.66 find
(a) its x-coordinate, (b) its y-coordinate. (c) Identify the point with its ordered pair of coordinates.

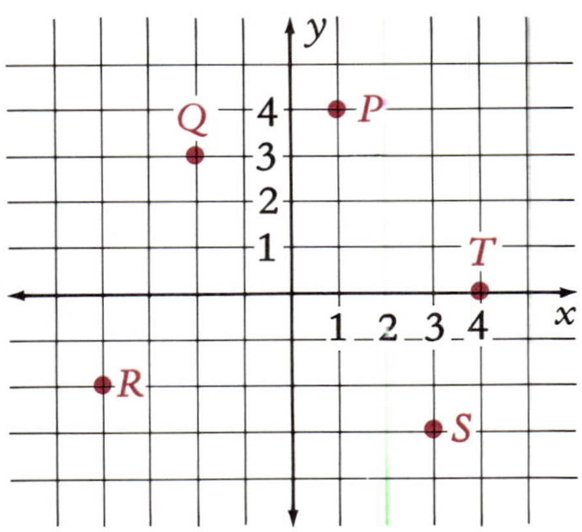

FIGURE 7.66

In Exercises 2–6 locate the following points on a rectangular coordinate system:

2. $(4, 0)$ **3.** $(0, -2)$ **4.** $(5, 3)$ **5.** $(-3, 4)$ **6.** $\left(\frac{-1}{2}, \frac{-3}{2}\right)$

7.2 *Graphs of Equations*

Graph the equations in Exercises 7-12.

7. $y = x + 2$ **8.** $y = 2x$ **9.** $y = 2 - x$

10. $y = \frac{x}{2}$ **11.** $y = 2$ **12.** $x = -2$

7.3 *Slope*

13. Let $P_1 = (3, 1)$ and $P_2 = (5, 6)$.
(a) Find the rise from P_1 to P_2.
(b) Find the run from P_1 to P_2.
(c) Find the slope of the line through P_1 to P_2.
(d) Does this line slope upward or downward to the right?

14. Draw the line through (2, 1) with slope $\frac{1}{2}$.

15. Indicate whether each of the following lines through P_1 and P_2 (*i*) is horizontal, (*ii*) is vertical, (*iii*) slopes upward to the right, (*iv*) slopes downward to the right.
(a) $P_1 = (4, 4)$, $P_2 = (6, 0)$ (b) $P_1 = (3, -3)$, $P_2 = (6, -3)$

7.4 *Equation of a Line*

16. Find (a) the slope and (b) the *y*-intercept of the line with equation

$$y = \frac{-x}{2} + 7$$

17. Find the equation of the line with slope 4 and *y*-intercept 0.

18. Find the equation of the line with x-intercept 3 and *y*-intercept -2.

19. Find the equation of the horizontal line through (3, 5).

20. (a) Find the equation in intercept form of the line drawn in Figure 7.67.

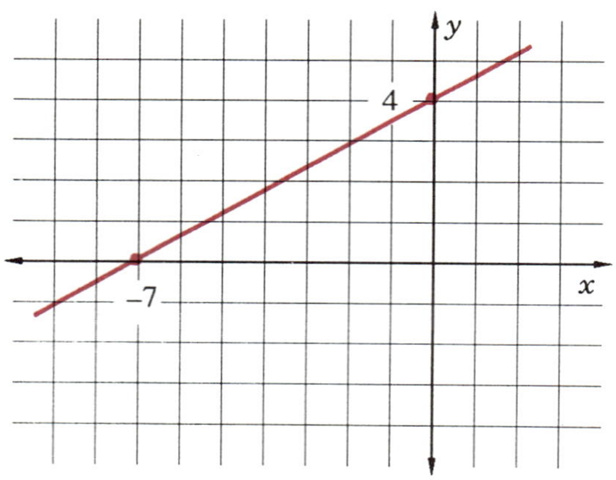

FIGURE 7.67

(b) Clear the equation of fractions.

7.5 *Intersection of Lines*

21. Graph the intersection of the lines given by:

$$L_1: y = 2x + 4 \qquad L_2: y = 7 - x$$

22. Show, graphically, that the indicated lines are parallel.

$$L_1: y = 4x - 3 \qquad L_2: y - 1 = 4(x + 1)$$

23. Graph the following lines to determine whether or not they intersect. If they do, find the intersection point.

$$L_1: y = x - 3 \qquad L_2: y = 2x + 2$$

24. Without graphing, indicate which two of these lines are parallel:

$$L_1\colon y = 5x \qquad L_2\colon y = -5x \qquad L_3\colon y = x + 5 \qquad L_4\colon y = x - 5$$

7.6 *Systems of Linear Equations*

25. Solve by substituting:

$$\begin{aligned} y &= 2x + 1 \\ y &= 6 - x \end{aligned}$$

26. Solve by substituting. Check your solution.

$$\begin{aligned} y + 1 &= 2x \\ \frac{y}{3} &= 3 - x \end{aligned}$$

27. Solve by the "adding" method.

$$\begin{aligned} 6a - 5b &= 7 \\ 3a - 2b &= 4 \end{aligned}$$

28. Solve by either substituting or adding:

$$\begin{aligned} 3x + 4y &= 2 \\ 2x + 3y &= -1 \end{aligned}$$

7.7 *Direct Variation*

29. Suppose y varies directly as x. Find the constant of variation if $y = 4$ when $x = -2$.

30. Assume y varies directly as x. Suppose $y = 10$ when $x = 4$. Find y when $x = 6$.

31. The number of pounds of potatoes prepared in a college cafeteria varies directly with the size of the student body. If 500 students consume 150 pounds of potatoes at a meal, how many pounds should be prepared for 600 students?

And These from Chapters 1–6

32. Find the value of $(-3)^3 - (-2)^4$

33. Find the value of $\dfrac{4(2 - 5)^2 - 3[1 - (5 - 7)]}{(12 - 9)^2}$

34. Multiply $4a^2(x - 2)\cdot 5ab^2(x - 2)^3$

35. Simplify $4(x - 2)^2 - 2(x + 1)^2$

36. Find the prime factorization of 288.

37. Find the *gcd* of 56, 98, and 112.

38. Simplify $\dfrac{\frac{3}{4}}{\frac{1}{8}}$

39. Solve the equation $8(x + 5) - 5(x - 8) = 8$

40. Solve the equation $\dfrac{4}{x} = \dfrac{6}{x + 2}$

41. Factor $4x^4 - 9x^2$

42. Factor $6x^2 + 11x - 10$

43. Factor $a^2x^3 + 2a^2x^2 + a^2x$

44. Solve for x:

$$5x - 2xy + 3ax = 10$$

45. A saleswoman works on a $7\frac{1}{2}\%$ rate of commission. What must her sales be in order for her to earn a commission of \$450?

46. Solve the inequality

$$-4 < 3y + 2 < 17$$

Practice Exam 1 on Chapter 7

1. For each point in Figure 7.68, find
(a) its x-coordinate, (b) its y-coordinate. (c) Identify the point with its ordered pair of coordinates.

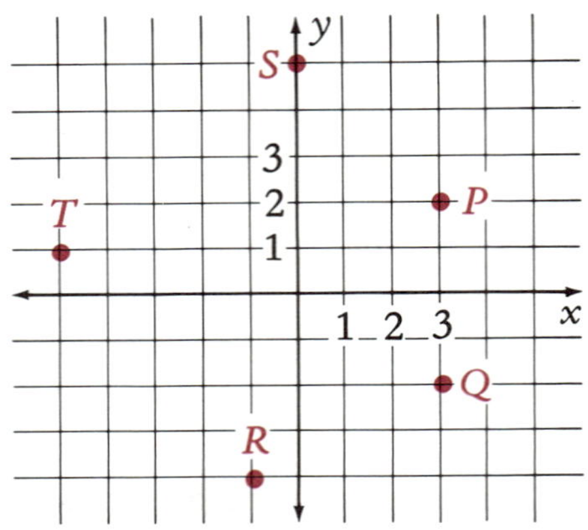

FIGURE 7.68

2. Graph the equation $y = 2x + 3$

3. Let $P_1 = (6, 3)$ and $P_2 = (3, 6)$.
(a) Find the rise from P_1 to P_2.
(b) Find the run from P_1 to P_2.
(c) Find the slope of the line through P_1 and P_2.
(d) Does this line slope upward or downward to the right?

4. Find (a) the slope and (b) the y-intercept of the line with equation $y = 4x - 3$.

5. Find the equation of the line with slope -2 and y-intercept 3.

6. Solve by either substituting or adding. *Check your solution.*

$$2x + y = 8$$
$$x - y = 1$$

7. Solve by either substituting or adding.

$$6x + 5y = 8$$
$$5x + 3y = 9$$

8. Assume y varies directly as x. Suppose $y = 3$ when $x = 2$. Find y when $x = \frac{1}{2}$.

Practice Exam 2 on Chapter 7

1. Identify each point in Figure 7.69 with its ordered pair of coordinates.

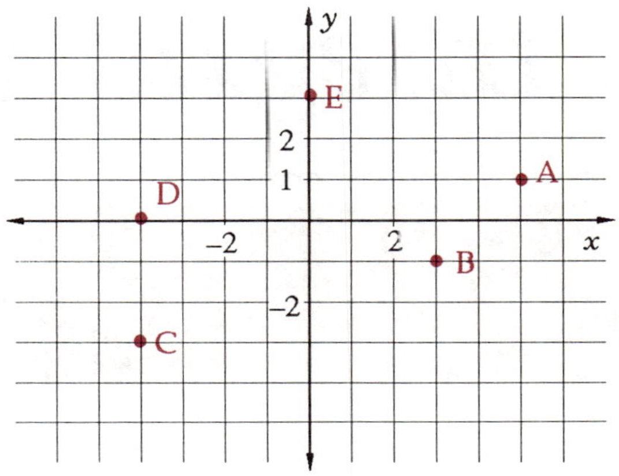

FIGURE 7.69

2. Graph the equation $y = 5 - 2x$.
3. Find the slope of the line that passes through each of the following pairs of points.

(a) (0, 5) and (3, 7) (b) (−4, 10) and (8, −2)

4. Find (a) the slope and (b) the y-intercept of the line with equation $3x + 2y = 4$.
5. Write the equation of the line that passes through (0, 8) and (−4, 0).
6. Solve the following system of equations by the substitution method. *Check your solution.*

$$x + 5 = 2y$$
$$3y - 4x = 5$$

7. Solve by the adding method.

$$6x + 5y = 14$$
$$3x - 2y = 16$$

8. If y varies directly as x, and $y = 20$ when $x = 25$,

(a) find the constant of variation,
(b) find y when $x = 10$, and
(c) find x when $y = 1$.

REVIEW

8

Roots and Integral Exponents

8.1 Square Roots

A. Roots of Numbers

You know that $$2^2 = 2 \cdot 2 = 4$$

Suppose you are asked,

What *positive* number times itself is 4?

In other words, you are asked to find a *positive* number such that

(the number) · (the number) = 4

Clearly 2 satisfies this condition. 2 is called the "positive square root of 4." Note that the *negative* number -2 also works because

$$(-2)(-2) = 4$$

-2 is called the "negative square root of 4."

DEFINITION

SQUARE ROOT. Let a be positive. Then b is called the **positive square root of a** if b is positive and if

$$b^2 = a$$

In this case, $-b$ is called the **negative square root of a.** Also the **square root of 0** is 0.

Write

$$b = \sqrt{a} \qquad \textit{Read: } b \text{ equals the positive square root of } a.$$

if b is the *positive* square root of a. Here the symbol $\sqrt{\ }$ is known as a **radical sign.** Thus

$$\sqrt{4} = 2 \qquad \text{The } \textit{positive} \text{ square root of 4 is 2.}$$

whereas $$-\sqrt{4} = -2 \qquad \text{The } \textit{negative} \text{ square root of 4 is } -2.$$

Example 1 ▶ (a) $\sqrt{9} = 3$ because 3 is positive and $3^2 = 9$.
(b) $\sqrt{25} = 5$ because 5 is positive and $5^2 = 25$. ◀

Example 2 ▶ Find (a) $\sqrt{100}$ (b) $\sqrt{144}$ (c) $-\sqrt{144}$

Solution:
(a) $\sqrt{100} = 10$ (b) $\sqrt{144} = 12$ (c) $-\sqrt{144} = -12$ ◀

Answers

1. (a) ______
(b) ______
(c) ______
(d) ______
2. (a) ______ (b) ______
(c) ______
3. ______

PRACTICE EXERCISES FOR TOPIC A

1. Find each square.

(a) 15^2 (b) $(-13)^2$ (c) 20^2 (d) 1000^2

2. Find each square root.

(a) $\sqrt{400}$ (b) $-\sqrt{225}$ (c) $\sqrt{169}$

3. Find the length of the side of a square whose area is 81 square inches.

Now try the exercises for Topic A on page 371.

B. Roots of Algebraic Expressions

Just as you have considered squares such as

$$a^2, \qquad (3x)^2, \qquad \text{and } (st)^2$$

so, too, you will consider square roots of algebraic expressions.

Example 3 ▶ Assume $a > 0$, $x > 0$, $s > 0$, and $t > 0$. Find:

(a) $\sqrt{a^2}$ (b) $-\sqrt{a^2}$ (c) $\sqrt{(3x)^2}$ (d) $\sqrt{(st)^2}$

Solution:

(a)
$$\sqrt{a^2} = a$$
because $a > 0$ and $a^2 = a \cdot a$

(b) By part (a),
$$\sqrt{a^2} = a \qquad \text{(The } positive \text{ square root of } a^2 \text{ is } a.)$$
Thus
$$-\sqrt{a^2} = -a \qquad \text{(The } negative \text{ square root of } a^2 \text{ is } -a.)$$

(c) Note that $3x \cdot 3x = (3x)^2$. Thus
$$\sqrt{(3x)^2} = 3x$$

(d)
$$\sqrt{(st)^2} = st$$

The product of positive numbers is positive. Thus in (c) and (d), $3x$ and st are each positive. ◀

Answers

4. (a) ______

(b) ______

(c) ______

PRACTICE EXERCISES FOR TOPIC B

4. Find each square root. Assume $a > 0$.

(a) $-\sqrt{57^2}$ (b) $\sqrt{(8a)^2}$ (c) $\sqrt{(a^3b^4)^2}$

Now try the exercises for Topic B on page 372.

C. Roots of Even Powers

Observe that
$$10\,000 = 100 \cdot 100 \quad \text{and} \quad 1\,000\,000 = 1000 \cdot 1000$$
Thus
$$\sqrt{10^4} = \sqrt{10\,000} = 100 = 10^2$$
and
$$\sqrt{10^6} = \sqrt{1\,000\,000} = 1000 = 10^3$$
Therefore
$$\sqrt{10^{2 \cdot 2}} = 10^2$$
and
$$\sqrt{10^{2 \cdot 3}} = 10^3$$

You can also consider square roots of *even* powers of variables.

Example 4 ▶ Let $b > 0$. Find:

(a) $\sqrt{a^4}$ (b) $\sqrt{b^6}$ (c) $\sqrt{c^{20}}$

Solution:

(a)
$$\sqrt{a^4} = a^2$$
because
$$a^4 = a^2 \cdot a^2$$
and a^2 is *nonnegative* (positive or 0).

(b)
$$\sqrt{b^6} = b^3$$
Note that
$$b^6 = b^3 \cdot b^3$$
Also $b^3 > 0$ because $b > 0$

(c) $$\sqrt{c^{20}} = c^{10}$$

because $$c^{20} = c^{10} \cdot c^{10}$$

Also $$c^{10} = c^5 \cdot c^5 = (c^5)^2$$

Thus c^{10} is nonnegative. ◀

In general, when k is a positive integer,

$$\sqrt{a^{2k}} = a^k$$

provided that when k is odd, then a is nonnegative, so that a^k is nonnegative. Thus

$$\sqrt{a^8} = a^4 \qquad \text{for all } a$$

but $$\sqrt{a^{10}} = a^5 \qquad \text{for nonnegative } a$$

Answers

5. (a) ______

(b) ______

(c) ______

PRACTICE EXERCISES FOR TOPIC C

5. Find each square root. Assume $a > 0$.

(a) $\sqrt{x^8}$ **(b)** $\sqrt{3^6}$ **(c)** $\sqrt{a^{22}}$

Now try the exercises for Topic C on page 372.

Use this area for work space.

EXERCISES

A. In Exercises 1–8 find each square. (*See Practice Exercise 1.*)

1. 14^2 **2.** $(-15)^2$ **3.** 17^2

4. 21^2 **5.** 25^2 **6.** 30^2

7. 100^2 **8.** 2000^2

In Exercises 9–20 find each square root. (*See Examples 1 and 2.*)

9. $\sqrt{16}$ **10.** $\sqrt{36}$ **11.** $\sqrt{49}$

12. $\sqrt{64}$ **13.** $-\sqrt{25}$ **14.** $-\sqrt{100}$

15. $\sqrt{81}$ **16.** $\sqrt{121}$ **17.** $-\sqrt{1\,000\,000}$

18. $-\sqrt{100\,000\,000}$ **19.** $\sqrt{0}$ **20.** $\sqrt{1}$

21. Find the length of the side of a square whose area is 64 square inches. (*See Practice Exercise 3.*)

22. Find the length of the side of a square whose area is 121 square inches.

23. The area of a square tablecloth is 16 square feet. Find the length of each side.

24. There are 12 inches in 1 foot. How many square inches are there in 1 square foot?

25. There are 3 feet in 1 yard. How many square feet are there in 1 square yard?

26. How many square inches are there in 1 square yard? (*See Exercises 24 and 25.*)

Use this area for work space.

B. In Exercises 27–33 find each square root. Assume a, b, c, x, y, and z are each positive or 0. (*See Example 3.*)

27. $\sqrt{x^2}$ **28.** $-\sqrt{z^2}$ **29.** $\sqrt{17^2}$ **30.** $-\sqrt{379^2}$

31. $\sqrt{(ab)^2}$ **32.** $\sqrt{(xyz)^2}$ **33.** $\sqrt{(ac^3)^2}$

34. Pick out the *error(s)*. Suppose a is *negative*. Then

(a) $\sqrt{a^2} = a$ (b) $\sqrt{a^2} = -a$ (c) $\sqrt{(-a)^2} = -a$

(d) $-\sqrt{a^2} = -a$ (e) $\sqrt{(-a)^2} = |a|$

C. In Exercises 35–50 find each square root. Assume a, b, c, x, y, and z are each positive or 0. (*See Example 4.*)

35. $\sqrt{c^4}$ **36.** $\sqrt{x^8}$ **37.** $\sqrt{y^{10}}$ **38.** $-\sqrt{a^{12}}$

39. $\sqrt{(xy)^4}$ **40.** $\sqrt{(bc)^6}$ **41.** $\sqrt{(ab^3)^4}$ **42.** $\sqrt{(7x^2y^3)^{10}}$

43. $\sqrt{x^{18}}$ **44.** $\sqrt{z^{64}}$ **45.** $\sqrt{a^{40}}$ **46.** $\sqrt{b^{200}}$

47. $\sqrt{2^4}$ **48.** $-\sqrt{2^6}$ **49.** $\sqrt{2^8}$ **50.** $\sqrt{2^{10}}$

8.2 Irrational Square Roots

A. Irrational Numbers and the Number Line

A rational number is a number that can be written in the form $\frac{N}{D}$, where N and D are integers and $D \neq 0$. Thus $\frac{2}{1}$ and $\frac{-4}{5}$ are rational numbers. You may be surprised to learn that there are points on the number line that do not represent rational numbers. Such points correspond to **irrational numbers.**

To obtain examples of irrational numbers, observe that

$$0 = 0^2, \quad 1 = 1^2, \quad 4 = 2^2, \quad 9 = 3^2, \ldots$$

Thus

$$\sqrt{0} = 0, \quad \sqrt{1} = 1, \quad \sqrt{4} = 2, \quad \sqrt{9} = 3, \ldots$$

But the square roots of integers such as

$$2, 3, 5, 6, 7, \ldots$$

are irrational. For example, it can be shown that $\sqrt{2}$ is irrational. Therefore $\sqrt{2}$ cannot be written as a (regular) decimal, which is a rational number. But $\sqrt{2}$ can be *approximated* by a decimal to as many places as desired. Observe how this can be done.

$$\begin{array}{r} 1.4 \\ \times 1.4 \\ \hline 5\,6 \\ 1\,4 \\ \hline 1.9\,6 \end{array} \qquad \begin{array}{r} 1.5 \\ \times 1.5 \\ \hline 7\,5 \\ 1\,5 \\ \hline 2.2\,5 \end{array}$$

$$(1.4)^2 = 1.96 \text{ and } (1.5)^2 = 2.25$$

Thus $$(1.4)^2 < 2 < (1.5)^2$$ *Read this:* $(1.4)^2 < 2$ and $2 < (1.5)^2$

Therefore it seems reasonable that

$$1.4 < \sqrt{2} < 1.5$$

Next observe:

$$\begin{array}{rr} 1.41 \\ \times 1.41 \\ \hline 1\,41 \\ 56\,4 \\ 1\,41 \\ \hline 1.98\,81 \end{array} \qquad \begin{array}{rr} 1.42 \\ \times 1.42 \\ \hline 2\,84 \\ 56\,8 \\ 1\,42 \\ \hline 2.01\,64 \end{array}$$

$$(1.41)^2 = 1.9881 \text{ and } (1.42)^2 = 2.0164$$

Thus $$(1.41)^2 < 2 < (1.42)^2$$

Therefore $$1.41 < \sqrt{2} < 1.42$$

To three decimal digits, it can be shown that

$$\sqrt{2} \approx 1.414$$ *Read:* $\sqrt{2}$ is approximately equal to 1.414.

(*See Figure 8.1.*) Similarly, it can be shown that

$$\sqrt{3} \approx 1.732$$

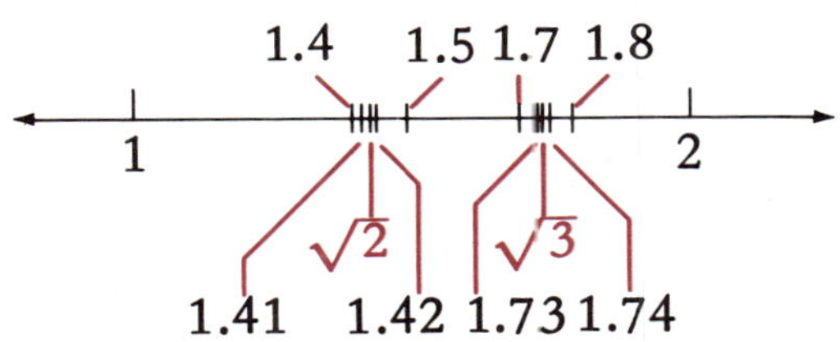

FIGURE 8.1. The irrational number $\sqrt{2}$ lies between the rational numbers 1.41 and 1.42. The irrational number $\sqrt{3}$ lies between the rational numbers 1.73 and 1.74.

The rational numbers and irrational numbers together comprise the **real numbers.** Every point on the number line corresponds to either a rational or an irrational real number.

We now know that every positive integer has a square root (rational or irrational) within the real number system. It seems natural to ask whether *negative* integers also have real numbers as square roots. Consider $\sqrt{-4}$. Note that -2 is *not* the square root of -4 because

$$(-2)(-2) = 4, \text{ rather than } -4$$

In fact, there is *no* real number—rational or irrational; positive, negative, or zero—whose square is -4. Thus

$$\sqrt{-4} \text{ is not defined (as a real number)}$$

and therefore there is no point on the number line that corresponds to $\sqrt{-4}$. Similarly, for any *negative number*, say $-a$,

$$\sqrt{-a} \text{ is not defined (as a real number)*}$$

Recall that every rational number can be expressed as either a (regular) decimal or as an infinite repeating decimal (in which one or more digits repeat). Thus

$$\frac{1}{4} = .25, \qquad \frac{2}{3} = .666\,666\ldots, \qquad \frac{4}{11} = .36\,36\,36\ldots$$

Every irrational number can be expressed as an infinite nonrepeating decimal. The digits do not repeat in any regular pattern. For example, the decimal representation of the irrational number $\sqrt{2}$ *to six decimal digits* is 1.414 214. Thus $\sqrt{2}$ can be *approximated* by the rational number 1.414 214.

A **fraction** is a number of the form $\frac{a}{b}$, where a and b are *any real numbers*, rational or irrational, provided $b \neq 0$. There are fractions, such as $\frac{3}{4}$, that are rational numbers because the numerator and denominator are *both* integers. There are also *irrational fractions*, such as $\frac{\sqrt{2}}{2}$, $\frac{1}{\sqrt{2}}$, and $\frac{\sqrt{2}}{\sqrt{5}}$, that cannot be expressed with *both* an integral numerator and an integral denominator.

Example 1 ▶ Indicate whether each of the following numbers is rational, irrational, or undefined (as a real number).

(a) $\sqrt{6}$ (b) $\frac{1}{\sqrt{3}}$ (c) $\frac{\sqrt{2}}{\sqrt{2}}$ (d) $\sqrt{5} - \sqrt{5}$ (e) $\sqrt{-5}$

Solution:

(a) irrational (b) irrational

(c) Divide numerator and denominator by $\sqrt{2}$. Thus

$$\frac{\overset{1}{\cancel{\sqrt{2}}}}{\underset{1}{\cancel{\sqrt{2}}}} = 1 \text{ (rational)}$$

(d) $\sqrt{5} - \sqrt{5} = 0$ (rational)

(e) The square root of a negative number is undefined (as a real number). ◀

PRACTICE EXERCISES FOR TOPIC A

Answers
1. (a)
(b)
(c)
(d)

1. Indicate whether each of the following is rational, irrational, or undefined (as a real number).

(a) $-\sqrt{36}$ (b) $\sqrt{-36}$ (c) $4\sqrt{6}$ (d) $\frac{7\sqrt{3}}{8\sqrt{3}}$

Now try the exercises for Topic A on page 378.

---

*In later courses you will develop the so-called "complex number system" in which $\sqrt{-a}$ is defined as a "new type of number."

B. Approximating Irrational Numbers

Example 2 ▶ Between which two consecutive integers is $\sqrt{92}$ located?

Solution:

81 and 100 are squares and

$$81 < 92 < 100$$

Thus $$\sqrt{81} < \sqrt{92} < \sqrt{100}$$

or $$9 < \sqrt{92} < 10$$

Therefore $\sqrt{92}$ lies between 9 and 10. (*See Figure 8.2.*) ◀

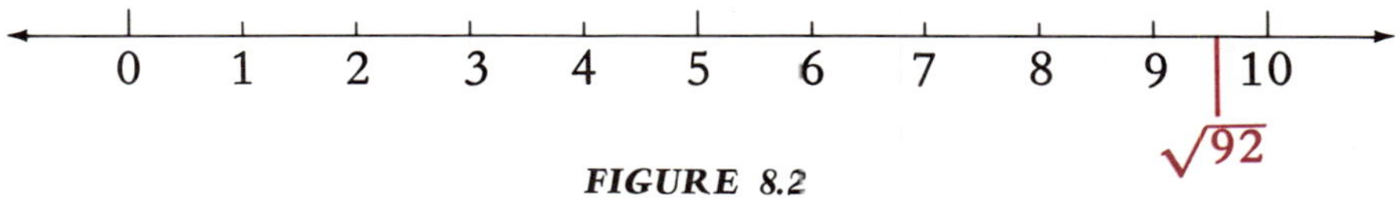

FIGURE 8.2

Example 3 ▶ Use $\sqrt{2} \approx 1.41$ to approximate

(a) $5\sqrt{2}$, (b) $5 + \sqrt{2}$, (c) $\frac{\sqrt{2}}{2}$

to one decimal digit.

Solution:

(a) $$5\sqrt{2} \approx 5(1.41) = 7.05$$

When the last digit is 5 or more, round off by increasing the next to last digit by 1. Thus to one decimal digit,

$$5\sqrt{2} \approx 7.1$$ ◀

(b) $$5 + \sqrt{2} \approx 5 + 1.41$$

Thus $$5 + \sqrt{2} \approx 6.41$$

Because the last digit is less than 5, to one decimal digit,

$$5 + \sqrt{2} \approx 6.4$$

(c) $$\frac{\sqrt{2}}{2} \approx \frac{1.41}{2} = .705$$

To one decimal digit,

$$\frac{\sqrt{2}}{2} \approx .7$$ ◀

PRACTICE EXERCISES FOR TOPIC B

2. Between which two consecutive integers is each of the following square roots located?

(a) $\sqrt{12}$ (b) $\sqrt{50}$ (c) $\sqrt{120}$ (d) $\sqrt{150}$

3. Approximate each of the following to one decimal digit.

(a) $\frac{3 + \sqrt{2}}{2}$ (Use $\sqrt{2} \approx 1.41$.)

(b) $4\sqrt{3}$ (Use $\sqrt{3} \approx 1.73$.)

Now try the exercises for Topic B on page 379.

Answers

2. (a) ____

(b) ____

(c) ____

(d) ____

3. (a) ____

(b) ____

C. Square Roots in Geometry

What is the geometric significance of $\sqrt{2}$? Consider the squares in Figure 8.3.

It is easy enough to draw the first and third squares, whose side lengths are rational. Whether or not you can draw the middle square, doesn't it seem reasonable that there should be a square whose area is 2 square inches? The length of a side is then $\sqrt{2}$ inches.

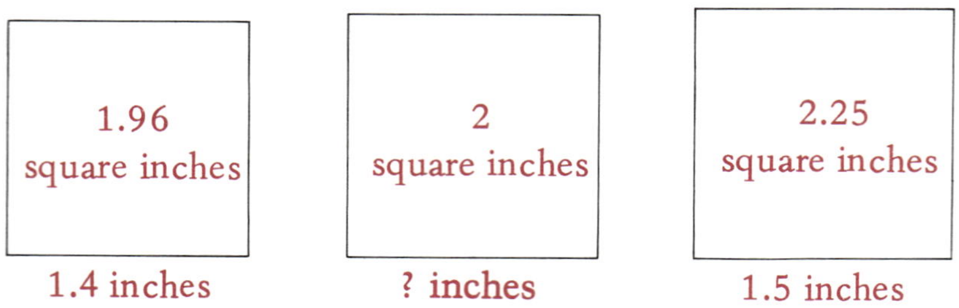

FIGURE 8.3

Another intuitive reason for introducing $\sqrt{2}$ relates to triangles, and will be explained in Example 6. First, let us consider some preliminary material on triangles.

A **right triangle** is a triangle in which one of the angles is 90°. This angle is called a **right angle.** The side opposite the right angle is called the **hypotenuse.** The ancient Greeks discovered that the sides of lengths a, b, and c of a right triangle are related by the formula

$$c^2 = a^2 + b^2$$

Here c is the length of the hypotenuse. This relationship is known as the **Pythagorean Theorem.** (*See Figure 8.4.*) Obtain the square root of each side of the above equation. Because $c > 0$,

$$c = \sqrt{a^2 + b^2}$$

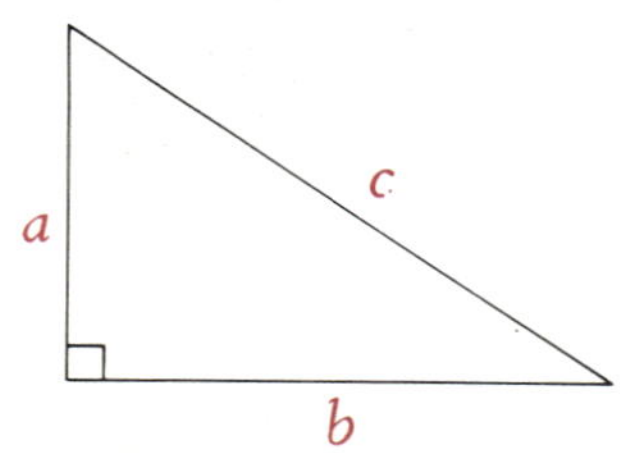

FIGURE 8.4. $\mathbf{c^2 = a^2 + b^2}$

Example 4 ▶ In the right triangle of Figure 8.5, find c if $a = 3$ inches and $b = 4$ inches.

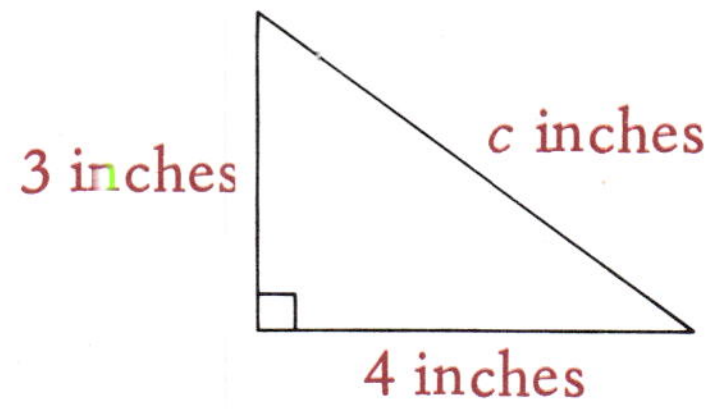

FIGURE 8.5

Solution:

$$c = \sqrt{a^2 + b^2}$$

Replace a by 3 and b by 4.

$$\begin{aligned} c &= \sqrt{3^2 + 4^2} \\ &= \sqrt{9 + 16} \\ &= \sqrt{25} \\ &= 5 \end{aligned}$$

The length of the hypotenuse is 5 inches. ◀

Example 5 ▶ In the right triangle of Figure 8.6, find b if $a = 5$ inches and $c = 13$ inches.

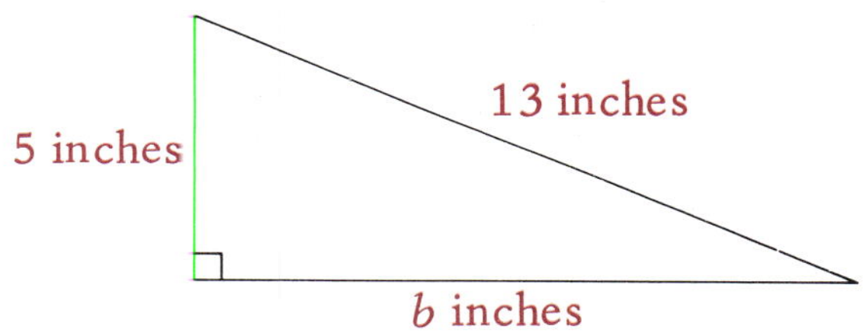

FIGURE 8.6

Solution:

$$\begin{aligned} a^2 + b^2 &= c^2 \\ b^2 &= c^2 - a^2 \\ b &= \sqrt{c^2 - a^2} \end{aligned}$$

Replace c by 13 and a by 5.

$$\begin{aligned} b &= \sqrt{13^2 - 5^2} \\ &= \sqrt{169 - 25} \\ &= \sqrt{144} \\ &= 12 \end{aligned}$$

◀

Next, an **isosceles triangle** is one in which two sides are of equal length.

Example 6 ▶ Consider the *isosceles right triangle* of Figure 8.7, in which the two equal sides are each of length 1 inch. What is the length of the hypotenuse? Express your result in radical form.

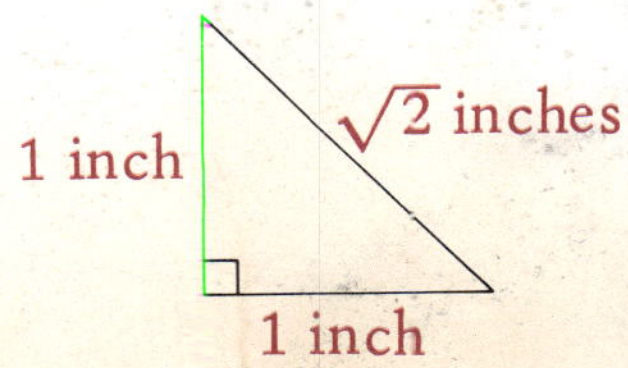

FIGURE 8.7

Solution:
By the Pythagorean Theorem,

$$c^2 = a^2 + b^2$$

Thus $$c^2 = 1^2 + 1^2 = 2$$

Therefore $$c = \sqrt{2}$$

The *exact* length of the hypotenuse is $\sqrt{2}$ inches. (The *approximate* length is 1.414 inches.) ◀

Example 7 ▶ In the right triangle of Figure 8.8, find c if $a = 3$ inches and $b = 5$ inches.

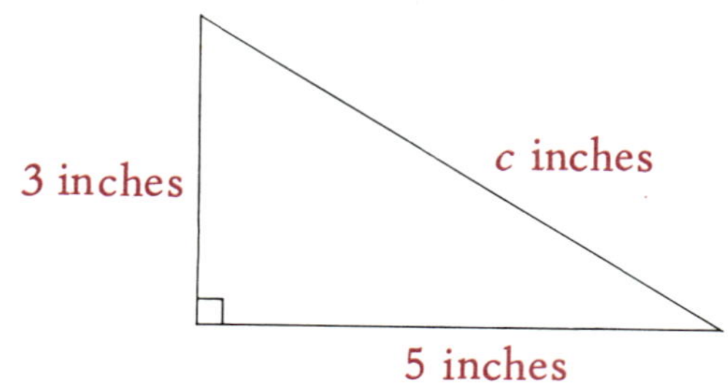

FIGURE 8.8

Solution:

$$c = \sqrt{a^2 + b^2}$$

Replace a by 3 and b by 5.

$$c = \sqrt{3^2 + 5^2} = \sqrt{9 + 25} = \sqrt{34}$$

The length of the hypotenuse is slightly less than 6 inches (because $\sqrt{36} = 6$). ◀

Answers

4. ______

5. ______

PRACTICE EXERCISES FOR TOPIC C

4. Approximate the length of the side of a square whose area is 80 square meters. (Use the next largest integer.)

5. Find the length of side a in the right triangle shown in Figure 8.9. (Leave the answer in radical form.)

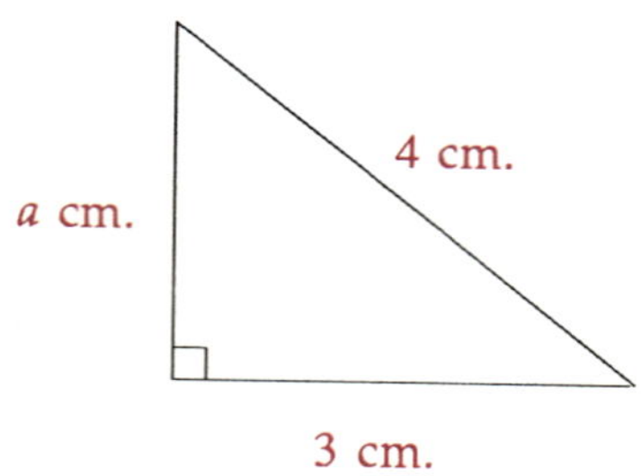

FIGURE 8.9

Now try the exercises for Topic C on page 379.

Use this area for work space.

EXERCISES

A. In Exercises 1–8 indicate whether each of the following is rational, irrational, or undefined (as a real number).

Use this area for work space.

1. $\sqrt{15}$ **2.** $\sqrt{16}$ **3.** $\dfrac{1}{\sqrt{7}}$ **4.** $\sqrt{25} - \sqrt{16}$

5. $\dfrac{\sqrt{5} - 5}{\sqrt{5}}$ **6.** $\dfrac{10}{\sqrt{5} - 5}$ **7.** $\dfrac{5\sqrt{5}}{6\sqrt{5}}$ **8.** $\dfrac{3\sqrt{5}}{6\sqrt{3}}$

B. In Exercises 9–20, between which two consecutive integers is each of the following square roots located?

9. $\sqrt{8}$ **10.** $\sqrt{14}$ **11.** $\sqrt{23}$

12. $\sqrt{32}$ **13.** $\sqrt{53}$ **14.** $\sqrt{65}$

15. $\sqrt{85}$ **16.** $-\sqrt{99}$ **17.** $\sqrt{125}$

18. $\sqrt{145}$ **19.** $\sqrt{401}$ **20.** $\sqrt{9999}$

In Exercises 21–26 use $\sqrt{2} \approx 1.41$ to approximate each of the following to one decimal digit. (*See Example 3.*)

21. $3\sqrt{2}$ **22.** $10 + \sqrt{2}$ **23.** $\sqrt{2} - 1$

24. $7 - \sqrt{2}$ **25.** $\dfrac{\sqrt{2}}{5}$ **26.** $1.5 + \sqrt{2}$

In Exercises 27–32 use $\sqrt{3} \approx 1.73$ to approximate each of the following to one decimal digit.

27. $2\sqrt{3}$ **28.** $100\sqrt{3}$ **29.** $\dfrac{\sqrt{3}}{2}$

30. $3.9 - \sqrt{3}$ **31.** $\dfrac{1 + \sqrt{3}}{2}$ **32.** $\dfrac{\sqrt{3}}{3} - 1$

In Exercises 33–38 use $\sqrt{2} \approx 1.41$ and $\sqrt{3} \approx 1.73$ to approximate each of the following to one decimal digit.

33. $\sqrt{2} + \sqrt{3}$ **34.** $\sqrt{3} - \sqrt{2}$ **35.** $\sqrt{2} + 3\sqrt{3}$

36. $\dfrac{\sqrt{2} - \sqrt{3}}{2}$ **37.** $\dfrac{\sqrt{3}}{5} + 10\sqrt{2}$ **38.** $\sqrt{2}\sqrt{3}$

C. **39.** Approximate the length of a side of a square whose area is 41 square inches. (Use the next *smallest* integer. *See Practice Exercise 4.*)

40. Approximate the length of a side of a square whose area is 118 square feet. (Use the next *largest* integer.)

In Exercises 41–48 (Figures 8.10–8.17) find the length of the indicated side of the right triangle. Leave your answer in radical form, when necessary. (*See Examples 4–7.*)

41.

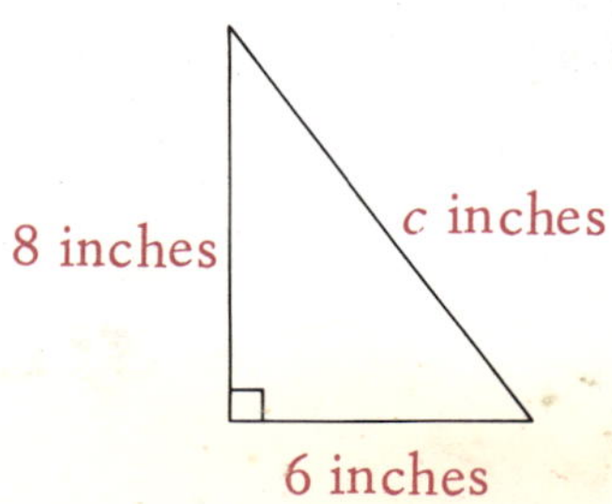

FIGURE 8.10

42.

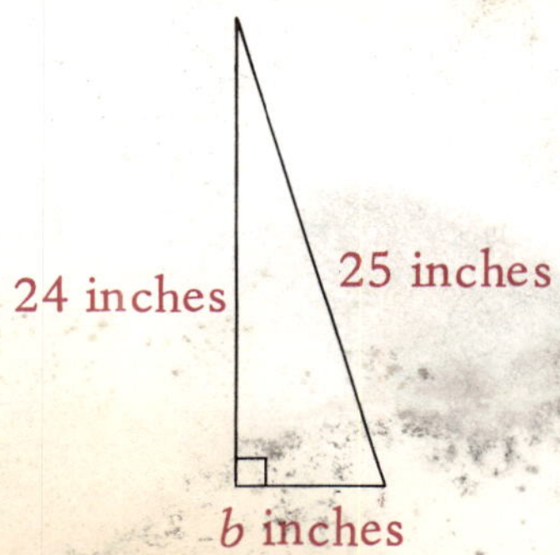

FIGURE 8.11

Use this area for work space.

43.

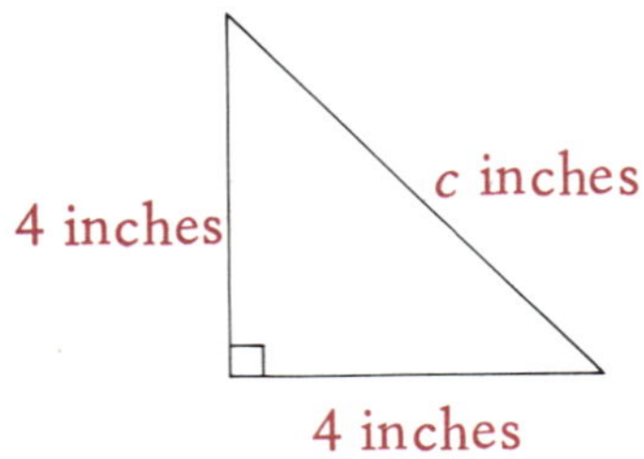

FIGURE 8.12

44.

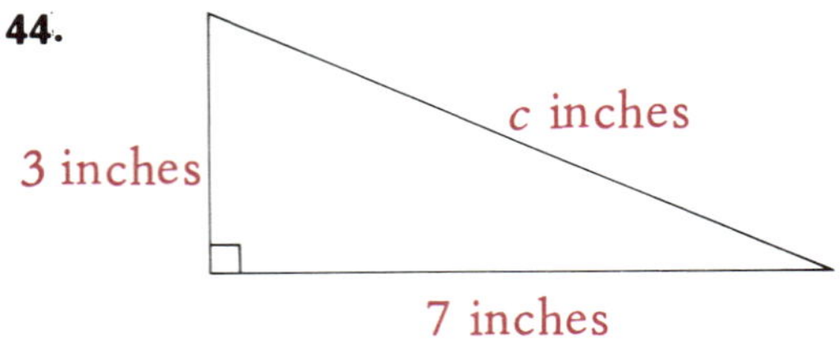

FIGURE 8.13

45.

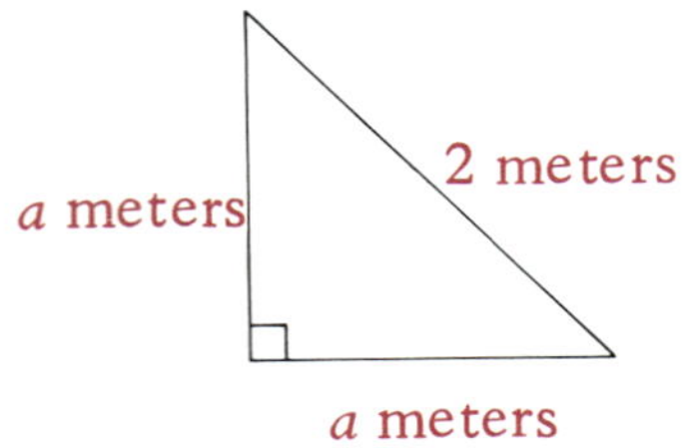

FIGURE 8.14

46.

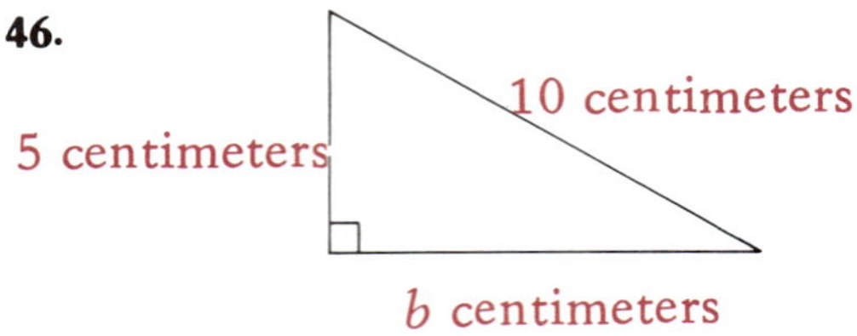

FIGURE 8.15

47.

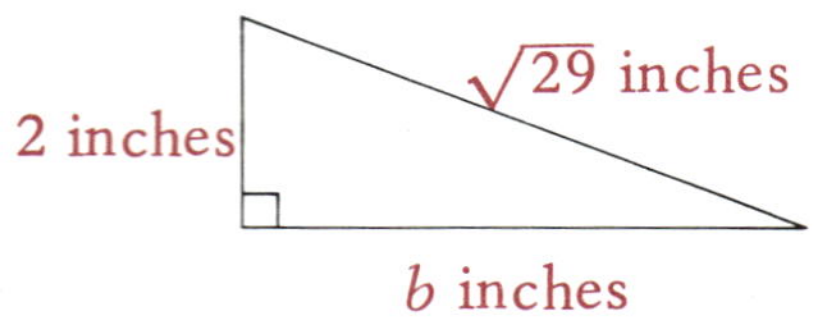

FIGURE 8.16

48.

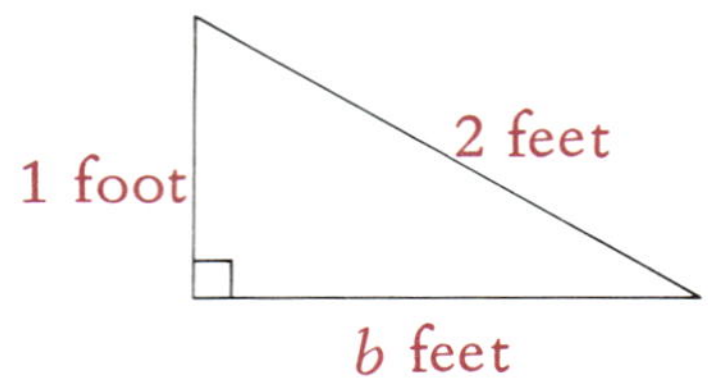

FIGURE 8.17

8.3 Roots of Products and Quotients

A. Roots of Products

Observe that $\sqrt{9 \cdot 4} = \sqrt{36} = 6$

Furthermore $\sqrt{9} \cdot \sqrt{4} = 3 \cdot 2 = 6$

It follows that $\sqrt{9 \cdot 4} = \sqrt{9} \cdot \sqrt{4}$

This suggests that (with suitable restrictions) *the square root of a product equals the product of the square roots.* More precisely,

$$\sqrt{ab} = \sqrt{a}\sqrt{b}, \qquad \text{where } a > 0 \text{ and } b > 0$$

This rule also applies to products of three or more factors. For example,

$$\sqrt{abc} = \sqrt{a}\sqrt{b}\sqrt{c}, \qquad \text{where } a > 0,\ b > 0, \text{ and } c > 0$$

Example 1 ► (a) $\sqrt{2500} = \sqrt{25 \cdot 100}$

$= \sqrt{25} \cdot \sqrt{100}$

$= 5 \cdot 10$

$= 50$

(b) $\sqrt{250\,000} = \sqrt{25\,00\,00}$

$= \sqrt{25 \cdot 100 \cdot 100}$

$= \sqrt{25} \cdot \sqrt{100} \cdot \sqrt{100}$

$= 5 \cdot 10 \cdot 10$

$= 500$ ◄

Example 2 ▶ Let $x > 0$ and $y > 0$. Then

(a) $\sqrt{9x^2} = \sqrt{9}\sqrt{x^2} = 3x$

(b) $\sqrt{25x^6y^{10}} = \sqrt{25}\sqrt{x^6}\sqrt{y^{10}} = 5x^3y^5$ ◀

Answers

1. (a) ______

(b) ______

(c) ______

PRACTICE EXERCISES FOR TOPIC A

1. Find each square root. Assume $x > 0$ and $y > 0$.

(a) $\sqrt{14\,400}$ (b) $\sqrt{16x^8}$ (c) $\sqrt{49x^2y^6}$

Now try the exercises for Topic A on page 385.

B. Simplifying by the Product Rule

The product rule enables you to simplify some irrational square roots, and often, to approximate them quickly, as in Example 3. You can also simplify square roots of odd powers of a variable, as in Example 4.

Example 3 ▶ (a) Simplify $\sqrt{8}$ (b) Estimate $\sqrt{8}$ to one decimal digit.

Solution:

(a)
$$8 = 4 \cdot 2$$
$$\sqrt{8} = \sqrt{4}\sqrt{2} = 2\sqrt{2}$$

The equivalent form $2\sqrt{2}$ is simpler than $\sqrt{8}$ in that the radical sign applies to a *smaller* positive integer. *Leave the answer in radical form.* Note that $2\sqrt{2}$ is the *exact* answer. In part (b) you will *approximate* $\sqrt{8}$.

(b)
$$\sqrt{2} \approx 1.41$$

Thus
$$\sqrt{8} = 2\sqrt{2} \approx 2(1.41) = 2.82 \approx 2.8$$ ◀

Example 4 ▶ Let $x > 0$. Simplify $\sqrt{x^9}$

Solution:

The "largest square factor" that can be removed from the radical symbol is x^8. Use $x^9 = x^8 \cdot x$. Then

$$\sqrt{x^9} = \sqrt{x^8 \cdot x} = \sqrt{x^8}\sqrt{x} = x^4\sqrt{x}$$ ◀

Answers

2. (a) ______

(b) ______

3. (a) ______

(b) ______

PRACTICE EXERCISES FOR TOPIC B

2. (a) Simplify $\sqrt{27}$ (b) Use $\sqrt{3} \approx 1.73$ to approximate $\sqrt{27}$ to one decimal digit.

3. Simplify. (Assume $x > 0$.) (a) $\sqrt{x^5}$ (b) $\sqrt{45x^3}$

Now try the exercises for Topic B on page 385.

C. $\sqrt{a^2}$

Observe that

$$\sqrt{5^2} = \sqrt{25} = 5$$

and $$\sqrt{(-5)^2} = \sqrt{(-5)\cdot(-5)} = \sqrt{25} = 5$$

In general, if p is *positive*, then

$$\sqrt{p^2} = p \qquad \text{and} \qquad \sqrt{(-p)^2} = p$$

Example 5 ▶ Find

(a) $\sqrt{4^2}$ (b) $\sqrt{(-4)^2}$ (c) $-\sqrt{4^2}$ (d) $\sqrt{-4}$

Solution:

(a) $\sqrt{4^2} = 4$ because 4 is positive.
(b) $\sqrt{(-4)^2} = \sqrt{16} = 4$
(c) $-\sqrt{4^2} = -4$, by part (a)
(d) $\sqrt{\underbrace{-4}_{\text{negative}}}$ is not defined (as a real number). ◀

Answers

4. (a) ______
(b) ______
(c) ______
(d) ______

PRACTICE EXERCISES FOR TOPIC C

4. Find:

(a) $\sqrt{18^2}$ (b) $\sqrt{(-18)^2}$ (c) $-\sqrt{18^2}$ (d) $\sqrt{-18^2}$

Now try the exercises for Topic C on page 386.

D. Roots of Quotients

Observe that $$\sqrt{\frac{100}{25}} = \sqrt{4} = 2$$

Also $$\frac{\sqrt{100}}{\sqrt{25}} = \frac{10}{5} = 2$$

It follows that $$\sqrt{\frac{100}{25}} = \frac{\sqrt{100}}{\sqrt{25}}$$

The square root of a quotient equals the quotient of the square roots. More precisely,

$$\sqrt{\frac{a}{b}} = \frac{\sqrt{a}}{\sqrt{b}}, \qquad \text{where } a > 0 \text{ and } b > 0$$

Thus $$\sqrt{\frac{4}{9}} = \frac{\sqrt{4}}{\sqrt{9}} = \frac{2}{3}$$

Example 6 ► (a) Let $b \neq 0$. Then

$$\sqrt{\frac{a^4}{b^8}} = \frac{\sqrt{a^4}}{\sqrt{b^8}} = \frac{a^2}{b^4}$$

(b) Let $x > 0$ and $z \neq 0$. Then

$$\sqrt{\frac{25x^2y^4}{81z^{12}}} = \frac{\sqrt{25x^2y^4}}{\sqrt{81z^{12}}} = \frac{\sqrt{25}\sqrt{x^2}\sqrt{y^4}}{\sqrt{81}\sqrt{z^{12}}} = \frac{5xy^2}{9z^6}$$ ◄

A fraction, such as $\frac{1}{\sqrt{5}}$, is often easier to work with when the denominator is rational. **Rationalize the denominator** of this fraction as follows.

$$\frac{1}{\sqrt{5}} = \frac{1 \cdot \sqrt{5}}{\sqrt{5} \cdot \sqrt{5}} = \frac{\sqrt{5}}{5}$$

Similarly,

$$\frac{4}{\sqrt{2}} = \frac{4 \cdot \sqrt{2}}{\sqrt{2} \cdot \sqrt{2}} = \frac{4\sqrt{2}}{2} = 2\sqrt{2}$$

Example 7 ► Let $c > 0$. Simplify $\sqrt{\frac{c^5}{32}}$

Solution:

$$\sqrt{\frac{c^5}{32}} = \frac{\sqrt{c^5}}{\sqrt{32}} = \frac{\sqrt{c^4}\sqrt{c}}{\sqrt{16}\sqrt{2}} = \frac{c^2\sqrt{c} \cdot \sqrt{2}}{4\sqrt{2} \cdot \sqrt{2}} = \frac{c^2\sqrt{2c}}{8}$$ ◄

Answers

5. (a) ______

(b) ______

6. (a) ______

(b) ______

PRACTICE EXERCISES FOR TOPIC D

5. Find each square root.

(a) $\sqrt{\frac{16}{81}}$ (b) $\sqrt{\frac{x^4}{25y^{12}}}$

6. Simplify. Leave answers in radical form with rational denominator. Assume $x > 0$.

(a) $\sqrt{\frac{2}{3x^2}}$ (b) $\sqrt{\frac{x^3}{10}}$

Now try the exercises for Topic D on page 386.

E. Roots of Sums and Differences

Let $a > 0$ and $b > 0$. Although

$$\sqrt{ab} = \sqrt{a}\sqrt{b} \qquad \text{and} \qquad \sqrt{\frac{a}{b}} = \frac{\sqrt{a}}{\sqrt{b}}$$

make sure that you realize that, in general,

$$\sqrt{a+b} \neq \sqrt{a} + \sqrt{b} \qquad \text{and} \qquad \sqrt{a-b} \neq \sqrt{a} - \sqrt{b}$$

For example, if $a = 25$ and $b = 4$, then

	$\sqrt{25+4} = \sqrt{29}$
whereas	$\sqrt{25} + \sqrt{4} = 5 + 2 = 7$
Note that	$\sqrt{29} < \sqrt{36} = 6 < 7$
so that	$\sqrt{25+4} \neq \sqrt{25} + \sqrt{4}$
Also	$\sqrt{25-4} = \sqrt{21}$
whereas	$\sqrt{25} - \sqrt{4} = 5 - 2 = 3$
Note that	$\sqrt{21} > \sqrt{16} = 4 > 3$
so that	$\sqrt{25-4} \neq \sqrt{25} - \sqrt{4}$

Answers

7. (a) ______

(b) ______

(c) ______

(d) ______

PRACTICE EXERCISES FOR TOPIC E

7. Pick out the *error(s)*.

(a) $\sqrt{100 - 64} = \sqrt{100} - \sqrt{64}$ (b) $\sqrt{\frac{64}{100}} = \frac{\sqrt{64}}{\sqrt{100}}$

(c) $\sqrt{2}\sqrt{3} = \sqrt{6}$ (d) $\sqrt{2} + \sqrt{3} = \sqrt{5}$

Now try the exercises for Topic E on page 386.

F. Roots of Decimals

Observe that

$$\sqrt{\frac{1}{100}} = \frac{\sqrt{1}}{\sqrt{100}} = \frac{1}{10} \qquad \text{and that} \qquad \sqrt{\frac{1}{10\,000}} = \frac{\sqrt{1}}{\sqrt{10\,000}} = \frac{1}{100}$$

Thus in decimal notation,

$$\sqrt{.01} = .1 \qquad \text{and} \qquad \sqrt{.0001} = .01$$

Example 8 ▶ (a) $\sqrt{.04} = \sqrt{4(.01)} = \sqrt{4}\sqrt{.01} = 2(.1) = .2$

(b) $\sqrt{.0025} = \sqrt{25(.0001)} = \sqrt{25}\sqrt{.0001} = 5(.01) = .05$

(c) For any x,

$\sqrt{.36x^4} = \sqrt{36(.01)x^4} = \sqrt{36}\sqrt{.01}\sqrt{x^4} = 6(.1)x^2 = .6x^2$ ◀

PRACTICE EXERCISES FOR TOPIC F

Answers

8. (a) ____
(b) ____
(c) ____

8. Find each square root.

(a) $\sqrt{.0049}$ (b) $\sqrt{1.21a^4}$ (c) $\sqrt{.000\,081}$

Now try the exercises for Topic F on page 386.

Use this area for work space.

EXERCISES

Assume $a > 0$, $b > 0$, $c > 0$, $d \neq 0$, $x > 0$, $y > 0$, and $z > 0$.

A. In Exercises 1–16 find each square root. (*See Examples 1 and 2.*)

1. $\sqrt{4900}$ **2.** $\sqrt{8100}$ **3.** $\sqrt{250\,000}$

4. $\sqrt{36\,000\,000}$ **5.** $\sqrt{400\,000\,000}$ **6.** $\sqrt{4x^2}$

7. $\sqrt{16a^4}$ **8.** $\sqrt{25b^8}$ **9.** $\sqrt{100a^{18}}$

10. $\sqrt{a^2b^2}$ **11.** $\sqrt{x^2y^4}$ **12.** $\sqrt{x^4y^4}$

13. $\sqrt{a^8x^{10}}$ **14.** $\sqrt{a^4b^2c^6}$ **15.** $\sqrt{36a^4x^2}$ **16.** $\sqrt{121y^2z^{12}}$

B. In Exercises 17–22 (a) Simplify each expression. (Leave your answer in radical form.) (b) Estimate the square root to the nearest tenth using $\sqrt{2} \approx 1.41$ and $\sqrt{3} \approx 1.73$. (*See Example 3.*)

Sample: $\sqrt{12}$	*Solution:*
	(a) $\sqrt{12} = \sqrt{4 \cdot 3} = \sqrt{4}\sqrt{3} = 2\sqrt{3}$
	(b) $\sqrt{12} \approx 2(1.73) = 3.46 \approx 3.5$

17. $\sqrt{18}$ **18.** $\sqrt{50}$ **19.** $\sqrt{72}$

20. $\sqrt{48}$ **21.** $\sqrt{75}$ **22.** $\sqrt{300}$

Use this area for work space.

In Exercises 23–28 simplify each expression. Leave answers in radical form. (*See Example 4.*)

23. $\sqrt{a^3}$ **24.** $\sqrt{c^7}$ **25.** $\sqrt{4b^5}$

26. $\sqrt{27b^9}$ **27.** $\sqrt{a^3b^5}$ **28.** $\sqrt{98a^4b^3}$

C. In Exercises 29–34 find each square root or indicate that it is undefined (as a real number). (*See Example 5.*)

29. $\sqrt{9^2}$ **30.** $\sqrt{(-9)^2}$ **31.** $\sqrt{-9^2}$

32. $-\sqrt{9^2}$ **33.** $\sqrt{-9}$ **34.** $-\sqrt{9}$

D. In Exercises 35–46 find each square root. (*See Example 6.*)

35. $\sqrt{\frac{9}{25}}$ **36.** $\sqrt{\frac{36}{49}}$ **37.** $\sqrt{\frac{81}{4}}$

38. $\sqrt{\frac{25}{144}}$ **39.** $\sqrt{\frac{100}{121}}$ **40.** $\sqrt{\frac{a^2}{c^2}}$

41. $\sqrt{\frac{x^2}{y^8}}$ **42.** $\sqrt{\frac{x^2}{4}}$ **43.** $\sqrt{\frac{25}{d^8}}$

44. $\sqrt{\frac{a^2b^6}{9}}$ **45.** $\sqrt{\frac{144x^4}{y^2z^{10}}}$ **46.** $\sqrt{\frac{25x^4y^8}{4z^{10}}}$

In Exercises 47–52 simplify each expression. Leave your answers in radical form and with rational denominators. (*See Example 7.*)

47. $\sqrt{\frac{a^5}{b^6}}$ **48.** $\sqrt{\frac{8x^7}{y^3}}$ **49.** $\sqrt{\frac{9a^9}{2b^2}}$

50. $\sqrt{\frac{48c^5}{27d^4}}$ **51.** $\sqrt{\frac{50a^3}{98b^8}}$ **52.** $\sqrt{\frac{175c^5}{108a^2b^4}}$

E. In Exercises 53–56 simplify, if possible.

53. $\sqrt{81} - \sqrt{4}$ **54.** $\sqrt{81 - 4}$ **55.** $\sqrt{a^2 + b^2}$ **56.** $\sqrt{a^2} + \sqrt{b^2}$

57. Pick out the *error(s)*. (*See Practice Exercise 7.*)

(a) $\sqrt{x^2 + y^2} = x + y$ (b) $\sqrt{x^2} - \sqrt{y^2} = x - y$

(c) $\sqrt{a^4} + \sqrt{9} = a^2 + 3$ (d) $\sqrt{x^2y^2} = xy$

(e) $\sqrt{4 + 9} = \sqrt{13}$ (f) $\sqrt{18} + \sqrt{7} = 5$

F. In Exercises 58–64 find each square root. (*See Example 8.*)

58. $\sqrt{.36}$ **59.** $\sqrt{.09}$ **60.** $\sqrt{.0064}$ **61.** $\sqrt{.000\,001}$

62. $\sqrt{.0144x^4}$ **63.** $\sqrt{.81x^8y^4}$ **64.** $\sqrt{.0121x^{12}y^{16}}$

65. Pick out the *error(s)*.

(a) $\sqrt{.1} = .1$ (b) $\sqrt{.0004} = .02$ (c) $\sqrt{.009} = .03$

(d) $\sqrt{1.44} = 1.2$ (e) $\sqrt{4.09} = 2.3$

8.4 Adding and Subtracting Roots

A. Simplifying Expressions With Radicals

Irrational square roots are added or subtracted as are terms. Thus

$$2\sqrt{3} + 3\sqrt{3} = 5\sqrt{3}$$

just as $$2x + 3x = 5x$$

But $$2\sqrt{3} + 3\sqrt{5}$$

cannot be further simplified, just as

$$2x + 3y$$

cannot be.

Example 1 ▶ Simplify $\sqrt{7} + 5\sqrt{7} - 3\sqrt{7}$

Solution:

Note that

$$\sqrt{7} = 1\sqrt{7}$$

Thus $$\sqrt{7} + 5\sqrt{7} - 3\sqrt{7} = (1 + 5 - 3)\sqrt{7} = 3\sqrt{7}$$ ◀

Example 2 ▶ Simplify $\sqrt{8} + \sqrt{18}$

Solution:

$$\sqrt{8} = \sqrt{4}\sqrt{2} = 2\sqrt{2}$$

$$\sqrt{18} = \sqrt{9}\sqrt{2} = 3\sqrt{2}$$

Therefore $$\sqrt{8} + \sqrt{18} = 2\sqrt{2} + 3\sqrt{2} = 5\sqrt{2}$$ ◀

Example 3 ▶ Let $a > 0$. Simplify $\sqrt{4a^5} + \sqrt{25a^5}$

Solution:

$$\sqrt{4a^5} = \sqrt{4}\sqrt{a^4}\sqrt{a} = 2a^2\sqrt{a}$$

$$\sqrt{25a^5} = \sqrt{25}\sqrt{a^4}\sqrt{a} = 5a^2\sqrt{a}$$

Thus $$\sqrt{4a^5} + \sqrt{25a^5} = 2a^2\sqrt{a} + 5a^2\sqrt{a} = 7a^2\sqrt{a}$$ ◀

Example 4 ▶ Let $a > 0$. Simplify $\sqrt{50a^3} + \sqrt{98a^3}$

Solution:

$$\sqrt{50a^3} = \sqrt{25}\sqrt{a^2}\sqrt{2a} = 5a\sqrt{2a}$$

$$\sqrt{98a^3} = \sqrt{49}\sqrt{a^2}\sqrt{2a} = 7a\sqrt{2a}$$

Thus $$\sqrt{50a^3} + \sqrt{98a^3} = 5a\sqrt{2a} + 7a\sqrt{2a} = 12a\sqrt{2a}$$ ◀

PRACTICE EXERCISES FOR TOPIC A

Answers

1. (a) ______
   (b) ______
   (c) ______

1. Simplify each expression.

(a) $\sqrt{6} + 10\sqrt{6} - 6\sqrt{6}$ (b) $\sqrt{125} - \sqrt{20}$ (c) $3a\sqrt{12a} - \sqrt{75a^3}$ (Assume $a > 0$.)

Now try the exercises for Topic A on page 388.

B. Combining Fractions

Irrational fractions are added and subtracted in the same way as rational fractions are. In Example 5 both fractions have the same denominator. In Example 6 the fractions have different denominators. You must first find equivalent fractions with the same denominator.

Example 5 ► Add $\frac{\sqrt{2}}{3} + \frac{2\sqrt{3}}{3}$

Solution:

$$\frac{\sqrt{2}}{3} + \frac{2\sqrt{3}}{3} = \frac{\sqrt{2} + 2\sqrt{3}}{3}$$ ◄

Example 6 ► Subtract $\frac{2}{3} - \frac{\sqrt{2}}{5}$

Solution:
The *lcd* is 15. Thus

$$\frac{2}{3} - \frac{\sqrt{2}}{5} = \frac{2\cdot 5}{3\cdot 5} - \frac{\sqrt{2}\cdot 3}{5\cdot 3} = \frac{10 - 3\sqrt{2}}{15}$$ ◄

Answers

2. (a) ____

(b) ____

(c) ____

PRACTICE EXERCISES FOR TOPIC B

2. Add or subtract as indicated.

(a) $\frac{3}{5} + \frac{\sqrt{3}}{5}$ (b) $2\sqrt{5} - \frac{\sqrt{5}}{4}$ (c) $\frac{\sqrt{7}}{9} + \frac{\sqrt{6}}{6}$

Now try the exercises for Topic B on page 389.

Use this area for work space.

EXERCISES

A. In Exercises 1–36 add or subtract, as indicated. Assume $a > 0$, $b > 0$, $x > 0$, and $y > 0$. (*See Examples 1–4.*)

1. $\sqrt{2} + \sqrt{2}$

2. $2\sqrt{3} + 4\sqrt{3}$

3. $5\sqrt{2} - \sqrt{2}$

4. $\sqrt{7} - 2\sqrt{7}$

5. $3\sqrt{6} + 2\sqrt{6}$

6. $7\sqrt{10} - 4\sqrt{10}$

7. $\sqrt{11} + 4\sqrt{11} + 3\sqrt{11}$

8. $6\sqrt{13} - \sqrt{13} + 4\sqrt{13}$

9. $2\sqrt{5} + 6\sqrt{5} - \sqrt{5}$

10. $\sqrt{7} - 2\sqrt{7} - 5\sqrt{7}$

11. $12\sqrt{2} - (7\sqrt{2} + 2\sqrt{2})$

12. $3\sqrt{7} - (2\sqrt{7} - 5\sqrt{7})$

13. $\sqrt{2} + \sqrt{3} + \sqrt{2} + 2\sqrt{3}$

14. $3\sqrt{5} - (\sqrt{2} + \sqrt{5})$

15. $\sqrt{7} - \sqrt{5} + 4\sqrt{7} + 3\sqrt{5}$

16. $\sqrt{11} - (\sqrt{2} + \sqrt{11} - \sqrt{2})$

17. $\sqrt{2} + \sqrt{8}$

18. $\sqrt{6} + \sqrt{24}$

19. $\sqrt{18} - 3\sqrt{2}$

20. $\sqrt{50} + \sqrt{32}$

21. $\sqrt{7} + \sqrt{28} - 2\sqrt{7}$

22. $\sqrt{18} - \sqrt{2} + \sqrt{50}$

Use this area for work space.

23. $\sqrt{300} + \sqrt{12} - \sqrt{75}$

24. $\sqrt{44} + \sqrt{99} - 5\sqrt{11}$

25. $\sqrt{4a} + \sqrt{9a}$

26. $\sqrt{64x} - \sqrt{49x}$

27. $\sqrt{100y^3} + \sqrt{81y^3}$

28. $\sqrt{a^5} - \sqrt{16a^5}$

29. $\sqrt{144ab^2} + \sqrt{121ab^2}$

30. $\sqrt{36x} + \sqrt{64x} - \sqrt{81x}$

31. $\sqrt{12a} + \sqrt{3a}$

32. $\sqrt{20b} - \sqrt{5b}$

33. $\sqrt{98x^3} + \sqrt{50x^3}$

34. $\sqrt{288yz^4} - \sqrt{72yz^4}$

35. $\sqrt{7ab^3} + \sqrt{28ab^3} + \sqrt{63ab^3}$

36. $\sqrt{125x^5y^3} - \sqrt{20x^5y^3} + \sqrt{45x^5y^3}$

B. In Exercises 37–48 add or subtract, as indicated.

37. $\frac{2}{5} + \frac{\sqrt{5}}{5}$

38. $\frac{1}{8} - \frac{\sqrt{2}}{8}$

39. $\frac{\sqrt{2}}{3} - \frac{\sqrt{5}}{3} + \frac{\sqrt{7}}{3}$

40. $\frac{\sqrt{2}}{2} + \frac{\sqrt{5}}{4}$

41. $\frac{\sqrt{2}}{5} - \frac{\sqrt{3}}{7}$

42. $\frac{6\sqrt{3}}{5} - \frac{12\sqrt{3}}{15}$

43. $\frac{\sqrt{2}}{4} + \frac{\sqrt{3}}{8} - \frac{2\sqrt{2}}{12}$

44. $\frac{\sqrt{2}}{3} + \frac{\sqrt{3}}{6} + \frac{1}{12}$

45. $\frac{\sqrt{5}}{8} - \left(\frac{\sqrt{3}}{6} + \frac{\sqrt{3}}{4}\right)$

46. $4\sqrt{7} - \frac{3\sqrt{7}}{2}$

47. $\frac{\sqrt{12}}{4} - \left(\frac{\sqrt{48}}{9} - \frac{\sqrt{75}}{3}\right)$

48. $\sqrt{\frac{21}{4}} - \sqrt{\frac{17}{16}}$

8.5 Multiplying and Dividing Roots

A. $\sqrt{a}\sqrt{b}$

Recall that the square root of a product equals the product of the square roots.

$$\sqrt{ab} = \sqrt{a}\sqrt{b}, \qquad \text{where } a > 0 \text{ and } b > 0$$

In Section 8.3 you used this to simplify the square root of a product such as $\sqrt{9x^2}$. Thus if $x > 0$,

$$\sqrt{9x^2} = \sqrt{9}\sqrt{x^2} = 3x$$

Now you will use the above law, in reverse, to simplify the product of square roots. Thus

$$\sqrt{2}\sqrt{3} = \sqrt{2 \cdot 3} = \sqrt{6}$$

And

$$\sqrt{5}\sqrt{5} = \sqrt{5 \cdot 5} = \sqrt{25} = 5$$

Example 1 ▶ Let $a > 0$ and $x > 0$. Multiply and simplify

(a) $\sqrt{a^3}\sqrt{a}$ (b) $\sqrt{2x}\sqrt{8x}$

Solution:

(a) $\sqrt{a^3}\sqrt{a} = \sqrt{a^3 \cdot a} = \sqrt{a^4} = a^2$

(b) $\sqrt{2x}\sqrt{8x} = \sqrt{(2x)(8x)} = \sqrt{16x^2} = \sqrt{16}\sqrt{x^2} = 4x$ ◀

Example 2 ▶ Let $a > 0$ and $b > 0$. Multiply and simplify

$$\sqrt{3a}\sqrt{27b}\sqrt{ab}$$

Solution:

$$\sqrt{3a}\sqrt{27b}\sqrt{ab} = \sqrt{(3a)(27b)(ab)} = \sqrt{81a^2b^2} = \sqrt{81}\sqrt{a^2}\sqrt{b^2} = 9ab$$ ◀

Answers

1. (a) ____

(b) ____

(c) ____

PRACTICE EXERCISES FOR TOPIC A

1. Multiply and simplify. Assume $x > 0$.

(a) $\sqrt{6}\sqrt{24}$ (b) $\sqrt{2x^3}\sqrt{18x}$ (c) $\sqrt{3x}\sqrt{7x}$

Now try the exercises for Topic A on page 394.

B. Distributive Laws

The Distributive Laws apply to products involving roots. Recall that

$$a(b + c) = ab + ac$$

Thus

$$\sqrt{2}(\sqrt{2} + \sqrt{5}) = \sqrt{2}\sqrt{2} + \sqrt{2}\sqrt{5} = \sqrt{2 \cdot 2} + \sqrt{2 \cdot 5} = 2 + \sqrt{10}$$

Example 3 ▶ Multiply (a) $(1 + \sqrt{7})(2 + \sqrt{7})$ (b) $(\sqrt{3} + \sqrt{2})(\sqrt{3} - \sqrt{2})$

Solution:

(a)
$$\begin{array}{l} 1 + \sqrt{7} \\ \underline{2 + \sqrt{7}} \\ 2 + 2\sqrt{7} \\ \underline{\quad + \sqrt{7} + \sqrt{7}\sqrt{7}} \\ 2 + 3\sqrt{7} + 7 \end{array}$$

($2 + 7 = 9$)

Thus $(1 + \sqrt{7})(2 + \sqrt{7}) = 9 + 3\sqrt{7}$

(b)
$$\begin{array}{r} \sqrt{3} + \sqrt{2} \\ \sqrt{3} - \sqrt{2} \\ \hline 3 + \sqrt{3}\sqrt{2} \qquad \\ - \sqrt{3}\sqrt{2} - 2 \\ \hline 3 \qquad\qquad - 2 \end{array}$$

1

Thus $(\sqrt{3} + \sqrt{2})(\sqrt{3} - \sqrt{2}) = 1$ ◀

Example 4 ▶ Multiply

$$\begin{array}{l} \quad 5\sqrt{7} + \sqrt{2} \\ \quad 2\sqrt{7} - 3\sqrt{2} \\ \hline 10 \cdot 7 + 2\sqrt{7}\sqrt{2} \\ \qquad\quad - 15\sqrt{7}\sqrt{2} - 3 \cdot 2 \\ \hline 70 \quad - 13\sqrt{14} - 6 \end{array}$$

64

Thus $(5\sqrt{7} + \sqrt{2})(2\sqrt{7} - 3\sqrt{2}) = 64 - 13\sqrt{14}$ ◀

Answers

2. (a) ______

(b) ______

PRACTICE EXERCISES FOR TOPIC B

2. Multiply:

(a) $\sqrt{2}(\sqrt{32} - 2\sqrt{3})$ (b) $(5 - \sqrt{5})(1 + 2\sqrt{5})$

Now try the exercises for Topic B on page 394.

C. $\dfrac{\sqrt{a}}{\sqrt{b}}$

The square root of a quotient equals the quotient of the square roots.

$$\sqrt{\frac{a}{b}} = \frac{\sqrt{a}}{\sqrt{b}}, \qquad \text{where } a > 0 \text{ and } b > 0$$

In Section 8.3 you simplified *the square root of a quotient*. Thus

$$\sqrt{\frac{4}{9}} = \frac{\sqrt{4}}{\sqrt{9}} = \frac{2}{3}$$

Now you will use the above law, together with the product law

$$\sqrt{ab} = \sqrt{a}\sqrt{b}, \qquad \text{where } a > 0 \text{ and } b > 0$$

to simplify *the quotient of square roots*. For example,

$$\frac{\sqrt{48}}{\sqrt{3}} = \sqrt{\frac{48}{3}} = \sqrt{16} = 4$$

Example 5 ▶ Let $x > 0$. Simplify

(a) $\dfrac{\sqrt{28x}}{\sqrt{7x}}$ (b) $\dfrac{\sqrt{3}\sqrt{14}}{\sqrt{6}}$

Solution:

(a) $\dfrac{\sqrt{28x}}{\sqrt{7x}} = \sqrt{\dfrac{28x}{7x}} = \sqrt{4} = 2$

(b) $\dfrac{\sqrt{3}\sqrt{14}}{\sqrt{6}} = \dfrac{\overset{1}{\cancel{\sqrt{3}}}\overset{1}{\cancel{\sqrt{2}}}\sqrt{7}}{\underset{1}{\cancel{\sqrt{3}}}\underset{1}{\cancel{\sqrt{2}}}} = \sqrt{7}$ ◀

PRACTICE EXERCISES FOR TOPIC C

Answers
3. (a)
(b)
(c)

3. Simplify:

(a) $\dfrac{\sqrt{60}}{\sqrt{15}}$ (b) $\dfrac{\sqrt{2}}{\sqrt{50}}$ (c) $\dfrac{\sqrt{75x}}{\sqrt{12x^3}}$ (Assume $x > 0$.)

Now try the exercises for Topic C on page 394.

D. Factoring and Dividing

You can often "isolate the common factor" of an expression containing radicals, just as you isolated the common factor of a polynomial. When this expression is the numerator or denominator of a fraction, you may be able to divide by factors common to the numerator and denominator.

Example 6 ▶ (a) Factor $2 + \sqrt{12}$ (b) Use part (a) to simplify the fraction $\dfrac{2 + \sqrt{12}}{4}$.

Solution:

(a) $2 + \sqrt{12} = 2 + \sqrt{4 \cdot 3}$
$= 2 + \sqrt{4}\sqrt{3}$
$= 2 + 2\sqrt{3}$
$= 2(1 + \sqrt{3})$

(b) $\dfrac{2 + \sqrt{12}}{4} = \dfrac{\overset{1}{\cancel{2}}(1 + \sqrt{3})}{\underset{2}{\cancel{4}}}$

$= \dfrac{1 + \sqrt{3}}{2}$ ◀

Example 7 ▶ Let $x > 0$ and $y > 0$.

(a) Factor $\sqrt{xy} - \sqrt{x^3}$

(b) Use part (a) to simplify the expression $\dfrac{\sqrt{xy} - \sqrt{x^3}}{\sqrt{x^5}}$.

Solution:

(a) $\sqrt{xy} - \sqrt{x^3} = \sqrt{x}\sqrt{y} - \sqrt{x}\sqrt{x^2} = \sqrt{x}(\sqrt{y} - x)$

(b) $\dfrac{\sqrt{xy} - \sqrt{x^3}}{\sqrt{x^5}} = \dfrac{\overset{1}{\cancel{\sqrt{x}}}(\sqrt{y} - x)}{\underset{1}{\cancel{\sqrt{x}}}\sqrt{x^4}} = \dfrac{\sqrt{y} - x}{x^2}$ ◀

Example 8 ► Let $a > 0$, $x > 0$, $y > 0$, and $x \neq y$.

(a) Factor $\sqrt{a^2x^2} - \sqrt{a^2y^2}$

(b) Use part (a) to simplify the expression $\dfrac{x^2 - y^2}{\sqrt{a^2x^2} - \sqrt{a^2y^2}}$.

Solution:

(a) $$\begin{aligned}\sqrt{a^2x^2} - \sqrt{a^2y^2} &= \sqrt{a^2}\sqrt{x^2} - \sqrt{a^2}\sqrt{y^2}\\ &= ax - ay\\ &= a(x - y)\end{aligned}$$

(b) $$\frac{x^2 - y^2}{\sqrt{a^2x^2} - \sqrt{a^2y^2}} = \frac{(x + y)(x - y)}{a(x - y)} = \frac{x + y}{a}$$ ◄

Answers

4. (a) ______

(b) ______

5. (a) ______

(b) ______

PRACTICE EXERCISES FOR TOPIC D

4. (a) Factor $8 - \sqrt{48}$ (b) Simplify $\dfrac{8 - \sqrt{48}}{20}$

5. Assume $x > 0$ and $y > 0$.

(a) Factor $\sqrt{9x^2} - \sqrt{9y^2}$ (b) Simplify $\dfrac{\sqrt{9x^2} - \sqrt{9y^2}}{9x + 9y}$

Now try the exercises for Topic D on page 394.

E. Multiplying and Dividing Fractions

Recall that to multiply fractions, first divide by factors common to the numerators and denominators. Then multiply the simplified numerators and denominators separately.

Example 9 ► $$\frac{\sqrt{12}}{5} \cdot \frac{\sqrt{6}}{4} = \frac{\overset{1}{\cancel{2}}\sqrt{3}}{5} \cdot \frac{\sqrt{2}\sqrt{3}}{\underset{2}{\cancel{4}}} = \frac{3\sqrt{2}}{10}$$ ◄

Recall that $$\frac{a}{b} \div \frac{c}{d} = \frac{a}{b} \cdot \frac{d}{c}$$

Thus *to divide fractions, invert the second fraction and multiply.*

Example 10 ► $$\frac{\sqrt{10}}{3} \div \frac{\sqrt{5}}{9} = \frac{\sqrt{10}}{\underset{1}{\cancel{3}}} \cdot \frac{\overset{3}{\cancel{9}}}{\sqrt{5}} = \overset{1}{\cancel{\sqrt{5}}}\sqrt{2} \cdot \frac{3}{\underset{1}{\cancel{\sqrt{5}}}} = 3\sqrt{2}$$ ◄

Answers

6. ______

7. ______

PRACTICE EXERCISES FOR TOPIC E

6. Multiply $\dfrac{\sqrt{30}}{20} \cdot \dfrac{\sqrt{5}}{\sqrt{3}}$

7. Divide $\dfrac{\sqrt{8}}{10} \div \dfrac{\sqrt{2}}{\sqrt{5}}$

Now try the exercises for Topic E on page 395.

Use this area for work space.

EXERCISES

Assume $a > 0$, $b > 0$, $c > 0$, $x > 0$, $y > 0$, and $z > 0$.

A. In Exercises 1–16 multiply and simplify. (*See Examples 1 and 2.*)

1. $\sqrt{2}\sqrt{5}$ **2.** $\sqrt{3}\sqrt{5}$ **3.** $\sqrt{7}\sqrt{8}$ **4.** $\sqrt{2}\sqrt{2}$

5. $\sqrt{7}\sqrt{7}$ **6.** $\sqrt{12}\sqrt{3}$ **7.** $\sqrt{50}\sqrt{2}$ **8.** $\sqrt{2}\sqrt{6}\sqrt{5}$

9. $\sqrt{a}\sqrt{a}$ **10.** $\sqrt{ax}\sqrt{ax}$ **11.** $\sqrt{b}\sqrt{b^5}$ **12.** $\sqrt{a}\sqrt{ab}\sqrt{b}$

13. $\sqrt{5x}\sqrt{20x}$ **14.** $\sqrt{2b^3}\sqrt{12b}$

15. $\sqrt{xyz^3}\sqrt{x^2y}\sqrt{zy^2}$ **16.** $\sqrt{3a^2b}\sqrt{2bc}\sqrt{6c^3}$

B. In Exercises 17–28 multiply.

17. $\sqrt{2}(\sqrt{2} + \sqrt{3})$ **18.** $\sqrt{a}(\sqrt{a} + \sqrt{b})$

19. $\sqrt{3}(\sqrt{27} - \sqrt{12})$ **20.** $2\sqrt{3}(\sqrt{3} + 12)$

21. $(7 + \sqrt{2})(7 - \sqrt{2})$ **22.** $(2\sqrt{5} + 3\sqrt{3})(2\sqrt{5} - 3\sqrt{3})$

23. $(2 + 3\sqrt{5})(4 + 2\sqrt{5})$ **24.** $(4\sqrt{7} - 2\sqrt{5})(2\sqrt{7} + 4\sqrt{5})$

25. $(\sqrt{5} - 3\sqrt{11})(2\sqrt{5} + \sqrt{11})$ **26.** $(6\sqrt{3} + 9\sqrt{5})(2\sqrt{3} + \sqrt{5})$

27. $(1 + \sqrt{2})^2$ **28.** $(2\sqrt{3} + \sqrt{5})^2$

C. In Exercises 29–40 simplify. If necessary, rationalize the denominator. (*See Example 5. See also Example 7 on page 383.*)

29. $\dfrac{\sqrt{27}}{\sqrt{3}}$ **30.** $\dfrac{\sqrt{2}}{\sqrt{8}}$ **31.** $\dfrac{-\sqrt{12}}{\sqrt{3}}$ **32.** $\dfrac{\sqrt{24}}{\sqrt{6}}$

33. $\dfrac{6}{\sqrt{3}}$ **34.** $\dfrac{\sqrt{98}}{\sqrt{2}}$ **35.** $\dfrac{\sqrt{18x}}{\sqrt{81x}}$ **36.** $\dfrac{\sqrt{9x}}{\sqrt{25z}}$

37. $\dfrac{\sqrt{48xyz}}{\sqrt{27xz^3}}$ **38.** $\dfrac{\sqrt{30}}{\sqrt{2}\sqrt{15}}$ **39.** $\dfrac{\sqrt{27}}{\sqrt{6}\sqrt{50}}$ **40.** $\dfrac{\sqrt{45}\sqrt{14}}{\sqrt{35}\sqrt{18}}$

D. In Exercises 41–50 (a) factor and (b) use part (a) to simplify the fraction or rational expression. (*See Examples 6–8.*)

41. (a) Factor $3 + \sqrt{18}$
(b) Simplify $\dfrac{3 + \sqrt{18}}{9}$

42. (a) Factor $2 - \sqrt{32}$
(b) Simplify $\dfrac{2 - \sqrt{32}}{8}$

43. (a) Factor $5 + \sqrt{50}$
(b) Simplify $\dfrac{5 + \sqrt{50}}{-15}$

44. (a) Factor $7 + \sqrt{98}$
(b) Simplify $\dfrac{7 + \sqrt{98}}{1 + \sqrt{2}}$

45. (a) Factor $\sqrt{2} + \sqrt{6}$
(b) Simplify $\dfrac{\sqrt{2} + \sqrt{6}}{\sqrt{8}}$

46. (a) Factor $\sqrt{6} - \sqrt{18}$
(b) Simplify $\dfrac{\sqrt{6} - \sqrt{18}}{3 - \sqrt{3}}$

47. (a) Factor $\sqrt{a^3b} + \sqrt{abc}$
(b) Simplify $\dfrac{\sqrt{a^3b} + \sqrt{abc}}{\sqrt{a^3b^3}}$

48. (a) Factor $\sqrt{4a^2} - \sqrt{4b^2}$
(b) Simplify $\dfrac{\sqrt{4a^2} - \sqrt{4b^2}}{10a - 10b}$

Use this area for work space.

49. (a) Factor $\sqrt{9a^2c^2} - \sqrt{16b^2c^2}$

(b) Simplify $\dfrac{15ac - 20bc}{\sqrt{9a^2c^2} - \sqrt{16b^2c^2}}$

50. (a) Factor $\sqrt{a^2c^4} + \sqrt{25a^2c^2}$

(b) Simplify $\dfrac{c^2 + 10c + 25}{\sqrt{a^2c^4} + \sqrt{25a^2c^2}}$

E. In Exercises 51–58 multiply or divide. Assume $s > 0$ and $t > 0$. *(See Examples 9 and 10.)*

51. $\dfrac{\sqrt{5}}{2} \cdot \dfrac{\sqrt{5}}{3}$

52. $\dfrac{\sqrt{8}}{5} \cdot \dfrac{\sqrt{3}}{\sqrt{2}}$

53. $\dfrac{\sqrt{12}}{\sqrt{5}} \cdot \dfrac{\sqrt{6}}{2} \cdot \dfrac{\sqrt{10}}{3}$

54. $\dfrac{\sqrt{3}}{\sqrt{2}} \cdot \dfrac{\sqrt{32}}{3} \cdot \dfrac{\sqrt{27}}{4}$

55. $\dfrac{\sqrt{2}}{3} \div \dfrac{\sqrt{8}}{6}$

56. $\dfrac{\sqrt{10}}{7} \div \dfrac{\sqrt{40}}{14}$

57. $\dfrac{\sqrt{st}}{\sqrt{48}} \div \dfrac{\sqrt{3}}{\sqrt{25s^2t}}$

58. $\dfrac{\sqrt{12}}{\sqrt{15}} \div \dfrac{\sqrt{45}}{\sqrt{144}}$

8.6 Zero and Negative Exponents

A. How to Define Zero and Negative Exponents

Zero and negative exponents can be defined so as to simplify our notation. Consider the pattern in Table 8.1, and fill in the blanks.

TABLE 8.1

$10^4 =$	10 000
$10^3 =$	1000
$10^2 =$	100
$10^1 =$	10
$10^0 =$	
$10^{-1} =$	
$10^{-2} =$	
$10^{-3} =$	
$10^{-4} =$	

Every time the exponent is decreased by 1, the corresponding number in the right-hand column is divided by 10. This suggests completing the table as follows:

$$10^1 = 10$$

$$10^0 = 1$$

$$10^{-1} = \frac{1}{10} = \frac{1}{10^1}$$

$$10^{-2} = \frac{1}{100} = \frac{1}{10^2}$$

$$10^{-3} = \frac{1}{1000} = \frac{1}{10^3}$$

$$10^{-4} = \frac{1}{10\,000} = \frac{1}{10^4}$$

Thus the following definition seems natural.

DEFINITION

Let $a \neq 0$ and let n be a positive integer (so that $-n$ is a negative integer). Define

$$a^{-n} = \frac{1}{a^n} \qquad \text{and} \qquad a^0 = 1$$

Example 1 ▶ (a) $2^{-1} = \frac{1}{2}$ (b) $2^{-2} = \frac{1}{2^2} = \frac{1}{4}$ (c) $2^{-3} = \frac{1}{2^3} = \frac{1}{8}$ (d) $2^{-4} = \frac{1}{2^4} = \frac{1}{16}$ ◀

Example 2 ▶ (a) $8^0 = 1$ (b) $x^0 = 1$, if $x \neq 0$
(c) $(10a)^0 = 1$, if $a \neq 0$, because any nonzero number raised to the 0th power equals 1. ◀

Observe that 0^{-m} is *undefined* because $\frac{1}{0^m}$ would involve division by 0 (which is undefined). *By convention,* 0^0 *is undefined.*

PRACTICE EXERCISES FOR TOPIC A

Answers
1. (a)
(b)
(c)
(d)
2. (a)
(b)

1. Find each value.

(a) 7^{-2} (b) 5^{-1} (c) $(-4)^{-3}$ (d) 2.6^0

2. Determine the exponent m.

(a) $8^m = \frac{1}{64}$ (b) $1 = 6^m$

Now try the exercises for Topic A on page 398.

B. Fractional Bases

The base a can be a fraction.

Example 3 ▶ (a) $\left(\frac{2}{3}\right)^{-1} = \frac{1}{\frac{2}{3}} = 1 \div \frac{2}{3} = \frac{3}{2}$

(b) $\left(\frac{2}{3}\right)^{-2} = \frac{1}{\left(\frac{2}{3}\right)^2} = \frac{1}{\frac{4}{9}} = \frac{9}{4}$ ◀

Example 4 ▶ Let $x \neq 0$.

$$\left(\frac{5}{x}\right)^{-3} = \frac{1}{\left(\frac{5}{x}\right)^3} = \frac{1}{\frac{125}{x^3}} = \frac{x^3}{125}$$ ◀

Examples 3 and 4 illustrate the general rule.

$$\left(\frac{a}{b}\right)^{-n} = \frac{b^n}{a^n}, \qquad \text{for } a \neq 0,\ b \neq 0, \text{ and } n \text{ a positive integer}$$

Answers
3. (a)
(b)
(c)

PRACTICE EXERCISES FOR TOPIC B

3. Find each value.

(a) $\left(\frac{5}{4}\right)^{-1}$ (b) $\left(\frac{-3}{8}\right)^{-2}$ (c) $(.2)^{-3}$

Now try the exercises for Topic B on page 398.

C. Rules for Exponents

Recall that for *positive integral exponents* m and n,

$$a^m \cdot a^n = a^{m+n} \quad \text{and} \quad \frac{a^m}{a^n} = a^{m-n}, \quad \text{where } m > n \text{ and } a \neq 0$$

These rules carry over to *zero and negative integral exponents.* Also the second rule now applies to $m = n$ and to $m < n$.

Example 5 ► Assume $a \neq 0$.

(a) $$a^5 \cdot a^{-2} = a^5 \cdot \frac{1}{a^2} = a^3$$

Thus $a^5 \cdot a^{-2} = a^{5+(-2)}$ *Here $m = 5$ and $n = -2$*

(b) $$\frac{a^5}{a^{-2}} = \frac{a^5}{\frac{1}{a^2}} = a^5 \cdot \frac{a^2}{1} = a^7$$

Thus $\frac{a^5}{a^{-2}} = a^{5-(-2)}$ *Here $m = 5$ and $n = -2$*

(c) $$a^{-5} \cdot a^{-2} = \frac{1}{a^5} \cdot \frac{1}{a^2} = \frac{1}{a^7} = a^{-7}$$

Thus $a^{-5} \cdot a^{-2} = a^{(-5)+(-2)}$ *Here $m = -5$ and $n = -2$* ◄

Example 6 ► Assume $x \neq 0$. Simplify

(a) $x^9 \cdot x^{-6}$ (b) $\frac{x^4}{x^{-7}}$ (c) $\frac{x^{-3}}{x^{-8}}$

Solution:

(a) $x^9 \cdot x^{-6} = x^{9+(-6)} = x^3$

(b) $\frac{x^4}{x^{-7}} = x^{4-(-7)} = x^{4+7} = x^{11}$

(c) $\frac{x^{-3}}{x^{-8}} = x^{-3-(-8)} = x^{-3+8} = x^5$ ◄

Answers

4. (a) ____
(b) ____
(c) ____
(d) ____

PRACTICE EXERCISES FOR TOPIC C

4. Simplify and write each answer without writing a fraction. Use negative exponents, if necessary. Assume $y \neq 0$.

(a) $\frac{y}{y^6}$ (b) $y \cdot y^{-3} \cdot y^2$ (c) $\frac{y^{-2}}{y^{-5}}$ (d) $\frac{y^3 \cdot y^{-4}}{y^7}$

Now try the exercises for Topic C on page 399.

Use this area for work space.

EXERCISES

A. In Exercises 1–22 find each value. (*See Examples 1 and 2.*)

1. 4^{-1} **2.** 1^{-1} **3.** $(-2)^{-1}$ **4.** 5^{-2}

5. 8^{-2} **6.** $(-4)^{-2}$ **7.** -4^{-2} **8.** 3^{-3}

9. $(-3)^{-3}$ **10.** 4^{-3} **11.** 5^{-3} **12.** 3^{-4}

13. 10^{-4} **14.** 10^{-5} **15.** 10^0 **16.** -10^0

17. $(-10)^0$ **18.** $2^{-3} \cdot 4^{-2}$ **19.** $\frac{2^{-2}}{4}$ **20.** $8 \cdot 2^{-4}$

21. $2^{-1} + 2^{-2}$ **22.** $2^{-1} + 3^{-1}$

In Exercises 23–30 determine the exponent m.

Sample: $\frac{1}{9} = 3^m$	*Solution:* $\frac{1}{9} = \frac{1}{3^2} = 3^{-2}$ Thus $m = -2$

23. $1 = 5^m$ **24.** $\frac{1}{5} = 5^m$ **25.** $\frac{1}{4} = 2^m$ **26.** $\frac{1}{8} = 2^m$

27. $\frac{1}{1000} = 10^m$ **28.** $\frac{1}{81} = 3^m$ **29.** $\frac{7}{10^2} = 7 \cdot 10^m$ **30.** $\frac{3}{10^5} = 3 \cdot 10^m$

31. Pick out the error(s).

(a) $5^{-1} = \frac{-1}{5}$ (b) $3^{-2} = \frac{1}{2^3}$ (c) $2^{-1} + 2^{-1} = 1$ (d) $2^{-4} = 4^{-2}$

(e) $2^{-1} - 3^{-1} = 2^{-1} \cdot 3^{-1}$

B. In Exercises 32–44 find each value. (*See Examples 3 and 4 and Practice Exercise 3.*)

32. $\left(\frac{1}{2}\right)^{-1}$ **33.** $\left(\frac{3}{5}\right)^{-1}$ **34.** $\left(\frac{-7}{6}\right)^{-1}$ **35.** $\left(\frac{1}{2}\right)^{-2}$

36. $\left(\frac{3}{2}\right)^{-2}$ **37.** $\left(\frac{4}{5}\right)^{-2}$ **38.** $\left(\frac{-3}{4}\right)^{-2}$ **39.** $\left(\frac{3}{4}\right)^{-3}$

40. $\left(\frac{-3}{4}\right)^{-3}$ **41.** $\left(\frac{1}{10}\right)^{-5}$ **42.** $\left(\frac{-1}{10}\right)^{-6}$ **43.** $(.3)^{-2}$

44. $(.5)^{-3}$

Use this area for work space.

In Exercises 45–49 determine the exponent m.

45. $25 = \left(\frac{1}{25}\right)^m$ **46.** $100 = \left(\frac{-1}{10}\right)^m$ **47.** $\frac{2}{3} = \left(\frac{3}{2}\right)^m$

48. $\frac{9}{4} = \left(\frac{2}{3}\right)^m$ **49.** $-8 = \left(\frac{-1}{2}\right)^m$

50. Pick out the *error(s)*.

(a) $\left(\frac{2}{5}\right)^{-1} = \frac{5}{2}$ (b) $\left(\frac{-3}{7}\right)^{-1} = \left(\frac{-7}{3}\right)$

(c) $\left(\frac{4}{5}\right)^{-1} \cdot \left(\frac{5}{4}\right)^{-1} = 1$ (d) $\left(\frac{1}{2}\right)^{-2} + \left(\frac{1}{3}\right)^{-1} = 7$

(e) $\left(\frac{3}{4}\right)^{-1} < \left(\frac{4}{3}\right)^{-1}$ (f) $\left(\frac{1}{.1}\right)^{-1} = .1$

C. In Exercises 51–62 assume $a \neq 0$, $b \neq 0$, and $y \neq 0$. Simplify and express your answer without writing a fraction. Use negative exponents, if necessary. (*See Examples 5 and 6.*)

51. $a^4 \cdot a^{-3}$ **52.** $a^3 \cdot a^{-4}$ **53.** $y^{-7} \cdot y^{-3}$

54. $b^{-1} \cdot b^{-6} \cdot b^{-8}$ **55.** $b^3 \cdot b^{-2} \cdot b^{-4}$ **56.** $\frac{a^6}{a^{-4}}$

57. $\frac{b^{-5}}{b^5}$ **58.** $\frac{x^{-3}}{x^{-6}}$ **59.** $\frac{y^{-4}}{y^{-4}}$

60. $\frac{x^{-1} \cdot x^{-7}}{x^{-9}}$ **61.** $\frac{a^5 \cdot a^{-4}}{a^{-3}}$ **62.** $\frac{b^{-5} \cdot b^{-4}}{b^{-7} \cdot b^2}$

8.7 Scientific Notation

Scientific notation is a convenient way of expressing large numbers, such as 420 000 000 000, and small positive numbers, such as .000 003. These numbers frequently occur in fields such as biology, chemistry, physics, engineering, and astronomy.

Consider the **powers of 10**—positive, zero, and negative:

$$10^1 = 10 \qquad 10^0 = 1 \qquad 10^{-1} = \frac{1}{10} = .1$$

$$10^2 = 100 \qquad 10^{-2} = \frac{1}{100} = .01$$

$$10^3 = 1000 \qquad 10^{-3} = \frac{1}{1000} = .001$$

$$10^4 = 10\,000 \qquad 10^{-4} = \frac{1}{10\,000} = .0001$$

$$10^5 = 100\,000 \qquad 10^{-5} = \frac{1}{100\,000} = .000\,01$$

$$10^6 = 1\,000\,000 \qquad 10^{-6} = \frac{1}{1\,000\,000} = .000\,001$$

A. Expressing Numbers in Scientific Notation

Every positive number N can be expressed as the product of

1. a number between 1 and 10 and
2. a power of 10

This form of writing a number is called **scientific notation.**

Example 1 ▶ To express 240 in scientific notation, observe that

$$240 = \frac{240}{100} \times 100 = 2.40 \times 10^2$$

To divide by 100 (or 10^2), move the decimal point **2** digits to the *left* (to obtain 2.40). Balance this by multiplying by $10^{\mathbf{2}}$. ◀

Example 2 ▶ To express .0051 in scientific notation, observe that

$$.0051 = (.0051 \times 1000) \times \frac{1}{1000} = 5.1 \times 10^{-3}$$

To multiply by 1000 (or 10^3), move the decimal point **3** digits to the *right* (to obtain 5.1). Balance this by multiplying by $10^{\mathbf{-3}}$. ◀

Refer to Examples 1 and 2 as you read the rule for writing a number in scientific notation.

Let N be a positive number, expressed as a decimal. To obtain the scientific notation,

$$N = M \times 10^m \quad \text{or} \quad N = M \times 10^{-m}$$

where m is a positive integer or 0:

1. Place the decimal point after the first nonzero digit of N to obtain M.
2. (a) If you moved the decimal point m digits to the *left*, you *divided* by 10^m. To balance this, multiply by 10^m. Thus

$$N = M \times 10^m$$

(b) If you moved the decimal point m digits to the *right*, you *multiplied* by 10^m. To balance this, multiply by 10^{-m}. Thus

$$N = M \times 10^{-m}$$

(c) If N is between 1 and 10, the scientific notation for N is N itself. Note that

$$N = N \times 10^0$$

Example 3 ▶ Express in scientific notation.
(a) 491.2 (b) .007 32
Solution:
(a) $N = 491.2$. Place the decimal point after the 4. Thus

$$M = 4.912$$

You moved the decimal point 2 digits to the *left:*

$$491.2 \text{ to } 4.912$$

Therefore $m = 2$ and

$$491.2 = 4.912 \times 10^2$$

(b) $N = .007\ 32$. The first *nonzero* digit is 7. Thus

$$M = 7.32$$

(The zeros to the left of 7 can be omitted.) You moved the decimal point 3 digits to the *right:*

$$.007\ 32 \text{ to } 007.\ 32$$

Therefore $m = -3$ and

$$.007\ 32 = 7.32 \times 10^{-3}$$

◀

Answers

1. (a) ______

(b) ______

(c) ______

PRACTICE EXERCISES FOR TOPIC A

1. Express in scientific notation.

(a) 63 070 000 (b) .000 902 (c) .008

Now try the exercises for Topic A on page 402.

B. Scientific Notation to Decimal Notation

To convert scientific notation,

$$M \times 10^m \quad \text{or} \quad M \times 10^{-m}, \ m \neq 0$$

back into the usual decimal notation, reverse the preceding process:

1. Move the decimal point m digits to the right if $m > 0$.
2. Move the decimal point m digits to the left if $m < 0$.

Example 4 ▶ Convert scientific notation back into the usual decimal notation.
(a) 3.52×10^5 (b) 1.74×10^{-2}
Solution:

(a) $$3.52 \times 10^5 = 352\,000$$

In order to move the decimal point 5 digits to the right, add 3 zeros after the 2. Because 352 000 is an integer, the decimal point can be omitted.

(b) $$1.74 \times 10^{-2} = .0174$$

In order to move the decimal point 2 digits to the left, add a zero before the 1.

◀

Answers

2. (a) ______

(b) ______

(c) ______

PRACTICE EXERCISES FOR TOPIC B

2. Convert scientific notation back into the usual decimal notation.

(a) 4×10^4 (b) 8.053×10^2 (c) 6.1×10^{-3}

Now try the exercises for Topic B on page 403.

C. Computing

Some arithmetic computations can be simplified by first expressing numbers in terms of powers of 10. Instead of using scientific notation, express each number as the product of an *integer* and a power of 10. Then write the final result in scientific notation.

Example 5 ▶ Find the following value.

$$\frac{75\,000 \times 2400}{2 \times 120\,000 \times .0125}$$

Solution:

$$\frac{75\,000 \times 2400}{2 \times 120\,000 \times .0125} = \frac{\overset{3}{\cancel{75}} \times 10^3 \times \overset{\overset{1}{\cancel{2}}}{\cancel{24}} \times 10^2}{\underset{1}{\cancel{2}} \times \underset{1}{\cancel{12}} \times 10^4 \times \underset{5}{\cancel{125}} \times 10^{-4}}$$

$$= \frac{3}{5} \times 10^{3+2-(4-4)}$$

$$= .6 \times 10^5$$

$$= 6 \times 10^4 \quad \text{in scientific notation}$$ ◀

Here are a few ways that scientific notation is used in the sciences:

The speed of light is 3×10^{10} centimeters per second. Thus in one minute light travels

$$60 \times 3 \times 10^{10} = 180 \times 10^{10} = 1.8 \times 10^2 \times 10^{10} = 1.8 \times 10^{12}$$

centimeters.

The earth's mass is approximately 6×10^{27} grams.

An electron has a charge of 4.8×10^{-10} electrostatic units.

All gases occupying equal volumes at the same temperature and pressure contain the same number of molecules **(Avogadro's Number)** which has been determined to be 6.023×10^{23} molecules per gram molecular weight.

Answers

3. (a) ____________

(b) ____________

PRACTICE EXERCISES FOR TOPIC C

3. Express the result of each computation in scientific notation.

(a) $\dfrac{10^{-4}}{10^2 \times 10^{-1}}$ (b) $\dfrac{3600 \times 5000}{.0045}$

Now try the exercises for Topic C on page 403.

Use this area for work space.

EXERCISES

A. In Exercises 1–16 express each number in scientific notation. *(See Examples 1–3.)*

1. 12.5 **2.** 465 **3.** 7028 **4.** 5.93

Use this area for work space.

5. .497 **6.** .005 31 **7.** .8 **8.** 6.66

9. 100 **10.** 7048.2 **11.** 361 000 **12.** 1 011 000

13. .000 005 **14.** .030 003 **15.** 6 **16.** .01

B. In Exercises 17–30 convert scientific notation back into the usual decimal notation. (*See Example 4.*)

17. 6.52×10^1 **18.** 3.41×10^3 **19.** 3.23×10^6

20. 3×10^4 **21.** 5×10^{-1} **22.** 3.7×10^{-2}

23. 9.01×10^{-4} **24.** 1.53×10^{-6} **25.** 8.78×10^8

26. 8.78×10^{-8} **27.** 2.001×10^2 **28.** 2.001×10^{-2}

29. 5.4314×10^4 **30.** $3.612\,54 \times 10^{-3}$

C. In Exercises 31–38 find each value. Express your result in scientific notation. (*See Example 5.*)

31. $10^4 \times 10^2 \times 10^{-3}$

32. $\dfrac{10^5 \times 10^8}{10^2 \times 10^6}$

33. $5000 \times 20\,000 \times 4000$

34. $8000 \times .0009$

35. $.0062 \times .000\,02$

36. $\dfrac{2000 \times 40\,000}{.05 \times 800}$

37. $\dfrac{2500 \times 70\,000}{.05 \times .014}$

38. $\dfrac{720\,000 \times .09}{1\,200\,000 \times .0003}$

39. A light-year is the distance over which light can travel in a year. One light year is approximately 6×10^{12} miles. A star is 180 light years away from earth. How many miles away is it?

40.

$$1 \text{ centimeter} \approx .3937 \text{ inch}$$

Express this number of inches in scientific notation.

41. One electron volt is the kinetic energy an electron acquires when it is accelerated in an electric field produced by a difference of potential of one volt.

$$1 \text{ electron volt} = 1.60 \times 10^{-19} \text{ joule}$$

How many joules equal one million electron volts?

42. The human eye is normally capable of responding to light waves whose lengths are in the range of 3.8×10^{-5} to 7.6×10^{-5} centimeter. Which of the following centimeter readings lie within this range?
(a) .000 54 centimeter (b) .000 054 centimeter
(c) 7×10^{-6} centimeter (d) $.39 \times 10^{-4}$ centimeter

43. The coefficient of linear expansion indicates the fractional change in the length of a rod per degree change of temperature. For tungsten, the coefficient is 4.4×10^{-6} per degree Celsius. What is the fractional change in the length of a tungsten rod when the temperature rises from 5° to 25° Celsius?

What Have You Learned in Chapter 8?

You have learned that if a is a positive number, then b is called the positive square root of a if b is positive and if $b^2 = a$. And $-b$ is then called the negative square root of a. Also the square root of 0 is 0.

You have seen that the square roots of integers such as 2, 3, and 5 are irrational. They can be *approximated* by rational numbers.

The square root of a product equals the product of the square roots, and the square root of a quotient equals the quotient of the square roots.

You have learned how to simplify expressions involving radicals.

You know that if $a \neq 0$ and if n is a positive integer, than a^{-n} is defined to be $\dfrac{1}{a^n}$ and a^0 is defined to be 1.

And you can express positive numbers in scientific notation, and can convert scientific notation back to the usual decimal notation.

Let's Review Chapter 8

Assume $a > 0$, $b > 0$, $c > 0$, $x > 0$, and $y > 0$.

8.1 Square Roots

In Exercises 1–4 find each square root.

1. $\sqrt{25}$

2. $\sqrt{64}$

3. $-\sqrt{121}$

4. $\sqrt{x^4}$

8.2 Irrational Square Roots

5. Between which two consecutive integers is $\sqrt{55}$ located?

6. Use $\sqrt{2} \approx 1.41$ to approximate $7\sqrt{2}$ to one decimal place.

7. Approximate the length of a side of a square whose area is 62 square inches. (Use the next *largest* integer.)

8. Find the length of the hypotenuse of the right triangle in Figure 8.18.

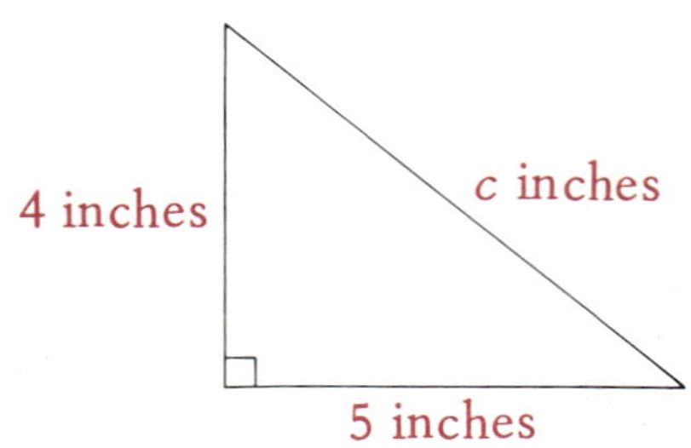

FIGURE 8.18

8.3 Roots of Products and Quotients

In Exercises 9–11 find each square root.

9. $\sqrt{25a^2}$

10. $\sqrt{a^4b^6}$

11. $\sqrt{\dfrac{81}{100}}$

12. Simplify $\sqrt{\dfrac{20a^4}{b^8}}$

8.4 *Adding and Subtracting Roots*

In Exercises 13–16 simplify each expression.

13. $5\sqrt{5} - (3\sqrt{5} - \sqrt{5})$

14. $\sqrt{9a} - \sqrt{4a}$

15. $\sqrt{64a^2b^4} + \sqrt{81a^2b^4}$

16. $\frac{4\sqrt{3}}{15} + \frac{2\sqrt{3}}{5}$

8.5 *Multiplying and Dividing Roots*

In Exercises 17–20 multiply as indicated.

17. $\sqrt{10}\sqrt{5}\sqrt{2}$

18. $\sqrt{xy}\sqrt{x^3y}$

19. $\sqrt{2}(\sqrt{8} - \sqrt{2})$

20. $\frac{\sqrt{8}}{\sqrt{3}} \cdot \frac{\sqrt{3}}{\sqrt{2}}$

21. Simplify $\frac{\sqrt{32}}{\sqrt{8}}$

22. Simplify $\frac{\sqrt{12x}}{\sqrt{75x}}$

23. (a) Factor $7 + \sqrt{98}$ (b) Simplify $\frac{7 + \sqrt{98}}{21}$

24. Divide $\frac{\sqrt{15}}{\sqrt{18}} \div \frac{\sqrt{5}}{\sqrt{32}}$

8.6 *Zero and Negative Exponents*

In Exercises 25–27 find each value.

25. $7^{-2} + 7^{-1}$

26. 8^0

27. $\left(\frac{2}{3}\right)^{-2}$

28. Assume $x \neq 0$. Express $\frac{x^{-4}}{x^3}$ without writing a fraction. Use a negative exponent.

8.7 *Scientific Notation*

In Exercises 29 and 30 express each number in scientific notation.

29. 265 000

30. .0257

31. Convert 3.45×10^{-4} back into the usual decimal notation.

32. Find the value of $\frac{16\,000 \times .06}{.004 \times 600}$. Express your result in scientific notation.

And These from Chapters 1–7

33. Multiply $\frac{81}{a^4} \cdot \frac{a^3}{27b^2c^3} \cdot \frac{bc^2}{2}$

34. Find $(.02)^2 + (.4)^3$

35. Add $\frac{3}{4} + \frac{5}{8} + \frac{7}{10}$

REVIEW

36. Fill in "$<$", "$=$", or "$>$".

(a) $2^2 \square \sqrt{2}$ (b) $1^2 \square \sqrt{1}$

(c) $\left(\frac{1}{4}\right)^2 \square \sqrt{\frac{1}{4}}$ (d) $(.09)^2 \square \sqrt{.09}$

37. Express 400 as the product of primes.

38. Solve $x^2 + 8x = -15$

39. Factor $x^2 + 8xy + 12y^2$

40. Find the *lcm* of $x^2 - 9$ and $x^2 + 4x + 3$

41. Simplify $\dfrac{\dfrac{a-b}{a^2-9}}{\dfrac{b^2-a^2}{a+3}}$

42. Solve $5x - 3ax + 2y = 1$ for x.

43. Solve $-2 \leq 5y - 1 \leq 14$

44. Graph the equation $y = 9 - 3x$.

45. Find the equation of the line with slope -2 and y-intercept 3.

46. Find the equation of the line with x-intercept 4 and y-intercept 5.

47. Find the equation of the horizontal line that passes through $(-2, -1)$.

48. Solve by either substituting or adding:

$$\begin{aligned} 6x + 2y &= 12 \\ 9x - 3y &= 0 \end{aligned}$$

Practice Exam 1 on Chapter 8

1. Find each square root.

(a) $\sqrt{144}$ (b) $\sqrt{100x^4}$ (c) $\sqrt{\frac{16}{49}}$

2. Simplify $\sqrt{\dfrac{32a^4b^8}{c^8}}$

3. Simplify $\dfrac{5\sqrt{2} - 3\sqrt{2}}{8}$

4. Assume $x > 0$, $y > 0$, and $z > 0$. Multiply $\sqrt{9x^3y}\sqrt{3xy^3z}\sqrt{12x^2z}$

5. Divide $\dfrac{\sqrt{20}}{\sqrt{63}} \div \dfrac{\sqrt{7}}{\sqrt{45}}$

6. (a) Factor $\sqrt{18} - 9$ (b) Simplify $\dfrac{\sqrt{18} - 9}{6}$

7. Use $\sqrt{2} \approx 1.41$ and $\sqrt{3} \approx 1.73$ to approximate $3\sqrt{2} + \dfrac{\sqrt{3}}{10}$ to one decimal digit.

8. Find each value.

(a) 4^{-3} (b) $(-6)^0$ (c) $\left(\frac{3}{5}\right)^{-2}$

9. Assume $a \neq 0$. Simplify

$$a^{-7} \cdot a^{10} \cdot a^{-2}$$

10. Express 48 700 in scientific notation.

11. Convert 2.87×10^{-2} back into the usual decimal notation.

12. Find the value of $\dfrac{.0081 \times 240}{.04 \times .009}$. Express your result in scientific notation.

Practice Exam 2 on Chapter 8

Assume $x > 0$, $y > 0$, and $n > 0$.

1. Find each square root.

(a) $\sqrt{225}$ (b) $\sqrt{\dfrac{49x^6}{100y^2}}$

2. Simplify: (a) $\sqrt{500}$ (b) $\sqrt{162x^3}$

3. Simplify: $\sqrt{\dfrac{9}{10}}$. Leave your answer in radical form and with rational denominator.

4. Multiply: (a) $\sqrt{3x}\sqrt{27x^3}$ (b) $\sqrt{5}(\sqrt{20} - 2\sqrt{3})$

(c) $(1 - \sqrt{2})(2 + 2\sqrt{2})$

5. Divide $\dfrac{5}{\sqrt{8}} \div \dfrac{\sqrt{75}}{\sqrt{6}}$

6. (a) Factor $12 - \sqrt{45}$ (b) Simplify $\dfrac{12 - \sqrt{45}}{15}$

7. Use $\sqrt{2} \approx 1.41$ to approximate $5 + 3\sqrt{2}$.

8. Evaluate: (a) 11^{-2} (b) $(-5)^0$ (c) $\left(\dfrac{-7}{10}\right)^{-2}$

9. Simplify: (a) $\dfrac{n^{-3}}{n^2}$ (b) $n^9 \cdot n^{-2} \cdot n^{-3}$

10. (a) Express .094 in scientific notation

(b) Express 8.71×10^3 in ordinary decimal notation.

11. Express the result of the following computation in scientific notation: $\dfrac{21\ 000}{200 \times .0003}$

12. Find the length of the hypotenuse of the right triangle in Figure 8.19.

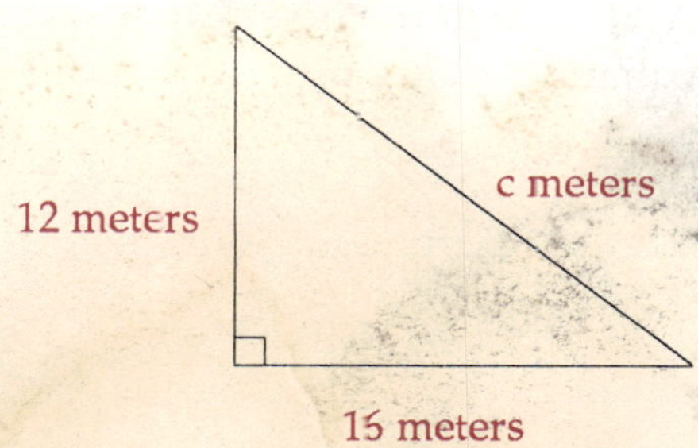

FIGURE 8.19

9

Quadratic Equations

9.1 Reducible Equations

In this chapter you will study techniques of solving equations such as

$$3x^2 + 7x + 2 = 0 \qquad \text{and} \qquad y^2 + y - 1 = 0$$

in which a variable appears to the second power.

A. Quadratic Equations

A **quadratic,** or **second-degree, equation** (in a single variable, x) is an equation *that can be written in the form*

$$Ax^2 + Bx + C = 0 \qquad \textit{where } A \neq 0$$

The quadratic equation

$$x^2 - 3x + 2 = 0$$

is already of this form with $A = 1$, $B = -3$, and $C = 2$. The quadratic equation

$$3x^2 + 5 = 2x - 3$$

can be transformed into the above form by adding $-2x + 3$ to both sides to obtain

$$3x^2 - 2x + 8 = 0$$

Here $A = 3$, $B = -2$, and $C = 8$

Quadratic equations have either 0, 1, or 2 (different) *real* roots.

Example 1 ► Show that

$$3(x^2 - 2x - 4) = 2x^2 + x - 3$$

is a quadratic equation.

Solution:

You must show that the given equation is equivalent to an equation of the form

$$Ax^2 + Bx + C = 0 \qquad \textit{where } A \neq 0$$

Use the Distributive Laws on the left side of the given equation.

$$3x^2 - 6x + 12 = 2x^2 + x - 3$$

Now bring all terms to the left side. Thus add $-2x^2 - x + 3$ to both sides.

$$3x^2 - 6x + 12 - 2x^2 - x + 3 = 2x^2 + x - 3 - 2x^2 - x + 3$$

$$x^2 - 7x + 15 = 0$$

This is of the required form with $A = 1$, $B = -7$, and $C = 15$. ◄

Example 2 ► Show that

$$2(x - 3)(x + 2) = x - 7$$

is a quadratic equation.

Solution:

$$2\underbrace{(x - 3)(x + 2)}_{x^2 - x - 6} = x - 7$$

$$2x^2 - 2x - 12 = x - 7$$

$$2x^2 - 2x - 12 - x + 7 = x - 7 - x + 7$$

$$2x^2 - 3x - 5 = 0$$

$$\begin{array}{r} x - 3 \\ x + 2 \\ \hline x^2 - 3x \quad\;\; \\ +\, 2x - 6 \\ \hline x^2 - \;\; x - 6 \end{array}$$

This is the form of a quadratic equation. Here $A = 2$, $B = -3$, and $C = -5$ ◄

Answers

1. (a) ____________

(b) ____________

PRACTICE EXERCISES FOR TOPIC A

1. Write each of the following equations in the form $Ax^2 + Bx + C = 0$, where $A \neq 0$. (Do *not* solve.)

(a) $\dfrac{3}{x + 1} = \dfrac{x}{2}$ (b) $(x + 1)^2 = x(2 - x) + 2$

Now try the exercises for Topic A on page 411.

B. Linear Equations

Example 3 ▶ Show that

$$x^2 - 4x + 6 = (x + 2)(x + 1)$$

is *not* a quadratic equation.
Solution:

$$x^2 - 4x + 6 = (x + 2)(x + 1)$$

$$x^2 - 4x + 6 = x^2 + 3x + 2$$

$$x^2 - 4x + 6 - x^2 - 3x - 2 = x^2 + 3x + 2 - x^2 - 3x - 2$$

$$-7x + 4 = 0$$

Thus the x^2 terms drop out ($A = 0$), and the given equation reduces to a first-degree, or linear, equation. You solve this, as usual, by bringing variables to one side, numerical terms to the other.

$$4 = 7x$$

$$\frac{4}{7} = x$$ ◀

In Example 3 all second-degree terms add up to 0. This equation reduces to a first-degree equation. Here is another example of a *reducible equation*.

Example 4 ▶ Show that

$$3t^2 + t(t + 1) + 9 = 2t(5 + 2t)$$

reduces to a first-degree equation. Solve it, and *check the root*.
Solution:

$$3t^2 + \underbrace{t(t + 1)}_{t^2 + t} + 9 = \underbrace{2t(5 + 2t)}_{10t + 4t^2}$$

$$3t^2 + t^2 + t + 9 = 10t + 4t^2$$

$$4t^2 + t + 9 - 4t^2 = 10t + 4t^2 - 4t^2$$

$$t + 9 = 10t$$

Thus all second-degree terms cancel. To solve, subtract t from each side.

$$9 = 9t$$

$$1 = t$$

Check:

$$3(1)^2 + 1(1 + 1) + 9 \stackrel{?}{=} 2(1)[5 + 2(1)]$$

$$3 + 2 + 9 \stackrel{?}{=} 2 \cdot 7$$

$$14 \stackrel{\checkmark}{=} 14$$ ◀

Answers

2. (a) ________

(b) ________

PRACTICE EXERCISES FOR TOPIC B

2. (a) Determine which of the following is a first-degree equation. (b) Solve it and (c) *check* the root.

i. $(x - 3)2x = 4$ *ii.* $4x^2 + 9x = (2x + 3)^2$

Now try the exercises for Topic B on this page.

Use this area for work space.

EXERCISES

A. In Exercises 1–12 write each equation in the form $Ax^2 + Bx + C = 0$, where $A \neq 0$. (Do *not* solve. *See Examples 1 and 2.*)

1. $2x^2 + 5x = 10$

2. $4x^2 - 7x = 9 - 3x$

3. $2x^2 - 9x + 1 = x^2 + 4x - 3$

4. $3(x^2 + 2x - 1) = 4(3 - x^2)$

5. $4(x + 3)(x + 2) = 3(x - 1)(x - 2)$

6. $2x(3x - 1) = x(5x + 4)$

7. $(x + 2)^2 = x(1 - x) - 1$

8. $(2x + 1)^2 = x(x + 1) + 2$

9. $\frac{x}{5} = \frac{x + 1}{x}$, $x \neq 0$

10. $x^2(x + 4) - 1 = x(x + 3)^2$

11. $\frac{x - 1}{4} = \frac{3}{x}$, $x \neq 0$

12. $\frac{x + 2}{5} = \frac{4}{x + 1}$, $x \neq -1$

B. In Exercises 13–18 show that each of the following reduces to a first-degree equation. (Do *not* solve. *See Example 3.*)

13. $x(x + 2) = (x + 1)(x + 3)$

14. $(2x + 1)(x - 1) = (x + 5)(2x - 3)$

15. $x^2 + 4x - 9 = x(x - 4)$

16. $4x^2 - (x + 1)(x - 2) = 3(x^2 - 2)$

17. $(x + 4)^2 = (x - 3)^2$

18. $x(x^2 - 2x + 4) = x(x - 1)^2 + 1$

19. Which of the following are quadratic equations?
(a) $x^2 + 5x = 5 + x^2$
(b) $3x^2 - 2x + 1 = 4x - 1 - 3x^2$
(c) $x^2 - x + 8 = x^3 + x - 10$
(d) $4x^2 + 1 - 2x^2 = 2x^2 - 3x + 5$

Use this area for work space.

20. Which of the following are quadratic equations?

(a) $\frac{x + 1}{x - 2} = \frac{5}{x - 1}$, $x \neq 1, 2$

(b) $\frac{x - 2}{x + 3} = \frac{x - 1}{x}$, $x \neq 0, -3$

(c) $x(x^2 + 1) = x(x^2 - 2x)$

(d) $x(x^2 - 4x + 1) = (x^2 + x)(x - 5)$

In Exercises 21–40 solve each equation. *Check* where indicated. (*See Example 4.*)

21. $x^2 + 5x = x^2 - 10$ (*Check.*)

22. $3x^2 + 6x - 2 = 2x^2 + 8x - (1 - x^2)$

23. $-5y + 2y^2 = 9y^2 + 35 - 7y^2$

24. $z^2 + 25 = (z - 5)^2$

25. $(x + 4)^2 = (x - 2)^2$ (*Check.*)

26. $x^2 - 3x = (x - 3)^2 + 1$

27. $(x - 4)(x + 2) = (x + 1)(x + 3)$

28. $(y - 2)^2 + 3 = (y + 1)^2$ (*Check.*)

29. $3 + (x - 1)^2 = (x + 1)^2 - 2$

30. $x(x - 4) + 10 = x(2x + 1) - x^2$

31. $(1 - x)^2 = 9 + x^2$

32. $2y - y^2 = 4y - (y + 2)^2$

33. $(z - 8)^2 + 6 = (z - 7)^2 + 1$ (*Check.*)

34. $(2x - 3)(3x - 2) = (6x + 1)(x - 2)$

35. $5u(u - 1) = 3(2u^2 - 6u) - (1 + u^2)$

36. $3x(4x - 1) = (6x + 1)(2x - 1)$

37. $4t(2t - 3) = 8(t + 1)(t - 1)$

38. $\frac{x - 1}{x} = \frac{x + 2}{x - 2}$, $x \neq 0, 2$

39. $\frac{y + 2}{y + 1} = \frac{y}{y - 3}$, $y \neq -1, 3$

40. $z(z + 1)^2 = (z - 1)z^2 + 3(z^2 + 1)$

41. Pick out the *error(s)*.

(a) $x^2 + 3x - 2 = 1 - x^2$ is a quadratic equation.

(b) $5x^2 + 7 = 6x^2 + x - x^2$ is a quadratic equation.

(c) The only root of the equation $x^2 - 3x + 1 = x^2 - 2$ is 1.

(d) The equation $x^2 - 2x + 5 = x^2 + 3 - 2x$ has no roots.

9.2 Factorable Equations

A. $x^2 + Bx + C = 0$

Factoring techniques were used to solve equations of the form

$$x^2 + Bx + C = 0$$

in Section 4.7. The basis of this method is the following fact:

A product is 0 if at least one of the factors is 0

And if none of its factors is 0, then a product is not 0

Thus

$$0 \cdot 5 = 0$$
$$-10 \cdot 0 = 0$$
$$0 \cdot 0 = 0$$
$$(-1) \cdot 3 \cdot (-2) \neq 0$$

To solve

$$x(x - 6) = 0$$

note that one of the factors x or x − 6 must be 0. Thus you solve two simpler equations

$$x = 0, \qquad x - 6 = 0$$

separately:

$$x = 0 \quad \Big| \quad \begin{aligned} x - 6 &= 0 \\ x &= 6 \end{aligned}$$

Clearly 0 and 6 are also the roots of the quadratic equation

$$x^2 - 6x = 0$$

because the left side can be factored as $x(x - 6)$, so that this equation is equivalent to

$$x(x - 6) = 0$$

Example 1 ▶ Solve $x^2 + 3x - 10 = 0$

Solution:

$$x^2 + 3x - 10 = (x + 5)(x - 2)$$

Thus

$$x^2 + 3x - 10 = 0$$

can be written as

$$(x + 5)(x - 2) = 0$$

$$\begin{aligned} x + 5 &= 0 \\ x &= -5 \end{aligned} \quad \Bigg| \quad \begin{aligned} x - 2 &= 0 \\ x &= 2 \end{aligned}$$

The roots are 2 and −5. ◀

In the next example, after you clear of "fractions," a quadratic equation emerges. You will be able to solve this by factoring.

Example 2 ▶ Solve $\frac{4}{x+1} - \frac{1}{x} = 1$

Solution: Multiply both sides by the *lcd*, x(x + 1).

$$\left(\frac{4}{x+1} - \frac{1}{x}\right) \cdot x(x+1) = 1 \cdot x(x+1) \qquad \textit{Use the Distributive Laws.}$$

$$\frac{4}{x+1} \cdot x(x+1) - \frac{1}{x} \cdot x(x+1) = x \cdot x + x \cdot 1$$

$$4x - (x+1) = x^2 + x$$

$$3x - 1 = x^2 + x$$

$$0 = x^2 - 2x + 1 \qquad \textit{Change sides.}$$

$$x^2 - 2x + 1 = 0$$

$$\underbrace{(x-1)}_{a}{}^2 = 0 \qquad a^2 = 0$$

Thus

$$\underbrace{x-1}_{a} = 0 \qquad a = 0$$

$$x = 1$$ ◀

PRACTICE EXERCISES FOR TOPIC A

Answers
1. ______
2. (a) ______
(b) ______
(c) ______

1. Find the roots of the equation

$$(x + 4)(x - 10) = 0$$

2. Solve each equation. *Check* where indicated.

(a) $x^2 + x - 30 = 0$ (*Check.*) (b) $\frac{10}{x-1} = 5x$ (c) $1 + \frac{6}{x^2} = \frac{7}{x}$

Now try the exercises for Topic A on page 417.

B. $Ax^2 + Bx + C = 0$

Example 3 ▶ Solve $2x^2 + 5x + 2 = 0$

Solution:

$$2x^2 + 5x + 2 = (2x + 1)(x + 2)$$

Thus $2x^2 + 5x + 2 = 0$

can be written as $(2x + 1)(x + 2) = 0$

$2x + 1 = 0$	$x + 2 = 0$
$2x = -1$	$x = -2$
$x = \frac{-1}{2}$	

The roots are $\frac{-1}{2}$ and -2. ◀

Example 4 ▶ Solve $6t^2 + 2t - 4 = 0$

Solution:

$$6t^2 + 2t - 4 = 2(3t^2 + t - 2) = 2(3t - 2)(t + 1)$$

Thus solve

$$2(3t - 2)(t + 1) = 0 \qquad \textit{Divide both sides by 2.}$$
$$(3t - 2)(t + 1) = 0$$

$3t - 2 = 0$	$t + 1 = 0$
$3t = 2$	$t = -1$
$t = \frac{2}{3}$	

The roots are $\frac{2}{3}$ and -1. ◀

Recall that if $a^2 = 0$, then $a = 0$; and if $a = 0$, then $a^2 = 0$.

Example 5 ▶ Solve $4x^2 - 20x + 25 = 0$

Solution:

$$4x^2 - 20x + 25 = (2x - 5)^2$$

Thus

$$\underbrace{(2x - 5)}_{a}{}^2 = 0 \qquad a^2 = 0$$

and consequently

$$2x - 5 = 0$$
$$2x = 5 \qquad a = 0$$
$$x = \frac{5}{2}$$

The only root is $\frac{5}{2}$. ◀

Answers

3. (a) ____

(b) ____

PRACTICE EXERCISES FOR TOPIC B

3. Solve each equation.

(a) $4x^2 - 5x + 1 = 0$ (b) $12x^2 + 8x - 4 = 0$

Now try the exercises for Topic B on page 418.

C. Difference of Squares

Recall the difference-of-squares factoring method:

$$x^2 - a^2 = (x + a)(x - a)$$

Example 6 ▶ Solve $9x^2 - 16 = 0$

Solution:

$$9x^2 - 16 = (3x + 4)(3x - 4)$$

Thus

$$(3x + 4)(3x - 4) = 0$$

You obtain the two equations

$3x + 4 = 0$	$3x - 4 = 0$
$3x = -4$	$3x = 4$
$x = \frac{-4}{3}$	$x = \frac{4}{3}$

The roots of the given equation are $\frac{4}{3}$ and $\frac{-4}{3}$. ◀

Answers

4. ______

PRACTICE EXERCISES FOR TOPIC C

4. Solve $49x^2 = 25$

Now try the exercises for Topic C on page 418.

D. Higher-Degree Equations

The same methods apply when the product of more than two factors equals 0. Set each of these factors equal to 0 to obtain simpler equations.

Example 7 ▶ Solve $4x^3 - 9x = 0$
Solution:

$$4x^3 - 9x = x(4x^2 - 9)$$
$$= x(2x + 3)(2x - 3)$$

Thus $$x(2x + 3)(2x - 3) = 0$$

You obtain three simpler equations:

$x = 0$	$2x + 3 = 0$	$2x - 3 = 0$
	$2x = -3$	$2x = 3$
	$x = \frac{-3}{2}$	$x = \frac{3}{2}$

The roots of the given equation are 0, $\frac{3}{2}$, and $\frac{-3}{2}$. ◀

Answers

5. (a) ______

(b) ______

PRACTICE EXERCISES FOR TOPIC D

5. Solve each equation.

(a) $8x^3 - 2x = 0$ (b) $x^4 - 4x^3 + 3x^2 = 0$

Now try the exercises for Topic D on page 418.

E. Given the Roots

Example 8 ▶ Find a quadratic equation with integral coefficients whose roots are -1 and $\frac{1}{2}$.

Solution:

If -1 and $\frac{1}{2}$ are roots of the equation, then

$$x = -1 \quad \text{or} \quad x = \frac{1}{2}$$

Therefore $$x + 1 = 0 \quad \text{or} \quad x - \frac{1}{2} = 0$$

so that $(x+1)\left(x-\frac{1}{2}\right)=0$

$$\begin{array}{r} x+1 \\ x-\frac{1}{2} \\ \hline x^2+x \\ -\frac{1}{2}x-\frac{1}{2} \\ \hline x^2+\frac{1}{2}x-\frac{1}{2} \end{array}$$

Consequently $x^2+\frac{1}{2}x-\frac{1}{2}=0$

is a quadratic equation whose roots are -1 and $\frac{1}{2}$. If you multiply both sides of this equation by 2, you obtain an equivalent equation with integral coefficients. Thus

$$2x^2+x-1=0$$

has roots -1 and $\frac{1}{2}$. ◀

In general, if a and b are the roots of a quadratic equation, then

$$x=a \quad \text{or} \quad x=b$$

Therefore $x-a=0$ or $x-b=0$

so that $(x-a)(x-b)=0$

Multiply the factors on the left side to see that

$$x^2-(a+b)x+ab=0$$

is a quadratic equation whose roots are a and b.

Answers

6. ______

PRACTICE EXERCISES FOR TOPIC E

6. Find a quadratic equation with integral coefficients whose roots are $\frac{2}{3}$ and -3.

Now try the exercises for Topic E on page 418.

Use this area for work space.

EXERCISES

A. In Exercises 1–4 find the roots of each equation. (*See Practice Exercise 1.*)

1. $(x-3)(x+7)=0$

2. $(x+9)^2=0$

3. $x\left(x+\frac{1}{4}\right)=0$

4. $\left(x-\frac{1}{2}\right)\left(x+\frac{3}{4}\right)=0$

In Exercises 5–20 solve each equation. *Check* where indicated. (*See Examples 1 and 2.*)

5. $x^2-5x+4=0$

6. $t^2-t-6=0$ (*Check.*)

7. $y^2+8y+16=0$

8. $z^2+4z-21=0$

Use this area for work space.

9. $2x^2 - 4x = 0$

10. $x^2 + 5x = 0$

11. $x^2 - 6x + 9 = 0$ *(Check.)*

12. $2t^2 + 20t + 50 = 0$

13. $\frac{4}{x} = x - 3,\ x \neq 0$

14. $\frac{15}{x - 2} = x,\ x \neq 2$

15. $1 + \frac{1}{x^2} - \frac{2}{x} = 0, \quad x \neq 0$

16. $\frac{4}{y} + y - 5 = 0, \quad y \neq 0$

17. $\frac{2}{t^2} - \frac{3}{t} + 1 = 0, \quad t \neq 0$

18. $1 + \frac{6}{x^2} = \frac{5}{x}, \quad x \neq 0$

19. $\frac{2}{1 - x} + \frac{x}{2 + x} = 0, \quad x \neq 1, -2$ *(Check.)*

20. $\frac{1}{t - 4} + \frac{1}{4} = \frac{-1}{t + 2}, \quad t \neq 4, -2$

B. In Exercises 21–42 solve each equation. *Check* where indicated. (*See Examples* 3–5.)

21. $(2x - 1)(3x + 1) = 0$

22. $(4x + 3)(5x - 8) = 0$

23. $2x^2 + 3x + 1 = 0$

24. $2x^2 + 7x + 3 = 0$

25. $2t^2 + 7t - 4 = 0$

26. $3y^2 + 7y + 2 = 0$ *(Check.)*

27. $4x^2 + 8x + 3 = 0$

28. $6y^2 + y - 1 = 0$

29. $12u^2 - 2u - 4 = 0$

30. $3x^2 - 10x + 3 = 0$

31. $2y^2 + 6 = 7y$

32. $2z^2 - 13z = 7$

33. $10t^2 - t - 3 = 0$

34. $12v^2 + 5v - 2 = 0$

35. $2x^2 - x = 0$

36. $3x^2 + 2x = 0$

37. $4y^2 - 12y + 9 = 0$

38. $9z^2 + 6z + 1 = 0$

39. $16x^2 + 24x + 9 = 0$ *(Check.)*

40. $100y^2 - 60y + 9 = 0$

41. $x^2 - \frac{x}{4} - \frac{1}{8} = 0$

42. $y^2 - \frac{2y}{3} + \frac{1}{9} = 0$ *(Check.)*

C. In Exercises 43–48 solve each equation. *Check* where indicated. (*See Example* 6.)

43. $x^2 - 25 = 0$

44. $9 - y^2 = 0$

45. $4t^2 = 49$

46. $36z^2 = 121$ *(Check.)*

47. $144 - 81x^2 = 0$

48. $169 - 100y^2 = 0$

D. In Exercises 49–54 solve each equation. *Check* where indicated. (*See Example* 7.)

49. $y^3 + 6y^2 + 9y = 0$

50. $u^3 - 6u^2 + 5u = 0$ *(Check.)*

51. $z^3 + 2z^2 - 15z = 0$

52. $x^3 + 20x = 9x^2$

53. $y^4 - y^2 = 0$

54. $2t^5 - 7t^4 + 6t^3 = 0$

E. In Exercises 55–64 find a quadratic equation with integral coefficients whose roots are as indicated. (*See Example* 8.)

55. 1 and 3

56. 2 and -4

57. 3 and 5

58. -4 and -5

59. 0 and 1

60. $\frac{1}{2}$ and 4

61. $\frac{1}{3}$ and -3

62. $\frac{2}{3}$ and $\frac{1}{4}$

63. only -3

64. only $-\frac{1}{2}$

Use this area for work space.

65. Pick out the *error(s)*. Assume A and B are any nonzero numbers.

(a) There is a quadratic equation whose roots are A and B.

(b) There is a quadratic equation whose only root is AB.

(c) There is a quadratic equation whose only root is $\frac{A}{B}$.

(d) There is a quadratic equation whose roots are both AB and $\frac{A}{B}$.

(e) The quadratic equation $Ax^2 + Bx = 0$ has $\frac{-B}{A}$ as its only root.

9.3 Equations of the Form $x^2 = a$

A. *a* Greater than Zero

Let n be a positive number, and let

	$\pm n$	stand for	both n and $-n$
For example,	± 2	stands for	both 2 and -2

If a is positive, there are two numbers whose square is a. These numbers are

$$\pm\sqrt{a}$$

because $\quad \sqrt{a}\sqrt{a} = a \quad$ and $\quad (-\sqrt{a})(-\sqrt{a}) = a$

Thus if a quadratic equation can be written in the form

$$x^2 = a$$

its roots are $\pm\sqrt{a}$.

Example 1 ▶ Solve $x^2 - 25 = 0$

Solution 1:

$$x^2 - 25 = 0$$
$$x^2 = 25$$
$$x = \pm 5$$

Solution 2:

$$x^2 - 25 = 0 \qquad \textit{Factor the left side.}$$
$$(x + 5)(x - 5) = 0$$

$$x + 5 = 0 \quad \Big| \quad x - 5 = 0$$
$$x = -5 \quad \Big| \quad x = 5$$

The roots are ± 5. ◀

Example 2 ▶ Solve $t^2 = 7$

Solution:

$$t^2 = 7$$
$$t = \pm\sqrt{7}$$ ◀

Example 3 ► Solve $x^2 - 24 = 0$
Solution:

$$x^2 - 24 = 0$$
$$x^2 = 24$$
$$x = \pm\sqrt{24}$$
$$x = \pm\sqrt{4 \cdot 6}$$
$$x = \pm\sqrt{4}\sqrt{6}$$
$$x = \pm 2\sqrt{6}$$ ◄

Example 4 ► Solve $(2u + 3)^2 = 25$ for u. *Check the roots.*
Solution:
This second-degree equation is of the form

$$x^2 = 25$$

Here $x = 2u + 3$

Thus $2u + 3 = \pm 5$

You obtain two first-degree equations:

$2u + 3 = 5$	$2u + 3 = -5$
$2u = 2$	$2u = -8$
$u = 1$	$u = -4$

The roots are 1 and -4.

Check for 1:	*Check* for -4:
$(2 \cdot 1 + 3)^2 \stackrel{?}{=} 25$	$[2(-4) + 3]^2 \stackrel{?}{=} 25$
$5^2 \stackrel{?}{=} 25$	$(-5)^2 \stackrel{?}{=} 25$
$25 \stackrel{\checkmark}{=} 25$	$25 \stackrel{\checkmark}{=} 25$

◄

Example 5 ► Solve $(3x - 1)^2 = 5$ for x. *Check the roots.*
Solution:

$$(3x - 1)^2 = 5$$
$$3x - 1 = \pm\sqrt{5}$$

$3x - 1 = \sqrt{5}$	$3x - 1 = -\sqrt{5}$
$3x = 1 + \sqrt{5}$	$3x = 1 - \sqrt{5}$
$x = \dfrac{1 + \sqrt{5}}{3}$	$x = \dfrac{1 - \sqrt{5}}{3}$

The roots differ only in the sign of $\sqrt{5}$. Thus the roots are

$$\frac{1 \pm \sqrt{5}}{3}$$

Check for $\dfrac{1+\sqrt{5}}{3}$:	*Check* for $\dfrac{1-\sqrt{5}}{3}$:
$\left[3\left(\dfrac{1+\sqrt{5}}{3}\right)-1\right]^2 \stackrel{?}{=} 5$	$\left[3\left(\dfrac{1-\sqrt{5}}{3}\right)-1\right]^2 \stackrel{?}{=} 5$
$[(1+\sqrt{5})-1]^2 \stackrel{?}{=} 5$	$[(1-\sqrt{5})-1]^2 \stackrel{?}{=} 5$
$(\sqrt{5})^2 \stackrel{?}{=} 5$	$(-\sqrt{5})^2 \stackrel{?}{=} 5$
$5 \stackrel{\checkmark}{=} 5$	$5 \stackrel{\checkmark}{=} 5$ ◀

Answers

1. (a) ______
(b) ______
(c) ______
(d) ______

PRACTICE EXERCISES FOR TOPIC A

1. Solve each equation.

(a) $x^2 = \dfrac{4}{25}$ (b) $x^2 - 45 = 0$ (c) $(5u - 1)^2 = 4$ (*Check.*)

(d) $(2u + 7)^2 = 6$

Now try the exercises for Topic A on page 422.

B. *a* Equal to Zero

Recall that

$$\text{if } a^2 = 0, \text{ then } a = 0$$

Thus if an equation has the form $a^2 = 0$, you can immediately set a equal to 0.

Example 6 ▶ Solve $(5z - 2)^2 = 0$ for z.

Solution:

This equation has the form

$$a^2 = 0$$

in which

$$a = 5z - 2$$

Thus from

$$(5z - 2)^2 = 0$$

you obtain

$$5z - 2 = 0$$
$$5z = 2$$

and finally,

$$z = \frac{2}{5}$$ ◀

Answers

2. (a) ______
(b) ______
(c) ______

PRACTICE EXERCISES FOR TOPIC B

2. Solve:

(a) $(5 - 3x)^2 = 0$ (b) $x(2x + 3)^2 = 0$ (c) $y^3 - 2y^2 + y = 0$

Now try the exercises for Topic B on page 423.

C. *a* Less than Zero

When a is *negative*, the equation

$$x^2 = a$$

does not have any real number as a root because the square of a real number is always *at least* 0. In more advanced courses, you will learn about other numbers, known as "complex numbers." If a is negative, the roots of the above equation are complex numbers. For the present, you need only consider real numbers.

To sum up, the equation

$$x^2 = a$$

has 2 (different) *real* roots if $a > 0$

1 *real* root if $a = 0$

0 *real* roots if $a < 0$

Answers

3. (a) ______

(b) ______

(c) ______

PRACTICE EXERCISES FOR TOPIC C

3. How many (different) *real* roots does each of the following equations have?

(a) $(2u + 9)^2 = 0$ (b) $(3u)^2 = -9$ (c) $(u - 4)^2 = 4$

Now try the exercises for Topic C on page 423.

Use this area for work space.

EXERCISES

A. In Exercises 1–36 solve each equation. *Check* where indicated. (*See Examples 1–5.*)

1. $x^2 = 16$

2. $x^2 - 36 = 0$

3. $y^2 - 49 = 0$

4. $t^2 - 100 = 0$

5. $x^2 = 6$

6. $x^2 - 11 = 0$

7. $y^2 - 13 = 0$ (*Check.*)

8. $z^2 - 15 = 0$

9. $x^2 - 8 = 0$

10. $t^2 - 12 = 0$

11. $u^2 - 18 = 0$

12. $y^2 - 20 = 0$

13. $2u^2 - 90 = 0$

14. $120 - 3x^2 = 0$

15. $x^2 - \frac{4}{9} = 0$

16. $u^2 - \frac{25}{16} = 0$ (*Check.*)

17. $3x^2 - 2 = 0$

18. $25t^2 - 12 = 0$

19. $(2u + 1)^2 = 49$ (*Check.*)

20. $(2x - 3)^2 = 1$ (*Check.*)

21. $(5x - 2)^2 = 16$

22. $\left(\frac{y}{2} + 1\right)^2 = 9$

23. $(3z + 4)^2 = 64$

24. $(5x - 2)^2 = 81$

25. $(3x - 2)^2 = 3$ (*Check.*)

26. $(2x + 1)^2 = 5$

Use this area for work space.

27. $(5t + 4)^2 = 7$

28. $(u - 4)^2 = 11$

29. $(3y + 5)^2 = 6$

30. $(4z - 1)^2 = 13$

31. $(2x - 1)^2 = 8$

32. $(2t - 5)^2 = 12$

33. $(z + 3)^2 = 32$

34. $\left(\frac{t}{2} + 1\right)^2 = 20$ (*Check.*)

35. $\left(\frac{x - 1}{2}\right)^2 = 98$

36. $\left(\frac{2x + 7}{3}\right)^2 = 80$

B. In Exercises 37–46 solve each equation. *Check* where indicated. (*See Example 6 and Practice Exercise 2.*)

37. $(x - 2)^2 = 0$

38. $(2t + 1)^2 = 0$

39. $(3 - 5z)^2 = 0$

40. $\left(\frac{4u + 5}{3}\right)^2 = 0$

41. $x(x + 4)^2 = 0$

42. $y(2y - 1)^2 = 0$ (*Check.*)

43. $x^2(5x + 3)^2 = 0$

44. $y^3(1 - 7y)^2 = 0$

45. $x^3 - 4x^2 + 4x = 0$

46. $y^4 + 6y^3 + 9y^2 = 0$

C. In Exercises 47–52 how many (different) *real* roots does each of the following equations have? (*See Practice Exercise 3.*)

47. $(4x - 5)^2 = 0$

48. $(7x - 2)^2 = 1$

49. $(3x + 8)^2 = -1$

50. $-(3x - 4)^2 = -7$

51. $x^2(4x - 9)^2 = 0$

52. $x(x - 1)(2x - 5)^2 = 0$

9.4 The Quadratic Formula

A. Applying the Formula

There is a mechanical procedure for finding the roots of *any quadratic equation*

$$Ax^2 + Bx + C = 0 \qquad \text{where } A \neq 0$$

This method is particularly useful when the previous methods do not easily apply. First write the equation in the above form. The roots are then given by the **quadratic formula:**

$$x = \frac{-B \pm \sqrt{B^2 - 4AC}}{2A}$$

Example 1 ▶ Use the quadratic formula to find the roots of the equation

$$x^2 + 5x + 3 = 0$$

Check the roots.

Solution:

$$x^2 + 5x + 3 = 0$$

is of the form

$$Ax^2 + Bx + C = 0$$

where $A = 1$, $B = 5$, and $C = 3$. Apply the quadratic formula,

$$\begin{aligned} x &= \frac{-B \pm \sqrt{B^2 - 4AC}}{2A} \\ &= \frac{-5 \pm \sqrt{5^2 - 4 \cdot 1 \cdot 3}}{2 \cdot 1} \\ &= \frac{-5 \pm \sqrt{25 - 12}}{2} \\ &= \frac{-5 \pm \sqrt{13}}{2} \end{aligned}$$

Check for $\frac{-5 + \sqrt{13}}{2}$:

$$\left(\frac{-5 + \sqrt{13}}{2}\right)^2 + 5\left(\frac{-5 + \sqrt{13}}{2}\right) + 3 \stackrel{?}{=} 0$$

$$\frac{25 - 10\sqrt{13} + 13}{4} + \frac{-25 + 5\sqrt{13}}{2} + 3 \stackrel{?}{=} 0$$

$$\frac{25 - 10\sqrt{13} + 13 - 50 + 10\sqrt{13} + 12}{4} \stackrel{?}{=} 0$$

$$\frac{\overbrace{(25 + 13 + 12 - 50)}^{0} + \overbrace{(-10\sqrt{13} + 10\sqrt{13})}^{0}}{4} \stackrel{?}{=} 0$$

$$0 \stackrel{\checkmark}{=} 0$$

Check for $\frac{-5 - \sqrt{13}}{2}$:

$$\left(\frac{-5 - \sqrt{13}}{2}\right)^2 + 5\left(\frac{-5 - \sqrt{13}}{2}\right) + 3 \stackrel{?}{=} 0$$

$$\frac{25 + 10\sqrt{13} + 13}{4} + \frac{-25 - 5\sqrt{13}}{2} + 3 \stackrel{?}{=} 0$$

$$\frac{25 + 10\sqrt{13} + 13 - 50 - 10\sqrt{13} + 12}{4} \stackrel{?}{=} 0$$

$$\frac{\overbrace{(25 + 13 + 12 - 50)}^{0} + \overbrace{(10\sqrt{13} - 10\sqrt{13})}^{0}}{4} \stackrel{?}{=} 0$$

$$0 \stackrel{\checkmark}{=} 0$$

◀

Example 2 ▶ Solve the quadratic equation

$$x^2 - 7x + 10 = 0$$

(a) by factoring the left side, and
(b) by applying the quadratic formula.

Solution:

(a)

$$x^2 - 7x + 10 = 0$$

$$(x - 2)(x - 5) = 0$$

$$x - 2 = 0 \quad \Big| \quad x - 5 = 0$$

$$x = 2 \quad \Big| \quad x = 5$$

The roots are 2 and 5.

(b)
$$x^2 - 7x + 10 = 0$$

Here $A = 1$, $B = -7$, and $C = 10$. Apply the quadratic formula:

$$x = \frac{-B \pm \sqrt{B^2 - 4AC}}{2A}$$
$$= \frac{-(-7) \pm \sqrt{(-7)^2 - 4 \cdot 1 \cdot 10}}{2 \cdot 1}$$
$$= \frac{7 \pm \sqrt{49 - 40}}{2}$$
$$= \frac{7 \pm \sqrt{9}}{2}$$
$$= \frac{7 \pm 3}{2}$$

$$x = \frac{7 + 3}{2} \quad\Big|\quad x = \frac{7 - 3}{2}$$
$$x = 5 \quad\Big|\quad x = 2$$

◀

As you see in Example 2, if you can factor the left side of an equation

$$Ax^2 + Bx + C = 0$$

you obtain the roots more easily than by applying the quadratic formula.

PRACTICE EXERCISES FOR TOPIC A

Answers
1. (a)
(b)
2. (a)
(b)

1. Apply the quadratic formula to solve:

(a) $4x^2 + 5x - 6 = 0$ (*Check.*) (b) $x^2 - 4x + 1 = 0$

2. (a) Apply the quadratic formula to solve the equation

$$6x^2 - 13x + 6 = 0$$

(b) Solve this equation by some other method.

Now try the exercises for Topic A on page 428.

B. The Discriminant

In the quadratic formula,

$$x = \frac{-B \pm \sqrt{B^2 - 4AC}}{2A}$$

the expression within the radical sign

$$B^2 - 4AC$$

is known as the **discriminant.** For the equation

$$2x^2 - 6x + 1 = 0$$

the discriminant is given by

$$\begin{aligned} B^2 - 4AC &= (-6)^2 - 4 \cdot 2 \cdot 1 \\ &= 36 - 8 \\ &= 28 \end{aligned}$$

When you evaluate the discriminant, you can tell the number of (different) *real* roots of the given equation.

$B^2 - 4AC$ (*Discriminant*)	*Number of Roots*
positive	2 *real* roots
0	1 *real* root
negative	no *real* roots

For, consider the quadratic formula,

$$x = \frac{-B \pm \sqrt{B^2 - 4AC}}{2A}$$

If the discriminant, $B^2 - 4AC$, is positive, you add and subtract its square root in the numerator and obtain 2 different roots.

If the discriminant is 0, then because $\sqrt{0} = 0$, the only root is $\frac{-B}{2A}$.

If the discriminant is negative, the equation has no *real* square root. (The roots obtained by applying the formula are not real numbers.)

Example 3 ► Consider the equation

$$x^2 + 7x + 1 = 0$$

(a) Evaluate the discriminant.
(b) How many (real) roots are there?
(c) Apply the quadratic formula to solve the equation.

Solution:

(a) $A = 1$, $B = 7$, and $C = 1$. The discriminant is given by

$$B^2 - 4AC = 49 - 4 = 45$$

(b) There are 2 roots because the discriminant is positive.

(c)
$$\begin{aligned} x &= \frac{-B \pm \sqrt{B^2 - 4AC}}{2A} \\ &= \frac{-7 \pm \sqrt{45}}{2} \\ &= \frac{-7 \pm 3\sqrt{5}}{2} \end{aligned}$$ ◄

Example 4 ▶ Consider the equation

$$9y^2 + 12y + 4 = 0$$

(a) Evaluate the discriminant.
(b) How many (real) roots are there?
(c) Apply the quadratic formula to solve the equation.
Solution:
(a) $A = 9$, $B = 12$, and $C = 4$. The discriminant is given by

$$\begin{aligned} B^2 - 4AC &= 12^2 - 4 \cdot 9 \cdot 4 \\ &= 144 - 144 \\ &= 0 \end{aligned}$$

(b) There is 1 root because the discriminant is 0.

(c)
$$\begin{aligned} x &= \frac{-B \pm \sqrt{B^2 - 4AC}}{2A} \\ &= \frac{-12 \pm \sqrt{0}}{2 \cdot 9} \\ &= \frac{-2}{3} \end{aligned}$$ ◀

Example 5 ▶ Consider the equation

$$2x^2 - 3x + 2 = 0$$

(a) Evaluate the discriminant.
(b) How many (real) roots are there?
Solution:
(a) $A = 2$, $B = -3$, and $C = 2$

$$\begin{aligned} B^2 - 4AC &= (-3)^2 - 4 \cdot 2 \cdot 2 \\ &= 9 - 16 \\ &= -7 \end{aligned}$$

(b) There are no (real) roots because the discriminant is negative. ◀

In later courses you will learn to apply the quadratic formula in the case where the discriminant is negative. You will also verify that the formula is correct in all cases.

PRACTICE EXERCISES FOR TOPIC B

Answers

3. (a) ____________

(b) ____________

(c) ____________

3. Consider the equation $4x^2 - 12x + 9 = 0$.

(a) Evaluate the discriminant.

(b) How many (different) *real* roots are there?

(c) If there are real roots, use the quadratic formula to solve the equation.

Now try the exercises for Topic B on page 428.

Use this area for work space.

EXERCISES

A. In Exercises 1–12 apply the quadratic formula to solve the equation. (*See Example 1.*) In Exercises 1–4 *check* the roots.

1. $x^2 + 5x - 2 = 0$ (*Check.*)

2. $y^2 + 9y + 1 = 0$ (*Check.*)

3. $3t^2 - 5t + 1 = 0$ (*Check.*)

4. $u^2 + 12u + 36 = 0$ (*Check.*)

5. $x^2 + 10x + 5 = 0$

6. $2z^2 - 9z + 6 = 0$

7. $u^2 + 3u - 1 = 0$

8. $u^2 + 3u + 1 = 0$

9. $2t^2 + 5t + 3 = 0$

10. $3x^2 - 7x + 4 = 0$

11. $3y^2 = 10y - 3$

12. $(y - 2)^2 = y + 3$

B. In Exercises 13–22: (a) Apply the quadratic formula to solve the equation. (b) Solve the equation by some other method. (*See Example 2.*)

13. $x^2 - 3x + 2 = 0$

14. $x^2 + 8x + 16 = 0$

15. $y^2 + y - 20 = 0$

16. $z^2 - 81 = 0$

17. $(t + 1)^2 = 5$

18. $2x^2 + 3x + 1 = 0$

19. $6y^2 - y = 1$

20. $10u^2 + 9u + 2 = 0$

21. $14t^2 - 11t + 2 = 0$

22. $(y + 2)^2 = 3$

In Exercises 23–38: (a) Evaluate the discriminant. (b) How many (different) *real* roots are there? (c) If there are real roots, use the quadratic formula to solve the equation. (*See Examples 3–5.*)

23. $x^2 + 4x + 2 = 0$

24. $x^2 + 6x + 9 = 0$

25. $2x^2 + 3 = 5x$

26. $5x^2 + 10x + 8 = 0$

27. $t^2 - 3t = 1$

28. $t^2 - 8t + 15 = 0$

29. $2y^2 - 6y + 3 = 0$

30. $z^2 + 20 = 9z$

31. $4x^2 - 4x + 1 = 0$

32. $2u^2 + 7u + 9 = 0$

33. $4z^2 - 20z + 5 = 0$

34. $u^2 + u + 1 = 0$

35. $4y^2 + 4y - 4 = 0$

36. $3x^2 - 6x + 3 = 0$

37. $3z^2 + z + 3 = 0$

38. $y^2 - 5y + 1 = 0$

39. Pick out the *error(s)*.

(a) If 1 is the only root of a quadratic equation, then its discriminant must be 0.

(b) If 2 is a root of a quadratic equation, then its discriminant must be positive.

(c) There is a quadratic equation whose roots are 1, 2, and 3.

(d) Every quadratic equation has *real* roots.

9.5 Applications

Stated problems can often be translated into quadratic equations. Your first task is to translate the problem into algebraic symbols. You then apply the techniques of this chapter to solve the corresponding quadratic equation.

A. Number Problems

Example 1 ▶ The sum of two numbers is 16 and the product is 48. Find these numbers.

Solution:

Let x be one of these numbers. Then the other is 16 − x because

the sum (of the two numbers) is 16.

$$x + (16 - x) = 16$$

The quadratic equation is obtained from the second piece of information:

The product (of the two numbers) is 48.

$$x(16 - x) = 48.$$

Multiply, as indicated, to obtain the quadratic equation

$$16x - x^2 = 48$$

or

$$x^2 - 16x + 48 = 0$$ *Factor the left side.*

$$(x - 4)(x - 12) = 0$$

$x - 4 = 0$	$x - 12 = 0$
$x = 4$	$x = 12$
$16 - x = 12$	$16 - x = 4$

In either case, the two numbers are 4 and 12.

PRACTICE EXERCISES FOR TOPIC A

Answers

1. ______

1. *Consecutive integers* are one unit apart. The sum of the squares of three consecutive positive integers is 110. Find these integers.

Now try the exercises for Topic A on page 433.

B. Geometry Problems

Example 2 ▶ The perimeter of a rectangle is 36 inches. Its area is 80 square inches. Find the dimensions.

Solution:

Let l and w be the dimensions of the rectangle. Then

the perimeter is 36 inches

$$2l + 2w = 36$$

the area is 80 square inches

$$lw = 80$$

(*See Figure 9.1.*)

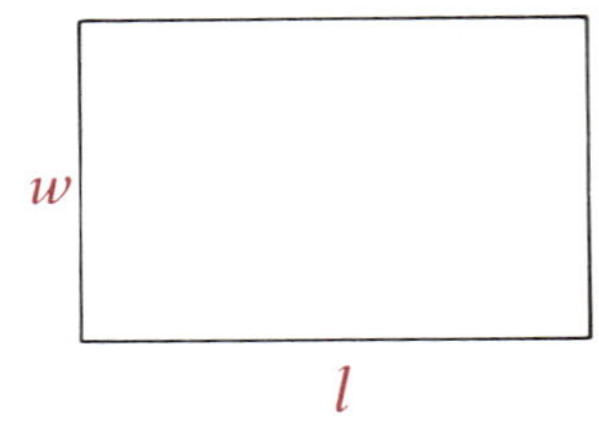

FIGURE 9.1

From the first equation,

$$l + w = 18$$
$$w = 18 - l$$

The second equation,

$$lw = 80$$

becomes

$$l(18 - l) = 80$$
$$18l - l^2 = 80$$

or

$$l^2 - 18l + 80 = 0 \quad \textit{Factor the left side.}$$
$$(l - 10)(l - 8) = 0$$

$l - 10 = 0$	$l - 8 = 0$
$l = 10$	$l = 8$
$w = 18 - l$	$w = 18 - l$
$w = 8$	$w = 10$

The rectangle is 8 inches by 10 inches. ◀

Answers

2. ______

PRACTICE EXERCISES FOR TOPIC B

2. The length of a rectangular rug is one foot longer than twice the width. The area of the rug is 36 square feet. Find the dimensions of the rug.

Now try the exercises for Topic B on page 433.

C. Motion Problems

Example 3 ▶ An object falls from a tower 320 feet high. (*See Figure 9.2.*) As it is falling, its height is given by

$$h = 320 - 16t^2 \text{ (feet)}$$

where t, the time falling, is measured in seconds. How long does it take for the object to hit the ground?

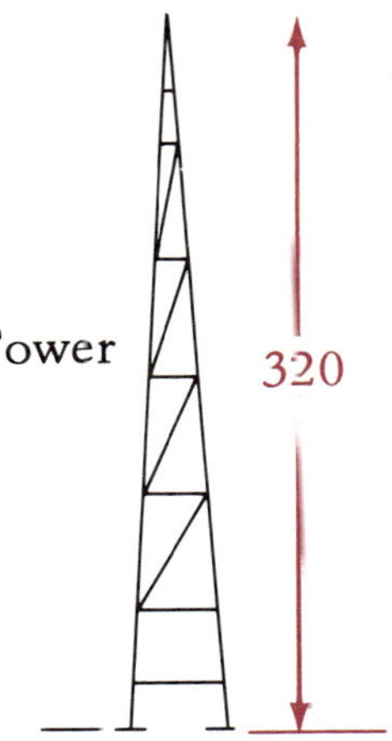

FIGURE 9.2

Solution:

$$h = 320 - 16t^2$$

Because h represents height, the object hits the ground when $h = 0$. Thus replace h by 0 and solve for t.

$$\begin{aligned} 0 &= 320 - 16t^2 \\ 16t^2 &= 320 \\ t^2 &= 20 \\ t &= \pm\sqrt{20} \\ &= \pm\sqrt{4 \cdot 5} \\ &= \pm\sqrt{4}\sqrt{5} \\ &= \pm 2\sqrt{5} \end{aligned}$$

Because of the physical nature of the problem, t *must be positive.*

$$\sqrt{5} \approx 2.24$$

$$2\sqrt{5} \approx 4.48$$

It takes slightly less than $4\frac{1}{2}$ seconds for the object to hit the ground. ◀

PRACTICE EXERCISES FOR TOPIC C

Answers
3. (a)
(b)

3. When an object is dropped from a roof 400 feet above the ground, its height, as it is falling, is given by the formula

$$h = 400 - 16t^2 \text{ (feet)}$$

where t is the time falling (in seconds).

(a) How high will an object be 2 seconds after it is dropped?

(b) How long will it take to reach the ground?

Now try the exercises for Topic C on page 433.

D. Profit Problems

The profit derived from manufacturing certain items can be described by means of a quadratic equation. Because of limited plant facilities, the profit will rise to a maximum and then taper off. This profit is often described by a curve called a **parabola,** as in Figure 9.3.

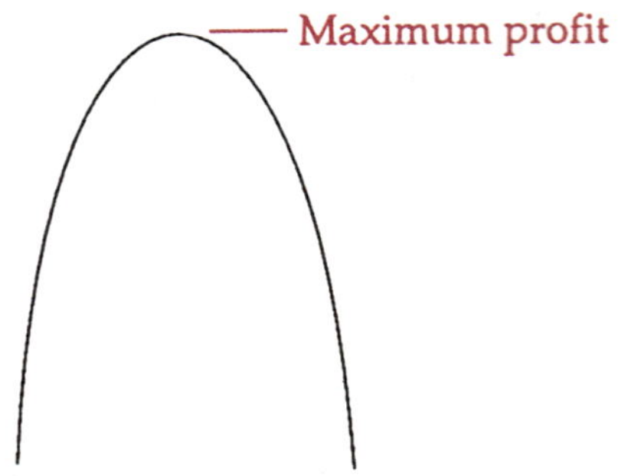

FIGURE 9.3

Example 4 ▶ The annual profit P (in thousands of dollars) derived from manufacturing x industrial machines is given by

$$P = 100 - x^2 + 20x$$

How many machines must be made in order to derive a profit of \$200 000?
Solution:
Note that P is measured in thousands of dollars. Thus let $P = 200$ in the equation.

$$P = 100 - x^2 + 20x$$

$$200 = 100 - x^2 + 20x$$

$$x^2 - 20x + 100 = 0$$

$$(x - 10)^2 = 0$$

$$x - 10 = 0$$

$$x = 10$$

Thus 10 machines must be produced. ◀

Answers

4. ________

PRACTICE EXERCISES FOR TOPIC D

4. Use the formula given in Example 4 to find out how many machines must be made to derive a profit of \$196 000.

Now try the exercises for Topic D on page 434.

Use this area for work space.

EXERCISES

In Exercises 1–14 set up an equation and then solve it.

A.

1. The sum of two numbers is 9 and the product is 20. Find these numbers. (*See Example 1.*)
2. The sum of two numbers is 20 and the product is 96. Find these numbers.
3. The sum of two numbers is 3 and the product is -70. Find these numbers.
4. The sum of two numbers is -14 and the product is 48. Find these numbers.
5. The sum of two numbers is 0 and the product is -36. Find these numbers.
6. The product of two *negative* integers is 72. The larger integer is twice the smaller one. Find these integers.
7. Four times a certain *positive* number is five less than its square. Find this number.
8. The square of a certain *negative* number is three more than twice the number. Find this number.
9. The sum of the squares of two consecutive positive integers is 145. Find these integers.
10. The sum of the squares of three consecutive negative integers is 77. Find these integers.

B.

11. The perimeter of a rectangle is 64 feet and the area is 240 square feet. Find the dimensions of the rectangle. (*See Example 2.*)
12. The perimeter of a rectangle is 70 inches and the area is 300 square inches. Find the dimensions of the rectangle.
13. The area of a garden is 1200 square feet. If the length of the garden were increased by 5 feet and the width decreased by 5 feet, the area would be decreased by 75 square feet. Find the dimensions of the garden.
14. The area of a Chinese rug is 360 square feet. A Moroccan rug, which is 4 feet longer but 3 feet narrower, has the same area. Find the dimensions of the Chinese rug.

C.

15. An object falls from a window 576 feet high. As it is falling, its height is given by

$$h = 576 - 16t^2 \text{ (feet)}$$

where t is the time falling (in seconds). How long does it take to hit the ground? (*See Example 3.*)

16. A flower pot falls from a window 192 feet high. As it is falling, its height is given by

$$h = 192 - 16t^2 \text{ (feet)}$$

where t is the time falling (in seconds). How long does it take to hit the ground?

17. A baseball is thrown straight up from ground level. Its height, h feet, after t seconds is given by

$$h = 128t - 16t^2$$

(a) How long does it take to rise 112 feet?
(b) At what time, t, is it 112 feet above the ground on its way down?

Use this area for work space.

D.

18. The monthly profit P (in *thousands* of dollars) derived from manufacturing prefabricated houses is given by

$$P = -x^2 + 16x - 14$$

where x is the number of houses constructed. How many houses must be constructed in order to make a monthly profit of $50 000? (*See Example 4.*)

19. In Exercise 18 how many houses must be constructed in order to make a monthly profit of $46 000?

20. In Exercise 16, because of limited plant facilities, it is unprofitable to build too many houses per month. *To the nearest integer,* what is the largest number of houses the firm can construct per month before it loses money? (*Hint:* Replace P by 0 in the equation of Exercise 16 and apply the quadratic formula.)

21. The cost C (in dollars) of building a bookcase is given by

$$C = \frac{l^2}{10} - 3l$$

where l is the total length of the shelves (in feet). Suppose there are to be 5 equally long shelves. How long will each shelf be if the cost is to be $100?

What Have You Learned in Chapter 9?

You have learned that a quadratic, or second-degree, equation (in the variable x) is an equation that can be written in the form $Ax^2 + Bx + C = 0$, where $A \neq 0$. There are, however, equations with second-degree terms, all of which drop out upon simplification. Such equations reduce to first-degree equations.

You have learned various methods of solving quadratic equations.

You know that when the discriminant,

$$B^2 - 4AC$$

of a quadratic equation

$$Ax^2 + Bx + C = 0 \qquad \text{where } A \neq 0$$

is positive, there are 2 (different) *real* roots; when the discriminant is 0, there is 1 *real* root, and when the discriminant is negative, there are no *real* roots. In the first two cases, you can apply the quadratic formula to obtain the roots:

$$x = \frac{-B \pm \sqrt{B^2 - 4AC}}{2A}$$

And you can apply these methods of solving quadratic equations to solve stated problems.

Let's Review Chapter 9

9.1 *Reducible Equations*

In Exercises 1–3 write each equation in the form $Ax^2 + Bx + C = 0$, where $A \neq 0$, or else indicate that the equation is reducible. (Do *not* solve.)

1. $4x^2 - 3x = 2x^2 + x - 3$

2. $3x^2 + 1 = (2x - 1)(x + 2)$

3. $(2x - 1)^2 = 4(x + 5)(x - 5)$

In Exercises 4–6 solve each equation.

4. $3x^2 + 8x - 2 = 2x^2 - (4 - x^2)$ (*Check.*)

5. $(x + 5)(x - 1) = (x + 6)^2 - 1$

6. $\dfrac{y + 4}{y} = \dfrac{y - 1}{y + 1}, \quad y \neq 0, -1$

9.2 *Factorable Equations*

In Exercises 7–11 solve each equation.

7. $x^2 - 7x + 12 = 0$

8. $3x^2 + 5x - 2 = 0$ (*Check.*)

9. $25t^2 - 30t + 9 = 0$

REVIEW

10. $\frac{1}{x} + \frac{2}{x+2} = 1, \quad x \neq 0, -2$

11. $y^3 - 8y^2 + 16y = 0$

12. Find a quadratic equation with integral coefficients whose roots are 2 and $\frac{1}{4}$.

9.3 *Equations of the Form $x^2 = a$*

Solve the equations in Exercises 13–16. Use the method:

If $x^2 = a$, where $a > 0$, then $x = \pm\sqrt{a}$

13. $x^2 - 17 = 0$

14. $x^2 - 80 = 0$

15. $(2u + 1)^2 = 9$ (*Check.*)

16. $(3y - 5)^2 = 2$

9.4 *The Quadratic Formula*

17. (a) Evaluate the discriminant of the equation

$$x^2 + 3x + 5 = 0$$

(b) Without solving, indicate how many (different) *real* roots there are.

In Exercises 18–20 apply the quadratic formula to solve the equation.

18. $x^2 + 5x + 5 = 0$ (*Check.*)

19. $4x^2 + 4x + 1 = 0$

20. $y^2 + 6y - 1 = 0$

9.5 *Applications*

21. Find two integers whose sum is 4 and whose product is -45.

22. The perimeter of a rectangle is 30 inches. The area is 36 square inches. Find the dimensions.

And These from Chapters 1–8

23. Factor $5x^2 - 4x - 1$

24. Factor $10x^2 + 9x + 2$

25. Solve $5(2t - 1) = 4(3t - 2)$

26. Solve $\frac{4}{x} = \frac{5}{x-2}$

27. Simplify $\sqrt{25a^2x^4} + \sqrt{81a^2x^4}, \quad a > 0$

28. (a) Factor $6 - \sqrt{108}$

(b) Simplify $\frac{6 - \sqrt{108}}{12}$

29. Simplify $\dfrac{\dfrac{x - y}{x^2 - 16}}{\dfrac{3x - 3y}{x^2 + 8x + 16}}$

30. How much money must be deposited at an annual interest rate of 8% in order to earn \$120 interest in one year?

31. Solve $-1 < 1 - 4t < 9$

32. Find the equation of the vertical line that passes through $(-2, 5)$.

33. Find the equation of the line with x-intercept 4 and y-intercept -4. Use intercept form.

34. Solve by either substituting or adding.

$$\begin{aligned} 4x - 3y &= 7 \\ 10x + 7y &= 3 \end{aligned}$$

35. Assume y varies directly as x. Suppose $y = -30$ when $x = 12$. Find y when $x = -18$.

36. (a) Express .004 32 in scientific notation.

(b) Convert 6.36×10^8 back into the usual decimal notation.

37. Multiply and simplify $(2 - \sqrt{3})^2$

38. (a) Find the *lcm* of $(x - 1)^2$, $4x - 4$, and $x^2 - 1$. (Leave your answer in factored form.)

(b) Find the *gcd* of these same polynomials.

39. A radio marked to sell for \$80 is reduced by 40%. What is the sale price?

40. Solve for y: $2x - 3y + xy = 10$

REVIEW

Practice Exam 1 on Chapter 9

1. Solve $(2x + 1)^2 = 4(x - 1)^2$

2. Solve by factoring: $x^2 - 5x - 14 = 0$

3. Solve $(2y - 1)^2 = 5$ by using the method:

If $x^2 = a$, where $a > 0$, then $x = \pm\sqrt{a}$

4. Solve by applying the quadratic formula: $x^2 + 9x + 3 = 0$

In Exercises 5–7, solve by any method.

5. $2y^2 - 3y - 5 = 0$ (*Check.*)

6. $(2u - 1)^2 - 19 = 0$

7. $x^2 + 2x - 2 = 0$

8. Set up an equation and then solve it:
The sum of two numbers is 17 and the product is 60. Find these numbers.

Practice Exam 2 on Chapter 9

1. (a) Solve $3(x + 1)^2 = (3x + 4)(x + 1)$

(b) Is this a first-degree or second-degree equation?

2. Solve by factoring: $x^2 - 11x - 26 = 0$ (*Check.*)

REVIEW

3. Solve $\frac{x}{3} = 2 - \frac{2}{x + 1}$

4. Solve $(5x - 4)^2 = 3$. Do *not* use the quadratic formula.

5. (a) Determine the discriminant of the equation

$$3x^2 + 4x + 1 = 0$$

(b) How many (different) *real* roots does this equation have?

(c) Use the quadratic formula to solve for x.

In Questions 6 and 7 solve by any appropriate method.

6. $10x^2 + 2 = 12x$

7. $(x + 2)^2 - 9 = 0$

8. Set up an equation and then solve it to find the dimensions of a rectangle whose perimeter is 15 feet and whose area is 14 square feet.

Answers

DIAGNOSTIC TEST

Chapter 1, p. xvii

1. $A = -6$, $B = -4$, $C = -2$, $D = 3$, $E = 8$

2.

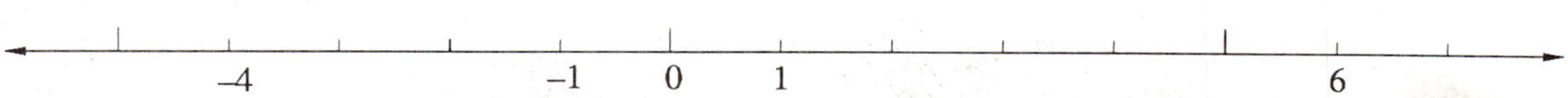

FIGURE D1A

3. (a) $-5 < -1$ (b) $7 > -10$

4. $-7 < -3 < 0 < 1 < 5$

5. (a) -12 (b) 7 (c) 0

6. (a) 16 (b) 2

7. -17

8. -7

9. -2

10. (a) -24 (b) 70

11. (a) -5 (b) 2 (c) 0 (d) undefined

12. (a) 64 (b) 36 (c) -1

13. (a) -5 (b) -35

14. 6

Chapter 2, p. xix

1. $2x + 5$
2. $4x^3y^2$
3. -8
4. (a) and (c)
5. 2
6. 24
7. (a) $5x$ (b) $5a + 2b$ (c) $6x^2 - 8x + 3$
8. (a) *iii* (b) *ii* (c) *i* (d) *iv*
9. $7a - 9b - 7$
10. $x - y$
11. $-2a + 3$
12. (a) $-56xy$ (b) $-18a + 6b - 12$
13. $6x + 4y - 1$
14. $3y$

Chapter 3, p. xx

1. (a) 2 is not a root. (b) -1 is a root.
2. $3x = x + 10$
3. (a) -3 (b) 5 (c) 72
4. (a) 6 (b) -4 (c) 28
5. (a) -3 (b) *Check:* $11 + 7(-3) \stackrel{?}{=} (-3) - 7$
   $11 + (-21) \stackrel{?}{=} -10$
   $-10 \stackrel{\checkmark}{=} -10$
6. (a) -1 (b) -5 (c) 2
7. $x + (2x + 1) = 22$; 7 feet, 15 feet
8. $150t$
9. 36 miles

Chapter 4, p. xxi

1. (a) $40x^6y^6$ (b) $3x^3y^3 - 6x^2y^2 + 3x^2y$
2. (a) $7x^3y$ (b) $4x^3 - 3x^2 + 1$
3. $4a^2b + 4ab + 2a^2$
4. (a) $3 \cdot 5 \cdot 11$ (b) $2^2 \cdot 3 \cdot 7$
5. 6
6. (a) $2(5x - 4y + 3)$ (b) $7xy(x + 3y - 2x^2y^2)$ (c) $a^2(a^2 + 3a - 6)$
7. (a) $3x^2 - 5x - 8$ (b) $4x^2 - 25$ (c) $x^3 - x^2 + x - 6$
8. (a) $(x + 10)(x - 10)$ (b) $3(a + 3b)(a - 3b)$ (c) $a(4 + x^2)(2 + x)(2 - x)$
9. (a) $(x + 5)(x - 6)$ (b) $(n + 8)(n - 4)$ (c) $(x + 9y)(x + y)$
10. (a) $0, -10$ (b) $8, -2$ (c) $5, -5$
11. (a) $(4x - 3)(x + 1)$ (b) $3(2x + 1)(x - 2)$ (c) $(3x - 2y)(2x + 5y)$

Chapter 5, p. xxii

1. Does $7x(2x^2 - 4x) = 14x^2(x - 2)$? Yes. $14x^3 - 28x^2 = 14x^3 - 28x^2$
2. (a) $\frac{2}{3}$ (b) $\frac{-9}{4}$ or $-\frac{9}{4}$ (c) $\frac{-1}{9}$ or $-\frac{1}{9}$
3. $\frac{1}{6}$
4. (a) $\frac{x^5}{y}$ (b) $\frac{5a}{8}$ (c) $\frac{2x}{3(x+3)}$
5. (a) $\frac{x+3}{3x}$ (b) $\frac{2x}{5x-4}$ (c) $\frac{x^2}{x-2}$
6. (a) $4x^3 + 3x^2 - x - 10$ (b) degree 3
7. (a) $x^2 + 2x$ (b) 1

8. (a) $\frac{-1}{24}$ (b) $\frac{4}{9(x+2)^2}$

9. (a) $\frac{12}{5}$ (b) $\frac{3a}{2(a-1)}$ or $\frac{3a}{2a-2}$

10. (a) $\frac{11}{12}$ (b) $\frac{4}{x}$ (c) 3

11. (a) 48 (b) $13x^2y$ (c) $x^2(x+1)(x-4)$

12. (a) $\frac{3}{4}$ (b) $\frac{5x+2}{2x^2}$ (c) $\frac{a^2+3a+11}{a^2+2a}$

13. (a) $\frac{5}{2}$ (b) $\frac{2}{x}$ (c) $\frac{4}{x^2+4}$

Chapter 6, p. xxiv

1. (a) −12 (b) 9

2. 250 pounds

3. (a) 12 (b) −8 (c) 4

4. (a) .65 (b) $\frac{11}{25}$

5. (a) 9.8 (b) .036

6. $150

7. $1200

8. $81.60

9. 15%

10. $35

11. $120

12. $y = x - 15$

13. $a = \frac{bc-b}{c+3}$

14. (a) true (b) false

15. (a) [0, 12] (b) (6, ∞) (c) (−9, 1)

16. (a) > (b) > (c) >

17. (a) (−∞, 3) (b) [−3, 2]

Chapter 7, p. xxvi

1.

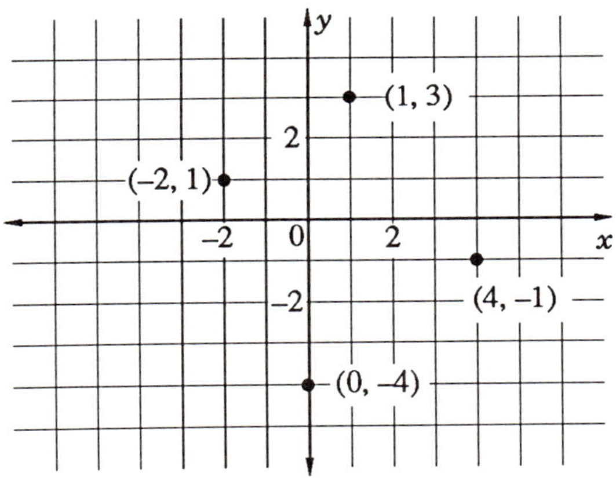

FIGURE D7A

2.
- *A*: (a) 2 (b) 2 (c) (2, 2)
- *B*: (a) 3 (b) −2 (c) (3, −2)
- *C*: (a) −1 (b) −3 (c) (−1, −3)
- *D*: (a) −2 (b) 0 (c) (−2, 0)

3. (a)

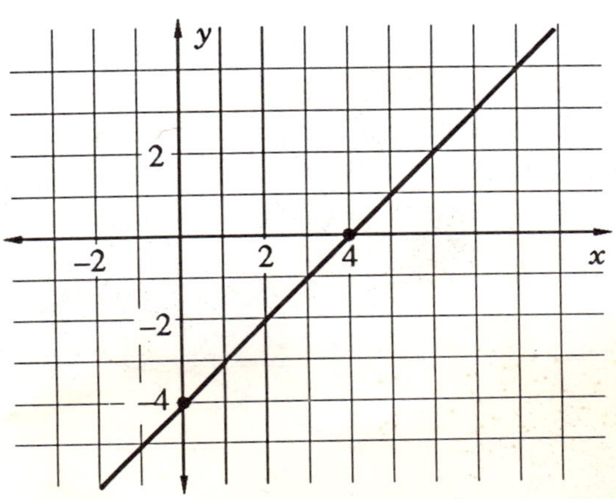

FIGURE D7B

(b)

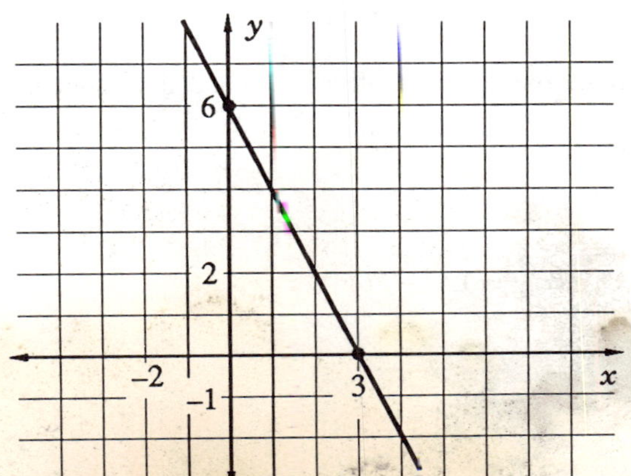

FIGURE D7C

(c)

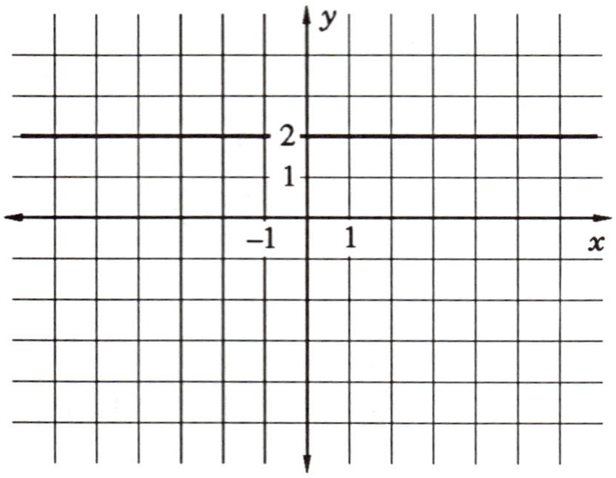

FIGURE D7D

4. (a) $\frac{-1}{2}$ (b) 0 (c) undefined

5.

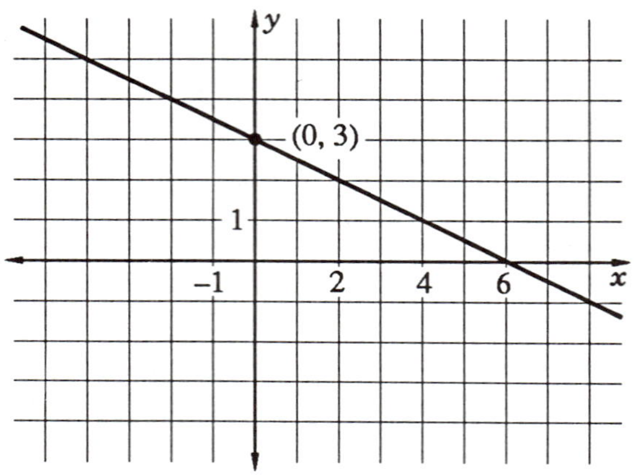

FIGURE D7E

6. (a) $\frac{1}{3}$ (b) -1

7. $y = -x + 8$

8. (a) $\frac{x}{3} + \frac{y}{-5} = 1$ or $\frac{x}{3} - \frac{y}{5} = 1$

(b) $-5x + 3y = -15$ or $5x - 3y = 15$

9.

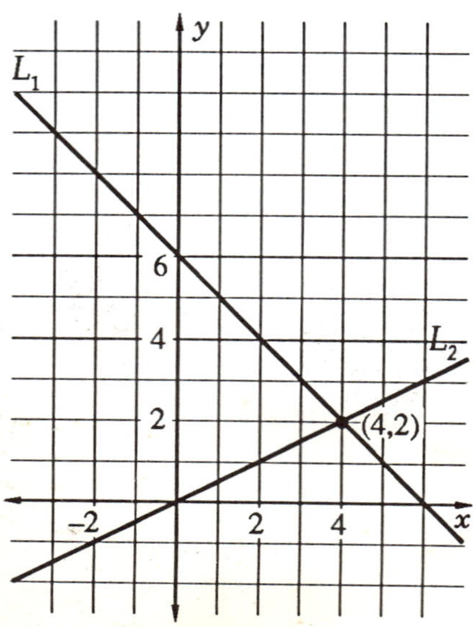

FIGURE D7F The point of intersection is (4, 2).

10. L_2 and L_4 are parallel because both have slope $\frac{1}{3}$.

11. (1, 6)

12. (7, 5)

13. (a) $\frac{8}{5}$, or 1.6 (b) 40

Chapter 8, p. xxviii

1. (a) 9 (b) -20 (c) x^4 (d) 3^8

2. $10 < \sqrt{116} < 11$

3. .5

4. $10\sqrt{3}$ inches

5. (a) $6x^8y^2$ (b) $\frac{4}{11}$ (c) $\frac{x^3}{3}$

6. (a) $2\sqrt{10}$ (b) $3x\sqrt{x}$ (c) $\frac{\sqrt{6}}{3}$

7. (a) $5\sqrt{5}$ (b) $\sqrt{15}$ (c) $5x\sqrt{2}$ (d) $\frac{3\sqrt{3}-2\sqrt{2}}{6}$

8. (a) 12 (b) $6-2\sqrt{3}$ (c) $24-7\sqrt{10}$

9. 5

10. (a) $3(4-\sqrt{2})$ (b) $\frac{4-\sqrt{2}}{2}$

11. (a) $\frac{1}{125}$ (b) 1 (c) $\frac{9}{16}$

12. (a) x^{-2} (b) x^{14}

13. (a) 6.31×10^{-4} (b) 7.25×10^8

14. 15 000

Chapter 9, p. xxix

1. (a) $4x^2-5x+2=0$ (b) $x^2-6x-16=0$

2. 1

3. (a) $0, -7$ (b) $5, \frac{-1}{2}$ (c) $9, -2$

4. $x^2-3x-10=0$

5. (a) $\pm 4\sqrt{2}$ (b) $1, -4$

6. (a) 16 (b) 2 real roots

7. (a) $\frac{3 \pm \sqrt{5}}{2}$ (b) $\frac{-2 \pm \sqrt{10}}{2}$

8. 5 and 12

9. 2 feet by 9 feet

EXERCISES

Answers are provided for the odd-numbered exercises of each section, and for all the exercises in the End-of-Chapter material.

CHAPTER 1

Section 1.1, p. 5

1.-7.

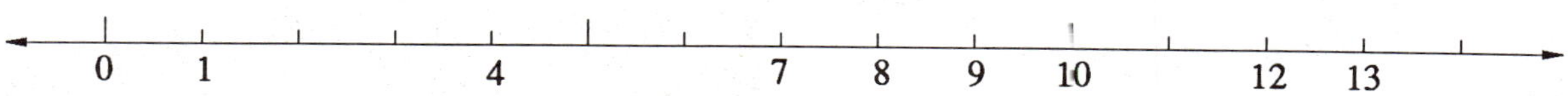

FIGURE 1A

9. not negative (positive)

11. not negative (positive)

13. not negative (0)

15. negative

17. -5

19. 0

21. -15

23.-30.

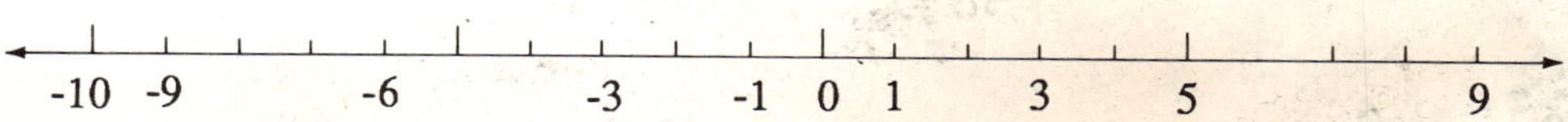

FIGURE 1B

31. $Q: 4$, $R: 7$, $S: -2$, $T: -5$, $U: -6$, $V: -7$
33. 19
35. 37
37. (c), (d), and (e) are the errors.

Section 1.2, p. 11

1. $5 < 7$
3. 8 is greater than 5.
5. <
7. >
9. >
11. right
13. right
15. $5 < 12 < 13 < 18$
17. $8 < 12 < 14 < 18 < 20$
19. $-8 < -6$
21. right
23. right
25. right
27. right
29. <
31. <
33. >
35. >
37. <
39. <
41. <
43. <
45. $-3 < -2 < -1$
47. $-6 < -3 < 3 < 6$
49. $-7 < -5 < -3 < 1 < 7$
51. $-18 < -15 < -10 < -4 < -3 < 15 < 18$
53. Marilyn
55. fallen
57. Joe

Section 1.3, p. 16

1. −2
3. −12
5. 2
7. 12
9. 0
11. −3
13. 0
15. 8
17. 17
19. 8
21. 17
23. 0
25. 100
27. −7
29. 7
31. 0
33. 12
35. 8 and −8
37. 11
39. −44
41. −1, 0, and 1
43. >
45. >
47. <
49. <
51. (b) is the error.

Section 1.4, p. 23

1. 17
3. 6
5. 0
7. −10
9. 3
11. 2
13. 20
15. −3
17. −5
19. −4
21. $3 + 3 = 6$
23. $(-3) + (-7) = -10$
25. $2 - (-6) = 8$
27. −41
29. 13 (Commutative Law of addition)
31. −2 (Commutative Law of addition)
33. −21 (Associative Law of addition)
35. −20
37. 7
39. −110
41. 15
43. −93
45. −1163
47. $164
49. 10 pages
51. -15° Celsius
53. 6°
55. $17

Section 1.5, p. 33

1. 20
3. 144
5. 0
7. −25
9. −30
11. 21 (Commutative Law of multiplication)

13. 40 (Associative Law of multiplication) **15.** $3 \cdot 3 = 9$ **17.** -6

19. 1 **21.** 0 **23.** 1 000 000 **25.** 2

27. 12 **29.** 6 **31.** $10 \div 2 = 5$ **33.** -8

35. 8 **37.** -1 **39.** 5 **41.** 0

43. undefined **45.** undefined **47.** 60 inches **49.** 800 centimeters

51. 480 minutes **53.** 16 000 **55.** 8 cents **57.** 7

59. \$12

Section 1.6, p. 39

1. 100 **3.** 9 **5.** 64 **7.** 9

9. 100 **11.** 144 **13.** 625 **15.** 1000

17. 8 **19.** 125 **21.** -1 **23.** -125

25. -8000 **27.** 81 **29.** 729 **31.** 64

33. 1 000 000 000 **35.** false **37.** \$32 **39.** 5

41. the second power **43.** 4 **45.** 36 **47.** 10 000

49. the second power **51.** the second power **53.** 64 **55.** 0

57. 1 **59.** 1 **61.** -128 **63.** 10 000 000

65. -243 **67.** -125 **69.** (b), (c), (d), and (e) are the errors.

Section 1.7, p. 45

1. 4 **3.** -5 **5.** -19 **7.** -17

9. (a) 3 (b) -1 **11.** 23 **13.** 16 **15.** 3

17. -16 **19.** 0 **21.** 4 **23.** 48

25. 18 **27.** 14 **29.** 1 **31.** -27

33. -16 **35.** 35 **37.** 0 **39.** 16

41. 1 **43.** 9 **45.** 29 **47.** 7

49. 20 **51.** 4 **53.** Both sides equal 96.

55. $4 \cdot 6 + 4 \cdot 1 = 24 + 4 = 28$ **57.** $8 \cdot 5 + 8 \cdot 3 = 40 + 24 = 64$

59. $9 \cdot 9 - 9 \cdot 6 = 81 - 54 = 27$ **61.** $17 \cdot 3 + 17 \cdot 7 = 17(3 + 7) = 17(10) = 170$

63. \$1.93 **65.** \$7000 **67.** 3 yards **69.** 10 weeks

Let's Review Chapter 1, p. 47

1. $Q = 3$, $R = 8$, $S = 10$, $T = -1$, $U = -6$ $V = -10$

2. (a) -12 (b) 12 **3.** $>$ **4.** left

5. $-10 < -3 < -1 < 0 < 1 < 4$
6. (a) -7 (b) 3 (c) 0
7. (a) 5 (b) 2 (c) 0
8. 10 and -10
9. -16
10. -7
11. -13
12. -15 (Associative Law of addition)
13. -6° Celsius
14. 12
15. (a) 3 (b) -1
16. (a) undefined (b) 0
17. 1440 minutes
18. (a) 49 (b) -27 (c) 16
19. the third power
20. 35
21. 29
22. 8
23. $92 \cdot 7 + 8 \cdot 7 = (92 + 8)7 = 100 \cdot 7 = 700$

CHAPTER 2

Section 2.1, p. 56

1. $x+5$
3. $x-5$
5. $x+3$
7. $x+12$
9. $2x$
11. $x-y$
13. $\frac{y}{x}$
15. $xy+1$
17. term
19. not a term
21. not a term
23. $t \cdot t$
25. $5 \cdot x \cdot y \cdot y$
27. $-x \cdot y$ [or $(-1) \cdot x \cdot y$]
29. $8 \cdot x \cdot x \cdot y$
31. u^2
33. x^2y^2
35. a^2bc^3
37. $-6xy^2z$
39. 7
41. 5
43. -3
45. 6
47. 20
49. 14
51. 0
53. polynomial
55. polynomial
57. not a polynomial
59. x and 4
61. x, $3y$, and $-2z$
63. $t^8, 2t^5, -3t^3, t^2, -t$, and 4
65. 10

Section 2.2, p. 61

1. 12
3. 7
5. 9
7. 36
9. 14
11. 13
13. 1
15. 3
17. -12
19. 10
21. 25
23. 32
25. 10
27. 8
29. 1 009 702
31. (a) 12 (b) 8 (c) 0
33. (a) 9 (b) 19 (c) 59
35. (a) 0 (b) 24 (c) 8
37. (a) 21 (b) 45 (c) 55
39. (a) 1 (b) 13 (c) 73
41. (a) 121 square inches (b) 225 square inches (c) 900 square inches
43. 880 square feet
45. 300 meters
47. (a) 376 square feet (b) 1120 square feet
49. (a) 300 miles (b) 900 miles (c) 3000 miles
51. (a) 80 feet (b) 128 feet (c) 0 feet, or ground level

Section 2.3, p. 68

1. like **3.** like **5.** unlike **7.** unlike

9. like **11.** $5 \mid 5x, 3x \mid 5y, 3y$ **13.** $x^2y, 3x^2y \mid xy^2, -xy^2, -3xy^2 \mid 3xy, 2xy \mid x^2y^2$

15. $3x$ **17.** x **19.** $4xy$ **21.** $5abc$

23. $6xy$ **25.** $5x$ **27.** 0 **29.** $8w$

31. $6xyz$ **33.** $14x$ **35.** $5a+3b+6$

37. $12-(3c-4d)$ or $12-3c+4d$ **39.** $6u-3v$ **41.** $5x$

43. $6a+12b$ **45.** $7ab-4cd$ **47.** $6a-3c$ **49.** $-2m+3n$

51. $13xyz-8abc$ **53.** $3y^2-2z^2$ **55.** 14 **57.** $4n-10p$

59. 5a **61.** $4x+8y+3z$ **63.** (a) $7x$ (b) $38x$ **65.** (b) and (e)

Section 2.4, p. 72

1. (i) **3.** (i) **5.** (iv) **7.** (iii)

9. (i) **11.** (i) **13.** $3x+5y$ **15.** $a+3$

17. $3a+2b+3c$ **19.** $5x$ **21.** $7a-2c$ **23.** $5r+2s+2u$

25. $22a+5b+5c+11d$ **27.** $7a-b+2c+6d+2e$ **29.** $11m+14n$ **31.** $10x+5y-z$

33. $3l+2w+3h+11$ **35.** $-x-2y-5z$ **37.** $-x^2+4x-7$ **39.** $2a+b$

41. $3m$ **43.** $21y-33z$ **45.** $7x-13y+8z$ **47.** $-2r+2t-4u+2v$

49. a **51.** $-4r+s+2t$ **53.** $6x-y$ **55.** $7+3a-6b+c$

57. $4x+y$ **59.** $7a+2b$

Section 2.5, p. 78

1. $15a$ **3.** $-28x$ **5.** $16xy$ **7.** $-12mn$

9. $24xyz$ **11.** $-36lmn$ **13.** $4x+4y$ **15.** $-10a-10b$

17. $2a-2b$ **19.** $-3m+3n$ **21.** $5x+5y$ **23.** $2ax+6y$

25. $4a+4b+4c$ **27.** $-10l+8m-6n$ **29.** $-12w+8x-4y+4z$ **31.** $2a$

33. $-3x$ **35.** $-3y$ **37.** $x+y$ **39.** $3x-4y$

41. $a+2b+3c$ **43.** $2x-3y+z-4$ **45.** $11x+3y$ **47.** $14a+17b$

49. x^2-10x **51.** $6x-6y-6z$ **53.** $3a+5b-4c$ **55.** $16x-12y$

57. $2x+y$ **59.** $2x+5y$ **61.** (c)

Let's Review Chapter 2, p. 81

1. $x+7$ **2.** $(-10)\cdot x\cdot x\cdot y\cdot y\cdot y$ **3.** -5 **4.** $x^2, -2x$, and 5

5. 50 **6.** 45 **7.** (a) 4 (b) -12 **8.** 400 square inches

9. yes **10.** no **11.** $10x$ **12.** $4uv$

13. $2m + 15n$ **14.** $5x + 2y - 3z$ **15.** $8a + 7b$ **16.** $4w + x + 4y - z$

17. $a + b + c$ **18.** $3x + 3z$ **19.** $13a$ **20.** $3m - 7n$

21. $30abc$ **22.** $8x^2 + 24y$ **23.** $-12x + 8y - 20z$ **24.** $5a - 2b + 3c$

25. $-3x^2 + 6x - 2$ **26.** $-14a - 6b$

And These From Chapter 1, p. 82

27. $>$ **28.** right **29.** (a) -2 (b) 2 **30.** 5 and -5

31. $<$ **32.** 1 **33.** 0

34. $3(6 + 2) = 3 \cdot 6 + 3 \cdot 2 = 18 + 6 = 24$ **35.** 10 miles west

CHAPTER 3

Section 3.1, p. 88

1. (a) $7x$ (b) 28 **3.** (a) $12 - 2x$ (b) 4 **5.** $x + 5 = 8$

7. $2x = -8$ **9.** $x - 6 = 10$ **11.** $x - 7 = 2$

13. One more than what number equals 5?

15. Ten times what number equals 40?

17. Three less than twice what number equals 13?

19. When three less than a number is divided by 4, the result equals 5. Find the given number.

21. $6 + 4 \stackrel{?}{=} 10$
$10 \stackrel{\checkmark}{=} 10$
6 is a root.

23. $7 - 2 \stackrel{?}{=} 5$
$5 \stackrel{\checkmark}{=} 5$
7 is a root.

25. $-3 + 5 \stackrel{?}{=} 2$
$2 \stackrel{\checkmark}{=} 2$
-3 is a root.

27. $5(-6) \stackrel{?}{=} 30$
$-30 \stackrel{x}{=} 30$
-6 is not a root.

29. $\frac{10}{2} \stackrel{?}{=} 5$
$5 \stackrel{\checkmark}{=} 5$
10 is a root.

31. $-7 + 1 \stackrel{?}{=} -6$
$-6 \stackrel{\checkmark}{=} -6$
-7 is a root.

33. $2(3) + 1 \stackrel{?}{=} 7$
$6 + 1 \stackrel{?}{=} 7$
$7 \stackrel{\checkmark}{=} 7$
3 is a root.

35. $3(1) + 1 \stackrel{?}{=} 4$
$3 + 1 \stackrel{?}{=} 4$
$4 \stackrel{\checkmark}{=} 4$
1 is a root.

37. $1 - 2 \stackrel{?}{=} -1$
$-1 \stackrel{\checkmark}{=} -1$
2 is a root.

39. $\frac{9 + 1}{5} \stackrel{?}{=} 2$
$\frac{10}{5} \stackrel{?}{=} 2$
$2 \stackrel{\checkmark}{=} 2$
9 is a root.

41. $2(-3) + 7 \stackrel{?}{=} 1$
$-6 + 7 \stackrel{?}{=} 1$
$1 \stackrel{\checkmark}{=} 1$
-3 is a root.

43. $-2(2) + 1 \stackrel{?}{=} 3$
$-4 + 1 \stackrel{?}{=} 3$
$-3 \stackrel{x}{=} 3$
2 is not a root.

45. $1 + 1 \stackrel{?}{=} 2(1)$
$2 \stackrel{\checkmark}{=} 2$
1 is a root.

47. $1 + 0 \stackrel{?}{=} \frac{0}{2}$
$1 \stackrel{x}{=} 0$
0 is not a root.

49. $\frac{1-5(2)}{3} \stackrel{?}{=} -3$

$\frac{1-10}{3} \stackrel{?}{=} -3$

$\frac{-9}{3} \stackrel{?}{=} -3$

$-3 \stackrel{\checkmark}{=} -3$

2 is a root.

51. $-1 \stackrel{?}{=} -(-1)$

$-1 \stackrel{x}{=} 1$

-1 is not a root.

53. (a) is the error.

Section 3.2, p. 95 *(The checks follow number 69.)*

1. yes
3. yes
5. (b), (c), (d), and (e)
7. (b) and (e)
9. yes
11. no
13. 5
15. 9
17. 0
19. 2
21. 10
23. -4
25. 7
27. 12
29. 1
31. 14
33. 81
35. 55
37. 3
39. 9
41. 5
43. 0
45. -10
47. -6
49. -2
51. -4
53. 15
55. 8
57. -10
59. -12
61. $x+4=6$; 2
63. $5+x=9$; 4
65. $\frac{x}{2}=6$; 12
67. $\frac{x}{-5}=10$; -50
69. $x+3=7$; 4

Checks:

25. $7-3 \stackrel{?}{=} 4$

$4 \stackrel{\checkmark}{=} 4$

27. $12-5 \stackrel{?}{=} 7$

$7 \stackrel{\checkmark}{=} 7$

29. $1+8 \stackrel{?}{=} 9$

$9 \stackrel{\checkmark}{=} 9$

53. $\frac{15}{3} \stackrel{?}{=} 5$

$5 \stackrel{\checkmark}{=} 5$

55. $2(8) \stackrel{?}{=} 16$

$16 \stackrel{\checkmark}{=} 16$

57. $\frac{-10}{2} \stackrel{?}{=} -5$

$-5 \stackrel{\checkmark}{=} -5$

59. $2(-12) \stackrel{?}{=} -24$

$-24 \stackrel{\checkmark}{=} -24$

Section 3.3, p. 101

1. 5
3. 5
5. 7
7. 3
9. (a) -7

(b) *Check:*

$-7-(-7) \stackrel{?}{=} 0$

$-7+7 \stackrel{?}{=} 0$

$0 \stackrel{\checkmark}{=} 0$

11. $7-x=0$; 7
13. 4
15. 2
17. -4
19. -1
21. 12
23. 12
25. (a) 4

(b) *Check:*

$3(4)+9 \stackrel{?}{=} 6(4)-3$

$12+9 \stackrel{?}{=} 24-3$

$21 \stackrel{\checkmark}{=} 21$

27. $2x+5=x-3$; -8
29. 75
31. 2
33. 2
35. -4
37. 6

39. (a) 6
(b) $2(6) + 4(6) \stackrel{?}{=} 27 + 9$
$12 + 24 \stackrel{?}{=} 36$
$36 \stackrel{\checkmark}{=} 36$

41. 4

43. 1

45. -5

47. -3

49. -10

51. 2

53. (a) -2
(b) *Check:*
$4(-2) - 3 \stackrel{?}{=} 7(-2) + 3$
$-8 - 3 \stackrel{?}{=} -14 + 3$
$-11 \stackrel{\checkmark}{=} -11$

55. (a) 1
(b) *Check:*
$6 - 2(1) \stackrel{?}{=} 9 - 5(1)$
$6 - 2 \stackrel{?}{=} 9 - 5$
$4 \stackrel{\checkmark}{=} 4$

57. $2x + 6 = 6x - 6$; 3

Section 3.4, p.105 *(The checks follow number 41.)*

1. 13 **3.** -3 **5.** -19 **7.** 32

9. -11 **11.** 4 **13.** 6 **15.** 4

17. -1 **19.** -4 **21.** 2 **23.** 1

25. 4 **27.** 1 **29.** 5 ounces **31.** 0

33. 3 **35.** 2 **37.** 1 **39.** -12

41. 2

Checks:

25. $4 - [12 - 2(4)] \stackrel{?}{=} 0$
$4 - 12 + 8 \stackrel{?}{=} 0$
$0 \stackrel{\checkmark}{=} 0$

27. $-3(1 + 2) \stackrel{?}{=} 1 - 10$
$-3(3) \stackrel{?}{=} -9$
$-9 \stackrel{\checkmark}{=} -9$

41. $4(2 + 3) \stackrel{?}{=} 20$
$4(5) \stackrel{?}{=} 20$
$20 \stackrel{\checkmark}{=} 20$

Section 3.5, p. 111 *(The check follows number 45.)*

1. (a) $x + 2$ (b) $x - 6$ (c) $3x$ (d) $x - 5$

3. (a) $\frac{t}{2}$ (b) $t + 4$ (c) $t - 10$ (d) $5\left[\frac{t}{2} + (t + 4) + (t - 10)\right]$

5. $x - 50$

7. $x + 8$

9. $x + 10 = 23$; 13

11. $x - 7 = 25$; 32

13. $3x - 1 = 23$; 8

15. $5x - 2 = 28$; 6

17. $(16 - x)$ feet

19. $x + (x + 6) = 36$, or $36 - x = x + 6$; 15 meters, 21 meters

21. $x + 2x + (2x - 5) = 35$; 8 feet, 11 feet, 16 feet

23. $x + (x - 10) = 2420$; 1215

25. $x + (x + 22) + [x + (x + 22)] = 2004$; Independent: 1002; Democrat: 512; Republican: 490

27. $x + (x + 40) = 3x - 160$; \$200

29. $3x + 500 = 5000$; 1500

31. (a) 10 meters
(b) 200 square meters

33. 60°

35. 45°

37. $30^\circ, 60^\circ$

39. $50^\circ, 55^\circ, 75^\circ$

41. $5x + 10(2x) = 75$; 3

43. $3x + 15(x + 1) = 87$; four 3-cent stamps and five 15-cent stamps

45. $x + (4x + 5) = 50$; Beethoven 9, Mozart 41

Check:

41. $5(3) + 10[2(3)] \stackrel{?}{=} 75$
$15 + 60 \stackrel{?}{=} 75$
$75 \stackrel{\checkmark}{=} 75$

Section 3.6, p. 120 *(The checks follow number 17.)*

1. 90 miles **3.** 240 miles **5.** (a) 350 miles per hour (b) 5 hours (c) $\frac{m}{350}$

7. 16 miles **9.** (a) 25 miles (b) 75 miles **11.** 690 miles

13. 6 kilometers **15.** 3 hours **17.** 18 miles

Checks:

11. $6(50) + 6(65) \stackrel{?}{=} 690$
$300 + 390 \stackrel{?}{=} 690$
$690 \stackrel{\checkmark}{=} 690$

13. $10(2) - 7(2) \stackrel{?}{=} 6$
$20 - 14 \stackrel{?}{=} 6$
$6 \stackrel{\checkmark}{=} 6$

15. $45(4) \stackrel{?}{=} 60(3)$
$180 \stackrel{\checkmark}{=} 180$

17. $\frac{18}{6} + \frac{18}{9} \stackrel{?}{=} 5$
$3 + 2 \stackrel{?}{=} 5$
$5 \stackrel{\checkmark}{=} 5$

Let's Review Chapter 3, p. 122

1. $3(4) + 2 \stackrel{?}{=} 14$
$12 + 2 \stackrel{?}{=} 14$
$14 \stackrel{\checkmark}{=} 14$
4 is a root.

2. $(-6)(-3) \stackrel{?}{=} 18$
$18 \stackrel{\checkmark}{=} 18$
-3 is a root.

3. $\frac{5(5) - 5}{5} \stackrel{?}{=} 5$
$\frac{25 - 5}{5} \stackrel{?}{=} 5$
$\frac{20}{5} \stackrel{?}{=} 5$
$4 \stackrel{x}{=} 5$
5 is not a root.

4. $2(0) \stackrel{?}{=} 0$
$0 \stackrel{\checkmark}{=} 0$
0 is a root.

5. $\frac{x}{3} = 10$ **6.** 12 **7.** -9 **8.** 48

9. (a) -18 (b) *Check*: $-18 + 12 \stackrel{?}{=} -6$
$-6 \stackrel{\checkmark}{=} -6$

10. $-3x = 18; -6$ **11.** -3

12. 7 **13.** 50 **14.** 5

15. (a) 1 (b) *Check*: $7 - 3(1) \stackrel{?}{=} 9(1) - 5$
$4 \stackrel{\checkmark}{=} 4$

16. 10 **17.** -8

18. 2 **19.** 6 **20.** 1 **21.** $3x + 5 = 80$; 25

22. $x + 2(x + 5) = 40$; 10 meters **23.** $x + (x + 2) + 2[x + (x + 2)] = 144$; 96, 25, 23

24. $180 - x = 3(90 - x) - 10$; 40° **25.** $25x + 5(2x) + 10(2x + 3) = 85$; 1 quarter, 2 nickels, and 5 dimes.

26. 525 miles **27.** 48 miles per hour **28.** 70 miles per hour

And These from Chapters 1 and 2, p. 124

29. the fifth power **30.** 50 **31.** 672 hours **32.** -8

33. 36 **34.** $x^2 - 1$ **35.** -7 **36.** 23

37. $6x - 4y + 1$ **38.** 200 *abc* **39.** $-3x + 2y - 5z$ **40.** $-12x^2 - 3x - 6$

CHAPTER 4

Section 4.1, p. 131

1. x^4
3. c^{10}
5. c^9
7. z^{12}
9. $4a^6$
11. x^3y^3
13. $10c^3d^3$
15. $30x^5y^2$
17. $6x^3y^4z^6$
19. $18x^2y^3$
21. x^2
23. a^2b
25. x^3yz
27. $-3b$
29. $-7st^6$
31. $6ac^2$
33. $a^3 + a^2b$
35. $-4y^3 - 4y^3z$
37. $3x^2yz^2 - 3xy^3z^2$
39. $50x^6y^2z^6 + 80x^3y^3z^5$
41. $2a^3b + 2a^2b^2 + 2ab^3$
43. $-x^3yz^2 - 2x^3y^2z^2 + 3x^3y^2z^3$
45. $a + 1$
47. $x - 1$
49. $x - a$
51. $b - 3$
53. $4a + 6 + 3b$
55. $6x^3y^4 + 9x^4y^3$
57. (a), (b), (e), and (f) are the errors.
59. $5a^2 + ab$
61. $2x^3 + 6x^2y$
63. $2x^3 + 2x^2 - 2x$
65. $4a + 7b$

Section 4.2, p. 137

1. 1, 2, 4
3. 1, 2, 3, 5, 6, 10, 15, 30
5. composite
7. prime
9. composite
11. prime
13. $2 \cdot 7$
15. $2 \cdot 3^2$
17. $-(2^3 \cdot 3)$
19. $2^2 \cdot 7$
21. $2^2 \cdot 3^2$
23. $2 \cdot 3 \cdot 7$
25. $5 \cdot 13$
27. $2 \cdot 5 \cdot 7$
29. $2^2 \cdot 3^3$
31. 13^2
33. 6
35. 2, 3, 5, 7, 11, 13, 17, 19, 23, 29, 31, 37, 41, 43, and 47
37. (a), (b), and (d) are the errors.
39. 9
41. 12
43. 24
45. 2
47. 8
49. 9
51. 6
53. 3
55. 5
57. 1
59. 6
61. 6

Section 4.3, p. 141

1. $4a + 10 = 2(2a + 5)$
3. $8y^2 + 12y - 20 = 4(2y^2 + 3y - 5)$
5. $5b^3 - b^2 = b^2(5b - 1)$
7. (a) $8a^2 + 12a = 4(2a^2 + 3a)$
   (b) $8a^2 + 12a = a(8a + 12)$
9. (a) $25a^2b^3 - 15ab^3 + 40ab^2 = 5(5a^2b^3 - 3ab^3 + 8ab^2)$
   (b) $25a^2b^3 - 15ab^3 + 40ab^2 = a(25ab^3 - 15b^3 + 40b^2)$
   (c) $25a^2b^3 - 15ab^3 + 40ab^2 = b^2(25a^2b - 15ab + 40a)$
11. (c)
13. $2(a + 1)$
15. $-6(x + 1)$
17. $5(x^2 + 2)$
19. $10(2x^2 + 5)$
21. $16a + 25$
23. $t(t + 1)$
25. $x(2x + 3)$
27. $y^2(y - 1)$
29. $a^7(a^3 - 1)$
31. $10x^2(x - 2)$
33. $4a^3(8a + 3)$
35. $xy(x + y)$
37. $xyz^2(x - 1)$
39. $2ab(ac + 2b)$
41. $4m^2n(4p + 5m^2n^3)$
43. $ab^3c^2(a^2b + a + 1)$
45. $2(x^2 + 2x + 1)$

47. $10x(x^2 - 2x + 5)$
49. $5a^2(a^2 - 2a + 3)$
51. $2x^3y(10x^3 - 12x + 15)$
53. $-18mn^3(m + 3n + 4)$
55. $2xy(x^4 + 6x^3y + 4x^2y^2 - 3xy^3 + 2y^4)$
57. (d) is the error.

Section 4.4, p. 148

1. $x^2 + 7x + 6$
3. $x^2 + 6x + 8$
5. $a^2 + a - 6$
7. $b^2 - 6b + 8$
9. $c^2 + 12c + 36$
11. $x^2 - 64$
13. $x^2 + 2xy + y^2$
15. (a), (c), and (d) are the errors.
17. $2y^2 - 13y + 15$
19. $12n^2 - 23n + 10$
21. $5z^2 + 5z - 10$
23. $-3b^2 + 9b - 6$
25. $x^4 + 8x^2 + 16$
27. $a^8 - 1$
29. $x^6 + 2x^3 + 1$
31. $x^4 + 5x^2 + 6$
33. $c^6 - 5c^3 + 4$
35. $m^2 + 3mn + 2n^2$
37. $m^2 - 2mn - 8n^2$
39. $-10a^2 + 36ab - 18b^2$
41. $x^3 + 3x^2 + 5x + 6$
43. $a^4 - 5a^3 - a^2 + 15a - 6$
45. $2x^4 + 10x^3 + 25x^2 + 29x + 15$
47. $x^4 + 8x^3 + 15x^2 - x + 12$
49. $2x^2 + 10x + 10$
51. $2y^2 + 15y + 11$
53. $-5a^2 + 8a - 9$
55. $2x^2 + 7x + 5$
57. $x^2 + 2x - 3$
59. 7
61. $s^2 + 6s + 9$

Section 4.5, p. 154

1. $(x + 3)(x - 3)$
3. $(x + 7)(x - 7)$
5. $(m + 10)(m - 10)$
7. $(y + 9)(y - 9)$
9. $(1 + x)(1 - x)$
11. $(x + y)(x - y)$
13. $s + 3$
15. $(2x + 1)(2x - 1)$
17. $(6x + 7)(6x - 7)$
19. $(1 + 3u)(1 - 3u)$
21. $(2x + y)(2x - y)$
23. $(3m + 7n)(3m - 7n)$
25. $(12y + 5x)(12y - 5x)$
27. $(20t + u)(20t - u)$
29. $2(x + 1)(x - 1)$
31. $3(s + 5)(s - 5)$
33. $3(10 + x)(10 - x)$
35. $2(x + 4y)(x - 4y)$
37. $8(a + 5b)(a - 5b)$
39. $6(m + 3n)(m - 3n)$
41. $6(2x + 7y)(2x - 7y)$
43. $a(a + 1)(a - 1)$
45. $a^2(a + 1)(a - 1)$
47. $x(x + y)(x - y)$
49. $m^2(m + 7n)(m - 7n)$
51. $2x(x + 1)(x - 1)$
53. $3x(x + y)(x - y)$
55. $2x(2x + 5y)(2x - 5y)$
57. (c)
59. (d)
61. (a), (c), and (d) are the errors.
63. $(x^2 + 100)(x + 10)(x - 10)$
65. $(9 + t^2)(3 + t)(3 - t)$
67. $5(c + d^2)(c - d^2)$
69. $(x^4 + 1)(x^2 + 1)(x + 1)(x - 1)$

Section 4.6, p. 161

1. $(x + 3)(x + 1)$
3. $(a + 7)(a + 1)$
5. $(z + 6)(z + 1)$
7. $(a + 2)(a - 1)$
9. $(b + 1)(b - 2)$
11. $(s + 3)(s - 4)$
13. $(y + 6)(y + 2)$
15. $(x - 7)(x - 2)$
17. $(a + 7)(a - 2)$

19. $(c+5)(c+4)$
21. $(x+8)(x+5)$
23. $(y+3)(y-10)$
25. $(x+12)(x-3)$
27. $(x^2+4)(x^2+2)$
29. (d)
31. $(b+1)^2$
33. $(x+3)^2$
35. $(a-6)^2$
37. $2(x+2)(x+1)$
39. $3(x+4)(x-2)$
41. $-(y-5)(y-4)$, or $(5-y)(y-4)$, or $(y-5)(4-y)$
43. $a(a+5)(a-2)$
45. $4(m+5)^2$
47. $3(6-x)(4+x)$
49. $2t(t-7)(t-5)$
51. $(x+a)^2$
53. $(x+4a)(x+a)$
55. $(m+2n)^2$
57. $(x+3y)^2$
59. $2(x+6y)^2$
61. $a(x+4y)(x+2y)$

Section 4.7, p. 166 *(The checks follow number 51.)*

1. 0, 4
3. 2, 3
5. $-3, 8$
7. 5
9. 0, 3
11. 1, 4
13. $-2, -4$
15. $4, -2$
17. $-3, -5$
19. $-3, -6$
21. $3, -7$
23. $5, -2$
25. -1
27. 5
29. 8
31. $1, -1$
33. $9, -9$
35. $11, -11$
37. $1, -9$
39. $1, -2$
41. $6, -6$
43. $0, 7, -7$
45. $0, 3, -3$
47. 0, 1, 9
49. 0, 1, 3
51. $2, -2$

Checks:

5. for -3:
$$2(-3+3)(-3-8) \stackrel{?}{=} 0$$
$$2(0)(-11) \stackrel{?}{=} 0$$
$$0 \stackrel{\checkmark}{=} 0$$
for 8:
$$2(8+3)(8-8) \stackrel{?}{=} 0$$
$$2(11)(0) \stackrel{?}{=} 0$$
$$0 \stackrel{\checkmark}{=} 0$$

13. for -2:
$$(-2)^2+6(-2)+8 \stackrel{?}{=} 0$$
$$4-12+8 \stackrel{?}{=} 0$$
$$0 \stackrel{\checkmark}{=} 0$$
for -4:
$$(-4)^2+6(-4)+8 \stackrel{?}{=} 0$$
$$16-24+8 \stackrel{?}{=} 0$$
$$0 \stackrel{\checkmark}{=} 0$$

39. for 1:
$$2(1)^2+2(1)-4 \stackrel{?}{=} 0$$
$$2(1)+2-4 \stackrel{?}{=} 0$$
$$2+2-4 \stackrel{?}{=} 0$$
$$0 \stackrel{\checkmark}{=} 0$$
for -2:
$$2(-2)^2+2(-2)-4 \stackrel{?}{=} 0$$
$$2(4)-4-4 \stackrel{?}{=} 0$$
$$8-4-4 \stackrel{?}{=} 0$$
$$0 \stackrel{\checkmark}{=} 0$$

Section 4.8, p. 171

1. $(3x+1)(x+1)$
3. $(7a+1)(a+1)$
5. $(3x-1)(x+1)$
7. $(5y-1)(y-1)$
9. $(2x+1)(x+3)$
11. $(2y+3)(y+1)$
13. $(2a+5)(a+1)$
15. $(2m-1)(m-2)$
17. $(2n-1)(n-3)$
19. $(2a-1)(a+4)$
21. $(2a+5)(a+5)$
23. $(3y-1)^2$
25. $(2a+1)^2$
27. $(3y-7)(y+1)$
29. $(2m+3)(2m+1)$
31. $(3x-2)(2x+3)$
33. $(7a+2)(a+1)$
35. $(5x+4)(2x-1)$
37. $(3x-13)(3x+1)$
39. $(5a-3)(a+4)$
41. $2(2x+1)(x+1)$

43. $(2b-1)(b+12)$ **45.** $3(2b+1)(b+1)$ **47.** $2(2a-1)(a-1)$
49. $2x(2x+3)(x+1)$ **51.** $(2x+y)^2$ **53.** $(2y+z)(y+z)$
55. $(2x+y)(x+2y)$ **57.** $(3a+2b)(3a-10b)$ **59.** $2(3x+y)(x+2y)$

Let's Review Chapter 4, p. 173 ***(The check follows number 100.)***

1. $6x^3y^4$ **2.** $8m^4n^4$ **3.** $2x^2y$
4. $8a^3$ **5.** x^3-5x^2y **6.** $8a^4b^2c+12a^3b^2c^2$
7. $2a+2b$ **8.** $4xy-3$ **9.** $5a^5$
10. 2, 3, 5, 7, 11, 13, 17, 19 **11.** (a) $2^2\cdot 3^2$ (b) $2^2\cdot 3\cdot 11$ **12.** 12
13. 4 **14.** $2(2x+3)$ **15.** $x(x^2+x-4)$
16. $3a^3(3a^3-5a+4)$ **17.** $x^2y^3z(x^3yz^2+xz^2+1)$ **18.** x^2+6x+5
19. $a^2-5a-14$ **20.** z^4-3z^2-4 **21.** $m^2-mn-6n^2$
22. $x^5-3x^4+6x^2-4x$ **23.** x^3-2x^2+x-11 **24.** $(a+5)(a-5)$
25. $(2y+7z)(2y-7z)$ **26.** $a(a+1)(a-1)$ **27.** $3x(y+2)(y-2)$
28. $(c+4)(c+3)$ **29.** $(m+4)(m-5)$ **30.** $t(t+3)^2$
31. $(x-3y)(x-2y)$ **32.** 0, 9 **33.** $-3, -4$
34. $6, -4$ **35.** $6, -6$ **36.** 0, 3, 6
37. $(4a+1)(a+1)$ **38.** $(2x+5)(x+1)$ **39.** $(2y-1)(y-3)$
40. $(4x-3y)(x+y)$ **41.** $5(m-2n)$ **42.** $(6+a)(6-a)$
43. $(x+7)(x+2)$ **44.** $7(a+2b)(a-2b)$ **45.** $x(x+2)(x-2)$
46. $x(3x+1)(x-2)$ **47.** $(m+5n)(m+3n)$ **48.** $4(m+2n^2)(m-2n^2)$
49. $s(s+7)(s-1)$ **50.** $t^2(t+4)(t-4)$ **51.** $(a-8)(a-2)$
52. $(5x+2)(x-2)$ **53.** $(x+18)(x-1)$ **54.** $(k+10)(k-4)$
55. $(7x+1)(x-1)$ **56.** $(n+5)(n-11)$ **57.** $(6x+5)(x+2)$
58. $(7+b^3)(7-b^3)$ **59.** $(x+2y)(x-10y)$ **60.** $a(a+5)(a-5)$
61. $8(3a+5)(3a-5)$ **62.** $2(5x-y)(x+y)$ **63.** $2a(9a+9-4b)$
64. $(9x+2)(9x-2)$ **65.** $(x+3)(x-10)$ **66.** $(b-2)(b-5)$
67. $4(x+2)(x+4)$ **68.** $(x+yz)(x-yz)$ **69.** $4x^2(x+5)(x-5)$
70. $2b^2(3a+2)(3a-2)$ **71.** $3a(a+4)(a+1)$ **72.** $(9x-y)(x+y)$
73. $6(2x+5y)(2x-5y)$ **74.** $3a(3a+4)(a-1)$ **75.** $(6x-5)(x-1)$
76. $30(2+n)(2-n)$ **77.** $5ab(2a+5b-1)$ **78.** $11k(1+2k)(1-2k)$
79. $2a(4+5b)$ **80.** $(2x-7)^2$ **81.** $a^2(x+3a)(x-3a)$
82. $10(a+7)(a-8)$ **83.** $7x(2x-1)(x+1)$ **84.** $6(3x^3+2y^2)(3x^3-2y^2)$

85. $(9+x^2)(3+x)(3-x)$
86. $5r(1+2q)(1-2q)$
87. $3x(3x+2)$
88. $a(a^2+1)(a+1)(a-1)$
89. $a^2x^2(x+a)(x-a)$
90. $5b^2(a+4)(a-1)$

And These from Chapters 1-3, p. 175

91. $2<7<17<23<29<41$
92. 15
93. 5
94. -72
95. 12
96. 148 feet
97. $6a^2+3ab+4b^2$
98. $-2x-9xy+3y$
99. $-5x^3+2xy+3y$
100. x^2+5x

Check:

34. for 6:
$6^2-2(6)-24 \overset{?}{=} 0$
$36-12-24 \overset{?}{=} 0$
$0 \overset{\checkmark}{=} 0$

for -4:
$(-4)^2-2(-4)-24 \overset{?}{=} 0$
$16+8-24 \overset{?}{=} 0$
$0 \overset{\checkmark}{=} 0$

CHAPTER 5

Section 5.1, p. 187

1. (a) 3 (b) 8
3. (a) -3 (b) 8
5. $\frac{-4}{3}$
7. (a) x^2-x+1 (b) $x+3$
9. (a) $-x$ (b) 3
11. $\frac{20}{11}$
13. -2
15. -1
17. $\frac{4}{5}$
19. 3
21. 5
23. $\frac{3}{10}$
25. (a) $\frac{-1}{2}$ (b) 0 (c) $\frac{-2}{3}$
27. 2
29. 9 (feet per second per second, that is, 9 feet per second every second)
31. 49
33. $\frac{6}{15}$
35. 14
37. $\frac{18}{6x}$
39. $4\cdot 4 \overset{?}{=} 1\cdot 16$
$16 \overset{\checkmark}{=} 16$
Thus $\frac{4}{16}=\frac{1}{4}$
41. Does $14\cdot 7 = 1\cdot 2$?
No. $98 \neq 2$
Thus $\frac{14}{2} \neq \frac{1}{7}$
43. Does $4\cdot 3t = 12\cdot t$?
Yes. $12t = 12t$
Thus $\frac{4}{t}=\frac{12}{3t}$
45. Does $(x-1)\,6x^2 = (x^3-x^2)\,6x$?
No. $6x^3-6x^2 \neq 6x^4-6x^3$
Thus $\frac{x-1}{6x} \neq \frac{x^3-x^2}{6x^2}$
47. (a) and (c)
49. $(-2)(-9) \overset{?}{=} 2(9)$
$18 \overset{\checkmark}{=} 18$
51. $\frac{-7}{8}$
53. $\frac{-x-1}{3}$
55. $\frac{1}{2}$
57. $\frac{1}{4}$
59. $\frac{-4}{5}$
61. $\frac{-2}{3}$
63. $\frac{3}{8}$
65. $\frac{1}{5}x^2+\frac{2}{5}x-\frac{4}{5}$
67. $\frac{1}{4}x^3-\frac{1}{2}x^2+\frac{2}{3}x-\frac{3}{4}$

Section 5.2, p. 195

1. $\frac{1}{3}$
3. 1
5. $\frac{1}{y^3}$
7. $\frac{1}{a}$
9. $\frac{m^7}{n}$
11. abc
13. $\frac{x^2y}{z^{15}}$
15. $\frac{b}{2}$
17. $\frac{-ab}{2c}$
19. $\frac{3}{5a^2c}$
21. $\frac{2}{m^2n}$
23. $\frac{3m^9p}{2n}$
25. (a) is the error.
27. $\frac{11}{24}$
29. $\frac{-16}{21}$
31. $\frac{15}{8}$
33. $\frac{11a}{4}$
35. $\frac{8}{3y^6}$
37. $\frac{2a^3}{3}$
39. $x+1$
41. $(a+b)^2$
43. 1
45. $\frac{1}{(a-b)^3}$
47. 2
49. $-6(a+b)^3$
51. $x+y$
53. $\frac{(x+y)^2}{(x-y)^2}$
55. $\frac{(a+b)(x+y)^6}{4(x-y)}$
57. $\frac{1}{25x(a+2)}$

Section 5.3, p. 200

1. $m+2n$
3. $x+y$
5. $\frac{2a-b}{4}$
7. $\frac{b+c^2}{a}$
9. $x+2y$
11. $3(a+2b)$
13. $\frac{x^2-2x}{2}$ or $\frac{x(x-2)}{2}$
15. $3(x+y-z)$
17. $\frac{a-2x+3ax}{4ax}$
19. $\frac{1}{a-b}$
21. $\frac{1}{2x+3y}$
23. $\frac{20xy}{2y-5x+10xy}$
25. $\frac{a+b}{x+y}$
27. $\frac{3(x+y)}{x-y}$
29. $\frac{x+y}{2(x-y)}$
31. $\frac{6}{5}$
33. $\frac{a}{b}$
35. $\frac{5}{a-b}$
37. $\frac{x-1}{2}$
39. $\frac{3}{5(m-n)}$
41. $\frac{a+2b}{2}$
43. $\frac{2x-3y}{6}$
45. $\frac{x+1}{x+3}$
47. $\frac{y-3}{y+1}$
49. $\frac{u-6}{u+1}$
51. $\frac{m+4}{m+2}$
53. $\frac{x-2}{x-4}$
55. $\frac{a-b}{2(a+2b)}$
57. $\frac{x+4}{2(x-2)}$
59. $\frac{2x+5}{x+3}$
61. -1
63. -2
65. (a), (d), (e), (f), and (g)

Section 5.4, p. 208 *(The checks follow number 59.)*

1. (a) x^2+2x+1 (b) x^2: degree 2; $2x$: degree 1; 1: degree 0 (c) 2
3. (a) $-u^4+u^3-u^2+u$ (b) $-u^4$: degree 4; u^3: degree 3; $-u^2$: degree 2; u: degree 1 (c) 4
5. (a) $z^{10}+z^7-z^4+z$ (b) z^{10}: degree 10; z^7: degree 7; $-z^4$: degree 4; z: degree 1 (c) 10
7. $x+4$
9. $x+5$
11. $t+5$
13. x^2+x+1
15. $3t^2-1$
17. $x+9$
19. z^2+1
21. $a+3$

23. a^2+5 **25.** $2a^2-a+2$ **27.** $x+2$ **29.** x^2+2x+1

31. (a) 4 (b) 3 (c) $\underbrace{\text{degree }(3)}_{0} < \underbrace{\text{degree }(2x)}_{1}$

33. (a) $2x$ (b) 1 (c) $\underbrace{\text{degree }(1)}_{0} < \underbrace{\text{degree }(x)}_{1}$

35. (a) $2x$ (b) 5 (c) $\underbrace{\text{degree }(5)}_{0} < \underbrace{\text{degree }(x^2-2)}_{2}$

37. (a) 3 (b) -1 **39.** (a) 5 (b) -19 **41.** (a) $x+3$ (b) -2 **43.** (a) x^2+2x+1 (b) 4

45. (a) $3x+5$ (b) $10x+2$ **47.** (a) $x+5$ (b) 5 (c) $x+5+\dfrac{5}{x+3}$

49. (a) $x+5$ (b) 25 (c) $x+5+\dfrac{25}{x-5}$ **51.** (a) $2x+2$ (b) 3 (c) $2x+2+\dfrac{3}{2x+1}$

53. (a) $4x+1$ (b) 9 (c) $4x+1+\dfrac{9}{3x+2}$ **55.** (a) x^2-1 (b) 0 (c) $x^2-1+\dfrac{0}{x^2+1}$

57. (a) x^2-4x+6 (b) -5 (c) $x^2-4x+6+\dfrac{-5}{x+1}$ **59.** (a) x^2-4x+9 (b) -16 (c) $x^2-4x+9+\dfrac{-16}{x+2}$

Checks:

7. $\begin{array}{r} x+4 \\ x+1 \\ \hline x^2+4x \\ x+4 \\ \hline x^2+5x+4 \end{array}$

9. $\begin{array}{r} x+5 \\ x+3 \\ \hline x^2+5x \\ 3x+15 \\ \hline x^2+8x+15 \end{array}$

11. $\begin{array}{r} t+5 \\ t+9 \\ \hline t^2+5t \\ 9t+45 \\ \hline t^2+14t+45 \end{array}$

13. $\begin{array}{r} x^2+x+1 \\ x+4 \\ \hline x^3+x^2+x \\ 4x^2+4x+4 \\ \hline x^3+5x^2+5x+4 \end{array}$

15. $\begin{array}{r} 3t^2-1 \\ t+4 \\ \hline 3t^3 \quad -t \\ 12t^2 \quad -4 \\ \hline 3t^3+12t^2-t-4 \end{array}$

37. (c) $\begin{array}{r} 3(x+1)=3x+3 \\ -1 \\ \hline 3x+2 \end{array}$

39. (c) $\begin{array}{r} 5(x+5)=5x+25 \\ -19 \\ \hline 5x+6 \end{array}$

41. (c) $\begin{array}{r} x+3 \\ x+4 \\ \hline x^2+3x \\ 4x+12 \\ \hline x^2+7x+12 \\ -2 \\ \hline x^2+7x+10 \end{array}$

43. (c) $\begin{array}{r} x^2+2x+1 \\ x+1 \\ \hline x^3+2x^2+x \\ x^2+2x+1 \\ \hline x^3+3x^2+3x+1 \\ +4 \\ \hline x^3+3x^2+3x+5 \end{array}$

45. (c) $\begin{array}{r} 3x+5 \\ x^2-x \\ \hline 3x^3+5x^2 \\ -3x^2-5x \\ \hline 3x^3+2x^2-5x \\ +10x+2 \\ \hline 3x^3+2x^2+5x+2 \end{array}$

Section 5.5, p. 215

1. $\dfrac{1}{6}$ **3.** $\dfrac{3}{2}$ **5.** $\dfrac{1}{12}$ **7.** $\dfrac{5}{14}$

9. $\dfrac{-4}{15}$ **11.** $\dfrac{2}{3}$ **13.** 24 **15.** $\dfrac{2}{3}$

17. $\dfrac{3}{8}$ **19.** $\dfrac{ax}{y}$ **21.** $2b$ **23.** $\dfrac{20xb}{ay}$

25. $\dfrac{48a^2}{x^2}$ **27.** $a(x-y)^3$ **29.** $\dfrac{2}{15}x^3y$ or $\dfrac{2x^3y}{15}$ **31.** $\dfrac{3}{5}x^6y^6z^3$ or $\dfrac{3x^6y^6z^3}{5}$

33. $3a$ **35.** $\dfrac{5y}{x}$ **37.** $\dfrac{2(a-2)}{3(a+2)}$ **39.** $\dfrac{3}{a+3}$

41. $\frac{4}{a}$ 43. 2 45. $\frac{-1}{3}$ 47. 25

49. 6 51. $\frac{22}{25}$ 53. $\frac{5}{2}$ 55. $\frac{b}{a}$

57. $\frac{1}{a}$ 59. $\frac{b^2}{ac^2}$ 61. $\frac{2}{x-a}$ 63. 6

65. $\frac{x+2}{a^2b}$ 67. a^2 69. $2a^2$ 71. $\frac{3}{x}$

73. (c) and (e) are the errors.

Section 5.6, p. 220

1. $\frac{2}{5}$ 3. 1 5. 1 7. $\frac{1}{2}$

9. $\frac{3}{4}$ 11. 1 13. $\frac{8}{9}$ 15. $\frac{3}{4}$

17. $\frac{3}{2}$ 19. 0 21. loss of $\frac{3}{10}$ of a point 23. $\frac{2}{x}$

25. $\frac{1}{y}$ 27. $\frac{11}{x+1}$ 29. $\frac{8a}{x+2}$ 31. $\frac{9}{a}$

33. $\frac{6}{x}$ 35. 1 37. 3 39. -1

41. $\frac{1}{x+2}$ 43. $\frac{b^2+1}{b+1}$ 45. $y+1$ 47. $\frac{4-2x}{x}$

49. 6 51. $\frac{2}{3}x$ 53. $\frac{9}{10}y$ 55. a

57. $\frac{1}{3}x+\frac{1}{7}y$ 59. $x-y$

Section 5.7, p. 225

1. 6 3. 4 5. 6 7. 25

9. 12 11. 30 13. 60 15. 108

17. 30 19. 12 21. 300 23. ax

25. $5x$ 27. x^2 29. $(x+1)^2$ 31. a^2x^2

33. $5x^2yz$ 35. $12a^2bxy$ 37. $(x+a)(x-a)$ or x^2-a^2

39. $(x+2)(x-2)$ or x^2-4 41. $(a+5)^2(a-5)$ 43. $8(x-1)^2$

45. $a^2b^2(x-y)$ 47. $a^2b^7x^4y^3z$ 49. $25(a+2)(a-2)$

Section 5.8, p. 233

1. (a) 3 (b) $\frac{1}{3}=\frac{1}{3}$, $1=\frac{3}{3}$

3. (a) 6 (b) $\frac{1}{2}=\frac{3}{6}$, $\frac{5}{6}=\frac{5}{6}$

5. (a) 20 (b) $\frac{5}{4}=\frac{25}{20}$, $\frac{7}{10}=\frac{14}{20}$

7. (a) 140 (b) $\frac{7}{20}=\frac{49}{140}$, $\frac{3}{28}=\frac{15}{140}$

9. (a) 120 (b) $\frac{1}{8}=\frac{15}{120}$, $\frac{-1}{12}=\frac{-10}{120}$, $\frac{2}{15}=\frac{16}{120}$

11. (a) a^2b (b) $\frac{1}{a^2}=\frac{b}{a^2b}$, $\frac{a}{b}=\frac{a^3}{a^2b}$

13. (a) $x^2y^4z^2$ (b) $\frac{x-a}{x^2y^3z^2} = \frac{xy-ay}{x^2y^4z^2}$, $\frac{a}{xy^4z} = \frac{axz}{x^2y^4z^2}$

15. (a) $(x+1)(x-1)^2$
(b) $\frac{1}{x^2-1} = \frac{x-1}{(x+1)(x-1)^2}$, $\frac{x}{(x-1)^2} = \frac{x^2+x}{(x+1)(x-1)^2}$

17. (a) $a(x+2)(x-2)$
(b) $\frac{1}{ax-2a} = \frac{x+2}{a(x+2)(x-2)}$, $\frac{a}{x^2-4} = \frac{a^2}{a(x+2)(x-2)}$, $\frac{-x}{ax+2a} = \frac{-x^2+2x}{a(x+2)(x-2)}$

19. 19 inches
21. $\frac{5}{6}$
23. $\frac{3}{4}$
25. $\frac{4}{3}$
27. $\frac{11}{9}$
29. $\frac{37}{50}$
31. $\frac{5}{192}$
33. $\frac{59}{250}$
35. $\frac{71}{50}$
37. $\frac{3}{2}$
39. $\frac{53}{8}$
41. $\frac{x+y}{xy}$
43. $\frac{ax+a}{x^2}$
45. $\frac{2+x^2y^2z^2}{xyz}$
47. $\frac{x+a+2}{(x+a)^2}$
49. $\frac{3x+16}{x^2+10x+25}$ or $\frac{3x+16}{(x+5)^2}$
51. $\frac{-5a}{(x+2)(x+3)(x-2)}$
53. $\frac{3x^2-12x+11}{(x-1)(x-2)(x-3)}$
55. $\frac{ax+4x+a}{a^3(a+4)(a-4)}$
57. 4
59. $\frac{2}{x+1}$
61. $\frac{2}{a^2}$
63. $\frac{2x^2+7x+2}{(x+4)(x+2)}$
65. $\frac{a^3+5a^2-1}{a^2(a-1)}$
67. $\frac{22}{15}x$
69. (a), (b), and (e) are the errors.

Section 5.9, p. 239

1. (a) $\frac{3}{4}$ (b) 7
3. (a) $\frac{x^2}{x+y}$ (b) $\frac{x-1}{(x+y)^2}$
5. $\left(9-\frac{x}{4}\right) \div \frac{x+1}{x+4}$
7. $\frac{1}{4}$
9. $\frac{-1}{4}$
11. 3
13. $\frac{-1}{6}$
15. $\frac{3}{4}$
17. 3
19. 6
21. 5
23. $\frac{20}{23}$
25. $\frac{25}{3}$
27. $\frac{a}{xy}$
29. $\frac{-2}{xy}$
31. $\frac{a^3}{b^3}$
33. $-b$
35. $\frac{x^2}{2y^3}$
37. $\frac{4ab^2}{3d^2}$
39. $\frac{x-1}{x-2}$
41. $\frac{1+a}{a^3}$
43. -1
45. $\frac{-(1+a-b)}{a}$
47. $\frac{x}{2}$ or $\frac{1}{2}x$
49. (a), (b), (c), and (d) are the errors.

Let's Review Chapter 5, p. 241 *(The checks follow number 64.)*

1. $3(-15) \stackrel{?}{=} (-9)5$
$-45 \stackrel{\checkmark}{=} -45$
Thus $\frac{3}{5} = \frac{-9}{-15}$
2. $\frac{2}{3}$
3. $\frac{-5}{6}$
4. $\frac{4}{5}$
5. 32
6. $\frac{3}{4}$
7. (a) -2 (b) -2 (c) $\frac{5}{2}$
8. $\frac{1}{4}x^2 - \frac{1}{6}x + \frac{1}{3}$
9. $\frac{a^2}{b^4}$
10. $\frac{-10}{9}$
11. $\frac{4a^3}{3c^2}$
12. $\frac{2(x+y)^2}{3(x-y)}$
13. $2(x-y)$
14. $\frac{a+b}{x}$
15. $\frac{a+2c}{3}$
16. $\frac{2}{1-5y}$

17. $\dfrac{a+b}{5(a+2b)}$

18. $\dfrac{m-3}{m+3}$

19. $\dfrac{1}{s}$

20. $\dfrac{2(x+y)}{x-y}$

21. (a) $7x^6+5x^5-x^4+2x^2$ (b) 6

22. $y+5$

23. x^2-x+1

24. a^2-1

25. (a) 5 (b) 2 (c) $\underbrace{\text{degree }(2)}_{0} < \underbrace{\text{degree }(x+1)}_{1}$

26. (a) $4x-15$ (b) 39

27. (a) $a-3$ (b) -11 (c) $a-3+\dfrac{-11}{a+7}$

28. (a) $x^2-5x+20$ (b) -79 (c) $x^2-5x+20+\dfrac{-79}{x+4}$

29. $\dfrac{4}{45}$

30. ab^2

31. $\dfrac{2y\,(x+3)}{z\,(x-3)}$

32. 8

33. axy

34. $(x-y)(x+1)$

35. $\dfrac{1}{3}$

36. $\dfrac{1}{4}$

37. 2

38. $\dfrac{1}{x+3}$

39. 30

40. 540

41. $4x^3y$

42. $(x+2)(x-2)(x-1)$

43. (a) 15 (b) $\dfrac{1}{3}=\dfrac{5}{15}, \dfrac{2}{5}=\dfrac{6}{15}$

44. $\dfrac{5}{8}$

45. $\dfrac{3x^2-x+1}{x^3}$

46. $\dfrac{x^2-ax+a}{x^2-a^2}$ or $\dfrac{x^2-ax+a}{(x+a)(x-a)}$

47. $\dfrac{3}{2}$

48. $\dfrac{4}{5}$

49. $\dfrac{ab}{c}$

50. $\dfrac{1}{(x-1)(x+a)}$

And These from Chapters 1-4, p. 244

51. $5x^6y^4-15x^4y^4z+10x^6y^3z^4$

52. $-a^3b^2+3bc^2-2a^2c$

53. $(3x+2y)(x+3y)$

54. $4a^2(x+3)(x-3)$

55. $2^3\cdot 3^3\cdot 5$

56. $-2x^2-11x+25$

57. $42x^2+3x-9$

58. 3

59. 225 square centimeters

60. $5x-1$

61. 2

62. 1

63. -24

64. 12

Checks:

22.
$$\begin{array}{r} y+5 \\ \underline{y+2} \\ y^2+5y \\ \underline{2y+10} \\ y^2+7y+10 \end{array}$$

26. (c)
$$\begin{array}{r} 4x-15 \\ \underline{x+2} \\ 4x^2-15x \\ \underline{8x-30} \\ 4x^2-7x-30 \\ \underline{+39} \\ 4x^2-7x+9 \end{array}$$

CHAPTER 6

Section 6.1, p. 251 *(The checks follow number 51.)*

1. $\dfrac{9}{2}$

3. $\dfrac{1}{3}$

5. -81

7. $\dfrac{3}{4}$

9. $\dfrac{2}{5}$

11. 6

13. 2

15. -6

17. \$81.00 **19.** 22 **21.** 120 **23.** 4

25. −3 **27.** 10 **29.** 5 **31.** 9

33. 2 **35.** 2 **37.** 0 **39.** −3

41. −3 **43.** $\frac{6}{5}$ **45.** 1 **47.** −16

49. 7 **51.** $\frac{-8}{7}$

Checks:

7. $\frac{3}{4} - \frac{1}{4} \stackrel{?}{=} \frac{1}{2}$

$\frac{2}{4} \stackrel{?}{=} \frac{1}{2}$

$\frac{1}{2} \stackrel{\checkmark}{=} \frac{1}{2}$

9. $1 + 5\left(\frac{2}{5}\right) \stackrel{?}{=} 3$

$1 + 2 \stackrel{?}{=} 3$

$3 \stackrel{\checkmark}{=} 3$

49. $\frac{7+3}{7-2} \stackrel{?}{=} 2$

$\frac{10}{5} \stackrel{?}{=} 2$

$2 \stackrel{\checkmark}{=} 2$

51. $\dfrac{2}{\frac{-8}{7} + 2} \stackrel{?}{=} \dfrac{5}{1 - \frac{-8}{7}}$

$2\left(1 + \frac{8}{7}\right) \stackrel{?}{=} 5\left(\frac{-8}{7} + 2\right)$

$2 + \frac{16}{7} \stackrel{?}{=} -\frac{40}{7} + 10$

$\frac{30}{7} \stackrel{\checkmark}{=} \frac{30}{7}$

Section 6.2, p. 260 *(The checks follow number 77.)*

1. .13 **3.** .09 **5.** .0009 **7.** 12 or 12.

9. $\frac{29}{100}$ **11.** $\frac{2}{5}$ **13.** $\frac{9}{25}$ **15.** $\frac{1}{250}$

17. (b) and (c) **19.** .2 **21.** .08 **23.** .666 666 . . .

25. .142857 142857 . . . **27.** 1.5674 **29.** 1.909 **31.** x

33. 238 **35.** .3 **37.** .000 06 **39.** 3.865

41. .4 **43.** .4 **45.** .28 **47.** 400

49. $.01x^4y^4$ **51.** 1 **53.** 5 **55.** .23

57. 3 **59.** 24 **61.** 10 **63.** 12

65. 5 **67.** 50 **69.** 200 **71.** 2

73. 22.375 **75.** 4 **77.** $\frac{3}{59}$

Checks:

61. $10 - .02 \stackrel{?}{=} 9.98$

$9.98 \stackrel{\checkmark}{=} 9.98$

63. $\frac{12}{.06} \stackrel{?}{=} 200$

$12 \stackrel{?}{=} 200(.06)$

$12 \stackrel{\checkmark}{=} 12.00$

Section 6.3, p. 268 *(The checks follow number 37.)*

1. .17 **3.** .42 **5.** .105 **7.** 2.5

9. 52% **11.** .4% **13.** 110% **15.** 600

17. 960

19. 325

21. \$84

23. 6%

25. \$600

27. \$1000 at $7\frac{1}{2}\%$

29. \$450

31. $\frac{1}{4}$ of a year

33. \$3000 at 8% for $\frac{3}{4}$ of a year

35. \$302.25

37. \$543.54

Checks:

23. $.06(750) \stackrel{?}{=} 45$
$45.00 \stackrel{\checkmark}{=} 45$

25. $.09(600) \stackrel{?}{=} 54$
$54.00 \stackrel{\checkmark}{=} 54$

Section 6.4, p. 274 (The checks follow number 25.)

1. \$198

3. 8%

5. \$2500

7. (a) \$2520
(b) \$39 480

9. \$440

11. (a) \$12
(b) \$14.40

13. \$175

15. 60%

17. (a) \$80
(b) \$320

19. 40%

21. \$15.40

23. (a) \$12.50
(b) \$2.50

25. 25%

Checks:

3. $.08(5000) \stackrel{?}{=} 400$
$400 \stackrel{\checkmark}{=} 400$

21. $\frac{20-14}{20} \stackrel{?}{=} \frac{22-15.40}{22}$
$\frac{6}{20} \stackrel{?}{=} \frac{6.60}{22}$
$\frac{3}{10} \stackrel{?}{=} \frac{660}{2200}$
$6600 \stackrel{\checkmark}{=} 6600$

Section 6.5, p. 279 (The checks follow number 51.)

1. $\frac{y}{5}$

3. $y-2$

5. $-y-4$

7. $\frac{-7t}{5}$

9. $1-y-z$

11. $\frac{-2z}{3}$

13. $\frac{5-2z}{3}$

15. $\frac{2u-6}{3}$

17. $\frac{12-b+3c}{2}$

19. (a) $\frac{y+20}{5}$ (b) $5x-20$

21. (a) $\frac{4c-1}{10}$ (b) $\frac{10b+1}{4}$

23. (a) $\frac{11z-1}{3}$ (b) $\frac{3y+1}{11}$

25. (a) b (b) a

27. $\frac{y+10}{2}$

29. $\frac{6}{u}$

31. $x+2z-8$

33. $\frac{10}{4-y}$

35. $\frac{5}{a+b}$

37. $\frac{1}{a-c}$

39. $\frac{3}{x-1}$

41. (a) $\frac{1-4y}{y}$ (b) $\frac{1}{x+4}$

43. (a) $\frac{10-2z}{5-y}$, $y \neq 5$ (b) $\frac{5x+2z-10}{x}$, $x \neq 0$

45. $\frac{A}{w}$

47. (a) $\frac{2A}{b}$ (b) 50 inches

49. (a) $\frac{d}{t}$ (b) 60 miles per hour

51. (a) $P = \frac{I}{RT}$ (b) \$10 000

Checks:

27. $2\left(\frac{y+10}{2}\right) - y \stackrel{?}{=} 10$
$y + 10 - y \stackrel{?}{=} 10$
$10 \stackrel{\checkmark}{=} 10$

29. $\frac{3}{u} \stackrel{?}{=} \frac{\frac{6}{u}}{2}$
$\frac{3}{u} \stackrel{?}{=} \frac{6}{u} \cdot \frac{1}{2}$
$\frac{3}{u} \stackrel{\checkmark}{=} \frac{3}{u}$

31. $x - (x + 2z - 8) + 2z \stackrel{?}{=} 8$
$x - x - 2z + 8 + 2z \stackrel{?}{=} 8$
$8 \stackrel{\checkmark}{=} 8$

Section 6.6, p. 285

1. true
3. false
5. true
7. true
9. $\leq$
11. $\geq$
13. $\leq$
15. both
17. $\leq$
19. $x < 4$
21. $x \leq -1$
23. (a)
25. (a)
27. (f)
29. (d)
31. (e)
33. (c)
35. (f)
37. true
39. true
41. true
43. true
45. false
47. true
49. true
51. $(0, 2)$
53. $(5, \infty)$
55. $[-1, 1]$
57. $(-\infty, 0)$
59. $(1, 2)$
61. (a) is the error.

Section 6.7, p. 292

1. $<$
3. $>$
5. $<$
7. $<$
9. $\leq$
11. $\leq$
13. $67 \leq x \leq 84$
15. $0 < x + y < 90$
17. $>$
19. $>$
21. $<$
23. $>$
25. $>$
27. $>$
29. $<$
31. $>$
33. $<$
35. $>$
37. $\geq$
39. $\geq$
41. $\leq$
43. $\geq$
45. (d) and (e) are the errors.

Section 6.8, p. 298

1. $(-\infty, 6)$
3. $(-\infty, 3)$
5. $(-\infty, 9)$
7. $(5, \infty)$
9. $(-\infty, 5)$
11. $(6, \infty)$
13. $(-\infty, -4)$
15. $(-\infty, 5]$
17. $\left(-\infty, \frac{-8}{3}\right)$
19. $(3, \infty)$
21. $\left[\frac{9}{2}, \infty\right)$
23. $\left(-\infty, \frac{3}{2}\right)$
25. $\left(\frac{12}{7}, \infty\right)$
27. $(-\infty, 18]$
29. $\left[\frac{-2}{5}, \infty\right)$
31. $(5, 15)$
33. $[2, 6]$
35. $[-2, 2]$
37. $[0, 64]$
39. $(-12, 4)$
41. $(-6, 6)$
43. $[-4, 2]$
45. $\left(\frac{-1}{2}, \frac{1}{2}\right)$
47. $[160, 240]$
49. $[20, 25]$

Let's Review Chapter 6, p. 300

1. -27
2. 14
3. 4
4. 8
5. 77
6. .35
7. (a) .0484 (b) 33.3
8. .12
9. .02
10. 798
11. \$58.50
12. \$1500 at 8%; \$1000 at 9%
13. \$202.50
14. \$2000
15. \$16.80
16. (a) \$160 (b) \$40
17. (a) 25% (b) 25%
18. $\dfrac{y+3}{2}$
19. $\dfrac{6+2x}{7}$
20. (a) $\dfrac{-2b}{3}$ (b) $\dfrac{-3a}{2}$
21. (a) $\dfrac{7u}{3}$
    (b) *Check:*
    $$7u - 3\left(\frac{7u}{3}\right) \stackrel{?}{=} 0$$
    $$7u - 7u \stackrel{?}{=} 0$$
    $$0 \stackrel{\checkmark}{=} 0$$
22. $\dfrac{1-by}{a-c}$, $a \neq c$
23. (a) $\geq$ (b) $\leq$ (c) both
24. (a) false (b) true (c) true
25. (a) $(4, 7)$ (b) $[-2, -1]$ (c) $(-\infty, 2)$
26. (a) $<$ (b) $>$ (c) $>$
27. (a) $\geq$ (b) $\geq$ (c) $\geq$
28. $(-3, \infty)$
29. $\left[\frac{-1}{3}, \infty\right)$
30. $[-6, \infty)$
31. $[2, 5]$
32. $[350, 700]$

And These from Chapters 1–5, p. 302

33. 0
34. 5
35. -17
36. $5x^2 - 14x + 5$
37. $4y^3 - 5y^2 - 6y$
38. $5a^3 - 2a + 3$
39. $\dfrac{10}{9}$
40. $\dfrac{13}{7}$
41. 68 miles
42. $x^2(3y+4)(3y-4)$
43. $(3x-5)(x-2)$
44. $5a^2(a+3b)(a-3b)$
45. 5
46. $\dfrac{1}{x(x-1)}$
47. $\dfrac{x+1}{4x^4(x-4)}$
48. $\dfrac{3y^2+4y-16}{2(y+2)(y-2)}$

CHAPTER 7

Section 7.1, p. 309

1.

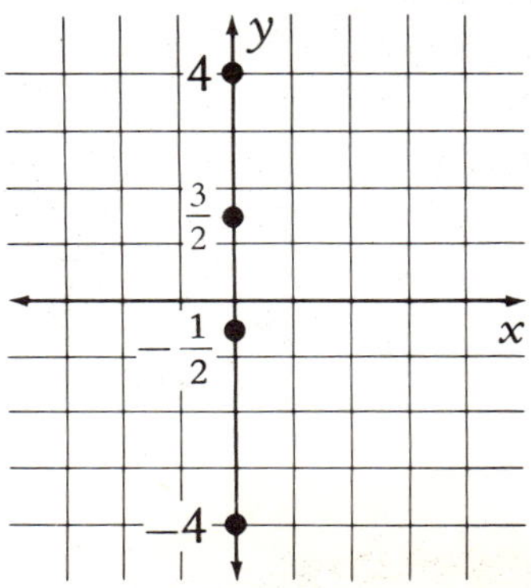

FIGURE 7A

3.

P:	(a) 5	(b) 1	(c) $P = (5, 1)$
Q:	(a) 2	(b) 8	(c) $Q = (2, 8)$
R:	(a) -3	(b) 2	(c) $R = (-3, 2)$
S:	(a) -3	(b) -2	(c) $S = (-3, -2)$
T:	(a) 3	(b) -2	(c) $T = (3, -2)$

5–27.

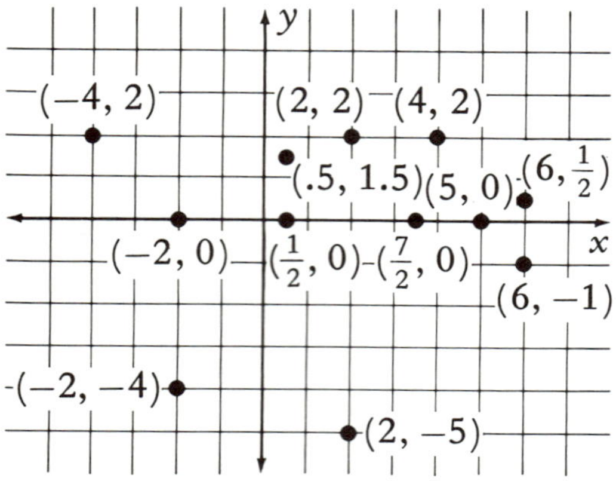

FIGURE 7B

Section 7.2, p. 318

1. (b), (c), and (d) are solutions.

3. (b), (c), (d), and (e) are solutions.

5. (a) $\frac{5}{3}$ (b) 5 (c) 2 (d) −1

7. *Table 7A*

x	$y = 7 - 2x$
0	7
1	5
2	3

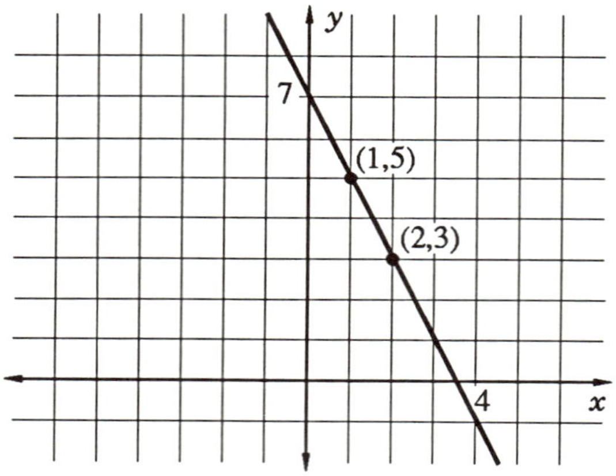

FIGURE 7C

9.

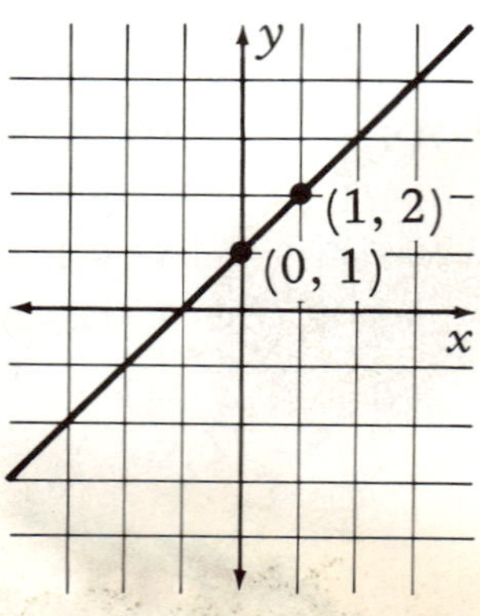

FIGURE 7D. The graph of $y = x + 1$

11.

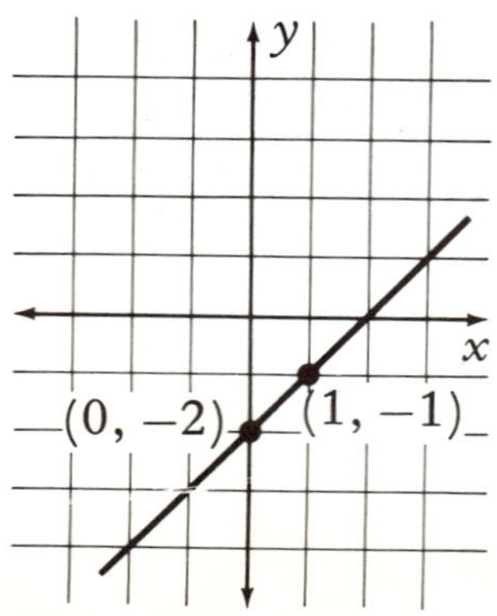

FIGURE 7E. The graph of $y = x - 2$

13.

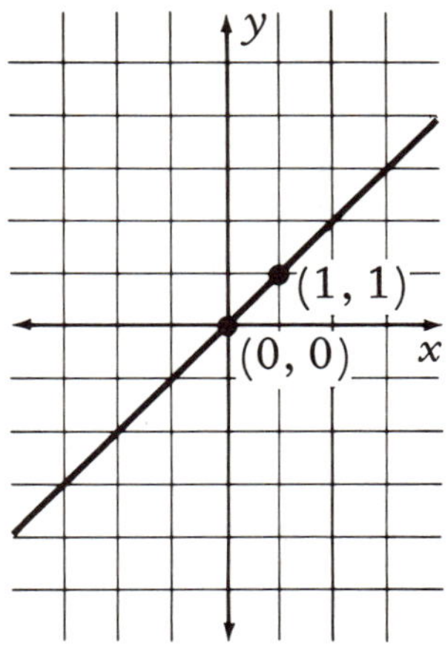

FIGURE 7F. The graph of y = x

15.

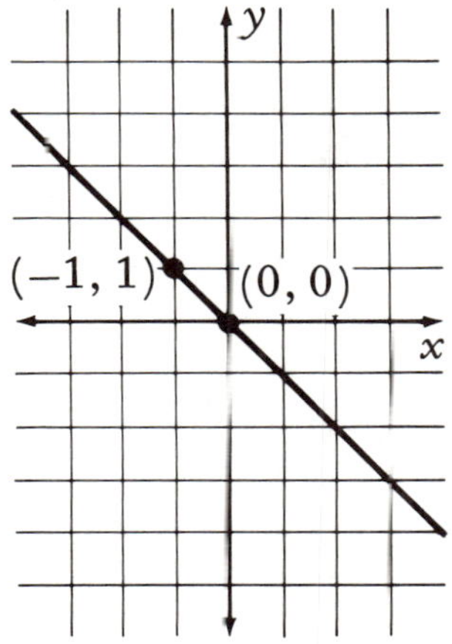

FIGURE 7G. The graph of y = −x

17.

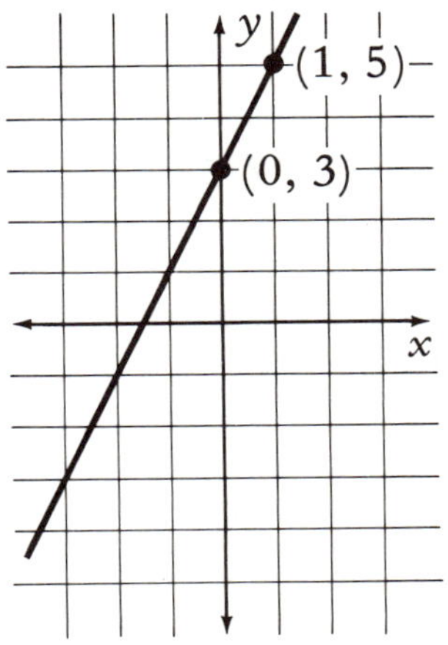

FIGURE 7H. The graph of y = 2x + 3

19.

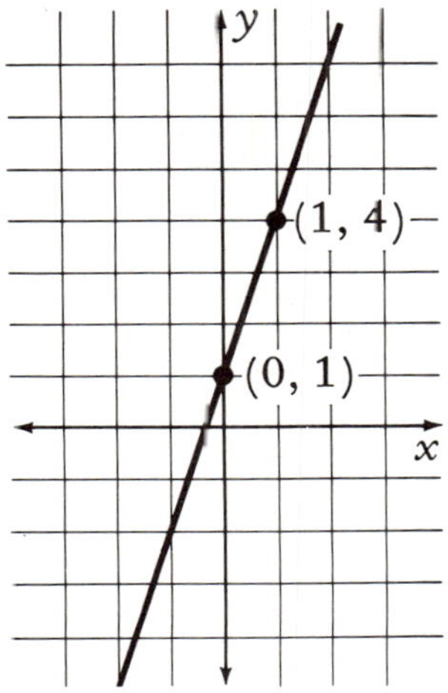

FIGURE 7I. The graph of y = 3x + 1

21.

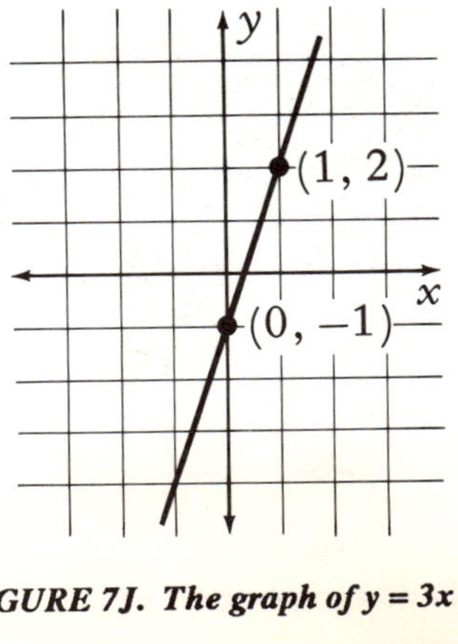

FIGURE 7J. The graph of y = 3x − 1

23.

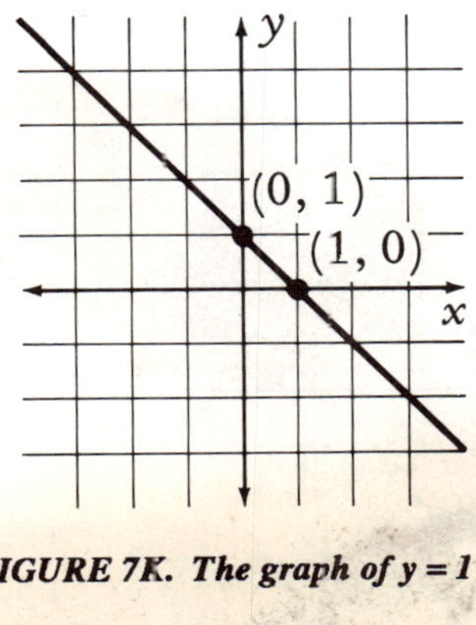

FIGURE 7K. The graph of y = 1 − x

25.

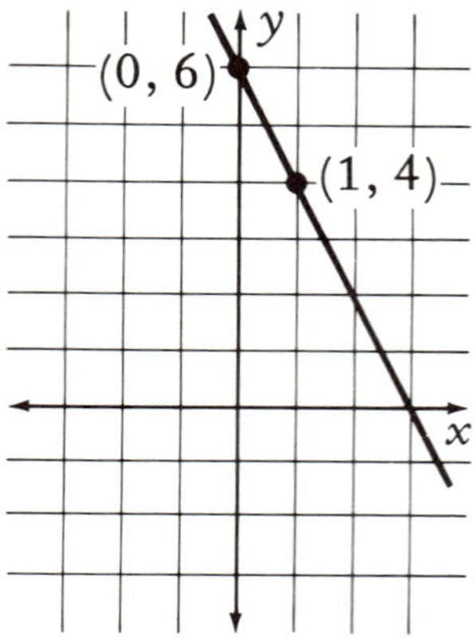

FIGURE 7L. The graph of $y = 6 - 2x$

27.

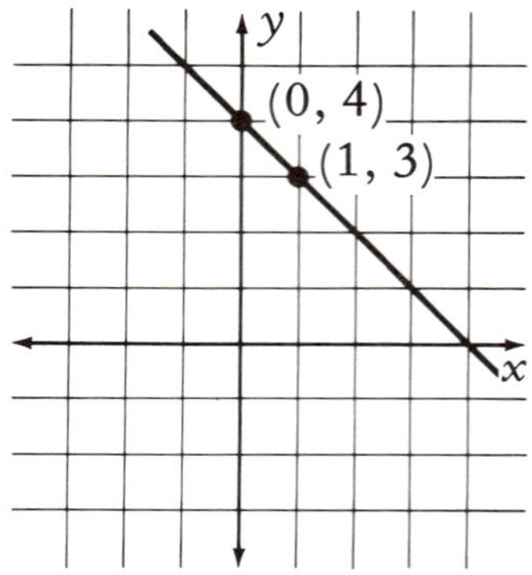

FIGURE 7M. The graph of $x + y = 4$

29.

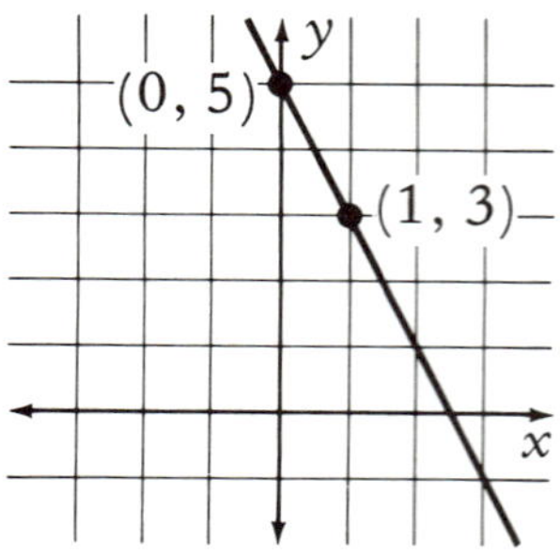

FIGURE 7N. The graph of $2x + y = 5$

31. (a) (See Figure 7O.) (b) \$8 (c) $7\frac{1}{2}$ items

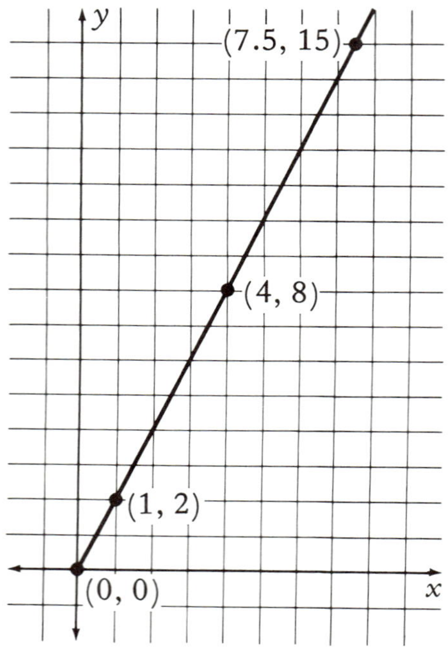

FIGURE 7O

33.

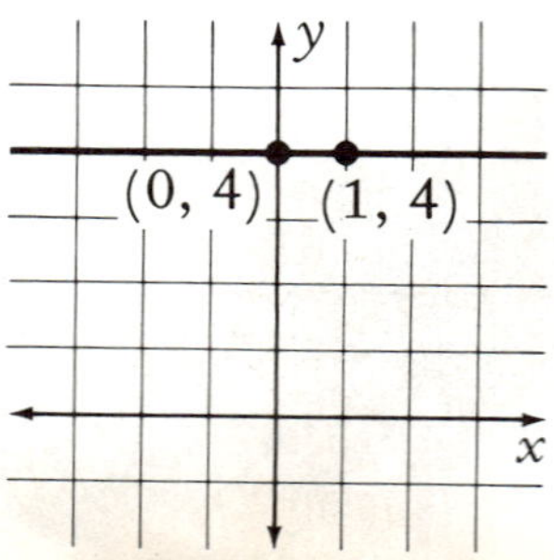

FIGURE 7P. The graph of $y = 4$

35.

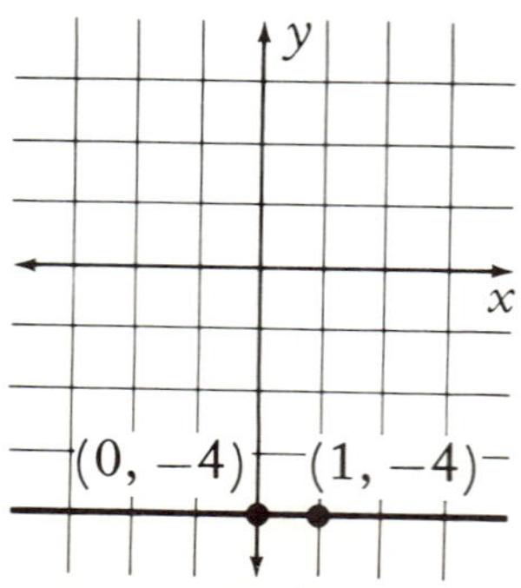

FIGURE 7Q. The graph of $y = -4$

37. (a) (*iii*), (b) (*v*), (c) (*ii*), (d) (*iv*)

Section 7.3, p. 327

1. (a) 1 (b) 1 (c) 1 (d) upward

3. (a) 2 (b) 1 (c) 2 (d) upward

5. (a) 1 (b) 2 (c) $\frac{1}{2}$ (d) upward

7. (a) −2 (b) 1 (c) −2 (d) downward

9. (a) −1 (b) −3 (c) $\frac{1}{3}$ (d) upward

11. (a) 2 (b) −4 (c) $\frac{-1}{2}$ (d) downward

13. (a) 1 (b) 4 (c) $\frac{1}{4}$ (d) upward

15. (a) $\frac{3}{2}$ (b) $\frac{1}{2}$ (c) 3 (d) upward

17. (a) 0 (b) 2 (c) 0 (d) neither

19. (a) −2 (b) 14 (c) $\frac{-1}{7}$ (d) downward

21. (a) $\frac{5-2}{2-0} = \frac{3}{2}$ (b) $\frac{11-2}{6-0} = \frac{9}{6} = \frac{3}{2}$

23.

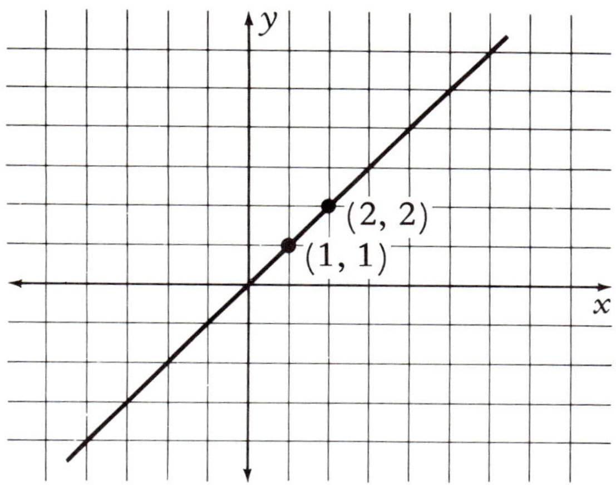

FIGURE 7R

25.

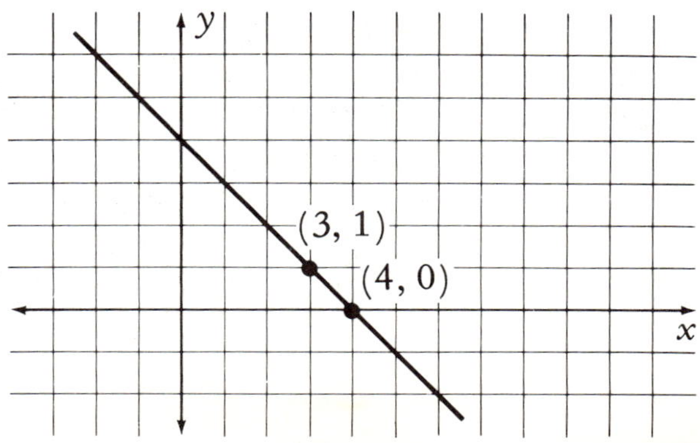

FIGURE 7S

27.

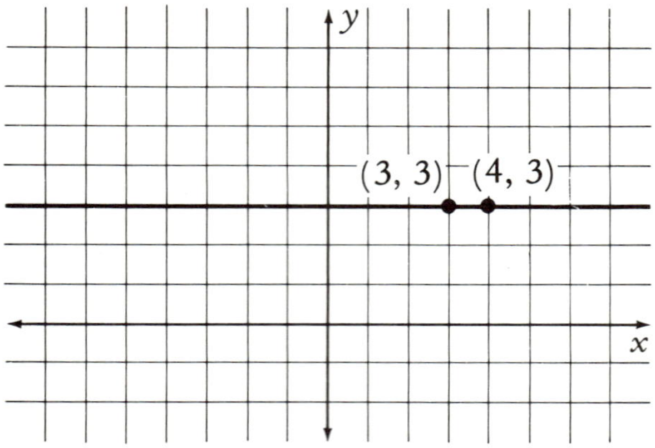

FIGURE 7T

29.

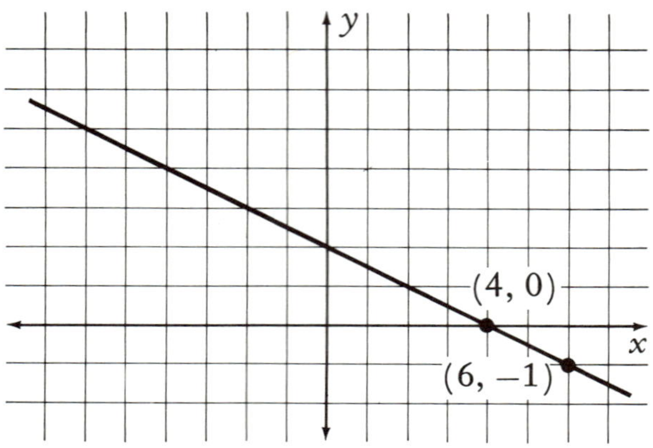

FIGURE 7U

31.

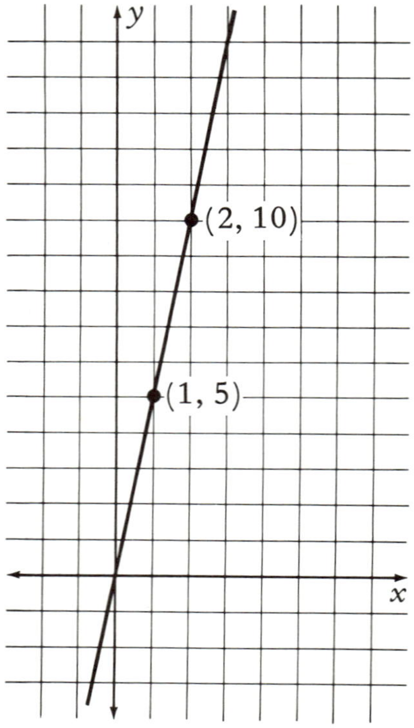

FIGURE 7V

33. (a) 4 (b) (*iii*) **35.** (a) 0 (b) (*i*) **37.** (a) 0 (b) (*i*) **39.** (a) −1 (b) (*iv*)

Section 7.4, p. 338

1. (a) 2 (b) 4 **3.** (a) $\frac{1}{2}$ (b) 5 **5.** (a) −1 (b) 2

7. (a) $\frac{1}{2}$ (b) 1 **9.** (a) 3 (b) 3 **11.** (a) 3 (b) −2

13. (a) $\frac{1}{2}$ (b) 4 **15.** (a) 2 (b) 10 **17.** (a) 0 (b) 5

19. $y = 4x + 1$

21. $y = -x + 5$

23. $y = \frac{1}{2}x + \frac{1}{4}$

25. $y = -2x + \frac{1}{2}$

27. (a) 2 (b) 1 (c) (*See Figure 7W.*)

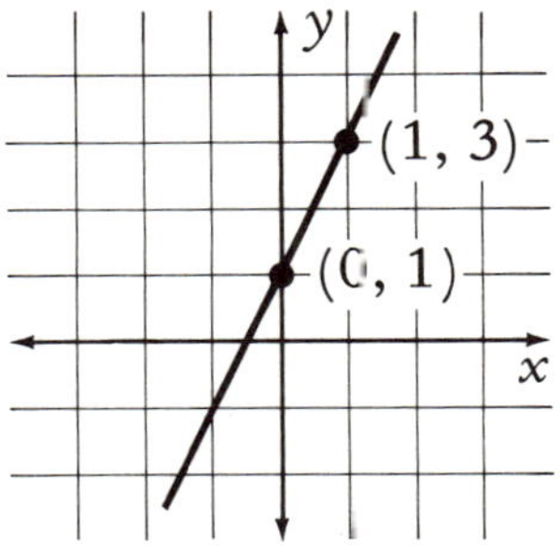

FIGURE 7W

29. (a) −2 (b) 1 (c) (*See Figure 7X.*)

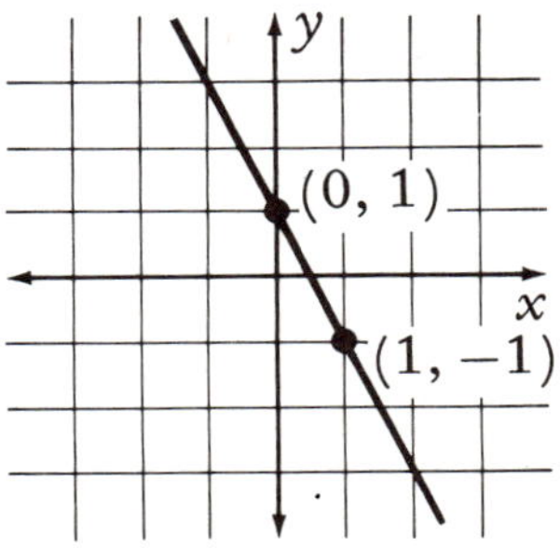

FIGURE 7X

31. (a) $\frac{1}{2}$ (b) $\frac{1}{2}$ (c) (*See Figure 7Y.*)

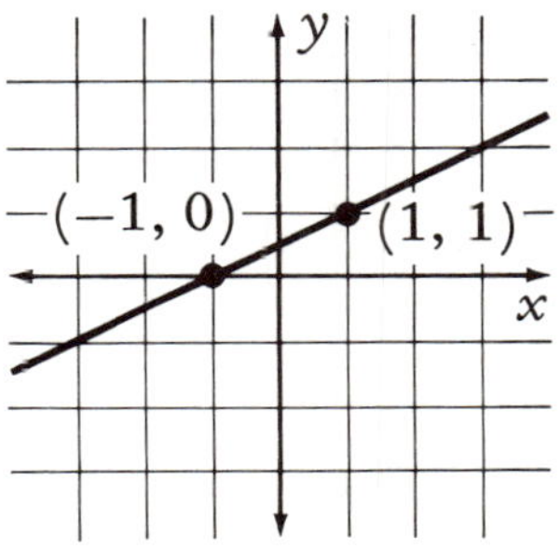

FIGURE 7Y

33. (a) $\frac{x}{2} + \frac{y}{4} = 1$ (b) $2x + y = 4$

35. (a) $\frac{x}{3} + \frac{y}{-3} = 1$ (b) $x - y = 3$

37. (a) $\frac{x}{-7} + y = 1$ (b) $7y - x = 7$

39. (a) $\frac{x}{3} + \frac{y}{3} = 1$ (b) $x + y = 3$ (c) (*See Figure 7Z.*)

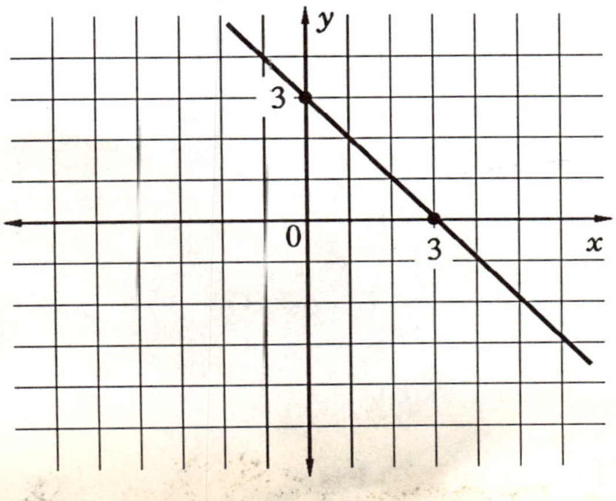

FIGURE 7Z

41. (a) $\frac{x}{2} + y = 1$ (b) $x + 2y = 2$ (c) *(See Figure 7AA.)*

43. $y = \frac{x}{2}$

45. $y = \frac{-3}{8}x$

47. $y = 3$

49. $y = 2$

51. $x = 2$

53. (a) $y = -3$ (b) *(See Figure 7BB.)*

55. (a) $x = 4$ (b) *(See Figure 7CC.)*

57. (a) $y = 0$ (b) *(See Figure 7DD.)*

59. $(7, 4)$

61. (a) and (c) are the errors.

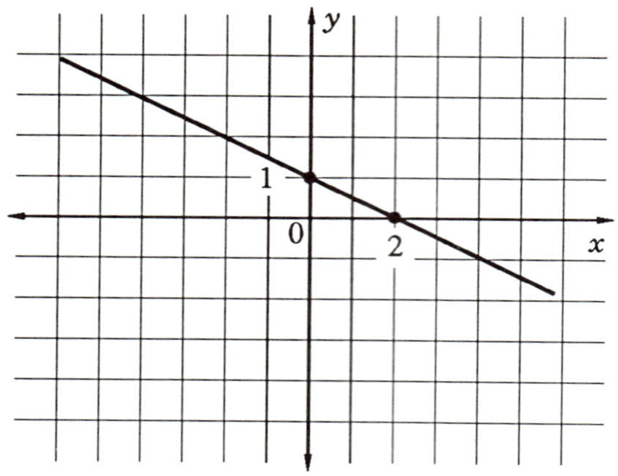

FIGURE 7AA

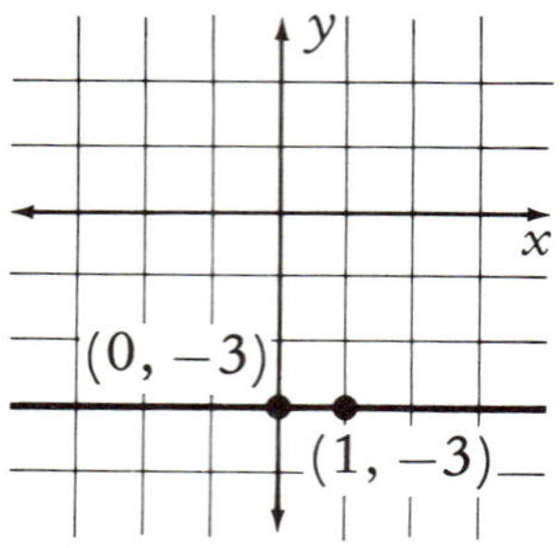

FIGURE 7BB

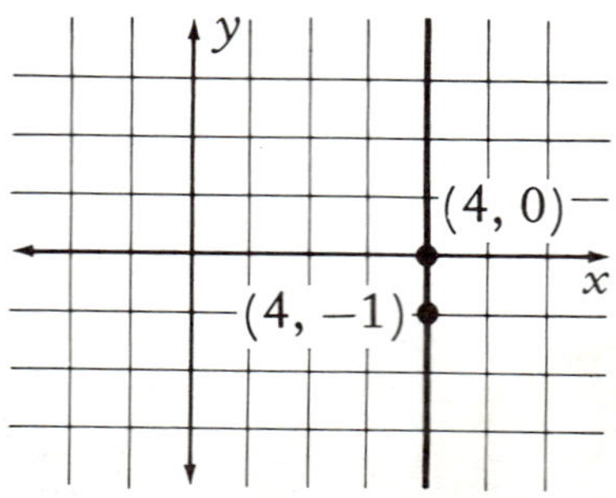

FIGURE 7CC

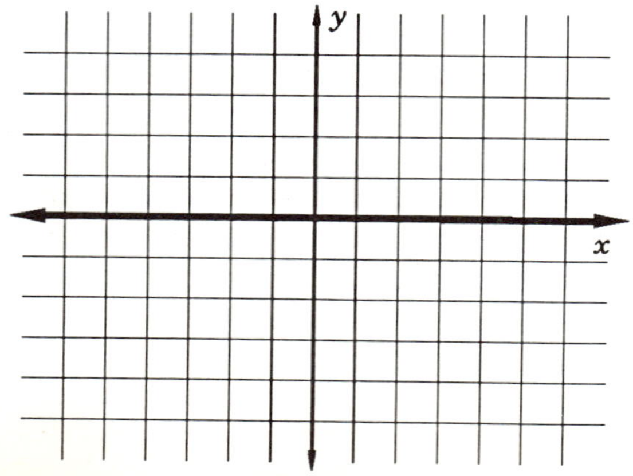

FIGURE 7DD

Section 7.5, p. 346 *(The checks follow number 33.)*

1.

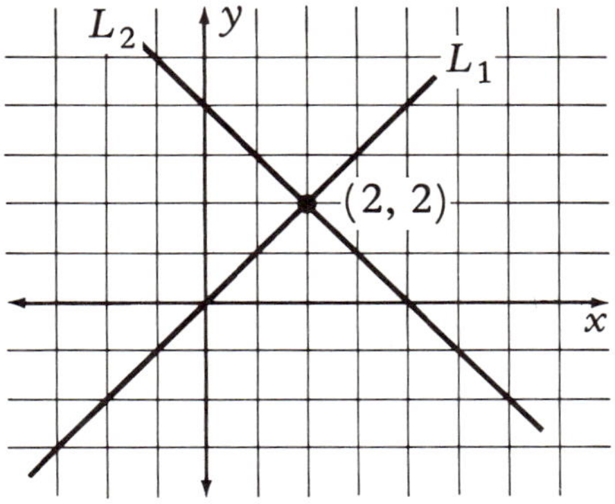

FIGURE 7EE

3.

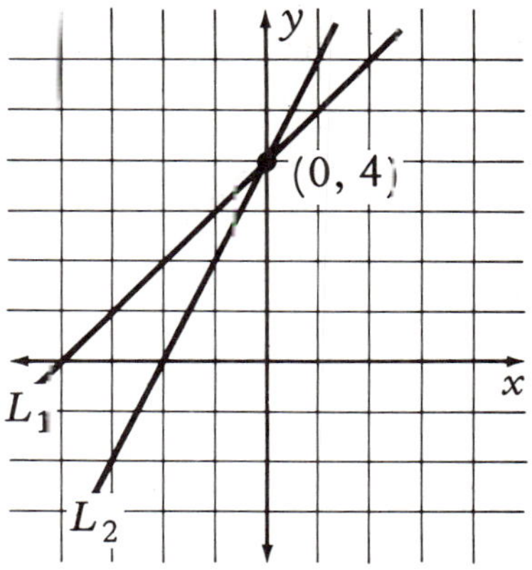

FIGURE 7FF

5.

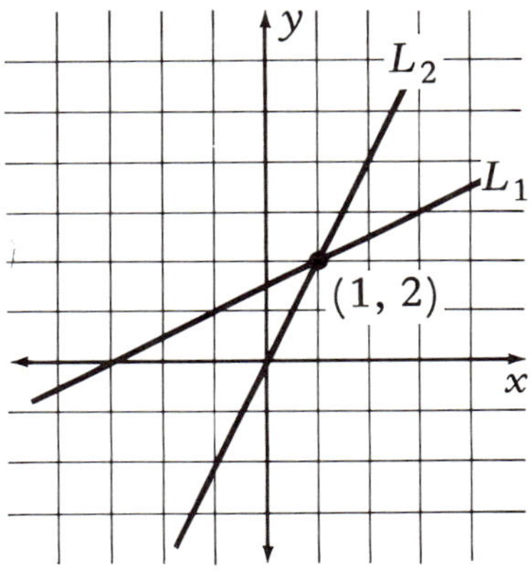

FIGURE 7GG

7.

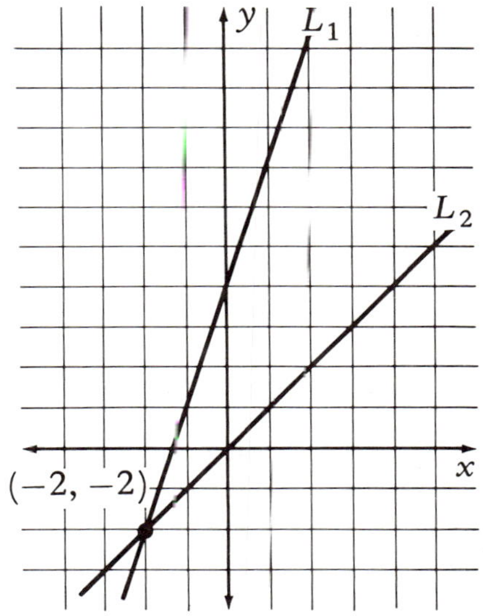

FIGURE 7HH

9.

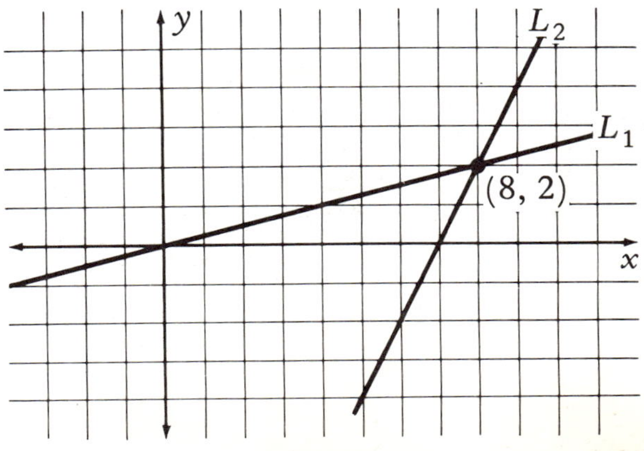

FIGURE 7II

11.

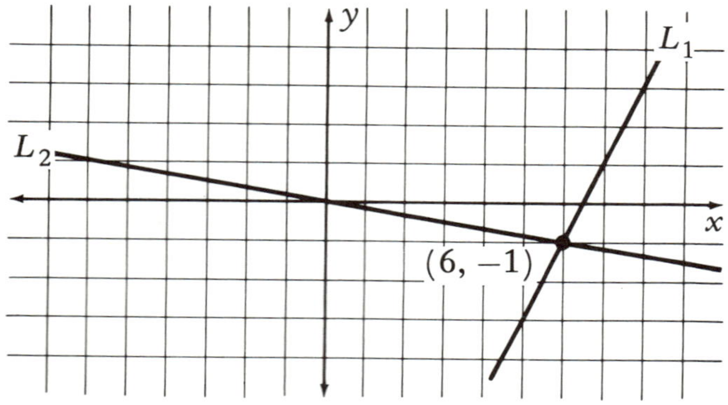

FIGURE 7JJ

13.

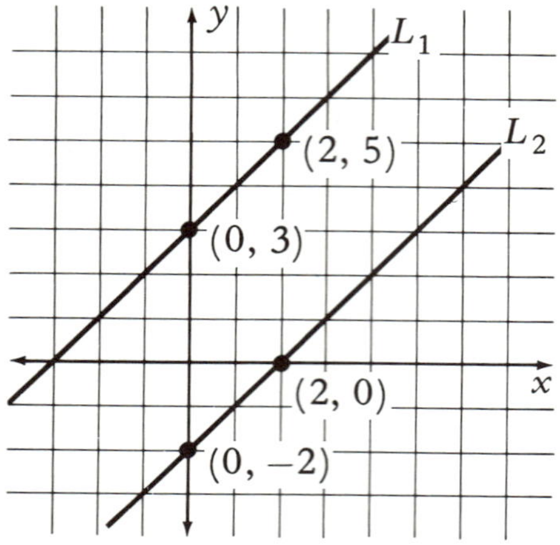

FIGURE 7KK. For any value of x, the y-values of the corresponding points on L_1 and L_2 are 5 units apart.

15.

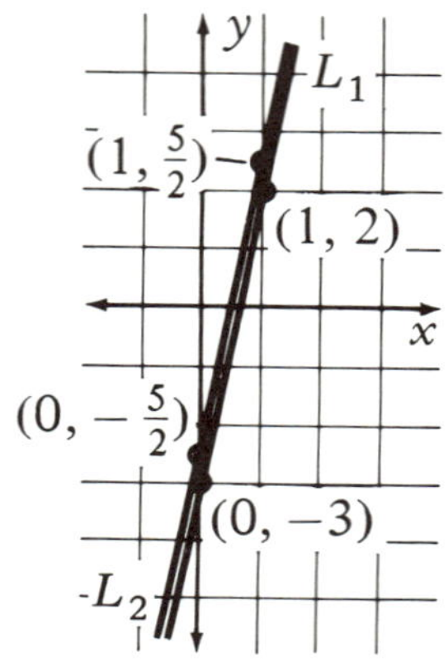

FIGURE 7LL. For any value of x, the y-coordinates of the corresponding points on L_1 and L_2 are $\frac{1}{2}$ unit apart.

17.

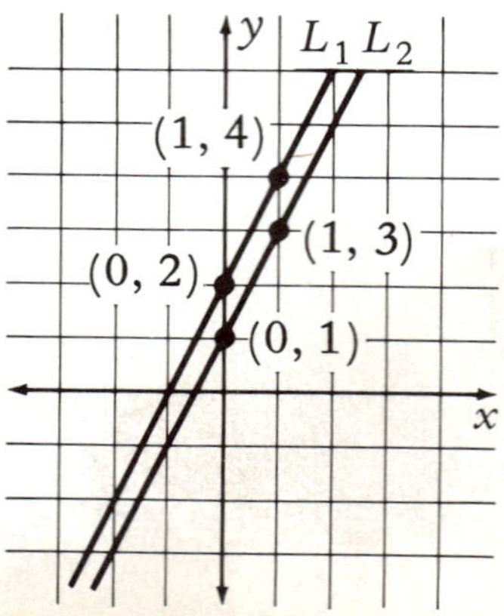

FIGURE 7MM. For any value of x, the y-coordinates of the corresponding points on L_1 and L_2 are 1 unit apart.

19.

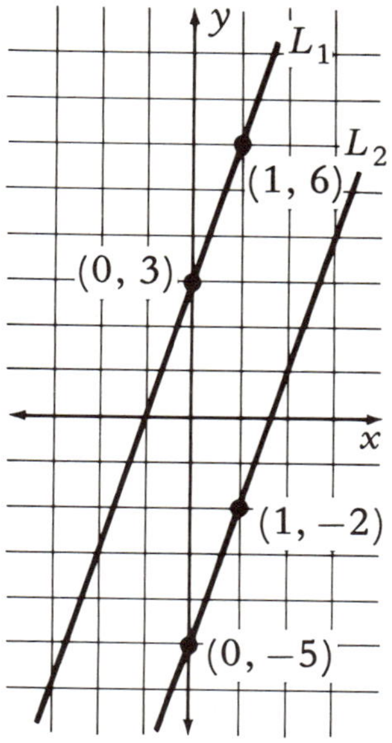

FIGURE 7NN. For any value of x, the y-coordinates of the corresponding points on L_1 and L_2 are 8 units apart.

21.

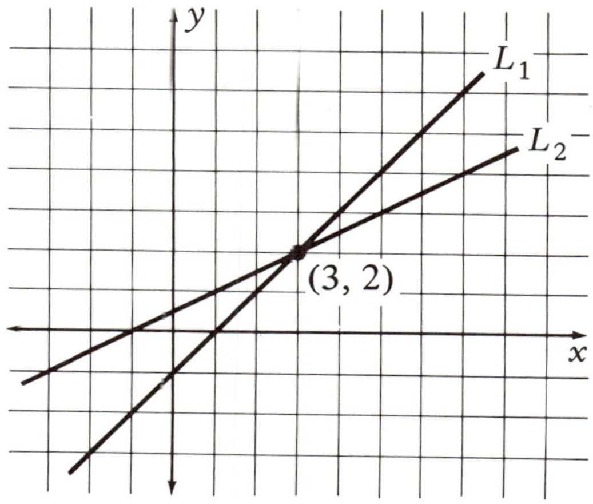

FIGURE 7OO

23.

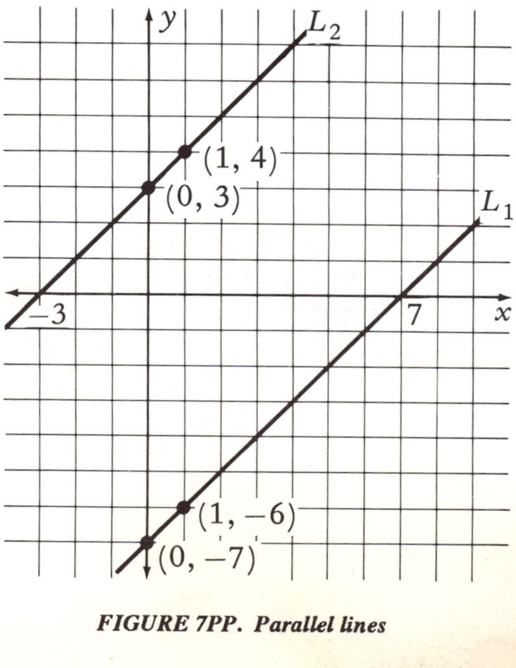

FIGURE 7PP. Parallel lines

25.

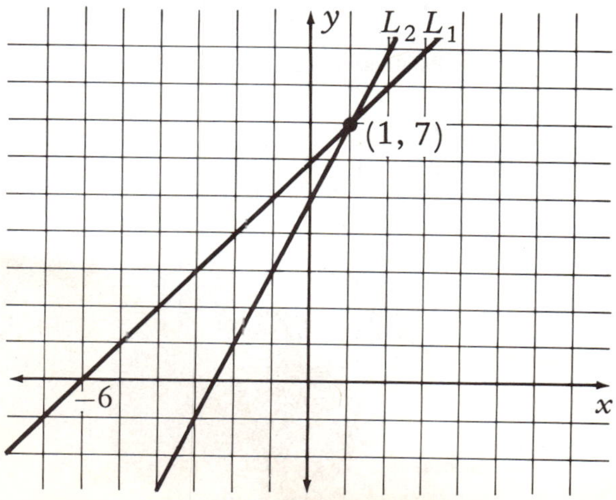

FIGURE 7QQ

27. 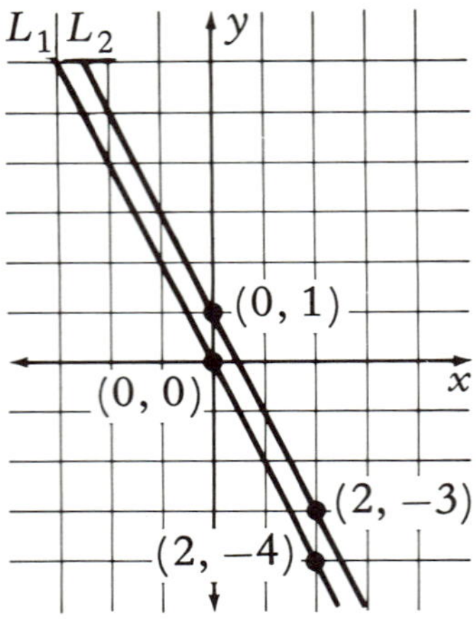

FIGURE 7RR. Parallel lines

29. L_1 and L_4

31. L_1 and L_3

33. (e) is the error.

Checks:

1. L_1: $2 \stackrel{\checkmark}{=} 2$ | L_2: $2 \stackrel{?}{=} 4 - 2$, $2 \stackrel{\checkmark}{=} 2$

3. L_1: $4 - 4 \stackrel{?}{=} 0$, $0 \stackrel{\checkmark}{=} 0$ | L_2: $4 - 2 \stackrel{?}{=} 2(0 + 1)$, $2 \stackrel{\checkmark}{=} 2$

Section 7.6, p. 356 *(The checks follow number 43.)*

1. $3 \stackrel{?}{=} 2 + 1$, $3 \stackrel{\checkmark}{=} 3$ | $3(3) \stackrel{?}{=} 4(2) + 1$, $9 \stackrel{\checkmark}{=} 9$

   Thus, (2, 3) is a solution of the system.

3. $3(-2) - 2(-4) \stackrel{?}{=} 2$, $-6 + 8 \stackrel{?}{=} 2$, $2 \stackrel{\checkmark}{=} 2$ | $5(-2) + (-4) \stackrel{?}{=} -6$, $-10 - 4 \stackrel{?}{=} -6$, $-14 \stackrel{x}{=} -6$

   Thus, (−2, −4) is not a solution of the system.

5. (3, 4)
7. (4, 5)
9. (1, 1)
11. $u = 5, v = 6$
13. $a = 5, b = 10$
15. (3, 1)
17. $a = 1, b = \frac{-3}{2}$
19. $c = 1, d = \frac{1}{5}$
21. $r = 2, s = 3$
23. (2, 1)
25. (4, 3)
27. (2, 2)
29. (18, 24)
31. $a = 3, b = 1$
33. (−3, −2)
35. (5, 1)
37. (10, 2)
39. 10 and 7
41. Each atlas is $1\frac{1}{2}$ inches thick; each dictionary is 2 inches thick.
43. Potatoes cost $.79 per pound; onions cost $.72 cents per pound.

Checks:

5. $4 \stackrel{?}{=} 3 + 1$, $4 \stackrel{\checkmark}{=} 4$ | $4 \stackrel{?}{=} 3(3) - 5$, $4 \stackrel{?}{=} 9 - 5$, $4 \stackrel{\checkmark}{=} 4$

7. $5 \stackrel{?}{=} 2(4) - 3$, $5 \stackrel{?}{=} 8 - 3$, $5 \stackrel{\checkmark}{=} 5$ | $5 \stackrel{?}{=} 3(4) - 7$, $5 \stackrel{?}{=} 12 - 7$, $5 \stackrel{\checkmark}{=} 5$

15. $3 + 1 \stackrel{?}{=} 4$, $4 \stackrel{\checkmark}{=} 4$ | $3 - 1 \stackrel{?}{=} 2$, $2 \stackrel{\checkmark}{=} 2$

17. $1 + 2\left(\frac{-3}{2}\right) \stackrel{?}{=} -2$, $1 - 3 \stackrel{?}{=} -2$, $-2 \stackrel{\checkmark}{=} -2$ | $3(1) - 2\left(\frac{-3}{2}\right) \stackrel{?}{=} 6$, $3 + 3 \stackrel{?}{=} 6$, $6 \stackrel{\checkmark}{=} 6$

Section 7.7, p. 361

1. 2
3. 3
5. −2
7. 2
9. 10
11. 10
13. 40
15. −144
17. 2
19. 6
21. $\frac{-1}{3}$
23. .35

25. .01 **27.** \$4.20 **29.** (a) 225 miles (b) 6 hours and 40 minutes

31. 3

Let's Review Chapter 7, p. 363 *(The check follows number 46.)*

1.

P:	(a) 1	(b) 4	(c) $P = (1, 4)$
Q:	(a) −2	(b) 3	(c) $Q = (-2, 3)$
R:	(a) −4	(b) −2	(c) $R = (-4, -2)$
S:	(a) 3	(b) −3	(c) $S = (3, -3)$
T:	(a) 4	(b) 0	(c) $T = (4, 0)$

2–6.

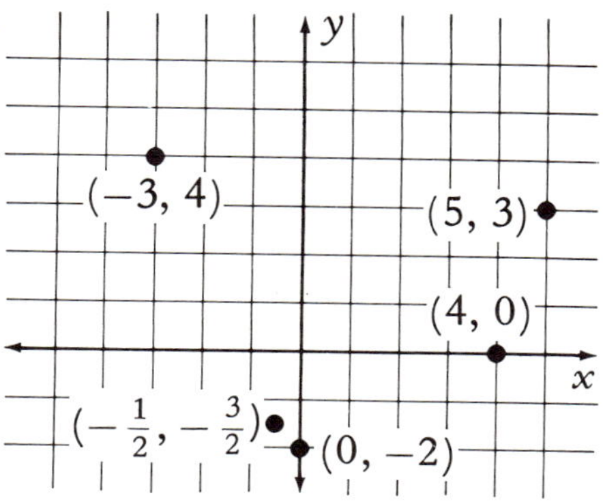

FIGURE 7SS

7.

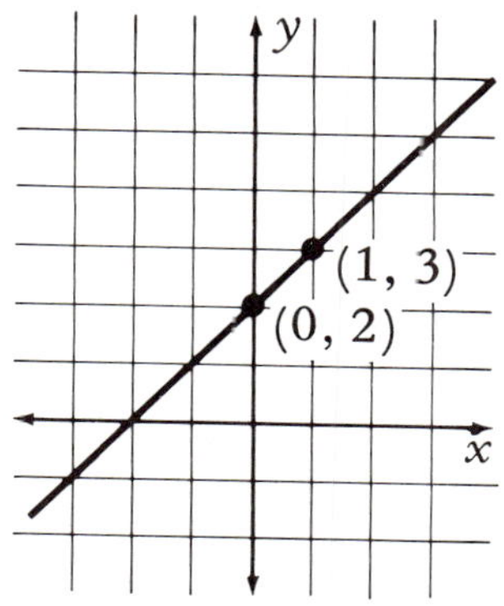

FIGURE 7TT. The graph of $y = x + 2$

8.

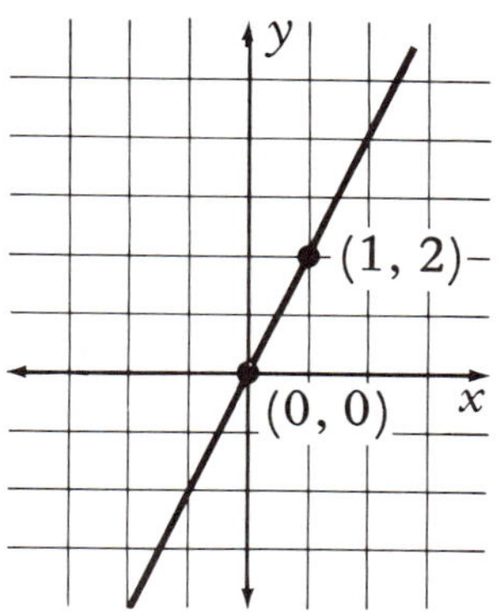

FIGURE 7UU. The graph of $y = 2x$

9.

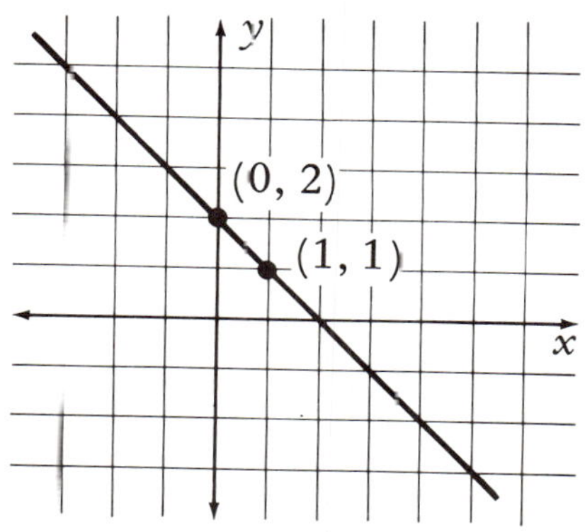

FIGURE 7VV. The graph of $y = 2 - x$

10.

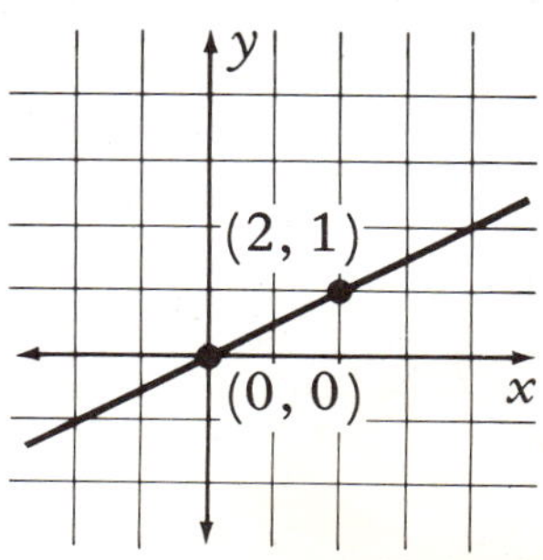

FIGURE 7WW. The graph of $y = \frac{x}{2}$

11.

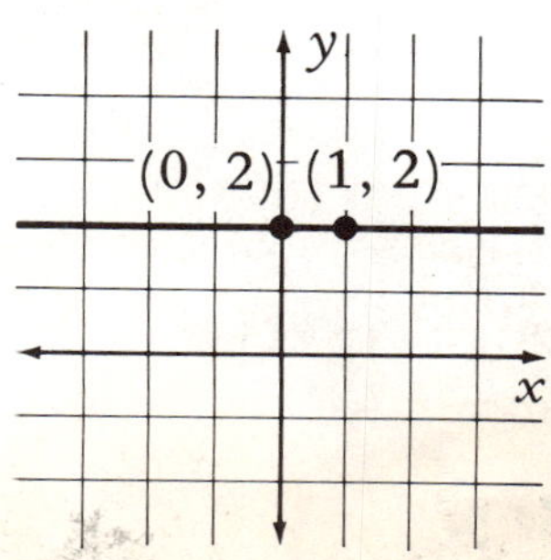

FIGURE 7XX. The graph of $y = 2$

12.

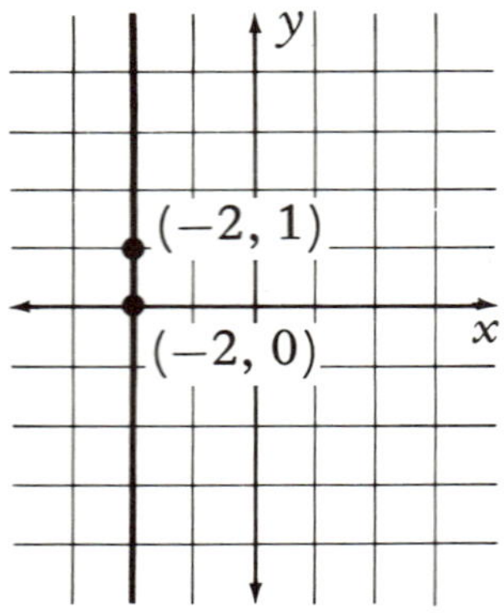

FIGURE 7YY. The graph of $x = -2$

13. (a) 5 (b) 2 (c) $\frac{5}{2}$ (d) upward to the right

14.

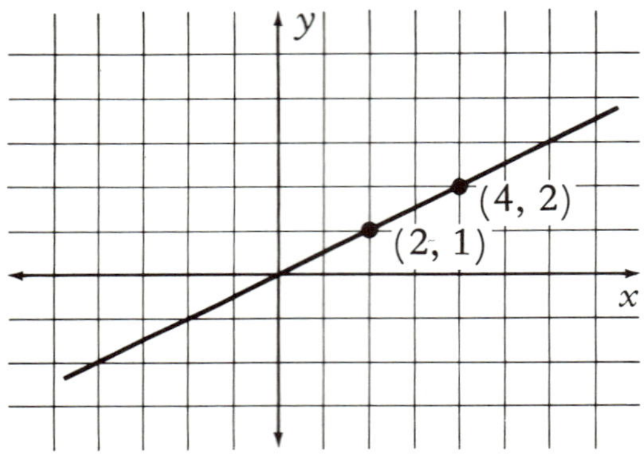

FIGURE 7ZZ

15. (a) (*iv*) (b) (*i*)

16. (a) $\frac{-1}{2}$ (b) 7

17. $y = 4x$

18. $\frac{x}{3} + \frac{y}{-2} = 1$, or $2x - 3y = 6$

19. $y = 5$

20. (a) $\frac{x}{-7} + \frac{y}{4} = 1$ (b) $-4x + 7y = 28$

21.

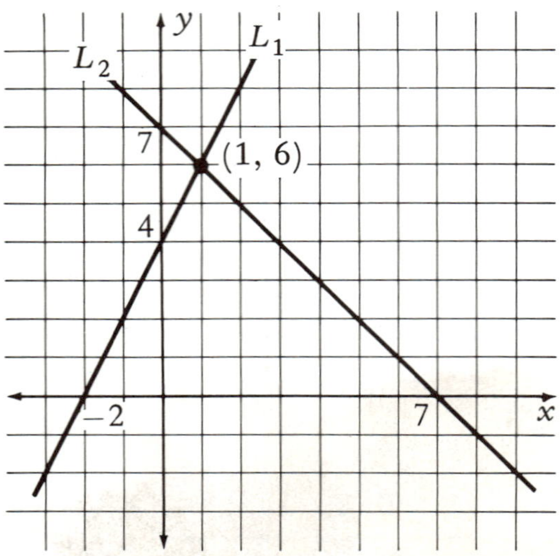

FIGURE 7AAA

22.

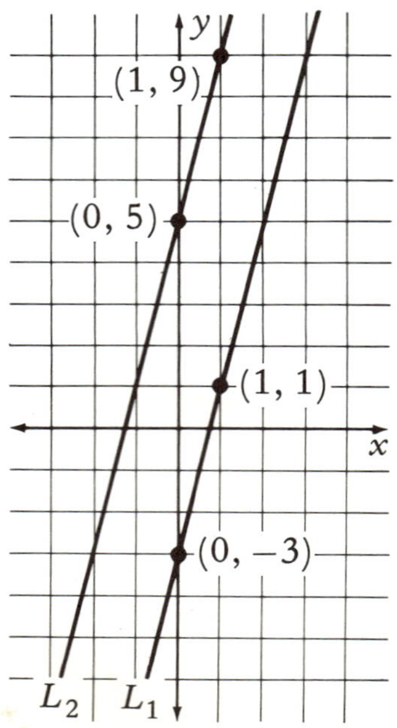

FIGURE 7BBB. For any value of x, the y-values of the corresponding points on L_1 and L_2 are 8 units apart.

23.

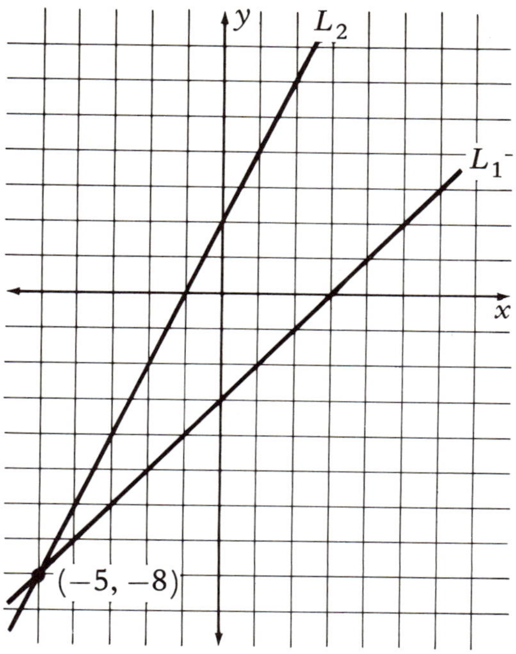

FIGURE 7CCC

24. L_3 and L_4

25. $\left(\frac{5}{3}, \frac{13}{3}\right)$

26. (2, 3)

27. (2, 1)

28. (10, −7)

29. −2

30. 15

31. 180 pounds

And These from Chapters 1–6, p. 365

32. −43

33. 3

34. $20a^2b^2(x-2)^4$

35. $2x^2 - 20x + 14$

36. $2^5 \cdot 3^2$

37. 14

38. 6

39. −24

40. 4

41. $x^2(2x+3)(2x-3)$

42. $(2x-5)(3x-2)$

43. $a^2x(x+1)^2$

44. $\dfrac{10}{5 - 2y + 3a}$

45. \$6000

46. (−2, 5)

Check:

26. $3 + 1 \stackrel{?}{=} 2(2)$ $\frac{3}{3} \stackrel{?}{=} 3 - 2$

$4 \stackrel{\checkmark}{=} 4$ $1 \stackrel{\checkmark}{=} 1$

CHAPTER 8

Section 8.1, p. 371

1. 196

3. 289

5. 625

7. 10 000

9. 4

11. 7

13. −5

15. 9

17. −1000

19. 0

21. 8 inches

23. 4 feet

25. 9 square feet

27. x

29. 17

31. ab

33. ac^3

35. c^2

37. y^5

39. $(xy)^2$

41. $(ab^3)^2$ or a^2b^6 43. x^9 45. a^{20} 47. 4

49. 16

Section 8.2, p. 378

1. irrational
3. irrational
5. rational
7. rational
9. $2 < \sqrt{8} < 3$
11. $4 < \sqrt{23} < 5$
13. $7 < \sqrt{53} < 8$
15. $9 < \sqrt{85} < 10$
17. $11 < \sqrt{125} < 12$
19. $20 < \sqrt{401} < 21$
21. 4.2
23. .4
25. .3
27. 3.5
29. .9
31. 1.4
33. 3.1
35. 6.6
37. 14.4
39. 6 inches
41. 10 inches
43. $4\sqrt{2}$ inches
45. $\sqrt{2}$ meters
47. 5 inches

Section 8.3, p. 385

1. 70
3. 500
5. 20 000
7. $4a^2$
9. $10a^9$
11. xy^2
13. a^4x^5
15. $6a^2x$
17. (a) $3\sqrt{2}$ (b) 4.2
19. (a) $6\sqrt{2}$ (b) 8.5
21. (a) $5\sqrt{3}$ (b) 8.7
23. $a\sqrt{a}$
25. $2b^2\sqrt{b}$
27. $ab^2\sqrt{ab}$
29. 9
31. undefined
33. undefined
35. $\frac{3}{5}$
37. $\frac{9}{2}$
39. $\frac{10}{11}$
41. $\frac{x}{y^4}$
43. $\frac{5}{d^4}$
45. $\frac{12x^2}{yz^5}$
47. $\frac{a^2\sqrt{a}}{b^3}$
49. $\frac{3a^4\sqrt{2a}}{2b}$
51. $\frac{5a\sqrt{a}}{7b^4}$
53. 7
55. $\sqrt{a^2+b^2}$
57. (a) and (f) are the errors.
59. .3
61. .001
63. $.9x^4y^2$
65. (a), (c), and (e) are the errors.

Section 8.4, p. 388

1. $2\sqrt{2}$
3. $4\sqrt{2}$
5. $5\sqrt{6}$
7. $8\sqrt{11}$
9. $7\sqrt{5}$
11. $3\sqrt{2}$
13. $2\sqrt{2} + 3\sqrt{3}$
15. $5\sqrt{7} + 2\sqrt{5}$
17. $3\sqrt{2}$
19. 0
21. $\sqrt{7}$
23. $7\sqrt{3}$
25. $5\sqrt{a}$
27. $19y\sqrt{y}$
29. $23b\sqrt{a}$
31. $3\sqrt{3a}$
33. $12x\sqrt{2x}$
35. $6b\sqrt{7ab}$
37. $\frac{2+\sqrt{5}}{5}$
39. $\frac{\sqrt{2}-\sqrt{5}+\sqrt{7}}{3}$
41. $\frac{7\sqrt{2}-5\sqrt{3}}{35}$
43. $\frac{2\sqrt{2}+3\sqrt{3}}{24}$
45. $\frac{3\sqrt{5}-10\sqrt{3}}{24}$
47. $\frac{31\sqrt{3}}{18}$

Section 8.5, p. 394

1. $\sqrt{10}$
3. $2\sqrt{14}$
5. 7
7. 10
9. a
11. b^3
13. $10x$
15. $xy^2z^2\sqrt{x}$
17. $2+\sqrt{6}$
19. 3
21. 47
23. $38+16\sqrt{5}$
25. $-23-5\sqrt{55}$
27. $3+2\sqrt{2}$
29. 3
31. -2
33. $2\sqrt{3}$
35. $\frac{\sqrt{2}}{3}$
37. $\frac{4\sqrt{y}}{3z}$
39. $\frac{3}{10}$
41. (a) $3(1+\sqrt{2})$ (b) $\frac{1+\sqrt{2}}{3}$
43. (a) $5(1+\sqrt{2})$ (b) $\frac{-(1+\sqrt{2})}{3}$
45. (a) $\sqrt{2}(1+\sqrt{3})$ (b) $\frac{1+\sqrt{3}}{2}$
47. (a) $\sqrt{ab}(a+\sqrt{c})$ (b) $\frac{a+\sqrt{c}}{ab}$
49. (a) $c(3a-4b)$ (b) 5
51. $\frac{5}{6}$
53. 2
55. 1
57. $\frac{5st\sqrt{s}}{12}$

Section 8.6, p. 398

1. $\frac{1}{4}$
3. $-\frac{1}{2}$
5. $\frac{1}{64}$
7. $-\frac{1}{16}$
9. $-\frac{1}{27}$
11. $\frac{1}{125}$
13. $\frac{1}{10\,000}$
15. 1
17. 1
19. $\frac{1}{16}$
21. $\frac{3}{4}$
23. 0
25. -2
27. -3
29. -2
31. (a) and (b) are the errors.
33. $\frac{5}{3}$
35. 4
37. $\frac{25}{16}$
39. $\frac{64}{27}$
41. 100 000
43. $\frac{100}{9}$
45. -1
47. -1
49. -3
51. a
53. y^{-10}
55. b^{-3}
57. b^{-10}
59. 1
61. a^4

Section 8.7, p. 402

1. 1.25×10^1
3. 7.028×10^3
5. 4.97×10^{-1}
7. 8×10^{-1}
9. 10^2
11. 3.61×10^5
13. 5×10^{-6}
15. 6
17. 65.2
19. 3 230 000
21 .5
23 .000 901
25. 878 000 000
27. 200.1
29. 54 314
31. 10^3

33. 4×10^{11} **35.** 1.24×10^{-7} **37.** 2.5×10^{11} **39.** 1.08×10^{15} miles

41. 1.60×10^{-13} joule **43.** 8.8×10^{-5}

Let's Review Chapter 8, p. 404

1. 5 **2.** 8 **3.** -11 **4.** x^2

5. $7 < \sqrt{55} < 8$ **6.** 9.9 **7.** 8 inches **8.** $\sqrt{41}$ inches

9. $5a$ **10.** a^2b^3 **11.** $\frac{9}{10}$ **12.** $\frac{2a^2\sqrt{5}}{b^4}$

13. $3\sqrt{5}$ **14.** $\sqrt{a}$ **15.** $17ab^2$ **16.** $\frac{2\sqrt{3}}{3}$

17. 10 **18.** x^2y **19.** 2 **20.** 2

21. 2 **22.** $\frac{2}{5}$ **23.** (a) $7(1+\sqrt{2})$ (b) $\frac{1+\sqrt{2}}{3}$

24. $\frac{4\sqrt{3}}{3}$ **25.** $\frac{8}{49}$ **26.** 1 **27.** $\frac{9}{4}$

28. x^{-7} **29.** 2.65×10^5 **30.** 2.57×10^{-2} **31.** .000 345

32. 4×10^2

And These from Chapters 1–7, p. 405

33. $\frac{3}{2abc}$ **34.** .0644 **35.** $\frac{83}{40}$

36. (a) > (b) = (c) < (d) < **37.** $2^4 \cdot 5^2$ **38.** $-5, -3$

39. $(x+6y)(x+2y)$ **40.** $(x+3)(x-3)(x+1)$ **41.** $\frac{-1}{(a-3)(a+b)}$

42. $\frac{1-2y}{5-3a}$, $a \neq \frac{5}{3}$ **43.** $\left[\frac{-1}{5}, 3\right]$

44. See Figure 8A. **45.** $y = -2x + 3$

46. $\frac{x}{4} + \frac{y}{5} = 1$ **47.** $y = -1$

48. (1, 3)

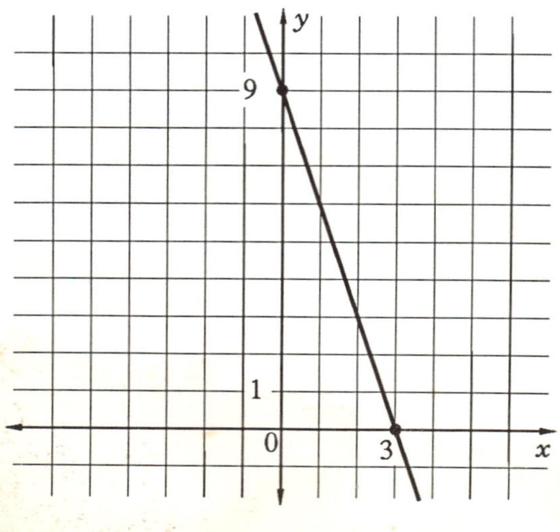

FIGURE 8A

CHAPTER 9

Section 9.1, p. 411 *(The checks follow number 41.)*

1. $2x^2 + 5x - 10 = 0$
3. $x^2 - 13x + 4 = 0$
5. $x^2 + 29x + 18 = 0$
7. $2x^2 + 3x + 5 = 0$
9. $x^2 - 5x - 5 = 0$
11. $x^2 - x - 12 = 0$
13. $x^2 + 2x = x^2 + 4x + 3$
 $2x = 4x + 3$
15. $x^2 + 4x - 9 = x^2 - 4x$
 $4x - 9 = -4x$
17. $x^2 + 8x + 16 = x^2 - 6x + 9$
 $8x + 16 = -6x + 9$
19. (b)
21. -2
23. -7
25. -1
27. $\frac{-11}{6}$
29. $\frac{5}{4}$
31. -4
33. 10
35. $\frac{-1}{13}$
37. $\frac{2}{3}$
39. -3
41. (b) is the error.

Checks:

21. $(-2)^2 + 5(-2) \stackrel{?}{=} (-2)^2 - 10$
$4 - 10 \stackrel{\checkmark}{=} 4 - 10$

25. $(-1+4)^2 \stackrel{?}{=} (-1-2)^2$
$3^2 \stackrel{?}{=} (-3)^2$
$9 \stackrel{\checkmark}{=} 9$

33. $(10-8)^2 + 6 \stackrel{?}{=} (10-7)^2 + 1$
$2^2 + 6 \stackrel{?}{=} 3^2 + 1$
$4 + 6 \stackrel{?}{=} 9 + 1$
$10 \stackrel{\checkmark}{=} 10$

Section 9.2, p. 417 *(The checks follow number 65.)*

1. $3, -7$
3. $0, \frac{-1}{4}$
5. $1, 4$
7. -4
9. $0, 2$
11. 3
13. $4, -1$
15. 1
17. $1, 2$
19. $4, -1$
21. $\frac{1}{2}, \frac{-1}{3}$
23. $\frac{-1}{2}, -1$
25. $\frac{1}{2}, -4$
27. $\frac{-1}{2}, \frac{-3}{2}$
29. $\frac{2}{3}, \frac{-1}{2}$
31. $\frac{3}{2}, 2$
33. $\frac{3}{5}, \frac{-1}{2}$
35. $0, \frac{1}{2}$
37. $\frac{3}{2}$
39. $\frac{-3}{4}$
41. $\frac{1}{2}, \frac{-1}{4}$
43. $5, -5$
45. $\frac{7}{2}, \frac{-7}{2}$
47. $\frac{4}{3}, \frac{-4}{3}$
49. $0, -3$
51. $0, 3, -5$
53. $0, 1, -1$
55. $x^2 - 4x + 3 = 0$
57. $x^2 - 8x + 15 = 0$
59. $x^2 - x = 0$
61. $3z^2 + 8x - 3 = 0$
63. $x^2 + 6x + 9 = 0$
65. (e) is the error.

Checks:

11. $3^2 - 6(3) + 9 \stackrel{?}{=} 0$
$9 - 18 + 9 \stackrel{?}{=} 0$
$0 \stackrel{\checkmark}{=} 0$

19. For 4:
$\frac{2}{1-4} + \frac{4}{2+4} \stackrel{?}{=} 0$
$\frac{2}{-3} + \frac{4}{6} \stackrel{?}{=} 0$
$\frac{-2}{3} + \frac{2}{3} \stackrel{?}{=} 0$
$0 \stackrel{\checkmark}{=} 0$

For -1:
$\frac{2}{1-(-1)} + \frac{-1}{2+(-1)} \stackrel{?}{=} 0$
$\frac{2}{2} + \frac{-1}{1} \stackrel{?}{=} 0$
$1 - 1 \stackrel{?}{=} 0$
$0 \stackrel{\checkmark}{=} 0$

39. $16\left(\frac{-3}{4}\right)^2 + 24\left(\frac{-3}{4}\right) + 9 \stackrel{?}{=} 0$

$16\left(\frac{9}{16}\right) - 18 + 9 \stackrel{?}{=} 0$

$9 - 18 + 9 \stackrel{?}{=} 0$

$0 \stackrel{\checkmark}{=} 0$

Section 9.3, p. 422 *(The checks follow number 51.)*

1. ± 4 **3.** ± 7 **5.** $\pm\sqrt{6}$ **7.** $\pm\sqrt{13}$

9. $\pm 2\sqrt{2}$ **11.** $\pm 3\sqrt{2}$ **13.** $\pm 3\sqrt{5}$ **15.** $\frac{\pm 2}{3}$

17. $\pm\sqrt{\frac{2}{3}}$ or $\frac{\pm\sqrt{6}}{3}$ **19.** $3, -4$ **21.** $\frac{6}{5}, \frac{-2}{5}$ **23.** $\frac{4}{3}, -4$

25. $\frac{2 \pm \sqrt{3}}{3}$ **27.** $\frac{-4 \pm \sqrt{7}}{5}$ **29.** $\frac{-5 \pm \sqrt{6}}{3}$ **31.** $\frac{1 \pm 2\sqrt{2}}{2}$

33. $-3 \pm 4\sqrt{2}$ **35.** $1 \pm 14\sqrt{2}$ **37.** 2 **39.** $\frac{3}{5}$

41. $0, -4$ **43.** $0, \frac{-3}{5}$ **45.** $0, 2$ **47.** 1

49. 0 **51.** 2

Checks:

7. For $\sqrt{13}$:

$(\sqrt{13})^2 - 13 \stackrel{?}{=} 0$

$13 - 13 \stackrel{?}{=} 0$

$0 \stackrel{\checkmark}{=} 0$

For $-\sqrt{13}$:

$(-\sqrt{13})^2 - 13 \stackrel{?}{=} 0$

$13 - 13 \stackrel{?}{=} 0$

$0 \stackrel{\checkmark}{=} 0$

19. For 3:

$[2(3) + 1]^2 \stackrel{?}{=} 49$

$7^2 \stackrel{?}{=} 49$

$49 \stackrel{\checkmark}{=} 49$

For −4:

$[2(-4) + 1]^2 \stackrel{?}{=} 49$

$[-7]^2 \stackrel{?}{=} 49$

$49 \stackrel{\checkmark}{=} 49$

25. For $\frac{2+\sqrt{3}}{3}$:

$\left[3\left(\frac{2+\sqrt{3}}{3}\right) - 2\right]^2 \stackrel{?}{=} 3$

$[2 + \sqrt{3} - 2]^2 \stackrel{?}{=} 3$

$[\sqrt{3}]^2 \stackrel{?}{=} 3$

$3 \stackrel{\checkmark}{=} 3$

For $\frac{2-\sqrt{3}}{3}$:

$\left[3\left(\frac{2-\sqrt{3}}{3}\right) - 2\right]^2 \stackrel{?}{=} 3$

$[2 - \sqrt{3} - 2]^2 \stackrel{?}{=} 3$

$[-\sqrt{3}]^2 \stackrel{?}{=} 3$

$3 \stackrel{\checkmark}{=} 3$

Section 9.4, p. 428 *(The checks follow number 39.)*

1. $\frac{-5 \pm \sqrt{33}}{2}$ **3.** $\frac{5 \pm \sqrt{13}}{6}$ **5.** $-5 \pm 2\sqrt{5}$ **7.** $\frac{-3 \pm \sqrt{13}}{2}$

9. $-1, \frac{-3}{2}$ **11.** $3, \frac{1}{3}$ **13.** $1, 2$ **15.** $4, -5$

17. $-1 \pm \sqrt{5}$ **19.** $\frac{1}{2}, \frac{-1}{3}$ **21.** $\frac{2}{7}, \frac{1}{2}$

23. (a) 8 (b) 2 (c) $-2 \pm \sqrt{2}$ **25.** (a) 1 (b) 2 (c) $1, \frac{3}{2}$

27. (a) 13 (b) 2 (c) $\dfrac{3 \pm \sqrt{13}}{2}$

29. (a) 12 (b) 2 (c) $\dfrac{3 \pm \sqrt{3}}{2}$

31. (a) 0 (b) 1 (c) $\dfrac{1}{2}$

33. (a) 320 (b) 2 (c) $\dfrac{5 \pm 2\sqrt{5}}{2}$

35. (a) 80 (b) 2 (c) $\dfrac{-1 \pm \sqrt{5}}{2}$

37. (a) -35 (b) 0

39. (b), (c), and (d) are the errors.

Checks:

1. For $\dfrac{-5+\sqrt{33}}{2}$:

$$\left(\frac{-5+\sqrt{33}}{2}\right)^2 + 5\left(\frac{-5+\sqrt{33}}{2}\right) - 2 \stackrel{?}{=} 0$$
$$\frac{58-10\sqrt{33}}{4} + \frac{-25+5\sqrt{33}}{2} - 2 \stackrel{?}{=} 0$$
$$\frac{58-10\sqrt{33}-50+10\sqrt{33}-8}{4} \stackrel{?}{=} 0$$
$$0 \stackrel{\checkmark}{=} 0$$

For $\dfrac{-5-\sqrt{33}}{2}$:

$$\left(\frac{-5-\sqrt{33}}{2}\right)^2 + 5\left(\frac{-5-\sqrt{33}}{2}\right) - 2 \stackrel{?}{=} 0$$
$$\frac{58+10\sqrt{33}}{4} + \frac{-25-5\sqrt{33}}{2} - 2 \stackrel{?}{=} 0$$
$$\frac{58+10\sqrt{33}-50-10\sqrt{33}-8}{4} \stackrel{?}{=} 0$$
$$0 \stackrel{\checkmark}{=} 0$$

3. For $\dfrac{5+\sqrt{13}}{6}$:

$$3\left(\frac{5+\sqrt{13}}{6}\right)^2 - 5\left(\frac{5+\sqrt{13}}{6}\right) + 1 \stackrel{?}{=} 0$$
$$3\left(\frac{38+10\sqrt{13}}{36}\right) + \frac{-25-5\sqrt{13}}{6} + 1 \stackrel{?}{=} 0$$
$$\frac{38+10\sqrt{13}-50-10\sqrt{13}+12}{12} \stackrel{?}{=} 0$$
$$0 \stackrel{\checkmark}{=} 0$$

For $\dfrac{5-\sqrt{13}}{6}$:

$$3\left(\frac{5-\sqrt{13}}{6}\right)^2 - 5\left(\frac{5-\sqrt{13}}{6}\right) + 1 \stackrel{?}{=} 0$$
$$3\left(\frac{38-10\sqrt{13}}{36}\right) + \frac{-25+5\sqrt{13}}{6} + 1 \stackrel{?}{=} 0$$
$$\frac{38-10\sqrt{13}-50+10\sqrt{13}+12}{12} \stackrel{?}{=} 0$$
$$0 \stackrel{\checkmark}{=} 0$$

Section 9.5, p. 433

1. $x(9-x) = 20$; 4 and 5

3. $x(3-x) = -70$; 10 and -7

5. $x(-x) = -36$; ± 6

7. $4x = x^2 - 5$; 5

9. $x^2 + (x+1)^2 = 145$; 8 and 9

11. $l(32-l) = 240$; 20 feet by 12 feet

13. $(x+5)\left(\dfrac{1200}{x} - 5\right) = 1125$; 40 feet by 30 feet

15. 6 seconds

17. (a) 1 second (b) 7 seconds

19. 6 or 10

21. 10 feet

Let's Review Chapter 9, p. 435 (The checks follow number 40.)

1. $2x^2 - 4x + 3 = 0$

2. $x^2 - 3x + 3 = 0$

3. reducible

4. $\dfrac{-1}{4}$

5. -5

6. $\dfrac{-2}{3}$

7. 3, 4

8. $-2, \dfrac{1}{3}$

9. $\dfrac{3}{5}$

10. $2, -1$

11. 0, 4

12. $4x^2 - 9x + 2 = 0$

13. $\pm\sqrt{17}$

14. $\pm 4\sqrt{5}$

15. $1, -2$

16. $\dfrac{5 \pm \sqrt{2}}{3}$

17. (a) -11 (b) 0

18. $\dfrac{-5 \pm \sqrt{5}}{2}$

19. $\dfrac{-1}{2}$

20. $-3 \pm \sqrt{10}$

21. 9 and -5

22. 12 inches by 3 inches

And These from Chapters 1–8, p. 436

23. $(5x+1)(x-1)$

24. $(5x+2)(2x+1)$

25. $\frac{3}{2}$

26. -8

27. $14ax^2$

28. (a) $6(1-\sqrt{3})$ (b) $\frac{1-\sqrt{3}}{2}$

29. $\frac{x+4}{3(x-4)}$

30. \$1500

31. $\left(-2, \frac{1}{2}\right)$

32. $x=-2$

33. $\frac{x}{4}+\frac{y}{-4}=1$

34. $(1, -1)$

35. 45

36. (a) 4.32×10^{-3} (b) 636 000 000

37. $7-4\sqrt{3}$

38. (a) $4(x-1)^2(x+1)$ (b) $x-1$

39. \$48

40. $\frac{10-2x}{x-3}$

Checks:

4.

$$3\left(\frac{-1}{4}\right)^2 + 8\left(\frac{-1}{4}\right) - 2 \stackrel{?}{=} 2\left(\frac{-1}{4}\right)^2 - \left(4 - \left(\frac{-1}{4}\right)^2\right)$$

$$3\left(\frac{1}{16}\right) - 2 - 2 \stackrel{?}{=} 2\left(\frac{1}{16}\right) - \left(4 - \frac{1}{16}\right)$$

$$\frac{3}{16} - 4 \stackrel{?}{=} \frac{2}{16} - \left(\frac{64-1}{16}\right)$$

$$\frac{3}{16} - \frac{64}{16} \stackrel{?}{=} \frac{2}{16} - \frac{63}{16}$$

$$\frac{-61}{16} \stackrel{\checkmark}{=} \frac{-61}{16}$$

8. For -2:

$$3(-2)^2 + 5(-2) - 2 \stackrel{?}{=} 0$$

$$3(4) - 10 - 2 \stackrel{?}{=} 0$$

$$12 - 10 - 2 \stackrel{?}{=} 0$$

$$0 \stackrel{\checkmark}{=} 0$$

For $\frac{1}{3}$:

$$3\left(\frac{1}{3}\right)^2 + 5\left(\frac{1}{3}\right) - 2 \stackrel{?}{=} 0$$

$$3\left(\frac{1}{9}\right) + \frac{5}{3} - 2 \stackrel{?}{=} 0$$

$$\frac{1}{3} + \frac{5}{3} - \frac{6}{3} \stackrel{?}{=} 0$$

$$0 \stackrel{\checkmark}{=} 0$$

15. For 1:

$$[2(1)+1]^2 \stackrel{?}{=} 9$$

$$(3)^2 \stackrel{?}{=} 9$$

$$9 \stackrel{\checkmark}{=} 9$$

For -2:

$$[2(-2)+1]^2 \stackrel{?}{=} 9$$

$$(-3)^2 \stackrel{?}{=} 9$$

$$9 \stackrel{\checkmark}{=} 9$$

18. For $\frac{-5+\sqrt{5}}{2}$:

$$\left(\frac{-5+\sqrt{5}}{2}\right)^2 + 5\left(\frac{-5+\sqrt{5}}{2}\right) + 5 \stackrel{?}{=} 0$$

$$\frac{30-10\sqrt{5}}{4} + \frac{-25+5\sqrt{5}}{2} + 5 \stackrel{?}{=} 0$$

$$\frac{30-10\sqrt{5}-50+10\sqrt{5}+20}{4} \stackrel{?}{=} 0$$

$$0 \stackrel{\checkmark}{=} 0$$

For $\frac{-5-\sqrt{5}}{2}$:

$$\left(\frac{-5-\sqrt{5}}{2}\right)^2 + 5\left(\frac{-5-\sqrt{5}}{2}\right) + 5 \stackrel{?}{=} 0$$

$$\frac{30+10\sqrt{5}}{4} + \frac{-25-5\sqrt{5}}{2} + 5 \stackrel{?}{=} 0$$

$$\frac{30+10\sqrt{5}-50-10\sqrt{5}+20}{4} \stackrel{?}{=} 0$$

$$0 \stackrel{\checkmark}{=} 0$$

Complete Solutions

to the Practice Exercises and Practice Exams

CHAPTER 1

Practice Exercises for Section 1.1, pages 2–5

1.

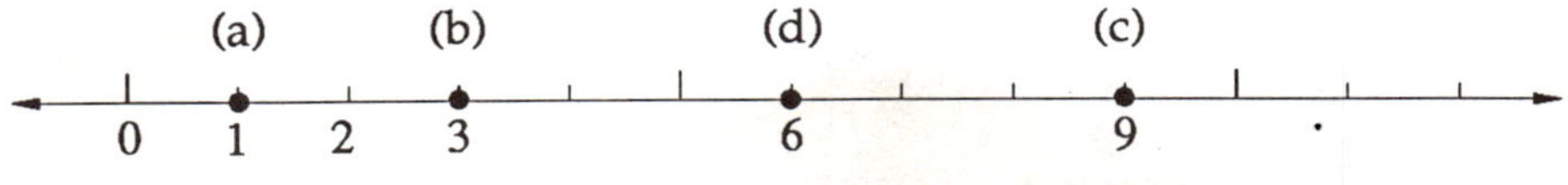

FIGURE 1A - S

2. (a) The point N corresponds to the integer 5; (b) P corresponds to 10; (c) R corresponds to 12.

3. The freezing temperatures are:
(a) $-10°$ C (*Read:* "minus 10 degrees Celsius" or "ten degrees below zero Celsius") and (d) $-2°$ C

4.

Positive Connotations	*Negative Connotations*
increase	decrease
profit	loss
increment	diminishment
grow	shrink

5. -4 **6.** right; left **7.** $A=-6$; $B=-4$; $C=-1$; $D=4$

8.

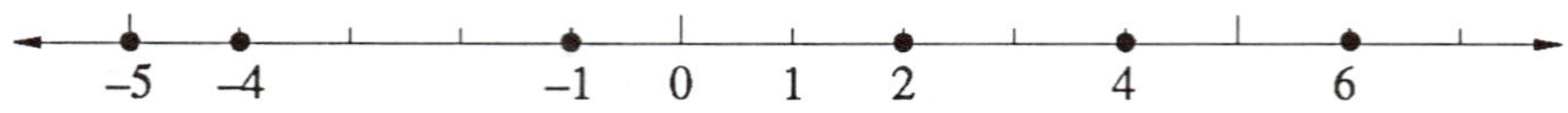

FIGURE 1B - S

Practice Exercises for Section 1.2, pages 8–11

1. $4 > 1$ **2.** (a) $21 > 17$ is read "21 is greater than 17." It is true because on the number line 21 lies to the right of 17.

(b) $0 < 2$ is read "zero is less than two." Note that zero is to the left of two on the number line.

3. (a) $99 < 100$ because 99 lies to the left of 100 on the number line.

(b) $40 > 37$ because 40 lies to the right of 37.

4. (a) –12 lies to the *left* of – 9.

(b) – 2 lies to the *right* of – 3. (See Figure 1C-S.)

FIGURE 1C - S

5. (a) $-12 < -9$ because of 4 (a).

(b) $-2 > -3$ because of 4 (b).

6. (a) $0 > -3$ because 0 lies to the right of -3 on the number line.

(b) $-3 < 0$ because -3 lies to the left of 0.

(c) $7 > -3$ because 7 lies to the right of -3.

(d) $-11 < 10$ because -11 lies to the left of 10.

7. $-10 < -8 < -5 < -4$

Of the integers listed, –10 occurs farthest to the left on the number line; – 4 occurs farthest to the right.

8. The temperature fell because –10° is a lower temperature than – 5°. That is, $-10 < -5$.

9. –10° is warmer than –15° because $-10 > -15$.

Practice Exercises for Section 1.3, pages 14–16

1. (a) The additive inverse of 19 is –19.

(b) The inverse of – 43 is 43.

(c) 0 is its own additive inverse.

(d) $-(-9) = 9$. The inverse is therefore – 9.

(e) The inverse of k is simply denoted as $-k$, regardless of whether k is positive, negative, or zero.

2. No. Every integer has exactly one additive inverse.

3. 6 and – 6 are each 6 units from the origin.

4. (a) $|10| = 10$ (b) $|-36| = 36$ (c) $|-7| = 7$

5. (a) $-|-4| = -4$ (b) $-(-13) = 13$ (c) $-|13| = -13$

6. (a) $4 < |-5|$ because $4 < 5$. (b) $|-5| > |-4|$ because $5 > 4$.

Practice Exercises for Section 1.4, pages 21–23

1. (a) 23

The sum of two positive numbers is positive.

(b) – 23

The sum of two negative numbers is negative. Simply add their absolute values and retain the negative sign. Thus, $-14 + (-9) = -(14 + 9) = -23$

(c) – 7

The sum of a positive number and a negative number will have the same sign as the number with the larger absolute value. Thus, $|-13| > |6|$. Subtract the smaller number from the larger one and prefix the result with a minus sign.

$$6 + (-13) = -(13 - 6) = -7$$

(d) 17

This sum is positive because the number with the larger absolute value is positive.

2. (a) 4

Think of subtraction as adding the inverse; then follow the rules for addition. Thus, $-3 - (-7) = -3 + 7 = 4$ (because the larger absolute value is positive).

(b) –10

$$-3 - 7 = -3 + (-7) = -10$$

(c) – 8

$$12 - 20 = 12 + (-20) = -8$$

(d) 30

$$24 - (-6) = 24 + 6 = 30$$

3. $-2+5=3$

Figure 1.27 on page 21 illustrates that when -2 is increased by 5, the result is 3.

4. (a) $-3+2=2+(-3)$ because both sides are equal to -1. (See Figures 1D-S and 1E-S.)
This is an illustration of the Commutative Law of addition.

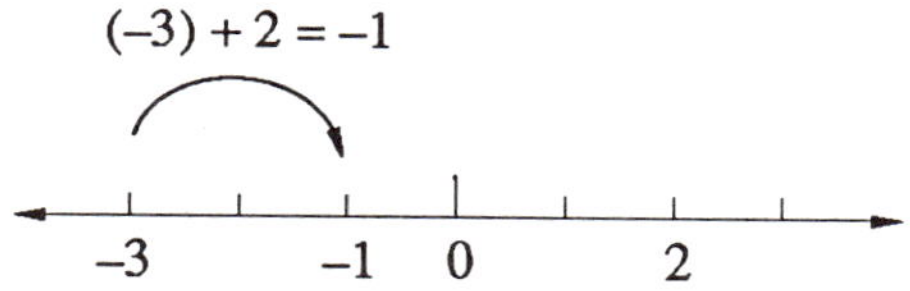

FIGURE 1D - S

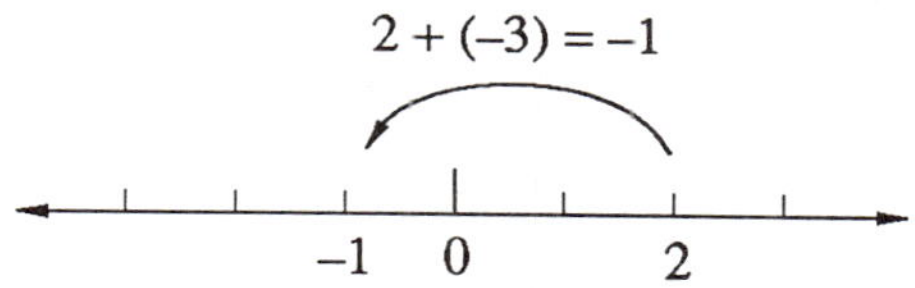

FIGURE 1E - S

(b) left side: $[7+(-4)]+5=3+5=8$
right side: $7+[(-4)+5]=7+1=8$

Both sums are equal to 8. This is an illustration of the Associative Law of addition.

5. (a) 1

The Commutative and Associative Laws permit us to find the sum of several numbers in various ways. Here are two possible solutions to 5 (a). Associate the first two numbers, then add the third:

$$7+(-12)+6=[7+(-12)]+6=-5+6=1$$

or change the order and add the numbers with like signs first, then add the total of the positive numbers to the total of the negative numbers.

$$7+(-12)+6=[7+6]+(-12)=13+(-12)=1$$

(b) 3

$$\begin{aligned}-22+30+(-5)&=[-22+(-5)]+30\\&=(-27)+30\\&=3\end{aligned}$$

(c) -5

$$\begin{aligned}15+(-8)+(-6)+4+(-10)&=[15+4]+[(-8)+(-6)+(-10)]\\&=19+(-24)\\&=-5\end{aligned}$$

6. The stock rises 5 points.

Add the points:

7	(Monday)
–10	(Tuesday)
8	(Wednesday)

The total is $7+(-10)+8=5$ for the three days.

Practice Exercises for Section 1.5, pages 29–33

1. (a) 96 (b) -36 (c) 20

2. (a) $(-5)\times 4=-20;\quad 4\times(-5)=-20$

Both products are equal to -20. This is an illustration of the Commutative Law of multiplication.

(b) left side: $2\times[7\times(-3)]=2\times(-21)=-42$
right side: $(2\times 7)\times(-3)=14\times(-3)=-42$

Both products are equal to -42. This is an illustration of the Associative Law of multiplication.

3. (a) -60 (b) 60 (c) -60 (d) 0

(a) $[10(-2)](3)=(-20)(3)=-60$ (b) $[(-10)(2)](-3)=(-20)(-3)=60$ (c) An odd number of negative factors makes the product negative.

(d) If any factor is zero, the product is zero.

4. (a) $42\div 3=14$ *Check:* $14\times 3=42$ (b) $\frac{91}{7}=13$ *Check:* $13\times 7=91$ (c) 12 *Check:* $12\times 9=108$

5. (a) -6 (b) -3 (c) 11 (d) 5 (e) -4

(d) $$\frac{100}{(-2)(-10)}=\frac{100}{20}=5$$

(e) $$\frac{6(-8)}{(-4)(-3)}=\frac{-48}{12}=-4$$

6. (a) 0 (b) undefined

7. 17 cents per pencil

To find the cost per pencil, divide the cost, 85 cents, by 5, the number of pencils.

$$\frac{85\text{ cents}}{5\text{ pencils}}=17\text{ cents per pencil}$$

Practice Exercises for Section 1.6, pages 36–39

1. (a) 49

$7^2 = 7 \times 7 = 49$

(b) 25

$(-5)^2 = (-5)(-5) = 25$

2. (a) 81

$9^2 = 9 \times 9 = 81$

(b) 36

$(-6)^2 = (-6)(-6) = 36$

(c) – 16

Apply the exponent first.

$-4^2 = -(4 \cdot 4) = -16$

3. (a) – 27

$(-3)^3 = (-3)(-3)(-3) = -27$

(b) 125

$5^3 = 5 \times 5 \times 5 = 125$

(c) – 216

$(-6)^3 = (-6)(-6)(-6) = -216$

(d) –1000

$-10^3 = -(10 \times 10 \times 10) = -1000$

4. (a) – 512

(b) 16

(c) 100 000

5. (a) False.

$5^2 = 25$, but $2^5 = 32$.

(b) True.

$4^2 = 4 \times 4 = 16$ and

$2^4 = 2 \times 2 \times 2 \times 2 = 16$.

6. (a) – 9 is the base.

(b) 3 is the exponent.

(c) The cube of – 9 is – 729.

7. (a) –1 (b) 1 (c) –128 (d) 0

(a) $(-1)^{13}$ has an odd number (13) of negative factors, and is therefore negative.

(b) $(-1)^{18}$ has an even number of negative factors, and is therefore positive.

(c) In $(-2)^7$, the number of negative factors is odd, making the result negative.

(d) Any (positive) power of zero is zero.

Practice Exercises for Section 1.7, pages 42–44

1. –11

$$\begin{aligned}(-18+3)-(7-16+5) &= (-15)-(-9+5)\\ &= -15-(-4)\\ &= -15+4\\ &= -11\end{aligned}$$

2. – 55

Evaluate the innermost grouping first, then the group in brackets. Finally, subtract that result from 50.

$$\begin{aligned}50-[100-(-14+9)] &= 50-[100-(-5)]\\ &= 50-[100+5]\\ &= 50-105\\ &= -55\end{aligned}$$

3. (a) 34 (b) –1 (c) 98 (d) 208

(a) $6+4(-3+10) = 6+4(7) = 6+28 = 34$

(b) $19-5\cdot 2^2 = 19-5\cdot 4 = 19-20 = -1$

(c) $$\begin{aligned}2\left(\frac{35-2}{3}-4\right)^2 &= 2\left(\frac{33}{3}-4\right)^2\\ &= 2(11-4)^2\\ &= 2(7)^2\\ &= 2(49)\\ &= 98\end{aligned}$$

(d) $$\begin{aligned}4\cdot 3^3+[5(-2)]^2 &= 4\cdot 27+[-10]^2\\ &= 108+100\\ &= 208\end{aligned}$$

4. 4

$3^2 - 5 = 9 - 5 = 4$

5. (a) Left side: $5(20-3) = 5(17) = 85$

Right side: $5\cdot 20 - 5\cdot 3 = 100 - 15 = 85$

(b) Left side: $(6+5)\cdot 9 = 11\cdot 9 = 99$

Right side: $6\cdot 9 + 5\cdot 9 = 54 + 45 = 99$

6. 126

$$\begin{aligned}7(10+8) &= 7\cdot 10 + 7\cdot 8 = 70 + 56\\ &= 126\end{aligned}$$

7. 90 cents

The total cost of the stamps is given by:

$12(25) + 5(22) = 300 + 110 = 410$ (cents)

Subtract this result from \$5 (= 500 cents).

$500 - 410 = 90$ cents (change)

Practice Exam 1 on Chapter 1, page 49

1. – 6 **2.** $-4 < -2$ **3.** $-12 < -2 < -1 < 0 < 1 < 12$ **4.** 19 **5.** 19

$|-19|$, the absolute value of –19, is 19 because –19 is 19 units from the origin.

6. –19

To add negative numbers, add their absolute values and prefix the result with a minus sign.

$(-8)+(-2)+(-9) = -(8+2+9) = -19$

7. -2

"Subtract" means "add the inverse." Thus,

$$(-5)-(-3)=(-5)+3$$

To add -5 and 3, find the difference of their absolute values and prefix the result with the sign of the larger absolute value.

$$(-5)+3=-2$$

8. 24

An even number of negative factors produces a positive product.

9. -10

$$\frac{-60}{(-2)(-3)}=\frac{-60}{6}=-10$$

With an odd number of negative factors in numerator and denominator, the result is negative.

10. 81

$(-3)^4=(-3)(-3)(-3)(-3)=81$

Even powers of negative numbers are positive.

11. The third power

-8 is the third power of -2 because $(-2)^3=-8$.

12. -2

$(-2+4)^2-2\times 3=2^2-6=4-6=-2$

13. 8

$$\frac{(8-2)^2}{9}+4=\frac{6^2}{9}+4=\frac{36}{9}+4=4+4=8$$

14. 32

Write:

$[7+(-3)]\cdot 2^3=4\cdot 8=32$

Practice Exam 2 on Chapter 1, page 49

1. -23

2. (a) $3>-5$ (b) $|-9|>|8|$

(a) 3 is greater than -5 because 3 lies to the right of -5 on the number line.

(b) $|-9|=9$ and $|8|=8$. Thus, $|-9|>|8|$.

3. $-7<-3<0<4<7$

4. 15

5. (a) 3 (b) 20 (c) -4

6. -12

7. -14

8. 21

9. -20

10. -90

11. 300

12. -7

13. -5

14. -64

15. 38

$$\begin{aligned}-10+3(2-6)^2&=-10+3(-4)^2\\&=-10+3(16)\\&=-10+48\\&=38\end{aligned}$$

16. -48

Write:

$(-2)^3+5(-8)=-8+(-40)=-48$

CHAPTER 2

Practice Exercises for Section 2.1, pages 53–56

1. (a) $z-5$ (b) $7z^2$ (c) $10+z^3$

2. (a) $10x^2y$, (c) -7, and (d) $x(-5)$, or $-5x$, are terms.

(b) $3+y^2$ and (e) $\frac{2}{x}$ are not terms because the definition does not allow for addition of numbers and variables or for division of a number by a variable.

3. (a) x^3 (b) x^2y^4 (c) $6ab^3$

4. (a) 9 (b) -1 (c) 24

In part (c), note that $(-4a)(-6b)=24ab$.

5. (a) $-12xy$ and (b) $3+2x^2$ are polynomials.

6. (a) $3x$ and 2 (b) y^3, $-y^2$, $5y$, and -3

Practice Exercises for Section 2.2, pages 59–61

1. -4

Replace x by -1. Thus,

$7(-1)+3=-7+3=-4$

2. -6

$3^2-5(3)=9-15=-6$

3. 6

Replace each occurrence of x by 5 and of y by -2. Thus,

$$\begin{aligned}&2(5)^2+4(5)(-2)-(-2)^2\\&=2(25)+20(-2)-(4)\\&=50+(-40)+(-4)\\&=6\end{aligned}$$

4. -86

When $a=2, b=1$, and $c=5$, then $a^2+2bc-10abc$ becomes

$$\begin{aligned}&(2)^2+2(1)(5)-10(2)(1)(5)\\&=4+10-100\\&=14-100\\&=-86\end{aligned}$$

5. 60 square meters

$$A=\frac{bh}{2}$$

$$A=\frac{15\cdot 8}{2}=\frac{120}{2}=60$$

6. 540 degrees

Replace n by 5 in the formula. Thus, $d=(5-2)\cdot 180=3\cdot 180=540$

Practice Exercises for Section 2.3, pages 65–67

1. (a) $10x$ and $-x$ are like terms because they differ only in their numerical coefficients, 10 and -1.

The pairs in (b) and (c) are not like terms.

(d) $5xy^2$ and y^2x are like terms.

2. $2xy$, $3yx$, and $-xy$ are like terms;
-2 and 5 are like terms;
$-2yx^2$ and $5x^2y$ are like terms;
$5x^2$ is unlike any of the other listed terms.

3. (a) $10x$

$9x+x=9x+1x=(9+1)x=10x$

(b) $2xy$

$-3xy+5xy=(-3+5)xy=2xy$

(c) $-4z^2$

$3z^2-7z^2=(3-7)z^2=-4z^2$

4. $3x+xy+(-y)$ or $3x+xy-y$

This expression cannot be simplified further because the terms are unlike terms.

5. $(9x+x^2)-8$, which equals x^2+9x-8

6. (a) $x-3$ (b) $12y-15$ (c) $-2x^2+5x$ (d) $2ab-3a$

Using the Commutative Law, you can rearrange in order to combine like terms.

(a) $6x-4-5x+1=(6x-5x)+(-4+1)=x-3$

(b) $9y-7+3y-8=(9y+3y)+(-7-8)=12y-15$

(c) $x^2+3x-3x^2+2x=(x^2-3x^2)+(3x+2x)=-2x^2+5x$

(d) You can also arrange in columns of like terms. Thus

$$\begin{array}{r}ab-b-2a\\ +\,ab+b-\ \ a\\ \hline 2ab+0-3a\end{array}=2ab-3a$$

Practice Exercises for Section 2.4, pages 70–72

1. (a) (*ii*) (b) (*iii*) (c) (*iv*) (d) (*i*)

(a) $2x-3y$ has two unlike terms and is therefore a binomial.

(b) $3x^2+5x-1$ is a trinomial because it has 3 unlike terms.

(c) $x^2-x^2=0$. Zero is not classified as a monomial.

(d) 2. A nonzero numerical term is a monomial.

2. $5x-2y$

3. $4x^2+6x-1$

Arrange in columns of like terms.

$$\begin{array}{rrr}3x^2 & & +\ 8\\ x^2 & -2x & -10\\ & 8x & +\ 1\\ \hline 4x^2 & +6x & -\ 1\end{array}$$

4. (a) $x-4y-3z$ (b) $-2x^3+x^2-8x+4$ (c) $-3x^2+3xy-y^2$

5. $2m+3mn-2n$

To subtract means to add the inverse.

Thus,

$$\begin{array}{r}9m+\ mn-n\\ -(7m-2mn+n)\\ \hline\end{array}\quad\xrightarrow{\text{Add:}}\quad\begin{array}{r}9m+\ mn-\ n\\ -7m+2mn-\ n\\ \hline 2m+3mn-2n\end{array}$$

6. $9x-7$

To simplify, work from innermost parentheses outward.
Note that $-(3x+4)$ indicates the inverse of $3x+4$, which is $-3x-4$.
Thus,

$$\begin{aligned}6x+1-[12-(3x+4)]&=6x+1-[12-3x-4]\\&=6x+1-[8-3x]\\&=6x+1-8+3x\\&=9x-7\end{aligned}$$

Practice Exercises for Section 2.5, pages 75–78

1. (a) $24x$ (b) $35x$ (c) $-12xy$ (d) $-180abc$

(a) $8(3x)=(8\cdot 3)\cdot x=24x$ (b) $(7x)(5)=7\cdot(x\cdot 5)=7\cdot(5\cdot x)=(7\cdot 5)\cdot x=35x$

It is important to recognize in (c) and (d) that multiplication is indicated, not addition or subtraction, as would be the case if the parentheses were missing. Thus, $6x-2y$ would indicate subtraction, but here, $(6x)(-2y)$ indicates multiplication.

(c) $(6x)(-2y)=(6)(-2)\cdot x\cdot y$
$=-12xy$

(d) $(-10a)(-9b)(-2c)=(-10)(-9)(-2)a\cdot b\cdot c$
$=-180abc$

2. (a) $30x+15y-45z$ (b) $8a-40b$ (c) $-21x+6$

(a) $15(2x+y-3z)=15(2x)+15(y)+15(-3z)=30x+15y-45z$

(b) $-8(-a+5b)=-8(-a)+(-8)(5b)=8a+(-40b)=8a=-40b$
Note that $-a=-1\cdot a$, so that $-8(-a)=(-8)(-1)\cdot a=8\cdot a$

(c) $(7x-2)(-3)=7x(-3)+(-2)(-3)=-21x+6$

3. (a) $3x^2$ (b) $5x-3$ (c) $-2x+3y-1$

$$\frac{6x^2}{2}=\frac{6}{2}x^2=3x^2$$

$$\frac{15x-9}{3}=\frac{15x}{3}+\frac{-9}{3}=5x+(-3)=5x-3$$

$$\frac{20x-30y+10}{-10}=\frac{20x}{-10}+\frac{-30y}{-10}+\frac{10}{-10}$$
$$=-2x+3y-1$$

4. (a) $4x-22y$ (b) $13a-9b$ (c) $4a+6$ (d) $6x-7$

Remove parentheses by using a Distributive Law, then combine like terms.

(a) $6(x-4y)+2(y-x)=6x-24y+2y-2x$
$=(6x-2x)+(-24y+2y)$
$=4x+(-22y)$
$=4x-22y$

(b) $5(3a-b)-2(a+2b)=5(3a-b)+(-2)(a+2b)$
$=15a-5b-2a-4b$
$=15a-2a-5b-4b$
$=13a-9b$

(c) $4-2[a-(3a+1)]=4-2[a-3a-1]$
$=4-2[-2a-1]$
$=4+4a+2$
$=4a+6$

(d) $\dfrac{12x+8}{4}-3(3-x)=\dfrac{12x}{4}+\dfrac{8}{4}+(-3)(3)+(-3)(-x)$
$=3x+2-9+3x$
$=6x-7$

5. $10a-4b$
The sum of $3a+b$ and $2a-3b$ multiplied by 2 is given by
$$2[(3a+b)+(2a-3b)]=2(3a+b+2a-3b)=2(5a-2b)=10a-4b$$

Practice Exam 1 on Chapter 2, page 83

1. -4

2. 45
Substitute 4 for each occurrence of x.
$(4)^2+8(4)-3=16+32-3$
$=45$

3. 325
Substitute 10 for x and 5 for y to obtain
$2(10)^2+5(5)^2=2(100)+5(25)$
$=200+125$
$=325$

4. 0
Rearrange to combine like terms.
$2a-3b+5a+3b-7a=(2a+5a-7a)+(-3b+3b)$
$=0+0$
$=0$

5. $12a+11b-13c+11d$

6. $3x+3y-z$
Change the signs of the bottom polynomial and add. Thus,

$$\begin{array}{r}5x-3y+9z\\ -(2x-6y+10z)\\ \hline\end{array}\longrightarrow\begin{array}{r}5x-3y+9z\\ -2x+6y-10z\\ \hline 3x+3y-z\end{array}$$

7. $-2m-n-3p$
Begin with innermost parentheses and work outward.
$4m-(3n+2p)-[6m-(2n-p)]=4m-(3n+2p)-[6m-2n+p]$
$=4m-3n-2p-6m+2n-p$
$=(4m-6m)+(-3n+2n)+(-2p-p)$
$=-2m+(-n)+(-3p)$
$=-2m-n-3p$

8. $-10a+20b-15c$

$$\begin{aligned}&(-5)(2a-4b+3c)\\&=(-5)(2a)+(-5)(-4b)+(-5)(3c)\\&=-10a+20b-15c\end{aligned}$$

9. $-3x^2+5x-9$

$$\begin{aligned}\frac{12x^2-20x+36}{-4}&=\frac{12x^2}{-4}+\frac{-20x}{-4}+\frac{36}{-4}\\&=-3x^2+5x-9\end{aligned}$$

10. $-8x+5y-12z$

$$\begin{aligned}&-2(3x-2y+4z)-\frac{4x-2y+8z}{2}\\&=-6x+4y-8z-(2x-y+4z)\\&=-6x+4y-8z-2x+y-4z\\&=(-6x-2x)+(4y+y)+(-8z-4z)\\&=-8x+5y-12z\end{aligned}$$

11. $-15a-5b-10$

First add $a+b-2$ and $2a+4$ to obtain $3a+b+2$; then multiply this sum by -5:

$$-5(3a+b+2)=-15a-5b-10$$

12. 80 cubic inches

$$\begin{aligned}V=lwh&=(8\text{ inches})(2\text{ inches})(5\text{ inches})\\&=(8\times2\times5)\text{ inches}^3\\&=80\text{ cubic inches}\end{aligned}$$

(Note that inches × inches × inches = inches^3, or cubic inches.)

Practice Exam 2 on Chapter 2, page 83

1. $5+2z$ or $2z+5$

Five more than twice the number

$$\underbrace{\text{Five}}_{5}\;\underbrace{\text{more than}}_{+}\;\underbrace{\text{twice}}_{2\cdot}\;\underbrace{\text{the number}}_{z}$$

2. -8

3. -5

Replace each occurrence of x by -1.

$$\begin{aligned}2x^2+7x&=2(-1)^2+7(-1)\\&=2\cdot1+(-7)\\&=2-7\\&=-5\end{aligned}$$

4. 3

Replace each occurrence of a by 2 and each occurrence of b by -3.

$$\begin{aligned}3a-ab-b^2&=3(2)-(2)(-3)-(-3)^2\\&=6-(-6)-9\\&=6+6+(-9)\\&=3\end{aligned}$$

5. No

Their variables have different exponents.

6. $3x^2+x+2$

Rearrange the terms to combine like terms.

$$\begin{aligned}x^2-3x+2x^2+4x+2&=(x^2+2x^2)+(-3x+4x)+2\\&=3x^2+x+2\end{aligned}$$

7. $xy-y$

8. $-20a^2+3a-2$

$$\begin{array}{r}-15a^2+a-3\\ \underline{-(5a^2-2a-1)}\end{array}\quad\xrightarrow{\text{Add:}}\quad\begin{array}{r}-15a^2+a-3\\ \underline{-5a^2+2a+1}\\ -20a^2+3a-2\end{array}$$

9. (a) $18xy$

$$\begin{aligned}(-3x)(-6y)&=(-3)(-6)xy\\&=18xy\end{aligned}$$

(b) $-14a+7b$

$$-7(2a-b)=-14a+7b$$

10. $-5x^2+x-3$

Divide each term in the numerator by -6.

$$\begin{aligned}\frac{30x^2-6x+18}{-6}&=\frac{30x^2}{-6}+\frac{-6x}{-6}+\frac{18}{-6}\\&=-5x^2+x-3\end{aligned}$$

11. $-x-6$

$$\begin{aligned}\frac{8x-12}{4}-3(x+1)&=\frac{8x}{4}-\frac{12}{4}-3(x)-3(1)\\&=2x-3-3x-3\\&=-x-6\end{aligned}$$

CHAPTER 3

Practice Exercises for Section 3.1, pages 86–88

1. (a) $3x-2$ is the left side. (b) 19 is the right side.

2. $10-x=3$

3. "What number added to 3 is equal to 9?"

$$\underbrace{\text{What number}}_{x}\;\underbrace{\text{added to}}_{+}\;\underbrace{3}_{3}\;\underbrace{\text{is equal to}}_{=}\;\underbrace{9}_{9}$$

4. (a) 12 is not a root.

Replace x by 12.

$$\begin{aligned}12-9&\overset{?}{=}-3\\3&\overset{\times}{=}-3\end{aligned}$$

(b) -3 is a root.

Replace each occurrence of y by -3.

$$\begin{aligned}4(-3)+1&\overset{?}{=}(-3)-8\\-12+1&\overset{?}{=}-3+(-8)\\-11&\overset{\checkmark}{=}-11\end{aligned}$$

(c) 1 is not a root.

Replace each occurrence of z by 1.

$$\begin{aligned}\frac{1-6}{5}&\overset{?}{=}1\\\frac{-5}{5}&\overset{?}{=}1\\-1&\overset{\times}{=}1\end{aligned}$$

Practice Exercises for Section 3.2, pages 91–95

1. (a), (c), (d), and (e) are all equivalent to $x = 20$ because they all have the same root, 20. That is, each results in a true statement when x is replaced by 20.

2. Yes, because both sides have been multiplied by the same quantity.

3. (a) 24

To isolate x, add 9 to both sides.

$$x-9=15$$
$$x-9+9=15+9$$
$$x=24$$

Check:

Replace x by 24.

$$24-9\stackrel{?}{=}15$$
$$15\stackrel{\checkmark}{=}15$$

(b) -5

To isolate x, add -7 to both sides (or subtract 7 from both sides).

$$x+7=2$$
$$x+7-7=2-7$$
$$x=-5$$

Check:

Replace x by -5.

$$-5+7\stackrel{?}{=}2$$
$$2\stackrel{\checkmark}{=}2$$

4. (a) 28

Multiply both sides by 7.

$$\frac{y}{7}=4$$
$$\frac{y}{7}\cdot 7=4\cdot 7$$
$$y=28$$

(b) 9

In this equation, k has been multiplied by 10. To *undo* multiplication, divide both sides by 10.

$$\frac{10k}{10}=\frac{90}{10}$$
$$k=9$$

(c) -40

To undo division by 8, multiply both sides by 8.

$$\frac{x}{8}\cdot 8=(-5)\cdot 8$$
$$x=-40$$

5. 9

Let x be the unknown number. Translating into algebraic symbols, "What number added to 12 is 21" becomes

$$x+12=21$$
$$x+12-12=21-12$$
$$x=9$$

6. 14

A certain number diminished by 8 is 6.

$$\underbrace{\text{A certain number}}_{x}\ \underbrace{\text{diminished by}}_{-}\ \underbrace{8}_{8}\ \underbrace{\text{is}}_{=}\ \underbrace{6.}_{6}$$

$$x-8=6$$
$$x-8+8=6+8$$
$$x=14$$

7. 5

$$12x=60$$
$$\frac{12x}{12}=\frac{60}{12}$$
$$x=5$$

Practice Exercises for Section 3.3, pages 98–101

1. 15

$$-x+15=0$$
$$-x+x+15=0+x$$
$$15=x$$

2. -4

$$8y=7y-4$$
$$8y-7y=7y-4-7y$$
$$y=-4$$

Check:

$$8(-4)\stackrel{?}{=}7(-4)-4$$
$$-32\stackrel{?}{=}-28-4$$
$$-32\stackrel{\checkmark}{=}-32$$

3. 5

$$5z+3=6z-2$$
$$5z+3-5z=6z-2-5z$$
$$3=z-2$$
$$3+2=z-2+2$$
$$5=z$$

4. 6

$$9n-7=5n+17$$
$$9n-7-5n=5n+17-5n$$
$$4n-7=17$$
$$4n-7+7=17+7$$
$$4n=24$$
$$\frac{4n}{4}=\frac{24}{4}$$
$$n=6$$

5. -30

$$\frac{m}{5}+8=2$$
$$\frac{m}{5}+8-8=2-8$$
$$\frac{m}{5}=-6$$
$$\frac{m}{5}\cdot 5=(-6)\cdot 5$$
$$m=-30$$

Check:

$$\frac{-30}{5}+8\stackrel{?}{=}2$$
$$-6+8\stackrel{?}{=}2$$
$$2\stackrel{\checkmark}{=}2$$

6. -3

$$7-x+1=x-5x-1$$
$$8-x=-4x-1$$
$$8-x+4x=-4x-1+4x$$
$$8+3x=-1$$
$$8+3x-8=-1-8$$
$$3x=-9$$
$$\frac{3x}{3}=\frac{-9}{3}$$
$$x=-3$$

7. 6

$$x+3-x+4=18-x-5$$
$$7=13-x$$
$$7+x=13$$
$$x=13-7$$
$$x=6$$

8. 3

To eliminate $-3x$ from the right side, we want to add $3x$; to eliminate -14 from the left side, we want to add 14. Both can be done in one step by adding $3x+14$ to both sides.

$$7x-14=16-3x$$
$$7x-14+3x+14=16-3x+3x-14$$
$$10x=30$$
$$\frac{10x}{10}=\frac{30}{10}$$
$$x=3$$

9. -2

The variable appears on both sides, but it appears with a larger coefficient on the right side. The most convenient strategy then is to isolate x on the right side, other terms on the left side. Add $-x$ and 7 to both sides.

$$x-21=8x-7$$
$$x-21-x+7=8x-7-x+7$$
$$-14=7x$$
$$\frac{-14}{7}=\frac{7x}{7}$$
$$-2=x$$

Check:

$$-2-21\stackrel{?}{=}8(-2)-7$$
$$-23\stackrel{?}{=}-16-7$$
$$-23\stackrel{\checkmark}{=}-23$$

Practice Exercises for Section 3.4, pages 104–105

1. 43

$$10(a+7)=4(3a-4)$$
$$10a+70=12a-16$$
$$10a+70-10a+16=12a-16-10a+16$$
$$86=2a$$
$$\frac{86}{2}=\frac{2a}{2}$$
$$43=a$$

2. -5

$$5y+9=y-(6-y)$$
$$5y+9=y-6+y$$
$$5y+9=2y-6$$
$$5y+9-2y-9=2y-6-2y-9$$
$$3y=-15$$
$$y=-5$$

3. 6

Because both sides contain a factor of 8, it will be convenient to divide both sides by -8 first.

$$-8(x-16)=80$$
$$\frac{-8(x-16)}{-8}=\frac{80}{-8}$$
$$x-16=-10$$
$$x-16+16=-10+16$$
$$x=6$$

Check:

$$-8(6-16)\stackrel{?}{=}80$$
$$-8(-10)\stackrel{?}{=}80$$
$$80\stackrel{\checkmark}{=}80$$

Practice Exercises for Section 3.5, pages 108–111

1. (a) $2x$ (b) $2x-8$

2. $y+50$

To say that the difference between two integers is 50 is just another way of saying that one number is 50 more than the other. So, if the smaller number is given by y, the larger is given by $50+y$, or $y+50$. Note that the difference between $y+50$ and y is 50. That is,

$$(y+50)-y=50$$

is a true statement regardless of the value of y.

3. $70-n$

Here n represents one integer; let's call the second one p. The sum of these two integers is 70. Thus,

$$n+p=70$$
$$p=70-n$$

4. cup of coffee: 35 cents; doughnut: 55 cents

Let x represent the cost of a cup of coffee, in cents. Because the doughnut costs 20 cents more, we can express its cost as $x+20$. We can now write an equation from the fact that the total cost of the two items is 90 cents. Thus, the cost of coffee + doughnut is 90 cents.

$$x+\overbrace{(x+20)}=90$$
$$2x+20=90$$
$$2x=70$$
$$x=35 \text{ (cost of cup of coffee)}$$
$$x+20=55 \text{ (cost of doughnut)}$$

5. Amy: \$90; Barbara: \$130; Chris: \$180

Let x = Amy's salary. Then Barbara's salary is $x+40$ and Chris's salary is $2x$. The total of all three salaries is \$400. Thus,

$$x+(x+40)+2x=400$$
$$4x+40=400$$
$$4x=360$$
$$x=90 \quad \text{(Amy)}$$
$$x+40=90+40=130 \quad \text{(Barbara)}$$
$$2x=2(90)=180 \quad \text{(Chris)}$$

It is easy to check that the total of these three salaries is \$400.

$$90+130+180=400$$

Practice Exercises for Section 3.6, pages 116–120

1. (a) 165 miles (b) 275 miles (c) $55t$ miles

rate · time = distance; miles per hour $= \frac{\text{miles}}{\text{hours}}$

(a) $\frac{55 \text{ miles}}{\text{hours}} \cdot 3 \text{ hours} = 165 \text{ miles}$

(b) $\frac{55 \text{ miles}}{\text{hours}} \cdot 5 \text{ hours} = 275 \text{ miles}$

2. (a) 11 miles per hour (b) 5 hours (c) $\frac{x}{11}$ hours

(a) $\text{rate} = \frac{\text{distance}}{\text{time}} = \frac{22 \text{ miles}}{2 \text{ hours}} = \frac{22}{2}$ miles per hour = 11 miles per hour

(b) $\text{time} = \frac{\text{distance}}{\text{rate}} = \frac{55}{11} = 5$ (hours)

(c) $t = \frac{d}{r}$; if the distance is x miles and the rate is 11 mph, then $t = \frac{x}{11}$ (hours).

3. 45 miles

	r	t	d
car	65	3	195
bus	50	3	150

They travel in the same direction.

$d = 195 - 150$
$d = 45$

4. 760 miles

	r	t	d
1st airplane	230	2	460
2nd airplane	150	2	300

They travel in opposite directions.

$d = 460 + 300$
$d = 760$

5. 300 miles; 5 hours going; 6 hours returning

	r	t	d
going	60	t	$60t$
returning	50	$t+1$	$50(t+1)$

To write an equation, consider the fact that the distance going is the same as the distance returning. Thus,

$$60t = 50(t+1)$$
$$60t = 50t + 50$$
$$10t = 50$$
$$t = 5 \text{ (hours)}$$
$$t + 1 = 5 + 1 = 6 \text{ (hours)}$$

The distance to Fairfield is $60 \frac{\text{miles}}{\text{hours}} \cdot 5 \text{ hours}$, or 300 miles.

Practice Exam 1 on Chapter 3, page 124

1. No

Replace x by -2 in the equation,

$$5x - 2 = -8$$
$$5(-2) - 2 \stackrel{?}{=} -8$$
$$-10 - 2 \stackrel{?}{=} -8$$
$$-12 \stackrel{\times}{=} -8$$

-2 is not a root of the equation.

2. Yes

Replace t by 3.

$$\frac{7(3) - 6}{3} \stackrel{?}{=} 5$$
$$\frac{21 - 6}{3} \stackrel{?}{=} 5$$
$$\frac{15}{3} \stackrel{?}{=} 5$$
$$5 \stackrel{\checkmark}{=} 5$$

3 is a root of the equation.

3. -9

$$y + 12 = 3$$
$$y + 12 - 12 = 3 - 12$$
$$y = -9$$

4. -6

$$\frac{x}{-3} = 2$$
$$\left(\frac{x}{-3}\right)(-3) = 2(-3)$$
$$x = -6$$

5. 5

$$3t - 1 = t + 9$$
$$3t - 1 - t + 1 = t + 9 - t + 1$$
$$2t = 10$$
$$\frac{2t}{2} = \frac{10}{2}$$
$$t = 5$$

6. 26

$$\frac{z-2}{4} = 6$$
$$\frac{z-2}{4} \cdot 4 = 6 \cdot 4$$
$$z - 2 = 24$$
$$z - 2 + 2 = 24 + 2$$
$$z = 26$$

7. 4

$$4x + 3(x - 1) = 25$$
$$4x + 3x - 3 = 25$$
$$7x = 28$$
$$x = 4$$

8. 2

$$5(2t - 1) = 3(7 - t)$$
$$10t - 5 = 21 - 3t$$
$$13t = 26$$
$$t = 2$$

Check: (Substitute 2 for each occurrence of t.)

$$5(2 \cdot 2 - 1) \stackrel{?}{=} 3(7 - 2)$$
$$5(4 - 1) \stackrel{?}{=} 3(5)$$
$$15 \stackrel{\checkmark}{=} 15$$

9. $10 - x = 12;\ -2$

10. 8 feet, 17 feet, 13 feet

Let x be the length of the shortest piece. The longest piece is 1 foot longer than twice the shortest, or $2x + 1$. The third piece is 4 feet shorter than the longest, or $2x + 1 - 4$, which equals $2x - 3$. The total length of the three pieces is 38 feet. Thus,

$$\begin{aligned} x + (2x + 1) + (2x - 3) &= 38 \\ 5x - 2 &= 38 \\ 5x &= 40 \\ x &= 8 \text{ (feet)} \\ 2x + 1 &= 17 \text{ (feet)} \\ 2x - 3 &= 13 \text{ (feet)} \end{aligned}$$

11. 18 degrees

The sum of the measures of the angles of a triangle is 180 degrees. A right triangle contains one 90-degree angle.
Let x be the measure of the smaller acute angle.
The larger acute angle is 4 times the smaller, or $4x$.
The equation we must solve is:

$$\begin{aligned} x + 4x + 90 &= 180 \\ 5x &= 90 \\ x &= 18 \text{ (degrees)} \end{aligned}$$

12. 430 miles

Because the cars are traveling in opposite directions, the distance between them is the sum of the distances they have traveled.

rate × time = distance

60 mph × 3 hours = 180 miles for the Plymouth.

50 mph × (3 + 2) hours = 250 miles for the Ford.

The distance between them is 180 + 250, or 430 (miles).

Practice Exam 2 on Chapter 3, page 125

1. No

-9 is not a root.
Replace x by -9 in the equation.

$$\begin{aligned} 4(-9-5) &\overset{?}{=} -16 \\ 4(-14) &\overset{?}{=} -16 \\ -56 &\overset{\times}{=} -16 \end{aligned}$$

2. Yes

Replace each occurrence of y by 8 in the equation.

$$\begin{aligned} 3(8) - 10 &\overset{?}{=} 22 - 8 \\ 24 - 10 &\overset{?}{=} 14 \\ 14 &\overset{\checkmark}{=} 14 \end{aligned}$$

3. -5

$$\begin{aligned} x + 15 &= 10 \\ x + 15 - 15 &= 10 - 15 \\ x &= -5 \end{aligned}$$

4. -24

$$\begin{aligned} \frac{n}{6} &= -4 \\ \frac{n}{6} \cdot 6 &= (-4) \cdot 6 \\ n &= -24 \end{aligned}$$

5. 18

$$\begin{aligned} 30 - \frac{a}{2} &= 21 \\ 30 - \frac{a}{2} - 21 + \frac{a}{2} &= 21 - 21 + \frac{a}{2} \\ 9 &= \frac{a}{2} \\ 9 \cdot 2 &= \frac{a}{2} \cdot 2 \\ 18 &= a \end{aligned}$$

6. 9

$$\begin{aligned} y - 8 &= 19 - 2y \\ y - 8 + 2y + 8 &= 19 - 2y + 2y + 8 \\ 3y &= 27 \\ \frac{3y}{3} &= \frac{27}{3} \\ y &= 9 \end{aligned}$$

7. 0

$$\begin{aligned} 4(2x + 1) &= 6 - (x + 2) \\ 8x + 4 &= 6 - x - 2 \\ 8x + 4 &= 4 - x \\ 9x &= 0 \\ x &= 0 \end{aligned}$$

Check:

$$\begin{aligned} 4(2 \cdot 0 + 1) &\overset{?}{=} 6 - (0 + 2) \\ 4(1) &\overset{?}{=} 6 - 2 \\ 4 &\overset{\checkmark}{=} 4 \end{aligned}$$

8. 360 miles

rate · time = distance. Therefore, $r = \frac{d}{t}$.

To find the rate of speed, divide distance by time.

$$r = \frac{135 \text{ miles}}{3 \text{ hours}} = 45 \text{ miles per hour}$$

To find the distance traveled in 8 hours, multiply this rate by 8.

$$d = 45(8) = 360 \text{ (miles)}$$

9. $x + 18 = 4x;\ 6$

10. $5x = x - 12;\ -3$

11. 9 girls

Let y be the number of girls in the class.
The number of boys can be represented by $y + 5$.
The total of boys and girls is 23. Thus,

$$\begin{aligned} y + y + 5 &= 23 \\ 2y &= 18 \\ y &= 9 \end{aligned}$$

12. 2 hours

	r	· t	= d
plane	175 mph	x	$175x$
bus	50 mph	$x + 5$	$50(x + 5)$

The plane and the bus covered the *same distance.*

$$\begin{aligned} 175x &= 50(x + 5) \\ 175x &= 50x + 250 \\ 125x &= 250 \\ x &= 2 \text{ (hours)} \end{aligned}$$

CHAPTER 4

Practice Exercises for Section 4.1, pages 128–131

1. (a) r^9 (b) n^{13}

(a) $r^8 \cdot r = r^8 \cdot r^1 = r^{8+1} = r^9$

(b) $n^4 \cdot n^7 \cdot n^2 = n^{4+7+2} = n^{13}$

2. (a) $-18a^8b^3$ (b) $21x^6y^3$

Use the Commutative and Associative Laws of multiplication to rearrange and group the factors.

(a) $9a^3(-2a^5b^3) = 9(-2)(a^3a^5)(b^3) = -18a^8b^3$

(b) $(-x^2y)(3xy^2)(-7x^3) = (-1)(3)(-7)(x^2 \cdot x^1 \cdot x^3)(y^1 \cdot y^2)$
$= 21x^{2+1+3}y^{1+2}$
$= 21x^6y^3$

3. (a) c^5 (b) ab^2 (c) $7x^3y$

$\dfrac{c^9}{c^4} = c^{9-4} = c^5$

$\dfrac{a^2b^4c}{ab^2c} = a^{2-1} \cdot b^{4-2} = a^1b^2$

(Notice that c divides out because $\dfrac{c}{c} = 1$.)

$\dfrac{28x^5y^2}{4x^2y} = \dfrac{28}{4} \cdot x^{5-2} \cdot y^{2-1} = 7x^3y$

4. $3x^7y^3 - 15x^6y^4$

$3x^3y(x^4y^2 - 5x^3y^3) = 3x^3y(x^4y^2) - 3x^3y(5x^3y^3)$
$= 3x^{3+4}y^{1+2} - 15x^{3+3}y^{1+3}$
$= 3x^7y^3 - 15x^6y^4$

5. $2a^6b^2 - ab + 3$

$\dfrac{10a^8b^3 - 5a^3b^2 + 15a^2b}{5a^2b} = \dfrac{10a^8b^3}{5a^2b} - \dfrac{5a^3b^2}{5a^2b} + \dfrac{15a^2b}{5a^2b}$
$= 2a^6b^2 - 1a^1b^1 + 3(1)$
$= 2a^6b^2 - ab + 3$

6. $4a^3 + 5ab$

$4a(a^2 + 2b) - 3ab = 4a^3 + 8ab - 3ab$
$= 4a^3 + 5ab$

7. $5a - 2b$

$\dfrac{9a^2b - 6ab^2}{3ab} + \dfrac{10a^2b^2}{5ab^2} = \dfrac{9a^2b}{3ab} - \dfrac{6ab^2}{3ab} + \dfrac{10a^2b^2}{5ab^2}$
$= 3a - 2b + 2a$
$= 5a - 2b$

Practice Exercises for Section 4.2, pages 135–136

1. 1, 2, 4, and 8

$8 = 8 \cdot 1$ and $8 = 2 \cdot 4$

Thus, the positive factors of 8 are 1, 2, 4, and 8.

2. (a) 35 is composite because $35 = 5 \cdot 7$.

(b) 29 is prime because its only factors are 1 and 29.

(c) 49 is composite because $49 = 7 \cdot 7$.

(d) 1 is neither prime nor composite.

3. (a) $87 = 3 \cdot 29$ (b) $200 = 2^3 \cdot 5^2$ (c) $-81 = -(3^4)$ (d) $117 = 3^2 \cdot 13$

4. (a) $gcd(30, 42) = 6$ (b) $gcd(18, 90) = 18$ (c) $gcd(32, 48, 60) = 4$

(a) $30 = 2 \cdot 3 \cdot 5$; $42 = 2 \cdot 3 \cdot 7$

The prime factors that appear in both 30 and 42 are 2 and 3. The product of these, 6, is the *gcd*.

(b) $18 = 2 \cdot 3^2$; $90 = 2 \cdot 3^2 \cdot 5$

$2 \cdot 3^2$ appears in the factorizations of both 18 and 90. Therefore, 18, which equals $2 \cdot 3^2$, is the *gcd*.

(c) $32 = 2^5$; $48 = 2^4 \cdot 3$; $60 = 2^2 \cdot 3 \cdot 5$

2 is the only prime appearing in all three factorizations. The smallest power of 2 that appears is 2^2. 4 is therefore the greatest common divisor of 32, 48, and 60.

5. (a) $gcd(25, 30) = 5$ (b) $gcd(60, -45, 15) = 15$

Practice Exercises for Section 4.3, pages 139–141

1. $\dfrac{12a + 8b - 20}{4} = 3a + 2b - 5.$

Therefore, $12a + 8b - 20 = 4(3a + 2b - 5)$.

This shows that 4 is a factor of the polynomial $12a + 8b - 20$.

2. $\dfrac{15x^3 + 10x^2}{5x^2} = 3x + 2.$

Therefore, $15x^3 + 10x^2 = 5x^2(3x + 2)$.

Thus, $5x^2$ is a factor of the polynomial $15x^3 + 10x^2$.

3. (a) 3 is a factor of $9x^2y^3z^3 - 12xy^2z^4$ because both terms of this polynomial are divisible by 3. That is,

$$\frac{9x^2y^3z^3}{3} - \frac{12xy^2z^4}{3} = 3x^2y^3z^3 - 4xy^2z^4 \text{ and}$$

therefore, $9x^2y^3z^3 - 12xy^2z^4 = 3(3x^2y^3z^3 - 4xy^2z^4)$

(b) x is a factor of the polynomial because x is a factor of each term. Thus,

$$9x^2y^3z^3 - 12xy^2z^4 = x(9xy^3z^3 - 12y^2z^4)$$

(c) y^2 is a factor of the polynomial because

$$9x^2y^3z^3 - 12xy^2z^4 = y^2(9x^2yz^3 - 12xz^4)$$

(d) z^3 is also a factor because

$$9x^2y^3z^3 - 12xy^2z^4 = z^3(9x^2y^3 - 12xy^2z)$$

4. (a) $7ab(3a+b)$

The common factors of the two terms of $21a^2b + 7ab^2$ are 7, a, and b. Thus, we divide by $7ab$ to find the other factor.

$$\frac{21a^2b + 7ab^2}{7ab} = 3a + b$$

Therefore,

$$21a^2b + 7ab^2 = 7ab(3a+b)$$

(b) $6x(6x^2 - 2x + 1)$

The greatest common factor of all three terms is $6x$.

$$\frac{36x^3 - 12x^2 + 6x}{6x} = 6x^2 - 2x + 1$$

Thus,

$$36x^3 - 12x^2 + 6x = 6x(6x^2 - 2x + 1)$$

Practice Exercises for Section 4.4, pages 143–148

1. $x^2 + 10x + 9$

$$\begin{array}{r} x + 1 \\ \underline{x + 9} \\ x^2 + x \\ \underline{9x + 9} \\ x^2 + 10x + 9 \end{array}$$

2. (a) $x^2 + 5x - 6$

Using the FOIL method:

First: $x \cdot x = x^2$
Outer: $x(-1) = -x$
Inner: $6(x) = 6x$
Last: $6(-1) = -6$

$(x+6)(x-1) = x^2 - x + 6x - 6$

Combining the Outer and Inner products, the result is $x^2 + 5x - 6$.

(b) $x^2 + 16x + 64$

Use the rule $(x+a)^2 = x^2 + 2 \cdot a \cdot x + a^2$ with $a = 8$.

$$\begin{aligned}(x+8)^2 &= x^2 + 2(8)x + 8^2 \\ &= x^2 + 16x + 64\end{aligned}$$

(c) $x^2 - 100$

$$(x+10)(x-10) = \underset{F}{x^2} - \underset{O}{10x} + \underset{I}{10x} + \underset{L}{(10)(-10)} = x^2 - 100$$

3. (a) $3x^2 + 14x - 5$

$$\begin{array}{r} 3x - 1 \\ \underline{x + 5} \\ 3x^2 - x \\ \underline{15x - 5} \\ 3x^2 + 14x - 5 \end{array}$$

(b) $3y^3 - 3y^2 - 6y$

When there are more than two factors, as we have here, the Associative Law says that it doesn't matter which two we choose to multiply first.

$$3y(y+1)(y-2) = 3y[(y+1)(y-2)] = 3y[y^2 - 2y + y - 2] = 3y(y^2 - y - 2) = 3y^3 - 3y^2 - 6y$$

4. (a) $x^3 - 2x^2 - 11x + 12$ (b) $a^4 + 3a - 2$

Multiply each term of one factor by each term of the other. Place the like terms of the product in columns so they may be combined easily.

(a)

$$\begin{array}{r} x^2 + 2x - 3 \\ \underline{x - 4} \\ x^3 + 2x^2 - 3x \\ \underline{-4x^2 - 8x + 12} \\ x^3 - 2x^2 - 11x + 12 \end{array}$$

(b)

$$\begin{array}{r} a^2 - a + 2 \\ \underline{a^2 + a - 1} \\ a^4 - a^3 + 2a^2 \\ a^3 - a^2 + 2a \\ \underline{- a^2 + a - 2} \\ a^4 + 0 + 0 + 3a - 2 \end{array}$$

5. $-12x + 12$

According to the rules for the order of operations, we must square each binomial before subtracting.

$$(x-4)^2 = x^2 - 8x + 16$$
$$(x+2)^2 = x^2 + 4x + 4$$

Thus,

$$\begin{aligned}(x-4)^2 - (x+2)^2 &= x^2 - 8x + 16 - (x^2 + 4x + 4) \\ &= x^2 - 8x + 16 - x^2 - 4x - 4 \\ &= -12x + 12\end{aligned}$$

6. The area of the rectangle on the left is given by

$$(x+2)(x-1) = x^2 - x + 2x - 2 = x^2 + x - 2$$

The area of the figure on the right is the area of the whole rectangle minus the area of the rectangle cut away. Thus,

$$x(x+1) - 1 \cdot 2 = x^2 + x - 2$$

Both areas are expressed by the same polynomial, $x^2 + x - 2$.

Practice Exercises for Section 4.5, pages 151–154

1. (a) $(k+8)(k-8)$

$k^2-64=k^2-8^2=(k+8)(k-8)$

(b) $(10+y)(10-y)$

$100-y^2=10^2-y^2=(10+y)(10-y)$

2. (a) $(1+7a)(1-7a)$

$1-49a^2=1^2-(7a)^2=(1+7a)(1-7a)$

(b) $(5y+3x)(5y-3x)$

$25y^2-9x^2$ is the difference of squares because $25y^2=(5y)^2$ and $9x^2=(3x)^2$. Thus,

$(5y)^2-(3x)^2=(5y+3x)(5y-3x)$

3. (a) $2(n+3)(n-3)$ (b) $10a(2+a)(2-a)$ (c) $3ab(2b+a)(2b-a)$

(a) $2n^2-18=2(n^2-9)=2(n+3)(n-3)$

(b) $40a-10a^3=10a(4-a^2)=10a(2+a)(2-a)$

(c) $12ab^3-3a^3b=3ab(4b^2-a^2)=3ab(2b+a)(2b-a)$

4. (a) $4x^2+9$ is the sum of squares and, as such, cannot be factored.

(e) $-9-4x^2$ can only be factored by isolating the common factor, –1. Thus,

$-9-4x^2=(-1)(9+4x^2)$

We cannot factor this further.

(b), (c), and (d) are each the difference of squares and thus can be factored.

5. (a) $(a^2+1)(a+1)(a-1)$

$a^4-1=(a^2)^2-1=(a^2+1)(a^2-1)$

The second factor in this factorization, a^2-1, can be factored further because it is the difference of squares. Thus,

$$(a^2+1)(a^2-1)=(a^2+1)[(a+1)(a-1)]$$
$$=(a^2+1)(a+1)(a-1)$$

(b) $(x^2+a)(x^2-a)$

$x^4-a^2=(x^2)^2-a^2=(x^2+a)(x^2-a)$

Neither of the factors on the right can be factored further.

Practice Exercises for Section 4.6, pages 156–161

1. $(t+11)(t+1)$

$t^2+12t+11$ is the product of two binomials whose first terms must yield the product t^2; thus, try $(t+\_)(t+\_)$. Choose for the second terms numbers whose product is 11 and whose sum is 12. Note that $11\cdot 1=11$ and $11+1=12$. Thus,

$t^2+12t+11=(t+11)(t+1)$

2. (a) $(a+2)(a+4)$ (b) $(x-8)(x-1)$ (c) $(y+2)(y-4)$

(a) To find the factors of a^2+6a+8, consider several points:

(*i*) The first term, a^2, is the product of a times a.

(*ii*) The last term, 8, may be the product of 2 and 4, or of 1 and 8.

(*iii*) The sign of the last term is "+", which indicates that its factors are either both positive or both negative.

(*iv*) Because the sign of the middle term, $6x$, is "+", the factors of 8 are both positive.

The possible factorizations therefore are $(a+1)(a+8)$ or $(a+2)(a+4)$.

Using the FOIL method, the Outer and Inner products in the first arrangement give the wrong middle term, $9a$. Multiplying $(a+2)(a+4)$ completely, we find that this is the correct factorization.

(b) To find the factors of x^2-9x+8, note that the last term is positive and the middle term is negative. This indicates that the factors of 8 are both negative. The possible factors of the polynomial are $(x-2)(x-4)$ or $(x-1)(x-8)$. Only $(x-1)(x-8)$ produces the proper middle term, $-9x$. Thus, $x^2-9x+8=(x-1)(x-8)$.

(c) Consider y^2-2y-8. Because the last term, -8, is negative, it must have one positive factor and one negative factor. Try various possibilities to get the correct middle term.

$(y+1)(y-8)$, $(y+4)(y-2)$, $(y+2)(y-4)$

The last pair works. Thus,

$y^2-2y-8=(y+2)(y-4)$

3. (a) $(x+6)^2$ (b) $(x-11)^2$

Both trinomials fit the pattern that occurs when a binomial is squared:

$$(x+a)^2 = x^2 + 2ax + a^2$$

In part (a), $a = 6$. That is,

$$x^2 + 12x + 36 = x^2 + 2(6)x + 6^2 = (x+6)^2$$

In part (b), $a = -11$. Thus,

$$x^2 - 22x + 121 = x^2 + 2(-11)x + (-11)^2 = (x-11)^2$$

4. (a) $7(x-2)(x-3)$ (b) $5x(x+3)(x-4)$

First isolate the common factor.

(a) $7x^2 - 35x + 42 = 7(x^2 - 5x + 6) = 7(x-2)(x-3)$

(b) $5x^3 - 5x^2 - 60x = 5x(x^2 - x - 12) = 5x(x+3)(x-4)$

5. (a) $(x+2y)(x+13y)$ (b) $4(x+6y)(x-y)$

In part (b), $4x^2 + 20xy - 24y^2 = 4(x^2 + 5xy - 6y^2)$
$= 4(x+6y)(x-y)$

Practice Exercises for Section 4.7, pages 163–166

1. (a) 0, -3 (b) 8, -1

A product is zero only when one of its factors is zero.
Thus, in part (a), $x(x+3) = 0$ only when $x = 0$ or when $x + 3 = 0$, that is, $x = -3$. The roots are therefore 0 and -3.
In part (b), $(x-8)(x+1) = 0$ when $x - 8 = 0$ or $x + 1 = 0$.
Solving each equation for x, $x = 8$ or $x = -1$.
The roots are 8 and -1.

2. (a) 5, -2

$x^2 - 3x - 10 = 0$
$(x+2)(x-5) = 0$
$x + 2 = 0 \mid x - 5 = 0$
$x = -2 \mid x = 5$

The roots are -2 and 5.

(b) -1, -9

$x^2 + 10x + 9 = 0$
$(x+1)(x+9) = 0$
$x + 1 = 0 \mid x + 9 = 0$
$x = -1 \mid x = -9$

The roots are -1 and -9.

3. (a) $x = 7$

$x^2 - 14x + 49 = 0$
$(x-7)(x-7) = 0$
$x - 7 = 0$
$x = 7$

(b) $t = -2$

$t^2 + 4t = -4$
$t^2 + 4t + 4 = 0$
$(t+2)^2 = 0$
$t + 2 = 0$
$t = -2$

4. (a) 10, -10

$x^2 - 100 = 0$
$(x+10)(x-10) = 0$
$x + 10 = 0 \mid x - 10 = 0$
$x = -10 \mid x = 10$

The roots are -10 and 10.

(b) 8, -8

$y^2 = 64$
$y^2 - 64 = 0$
$(y+8)(y-8) = 0$
$y + 8 = 0 \mid y - 8 = 0$
$y = -8 \mid y = 8$

The roots are -8 and 8.

5. (a) 0, -1, and -2

$3x^3 + 9x^2 + 6x = 0$
$3x(x^2 + 3x + 2) = 0$
$3x(x+1)(x+2) = 0$
$3x = 0 \mid x + 1 = 0 \mid x + 2 = 0$
$x = 0 \mid x = -1 \mid x = -2$

There are three roots: 0, -1, and -2.

(b) 4, -4

$5u^2 = 80$
$u^2 = 16$
$u^2 - 16 = 0$
$(u+4)(u-4) = 0$
$u + 4 = 0 \mid u + 4 = 0$
$u = -4 \mid u = -4$

The roots are -4 and 4.

Practice Exercises for Section 4.8, pages 169–171

1. (a) $(5x+2)(x-1)$

To factor $5x^2 - 3x - 2$, consider that
$5x^2$ is $5x \cdot x$ and -2 is $2(-1)$ or $-2(1)$.
Try various combinations to produce the middle term, $-3x$.

$5x+1$	$5x-1$	$5x-2$	$5x+2$
$x-2$	$x+2$	$x+1$	$x-1$

Only the last arrangement produces the correct middle term, $2x - 5x = -3x$.

(b) $(3x-1)(2x-1)$

Consider $6x^2 - 5x + 1$. Last term positive, middle term negative indicates that the +1 must be the product of $(-1)(-1)$.
The only possible factors are $(6x-1)(x-1)$ or $(3x-1)(2x-1)$.

$6x - 1$	$3x - 1$
$x - 1$	$2x - 1$
$6x^2 - x$	$6x^2 - 2x$
$-6x + 1$	$-3x + 1$
$6x^2 - 7x + 1$ (Wrong)	$6x^2 - 5x + 1$ (Right)

2. (a) $3(3x+2)(x-1)$

Isolate the common factor first.

$9x^2-3x-6=3(3x^2-x-2)=3(3x+2)(x-1)$

(b) $-a(2x+5)(x-1)$

Trinomials such as this are much easier to factor when the first term is positive. We can accomplish this by factoring out -1 as a common factor at the same time as we isolate the common factor, a. Thus,

$$-2ax^2-3ax+5a=-a(2x^2+3x-5)$$

Now factor the trinomial by trial and error. Check the possibilities.

$$\begin{array}{r}2x+1\\ x-5\\ \hline\end{array}\qquad \begin{array}{r}2x-1\\ x+5\\ \hline\end{array}\qquad \begin{array}{r}2x+5\\ x-1\\ \hline 2x^2+5x\\ -2x-5\\ \hline 2x^2+3x-5\ \text{(Right)}\end{array}\qquad \begin{array}{r}2x-5\\ x+1\\ \hline\end{array}$$

The complete factorization is $-a(2x+5)(x-1)$.

3. $(7x+y)(2x+y)$

The only possible combinations are:

$(14x+y)(x+y)=14x^2+15xy+y^2$ (Wrong)

or $(7x+y)(2x+y)=14x^2+9xy+y^2$ (Right)

Practice Exam 1 on Chapter 4, page 176

1. $-12x^3y^8$

$$\begin{aligned}(4x^2y^5)(-3xy^3)&=4(-3)(x^2x)(y^5y^3)\\&=-12x^{2+1}y^{5+3}\\&=-12x^3y^8\end{aligned}$$

2. $8x^2+14x-15$

$$\begin{array}{r}4x-3\\ 2x+5\\ \hline 8x^2-6x\\ 20x-15\\ \hline 8x^2+14x-15\end{array}$$

3. $a^5-6a^3+5a^2+8a-10$

$$\begin{array}{r}a^3-4a+5\\ a^2-2\\ \hline a^5-4a^3+5a^2\\ -2a^3\qquad+8a-10\\ \hline a^5-6a^3+5a^2+8a-10\end{array}$$

4. $3x-5y^2$

$$\begin{aligned}\frac{15x^2y^2-25xy^4}{5xy^2}&=\frac{15x^2y^2}{5xy^2}-\frac{25xy^4}{5xy^2}\\&=3x^{2-1}-5y^{4-2}\\&=3x-5y^2\end{aligned}$$

5. $2^2\cdot3^3$

$108=4\cdot27=2^2\cdot3^3$

6. $2x(3x-4)$

7. $(3a+4b)(3a-4b)$

$9a^2-16b^2=(3a)^2-(4b)^2=(3a+4b)(3a-4b)$

8. $(x+3)(x+6)$

9. $(2y+1)(y-1)$

10. $0,-5$

$$\begin{aligned}x^2+5x&=0\\x(x+5)&=0\end{aligned}$$

$x=0 \mid x+5=0$, $x=-5$

The roots are 0 and -5.

Check:

$(x=0)$	$(x=-5)$
$0^2+5\cdot0\stackrel{?}{=}0$	$(-5)^2+5(-5)\stackrel{?}{=}0$
$0+0\stackrel{?}{=}0$	$25+(-25)\stackrel{?}{=}0$
$0=0$	$0=0$

11. $0,2,4$

$$\begin{aligned}x^3-6x^2+8x&=0\\x(x^2-6x+8)&=0\\x(x-2)(x-4)&=0\end{aligned}$$

$x=0 \mid x-2=0$, $x=2 \mid x-4=0$, $x=4$

The roots are 0, 2, and 4.

12. -12

$$\begin{aligned}x^2+24x+144&=0\\(x+12)^2&=0\\x+12&=0\\x&=-12\end{aligned}$$

Practice Exam 2 on Chapter 4, page 176

1. k^{13}

$k^3\cdot k^9\cdot k=k^{3+9+1}=k^{13}$

2. $-21x^6y^4$

$$\begin{aligned}3x^4y(-7x^2y^3)&=3(-7)x^4\cdot x^2\cdot y^1\cdot y^3\\&=-21x^{4+2}y^{1+3}\\&=-21x^6y^4\end{aligned}$$

3. $5a^5b^2-40ab^5$

$$\begin{aligned}5ab^2(a^4-8b^3)&=5ab^2(a^4)-5ab^2(8b^3)\\&=5a^5b^2-40ab^5\end{aligned}$$

4. $18x^2-5x-2$

$$\begin{array}{r} 9x+2 \\ \underline{2x-1} \\ 18x^2+4x \\ \underline{-9x-2} \\ 18x^2-5x-2 \end{array}$$

5. $y^3-y^2-10y+6$

$$\begin{array}{r} y^2-4y+2 \\ \underline{y+3} \\ y^3-4y^2+2y \\ \underline{3y^2-12y+6} \\ y^3-y^2-10y+6 \end{array}$$

6. $3x^4y+xy-2$

$$\frac{18x^6y+6x^3y-12x^2}{6x^2}=\frac{18x^6y}{6x^2}+\frac{6x^3y}{6x^2}-\frac{12x^2}{6x^2}$$
$$=3x^{6-2}y+1x^{3-2}y-2(1)$$
$$=3x^4y+xy-2$$

7. $2\cdot5^2\cdot7$

$350=35\cdot10=5\cdot7\cdot2\cdot5=2\cdot5^2\cdot7$

8. $11a^2(7a+1)$ **9.** $3(2x+3)(2x-3)$ **10.** $(x-5)(x-7)$ **11.** $(x+8y)(x-5y)$ **12.** $(2x-3)(x-1)$

13. 1, 19

$$x^2+19=20x$$
$$x^2-20x+19=0$$
$$(x-19)(x-1)=0$$
$$\begin{array}{r|r} x-19=0 & x-1=0 \\ x=19 & x=1 \end{array}$$

14. 0, −2, 2

$$x^3-4x=0$$
$$x(x^2-4)=0$$
$$x(x+2)(x-2)=0$$
$$\begin{array}{r|r|r} x=0 & x+2=0 & x-2=0 \\ & x=-2 & x=2 \end{array}$$

CHAPTER 5

Practice Exercises for Section 5.1, pages 179–187

1. (a) 23 is the numerator. (b) 15 is the denominator.

2. $\frac{2}{7}$

$2\div7=\frac{2}{7}$

3. Divide the line segment between 0 and 7 into 4 equal parts. The first point of division to the right of 0 is $\frac{7}{4}$.

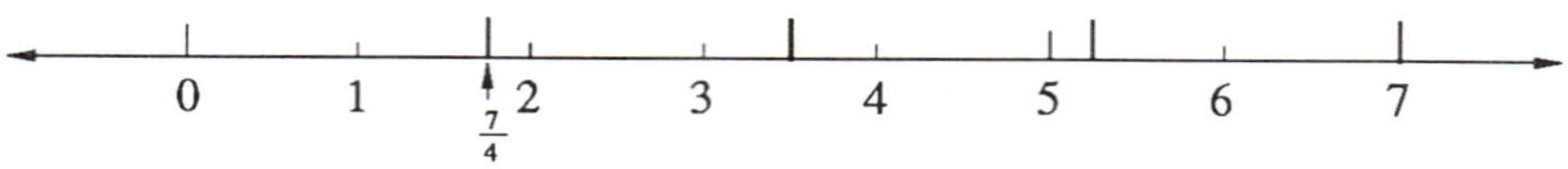

FIGURE 5A - S

4. (a) numerator: x^2+3; denominator: $4x$
(b) numerator: 1; denominator: $2x$
(c) numerator: x^2-2y; denominator: 1

5. (a) $\frac{1}{8}$

Replace each occurrence of x by 2.

$$\frac{(2)^2-3}{3(2)+2}=\frac{4-3}{6+2}=\frac{1}{8}$$

(b) $\frac{-3}{2}$

Replace each occurrence of x by 0.

$$\frac{0^2-3}{3(0)+2}=\frac{-3}{2}$$

6. $\frac{15}{7}$

Replace each occurrence of x by 5 and of y by −1.

$$\frac{(5)(-1)+4(5)}{5-2(-1)}=\frac{-5+20}{5-(-2)}=\frac{15}{7}$$

7. (a) $\frac{12}{28}$ (b) 15

We can obtain an equivalent fraction by multiplying both numerator and denominator by the same quantity.

In part (a) both 3 and 7 are multiplied by 4 to obtain $\frac{12}{28}$, which is equivalent to $\frac{3}{7}$.

In part (b) the denominator, 6, has been multiplied by 3 to obtain 18. We must therefore multiply the numerator by 3 as well. Thus,

$$\frac{5}{6}=\frac{5\cdot3}{6\cdot3}=\frac{15}{18}$$

8. (a) $\frac{3}{5} = \frac{15}{25}$ (b) $\frac{-8}{6} = \frac{12}{-9}$ (c) $\frac{3}{t-3} \neq \frac{9}{t-9}$

Check the cross products.

$3 \cdot 25 \stackrel{?}{=} 15 \cdot 5$ $(-8)(-9) \stackrel{?}{=} 12 \cdot 6$ $3(t-9) \stackrel{?}{=} 9(t-3)$

$75 \stackrel{\checkmark}{=} 75$ $72 \stackrel{\checkmark}{=} 72$ $3t - 27 \stackrel{\times}{=} 9t - 27$

9. $\frac{-3}{5} = \frac{3}{-5}$ because the cross products are equal. Thus, $(-3)(-5) = 3(5) = 15$

10. (a) $\frac{-9}{10}$ (b) $\frac{4}{5}$ (c) $\frac{-x}{2}$ (d) $\frac{-x+4}{3}$

11. (a) $\frac{4}{7}$ (b) $\frac{6}{5}$ (c) $\frac{-2}{3}$

Divide by factors common to the numerator and denominator.

$\frac{12}{21} = \frac{4 \cdot 3}{7 \cdot 3} = \frac{4}{7}$ $\frac{48}{40} = \frac{6 \cdot 8}{5 \cdot 8} = \frac{6}{5}$ $\frac{28}{-42} = \frac{2 \cdot 2 \cdot 7}{-3 \cdot 2 \cdot 7} = \frac{2}{-3} = \frac{-2}{3}$

12. $\frac{1}{5}x^2 + \frac{5}{2}x - 3$

$$\frac{2x^2 + 25x - 30}{10} = \frac{2x^2}{10} + \frac{25x}{10} - \frac{30}{10}$$
$$= \frac{x^2}{5} + \frac{5x}{2} - \frac{3}{1}$$
$$= \frac{1}{5}x^2 + \frac{5}{2}x - 3$$

Practice Exercises for Section 5.2, pages 194–195

1. (a) $\frac{1}{k^5}$ (b) 10^4 or 10 000 (c) $\frac{6y^2}{5x}$

$\frac{k^4}{k^9} = \frac{1 \cdot \cancel{k^4}^{\,1}}{k^5 \cdot \cancel{k^4}_{\,1}} = \frac{1}{k^5}$ $\frac{10^{12}}{10^8} = 10^{12-8} = 10^4$ $\frac{18x^3y^7}{15x^4y^5} = \frac{3 \cdot 6y^{7-5}}{3 \cdot 5x^{4-3}} = \frac{6y^2}{5x}$

2. (a) $\frac{5}{3}$

$\frac{225}{135} = \frac{3^2 \cdot 5^2}{3^3 \cdot 5} = \frac{5^{2-1}}{3^{3-2}} = \frac{5}{3}$

(b) $\frac{3z^4}{5xy}$

$\frac{420xy^2z^5}{700x^2y^3z} = \frac{2^2 \cdot 3 \cdot 5 \cdot 7 \cdot z^{5-1}}{2^2 \cdot 5^2 \cdot 7 \cdot x^{2-1}y^{3-2}} = \frac{3z^4}{5xy}$

3. (a) $\frac{1}{x-3}$ (b) $\frac{3(x+2y)^2}{x}$

$\frac{(x-3)^2}{(x-3)^3} = \frac{1}{(x-3)^{3-2}} = \frac{1}{x-3}$ $\frac{18x(x+2y)^5}{6x^2(x+2y)^3} = \frac{3 \cdot 6(x+2y)^{5-3}}{6x^{2-1}} = \frac{3(x+2y)^2}{x}$

Practice Exercises for Section 5.3, pages 198–200

1. (a) $\frac{3-2x}{2}$ (b) $y(4x^3+1)$

First, isolate the common factor in the numerator.

(a) $\frac{9x^2 - 6x^3}{6x^2} = \frac{3x^2(3-2x)}{2 \cdot 3x^2} = \frac{3-2x}{2}$

(b) $\frac{8x^6y^2 + 2x^3y^2}{2x^3y} = \frac{2x^3y^2(4x^3+1)}{2x^3y} = y(4x^3+1)$

2. $\frac{3n}{3n+2}$

Isolate the common factor in the denominator.

$\frac{9n^2}{9n^2+6n} = \frac{9n^2}{3n(3n+2)} = \frac{3n}{3n+2}$

3. (a) $\dfrac{2(x+1)}{x(x-1)}$ or $\dfrac{2x+2}{x^2-x}$ (b) $\dfrac{x}{2}$ (c) $\dfrac{-3}{a+2}$

Factor both numerator and denominator, wherever possible.

(a) $\dfrac{6x^2+6x}{3x^3-3x^2} = \dfrac{6x(x+1)}{3x^2(x-1)} = \dfrac{2(x+1)}{x(x-1)}$ or $\dfrac{2x+2}{x^2-x}$

(b) $\dfrac{x^2-3x}{2x-6} = \dfrac{x\cancel{(x-3)}^{1}}{2\cancel{(x-3)}_{1}} = \dfrac{x}{2}$

(c) $\dfrac{12-6a}{2a^2-8} = \dfrac{6(2-a)}{2(a^2-4)}$ [(a^2-4) can be factored further.]

$= \dfrac{3(2-a)}{(a+2)(a-2)}$ [Note that $2-a=(-1)(a-2)$.]

$= \dfrac{3(-1)\cancel{(a-2)}}{(a+2)\cancel{(a-2)}}$

$= -\dfrac{3}{a+2}$

Practice Exercises for Section 5.4, page 202–208

1. (a) The degree of $-4x^5$ is 5. (b) The degree of $3x^1$ is 1. (c) The degree of the numerical term 12 is zero.

2. (a) $-3y^4+y^3-6y+8$ (b) The degree of this polynomial is 4, the highest degree of any of its terms.

3. $3x+4$

$$
\begin{array}{r}
3x+4 \\
x-2\overline{)\,3x^2-2x-8} \\
\underline{\mp 3x^2 \pm 6x} \\
4x-8 \\
\underline{\mp 4x \pm 8} \\
0
\end{array}
$$

4. x^2+2x-3

The dividend has no 1st-degree term. Leave a space for that missing term by writing $0x$.

$$
\begin{array}{r}
x^2+2x-3 \quad \longleftarrow \text{quotient} \\
2x+3\overline{)\,2x^3+7x^2+0x-9} \\
\underline{\mp 2x^3 \mp 3x^2} \\
4x^2+0x \\
\underline{\mp 4x^2 \mp 6x} \\
-6x-9 \\
\underline{\pm 6x \pm 9} \\
0
\end{array}
$$

To check, multiply the quotient by the divisor. You should obtain the dividend.

$$
\begin{array}{r}
x^2+2x-3 \quad \longleftarrow \text{quotient} \\
\underline{\times\ 2x+3} \quad \longleftarrow \text{divisor} \\
2x^3+4x^2-6x \\
\underline{3x^2+6x-9} \\
2x^3+7x^2 \qquad -9 \quad \longleftarrow \text{dividend}
\end{array}
$$

5. quotient: x^2-2; remainder: 3

$$
\begin{array}{r}
x^2 \qquad -2 \quad \longleftarrow \text{quotient} \\
x+5\overline{)\,x^3+5x^2-2x-7} \\
\underline{\mp x^3 \mp 5x^2} \\
-2x-7 \\
\underline{\pm 2x \pm 10} \\
3 \quad \longleftarrow \text{remainder}
\end{array}
$$

To check, multiply the quotient by the divisor and add the remainder. You should obtain the dividend.

$$
\begin{array}{r}
x^2-2 \quad \longleftarrow \text{quotient} \\
\underline{\times\ x+5} \quad \longleftarrow \text{divisor} \\
x^3 \qquad -2x \\
\underline{5x^2 \qquad -10} \\
x^3+5x^2-2x-10 \\
\underline{+ \qquad 3} \quad \longleftarrow \text{remainder} \\
x^3+5x^2-2x-7 \quad \longleftarrow \text{dividend}
\end{array}
$$

6. $4x^2-x+1+\dfrac{-2}{x+3}$

$$
\begin{array}{r}
4x^2-x+1 \quad \longleftarrow \text{quotient} \\
x+3\overline{)\,4x^3+11x^2-2x+1} \\
\underline{\mp 4x^3 \mp 12x^2} \\
-x^2-2x \\
\underline{\pm x^2 \pm 3x} \\
x+1 \\
\underline{\mp x \mp 3} \\
-2 \quad \longleftarrow \text{remainder}
\end{array}
$$

Thus, $\dfrac{4x^3+11x^2-2x+1}{x+3} = 4x^2-x+1+\dfrac{-2}{x+3}$

Practice Exercises for Section 5.5, pages 212–215

1. (a) $\frac{5}{14}$ (b) $\frac{20}{3}$ (c) $\frac{1}{3}$

(a)
$$\frac{6}{7}\cdot\frac{5}{12}=\frac{6\cdot 5}{7\cdot 12}=\frac{30}{84}=\frac{5}{14}$$

This procedure is greatly simplified if we divide by the common factor 6 before multiplying. Thus,

$$\frac{\overset{1}{\cancel{6}}}{7}\cdot\frac{5}{\underset{2}{\cancel{12}}}=\frac{1\cdot 5}{7\cdot 2}=\frac{5}{14}$$

(b) Write the integer as a rational number with its denominator, 1. Divide by the common factor, 7.

$$\frac{\overset{4}{\cancel{28}}}{1}\cdot\frac{5}{\underset{3}{\cancel{21}}}=\frac{4\cdot 5}{1\cdot 3}=\frac{20}{3}$$

(c)
$$\frac{\overset{1}{\cancel{2}}\;\cancel{6}}{\underset{1}{\cancel{5}}}\cdot\frac{\overset{1}{\cancel{8}}}{\underset{3}{\cancel{9}}}\cdot\frac{\overset{1}{\cancel{5}}}{\underset{1}{\cancel{16}\;\cancel{2}}}=\frac{1\cdot 1\cdot 1}{1\cdot 3\cdot 1}=\frac{1}{3}$$

2. (a) $\frac{2(x+1)^2}{15ax^2}$ (b) $3a^2(a-2)$

(a)
$$\frac{2(x+1)}{3ax}\cdot\frac{x+1}{5x}=\frac{2(x+1)(x+1)}{3ax(5x)}=\frac{2(x+1)^2}{15ax^2}$$

(b) First, divide by the factors that are common to the numerator and denominator, namely, a, $a-2$, and 3.

$$\frac{\overset{a^2}{\cancel{a^3}}}{\underset{1}{\cancel{a-2}}}\cdot\frac{\overset{3\ (a-2)}{\cancel{9(a-2)}}}{\underset{1\cdot 1}{\cancel{3a}}}=3a^2(a-2)$$

3. $\frac{a^6b^6}{6}$

$$\frac{2}{5}a^3b\cdot\frac{10}{9}ab^4\cdot\frac{3}{8}a^2b=\frac{\overset{1}{\cancel{2}}}{\underset{1}{\cancel{5}}}\cdot\frac{\overset{1}{\cancel{2}}\;\cancel{10}}{\underset{3}{\cancel{9}}}\cdot\frac{\overset{1}{\cancel{3}}}{\underset{2}{\cancel{8}\;\cancel{4}}}\cdot a^3aa^2\cdot bb^4b$$
$$=\frac{1\cdot 1\cdot 1}{1\cdot 3\cdot 2}a^{3+1+2}\cdot b^{1+4+1}$$
$$=\frac{1}{6}a^6b^6$$
$$=\frac{a^6b^6}{6}$$

4. $\frac{x^2}{2(x-1)}$

$$\frac{x^2-2x-3}{6x-18}\cdot\frac{3x^3}{x^3-x}=\frac{(x-3)(x+1)}{6(x-3)}\cdot\frac{3x^3}{x(x+1)(x-1)}=\frac{x^2}{2(x-1)}$$

5. (a) 9 (b) $\frac{-3}{4}$ (c) $\frac{1}{10}$

(a)
$$6\div\frac{2}{3}=\frac{\overset{3}{\cancel{6}}}{1}\cdot\frac{3}{\underset{1}{\cancel{2}}}=\frac{9}{1}=9$$

(b)
$$\frac{-9}{10}\div\frac{6}{5}=\frac{\overset{-3}{\cancel{-9}}}{\underset{2}{\cancel{10}}}\cdot\frac{\overset{1}{\cancel{5}}}{\underset{2}{\cancel{6}}}=\frac{-3}{4}$$

(c)
$$\frac{2}{5}\div 4=\frac{\overset{1}{\cancel{2}}}{5}\cdot\frac{1}{\underset{2}{\cancel{4}}}=\frac{1}{10}$$

6. (a) $\frac{50}{x^4(x+5)}$ (b) $\frac{a}{4}$

To divide rational expressions, invert the divisor and multiply.

(a)
$$\frac{5x}{x+5}\div\frac{x^5}{10}=\frac{5x}{x+5}\cdot\frac{10}{x^5}=\frac{50}{x^4(x+5)}$$

(b)
$$\frac{9a-6}{12a}\div\frac{9a^2-4}{3a^3+2a^2}=\frac{9a-6}{12a}\cdot\frac{3a^3+2a^2}{9a^2-4}=\frac{\cancel{3}\cancel{(3a-2)}}{\cancel{3}\cdot 4\cancel{a}}\cdot\frac{\overset{a}{\cancel{a^2}}\cancel{(3a+2)}}{\cancel{(3a+2)}\cancel{(3a-2)}}=\frac{a}{4}$$

Practice Exercises for Section 5.6, pages 218–220

1. (a) 2 (b) $\frac{1}{3}$ (c) $\frac{3}{4}$

To add rational expressions with the same denominator, retain the denominator and add the numerators. Simplify, if possible.

$$\frac{7}{5}+\frac{3}{5}=\frac{7+3}{5}=\frac{10}{5}=2$$

$$\frac{5}{9}-\frac{2}{9}=\frac{5-2}{9}=\frac{3}{9}=\frac{1}{3}$$

$$\frac{11}{12}+\frac{5}{12}-\frac{7}{12}=\frac{11+5-7}{12}=\frac{9}{12}=\frac{3}{4}$$

2. (a) $\frac{5}{a}$ (b) $\frac{3(x-1)}{x-3}$ (c) $x-1$ (d) $\frac{x+1}{x}$

(a) $\frac{7}{2a}+\frac{3}{2a}=\frac{7+3}{2a}=\frac{10}{2a}=\frac{5}{a}$

(b) $\frac{3x}{x-3}-\frac{3}{x-3}=\frac{3x-3}{x-3}=\frac{3(x-1)}{x-3}$

(c) $\frac{x^2}{x+4}+\frac{3x}{x+4}-\frac{4}{x+4}=\frac{x^2+3x-4}{x+4}=\frac{\overset{1}{\cancel{(x+4)}}(x-1)}{\underset{1}{\cancel{x+4}}}=x-1$

(d) $\frac{5x-2}{5x}+\frac{7}{5x}=\frac{5x-2+7}{5x}=\frac{5x+5}{5x}=\frac{5(x+1)}{5x}=\frac{x+1}{x}$

3. $2x-\frac{1}{2}y$

$\frac{2x}{3}+\frac{y}{8}+\frac{4x}{3}-\frac{5y}{8}$. Note that this expression might be written as $\frac{2}{3}x+\frac{1}{8}y+\frac{4}{3}x-\frac{5}{8}y$.

Like terms are combined by adding their coefficients. Thus,

$$\left(\frac{2}{3}x+\frac{4}{3}x\right)+\left(\frac{1}{8}y-\frac{5}{8}y\right)=\left(\frac{2}{3}+\frac{4}{3}\right)x+\left(\frac{1}{8}-\frac{5}{8}\right)y$$
$$=\frac{6}{3}x+\frac{-4}{8}y$$
$$=2x+\left(\frac{-1}{2}\right)y$$
$$=2x-\frac{1}{2}y$$

Practice Exercises for Section 5.7, pages 224–225

1. 72

$8=2^3$; $18=2\cdot 3^2$. In the factorizations of 8 and 18, the highest power of 2 that occurs is 2^3; the highest power of 3 is 3^2. The product of these, $2^3\cdot 3^2$ or 72, is the least common multiple.

2. 378

$14=2\cdot 7$; $54=2\cdot 3^3$; $63=3^2\cdot 7$

The highest power of 2 here is 2; the highest power of 3 is 3^3 (in the factorization of 54); and the highest power of 7 is 7.
The *lcm* is $2\cdot 3^3\cdot 7=378$.

3. (a) $6xy$ (b) $36a^2b^3$ (c) $2x(x+5)$ (d) $6x(x+2)(x-2)$

(a) $lcm(3x, 2y)=3\cdot x\cdot 2\cdot y=6xy$

(b) $12a^2b=2^2\cdot 3a^2b$; $9ab^3=3^2\cdot ab^3$
The *lcm* is the product of the highest powers of all factors that occur.
Thus, the *lcm* is $2^2\cdot 3^2\cdot a^2b^3$, or $36a^2b^3$.

(c) $2x+10=2(x+5)$; $x^2+5x=x(x+5)$
$lcm(2x+10, x^2+5x)=2(x+5)\cdot x=2x(x+5)$

(d) $6x=2\cdot 3x$; $3x+6=3(x+2)$; $x^2-4=(x+2)(x-2)$
$lcm(6x, 3x+6, x^2-4)=2\cdot 3\cdot x(x+2)(x-2)=6x(x+2)(x-2)$

Practice Exercises for Section 5.8, pages 228–233

1. (a) $lcd\left(\frac{4}{9},\frac{7}{15}\right)=45$ (b) $\frac{4}{9}=\frac{20}{45}$, $\frac{7}{15}=\frac{21}{45}$

(a) $9=3^2$; $15=3\cdot 5$. Thus,

$$lcd\left(\frac{4}{9},\frac{7}{15}\right)=lcm(9,15)=3^2\cdot 5=45$$

(b) To express $\frac{4}{9}$ as an equivalent fraction with the denominator 45, multiply both numerator and denominator by 5. Thus, $\frac{4}{9}=\frac{4\cdot 5}{9\cdot 5}=\frac{20}{45}$.

To express $\frac{7}{15}$ with denominator 45, multiply numerator and denominator by 3. Thus, $\frac{7}{15}=\frac{7\cdot 3}{15\cdot 3}=\frac{21}{45}$.

2. (a) $lcd=15a^2b$ (b) $\frac{3b}{5a^2}=\frac{9b^2}{15a^2b}$, $\frac{2}{15ab}=\frac{2a}{15a^2b}$

(a) $lcd\left(\frac{3b}{5a^2},\frac{2}{15ab}\right)=lcm(5a^2, 15ab)=15a^2b$

(b) This *lcd* contains all the factors that appear in either denominator. Convert each denominator into the *lcd* by supplying whatever factors it lacks. Thus, $\frac{3b\cdot 3b}{5a^2\cdot 3b}=\frac{9b^2}{15a^2b}$ and $\frac{2\cdot a}{15ab\cdot a}=\frac{2a}{15a^2b}$

3. (a) $lcd = 10x^2(x-2)$ (b) $\frac{3}{10x^2} = \frac{3x-6}{10x^2(x-2)}$; $\frac{7}{2x^2-4x} = \frac{35x}{10x^2(x-2)}$

(a) $10x^2 = 2\cdot 5\cdot x^2$; $2x^2-4x = 2x(x-2)$

$lcm(10x^2,\ 2x^2-4x) = 2\cdot 5\cdot x^2(x-2) = 10x^2(x-2)$

(b) $\frac{3\cdot(x-2)}{10x^2\cdot(x-2)} = \frac{3x-6}{10x^2(x-2)}$; $\frac{7\cdot 5x}{2x(x-2)5x} = \frac{35x}{10x^2(x-2)}$

4. (a) $\frac{13}{15}$ (b) $\frac{13}{18}$

$lcd\left(\frac{7}{10},\ \frac{1}{6}\right) = lcm(10,\ 6) = 30$

Express both fractions with the denominator 30.

$\frac{7\cdot 3}{10\cdot 3} = \frac{21}{30}$; $\frac{1\cdot 5}{6\cdot 5} = \frac{5}{30}$

Thus,

$\frac{7}{10} + \frac{1}{6} = \frac{21}{30} + \frac{5}{30} = \frac{21+5}{30} = \frac{26}{30} = \frac{13}{15}$

$lcm(9, 12, 4) = 2^2\cdot 3^2 = 36$

$\frac{8\cdot 4}{9\cdot 4} = \frac{32}{36}$; $\frac{7\cdot 3}{12\cdot 3} = \frac{21}{36}$; $\frac{3\cdot 9}{4\cdot 9} = \frac{27}{36}$

Thus,

$\frac{8}{9} + \frac{7}{12} - \frac{3}{4} = \frac{32}{36} + \frac{21}{36} - \frac{27}{36} = \frac{32+21-27}{36} = \frac{26}{36} = \frac{13}{18}$

5. (a) $\frac{22}{5}$ (b) $-\frac{23}{8}$

(a) $4\frac{2}{5} = 4 + \frac{2}{5} = \frac{4\cdot 5}{1\cdot 5} + \frac{2}{5} = \frac{20}{5} + \frac{2}{5} + \frac{20+2}{5} = \frac{22}{5}$

(b) $-2\frac{7}{8} = -\left(2+\frac{7}{8}\right) = -\left(\frac{2}{1}\cdot\frac{8}{8}+\frac{7}{8}\right) = -\left(\frac{16}{8}+\frac{7}{8}\right) = -\left(\frac{16+7}{8}\right) = -\frac{23}{8}$

6. (a) $\frac{25y^2+6x}{20x^2y}$ (b) $\frac{x^2+7x-7}{x(x-1)}$ (c) $\frac{-2a^2+10a-10}{a(a+1)(a-1)}$

(a) $lcm(4x^2,\ 10xy) = 20x^2y$

$\frac{5y}{4x^2} + \frac{3}{10xy} = \frac{5y\cdot 5y}{4x^2\cdot 5y} + \frac{3\cdot 2x}{10xy\cdot 2x} = \frac{25y^2+6x}{20x^2y}$

(b) $lcm(x,\ x-1) = x(x-1)$

$\frac{7}{x} + \frac{x}{x-1} = \frac{7\cdot(x-1)}{x\cdot(x-1)} + \frac{x\cdot x}{(x-1)\cdot x} = \frac{7x-7}{x(x-1)} + \frac{x^2}{x(x-1)} = \frac{x^2+7x-7}{x(x-1)}$

(c) $a^2+a = a(a+1)$; $a^2-1 = (a+1)(a-1)$

$lcm(a^2+a,\ a^2-1) = a(a+1)(a-1)$

$\frac{10}{a^2+a} - \frac{2a}{a^2-1} = \frac{10\cdot(a-1)}{a(a+1)\cdot(a-1)} - \frac{2a\cdot a}{(a+1)(a-1)\cdot a} = \frac{10a-10}{a(a+1)(a-1)} - \frac{2a^2}{a(a+1)(a-1)} = \frac{-2a^2+10a-10}{a(a+1)(a-1)}$

Practice Exercises for Section 5.9, pages 236–239

1. (a) numerator: $-a$ (b) denominator: $a+\frac{1}{2}$

2. (a) numerator: $\frac{a+4}{a^2-1}$ (b) denominator: $\frac{a}{a^2-1}$

3. (a) $\frac{x}{2}\div 5$ (b) $\frac{a}{a-2}\div\frac{a^2}{a+2}$

4. (a) $\frac{25}{2}$ (b) 6

$\frac{10}{\frac{4}{5}} = 10\div\frac{4}{5} = \frac{\overset{5}{\cancel{10}}}{1}\cdot\frac{5}{\underset{2}{\cancel{4}}}$

$= \frac{25}{2}$

Before dividing, express $4\frac{1}{5}$ as a rational number.

$\frac{4\frac{1}{5}}{\frac{7}{10}} = \left(4+\frac{1}{5}\right)\div\frac{7}{10} = \frac{\overset{3}{\cancel{21}}}{\underset{1}{\cancel{5}}}\cdot\frac{\overset{2}{\cancel{10}}}{\underset{1}{\cancel{7}}} = \frac{6}{1} = 6$

5. (a) $\dfrac{2x^2}{x+2}$

$$\frac{\frac{x}{3}}{\frac{x+2}{6x}} = \frac{x}{3} \div \frac{x+2}{6x} = \frac{x}{\cancel{3}_{1}} \cdot \frac{\cancel{6x}^{2}}{x+2} = \frac{2x^2}{x+2}$$

(b) $4a^2$

Combine the terms of the denominator:

$$\frac{2}{a} + \frac{1}{4} = \frac{2 \cdot 4}{a \cdot 4} + \frac{1 \cdot a}{4 \cdot a} = \frac{8}{4a} + \frac{a}{4a} = \frac{a+8}{4a}$$

Thus,

$$\frac{a^2+8a}{\frac{2}{a}+\frac{1}{4}} = (a^2+8a) \div \frac{a+8}{4a} = \frac{a\cancel{(a+8)}^{1}}{1} \cdot \frac{4a}{\cancel{a+8}_{1}} = 4a^2$$

Practice Exam 1 on Chapter 5, page 244

1. $\dfrac{6(x+1)^3}{7x}$

$$\frac{36x^2(x+1)^4}{42x^3(x+1)} = \frac{36(x+1)^{4-1}}{42x^{3-2}} = \frac{6(x+1)^3}{7x}$$

2. $\dfrac{x^2-2}{3x}$

Factor the numerator.

$$\frac{5x^4-10x^2}{15x^3} = \frac{5x^2(x^2-2)}{5x^2 \cdot 3x} = \frac{x^2-2}{3x}$$

3. $\dfrac{x-3}{x+2}$

Factor the numerator and denominator.

$$\frac{x^2-9}{x^2+5x+6} = \frac{(x+3)(x-3)}{(x+3)(x+2)} = \frac{x-3}{x+2}$$

4. $\dfrac{2}{3}$

Replace each occurrence of x by 2 and each occurrence of y by -2.

$$\frac{5(2)+3(-2)}{2-2(-2)} = \frac{10-6}{2+4} = \frac{4}{6} = \frac{2}{3}$$

5. (a) $\dfrac{3}{2}$

$x=1$:

$$\frac{1^2-2(1)+4}{1^4+1} = \frac{1-2+4}{1+1} = \frac{3}{2}$$

(b) $\dfrac{7}{2}$

$x=-1$:

$$\frac{(-1)^2-2(-1)+4}{(-1)^4+1} = \frac{1+2+4}{1+1} = \frac{7}{2}$$

(c) $\dfrac{4}{17}$

$x=2$:

$$\frac{2^2-2(2)+4}{2^4+1} = \frac{4-4+4}{16+1} = \frac{4}{17}$$

6. (a) x^2+5x-8

$$\begin{array}{r l}
x^2+5x-8 & \leftarrow \text{quotient} \\
x+2\,\overline{)\,x^3+7x^2+2x} & \\
\mp x^3 \mp 2x^2 & \\
5x^2+2x & \\
\mp 5x^2 \mp 10x & \\
-8x & \\
\pm 8x \pm 16 & \\
16 & \leftarrow \text{remainder}
\end{array}$$

(b) 16

(c) *Check:*

$$\begin{array}{r l}
x^2+5x-8 & \leftarrow \text{quotient} \\
\times x+2 & \leftarrow \text{divisor} \\
\hline
x^3+5x^2-8x & \\
2x^2+10x-16 & \\
\hline
x^3+7x^2+2x-16 & \\
+ \qquad 16 & \leftarrow \text{remainder} \\
\hline
x^3+7x^2+2x & \leftarrow \text{dividend}
\end{array}$$

7. $\dfrac{8}{7}$

$$\frac{\cancel{3}^{1}}{\cancel{4}_{1}} \cdot \frac{\cancel{-8}^{-2}}{\cancel{9}_{\cancel{3}_{1}}} \cdot \frac{\cancel{12}^{4}}{-7} = \frac{-8}{-7} = \frac{8}{7}$$

8. $x(x-1)$

To divide, invert the divisor and multiply.

$$\frac{x^2}{x+1} \div \frac{x}{x^2-1} = \frac{x^2}{x+1} \cdot \frac{x^2-1}{x} = \frac{x^2}{x+1} \cdot \frac{(x+1)(x-1)}{x} = \frac{x(x-1)}{1} = x(x-1)$$

9. $\dfrac{2ax+x-2}{2x^2}$

$lcm\,(x, 2x, x^2) = 2x^2$

$$\frac{a}{x}+\frac{1}{2x}-\frac{1}{x^2}=\frac{a\cdot 2x}{x\cdot 2x}+\frac{1\cdot x}{2x\cdot x}-\frac{1\cdot 2}{x^2\cdot 2}$$
$$=\frac{2ax}{2x^2}+\frac{x}{2x^2}-\frac{2}{2x^2}$$
$$=\frac{2ax+x-2}{2x^2}$$

10. $\dfrac{3}{x-4}$

$lcm\,(x+4, x-4, x^2-16) = (x+4)(x-4)$

$$\frac{1}{x+4}+\frac{2}{x-4}+\frac{8}{x^2-16}=\frac{1\cdot(x-4)}{(x+4)\cdot(x-4)}+\frac{2\cdot(x+4)}{(x-4)\cdot(x+4)}+\frac{8}{(x+4)(x-4)}$$
$$=\frac{x-4+2(x+4)+8}{(x+4)(x-4)}$$
$$=\frac{x-4+2x+8+8}{(x+4)(x-4)}$$
$$=\frac{3x+12}{(x+4)(x-4)}$$
$$=\frac{3(x+4)}{(x+4)(x-4)}$$
$$=\frac{3}{x-4}$$

11. $\dfrac{1}{a(b-a)}$

$$\frac{\dfrac{a^2}{b^2-a^2}}{\dfrac{a^3}{a+b}}=\frac{a^2}{b^2-a^2}\div\frac{a^3}{a+b}=\frac{a^2}{(b+a)(b-a)}\cdot\frac{a+b}{a^3}=\frac{1}{a(b-a)}$$

Practice Exam 2 on Chapter 5, page 245

1. $\dfrac{-1}{4}$

Replace each occurrence of x by -4.

$$\frac{-4-9}{3(-4)^2-(-4)}=\frac{-13}{3(16)+4}=\frac{-13}{52}=\frac{-1}{4}$$

2. $\dfrac{9}{13}$

$$\frac{(-2)^3-2(5)}{3(-2)(5)+4}=\frac{-8-10}{-30+4}=\frac{-18}{-26}=\frac{9}{13}$$

3. $\dfrac{5x^3}{3(x-2)}$

$$\frac{30x^4(x-2)^2}{18x(x-2)^3}=\frac{6\cdot 5x^{4-1}}{6\cdot 3(x-2)^{3-2}}=\frac{5x^3}{3(x-2)}$$

4. $\dfrac{4x^2}{x+4}$

$$\frac{12x^3}{3x^2+12x}=\frac{3x\cdot 4x^2}{3x(x+4)}=\frac{4x^2}{x+4}$$

5. $\dfrac{8}{x(x-2)}$

Factor both numerator and denominator to determine whether they have factors in common that can be cancelled.

$$\frac{8x^3+16x^2}{x^5-4x^3}=\frac{8x^2(x+2)}{x^3(x^2-4)}=\frac{8(x+2)}{x(x+2)(x-2)}=\frac{8}{x(x-2)}$$

6. (a) $x^2-5x+10$ (b) -10 (c) *Check:*

(a)

$$\begin{array}{rl}
x^2-5x+10 & \longleftarrow \text{quotient}\\
x+5\,\overline{)\,x^3+0x^2-15x+40} & \\
\mp x^3 \mp 5x^2 & \\
-5x^2-15x & \\
\pm 5x^2 \pm 25x & \\
10x+40 & \\
\mp 10x \mp 50 & \\
-10 & \longleftarrow \text{remainder}
\end{array}$$

(c) *Check:*

$$\begin{array}{rl}
x^2-5x+10 & \longleftarrow \text{quotient}\\
\times\; x+5 & \longleftarrow \text{divisor}\\
x^3-5x^2+10x & \\
+5x^2-25x+50 & \\
x^3 \quad -15x+50 & \\
+ \qquad\quad -10 & \longleftarrow \text{remainder}\\
x^3 \quad -15x+40 & \longleftarrow \text{dividend}
\end{array}$$

7. $\dfrac{x-1}{x^2}$

Factor and divide by factors common to the numerator and denominator.

$$\frac{x^2-4x+3}{x^3}\cdot\frac{5x}{5x-15}=\frac{(x-3)(x-1)}{x^3}\cdot\frac{5x}{5(x-3)}=\frac{x-1}{x^2}$$

8. $\dfrac{5ab^4(a-3)}{2}$

To divide, invert the divisor and multiply.

$$\frac{a^4-3a^3}{10b}\div\frac{a^2}{25b^5}=\frac{a^3(a-3)}{10b}\cdot\frac{25b^5}{a^2}=\frac{5ab^4(a-3)}{2}$$

9. $\dfrac{a+2}{a(a-2)}$

$$lcd\left(\frac{2}{a-2},\frac{1}{a}\right)=a(a-2)$$

$$\frac{2}{a-2}-\frac{1}{a}=\frac{2\cdot a}{(a-2)\cdot a}-\frac{1\cdot(a-2)}{a\cdot(a-2)}=\frac{2a}{a(a-2)}-\frac{a-2}{a(a-2)}=\frac{2a-(a-2)}{a(a-2)}=\frac{a+2}{a(a-2)}$$

10. $\dfrac{x^2-x+2}{2(x+3)(x-3)}$

$$x^2-9=(x+3)(x-3);\quad 2x+6=2(x+3)$$

$$lcm(x^2-9,\ 2x+6)=2(x+3)(x-3)$$

$$\frac{x+1}{x^2-9}+\frac{x}{2x+6}=\frac{x+1\cdot 2}{(x+3)(x-3)\cdot 2}+\frac{x\cdot(x-3)}{2(x+3)\cdot(x-3)}=\frac{2x+2+x^2-3x}{2(x+3)(x-3)}=\frac{x^2-x+2}{2(x+3)(x-3)}$$

11. $\dfrac{5x+y^2-y-2x^2}{x^2y}$

$$\frac{5}{xy}+\frac{y-1}{x^2}-\frac{2}{y}=\frac{5\cdot x}{xy\cdot x}+\frac{(y-1)\cdot y}{x^2\cdot y}-\frac{2\cdot x^2}{y\cdot x^2}=\frac{5x+y^2-y-2x^2}{x^2y}$$

12. 16

$$\frac{3\frac{1}{3}}{\frac{3}{8}-\frac{1}{6}}=3\tfrac{1}{3}\div\left(\frac{3}{8}-\frac{1}{6}\right)=\left(3+\frac{1}{3}\right)\div\left(\frac{9}{24}-\frac{4}{24}\right)=\left(\frac{9}{3}+\frac{1}{3}\right)\div\frac{5}{24}=\frac{\overset{2}{\cancel{10}}}{\underset{1}{\cancel{3}}}\cdot\frac{\overset{8}{\cancel{24}}}{\underset{1}{\cancel{5}}}=\frac{16}{1}=16$$

13. $\dfrac{1}{x(x-1)}$

Combine the terms of the denominator into a single rational expression.

$$x-\frac{1}{x}=\frac{x^2}{x}-\frac{1}{x}=\frac{x^2-1}{x}$$

Thus,

$$\frac{\frac{x+1}{x^2}}{x-\frac{1}{x}}=\frac{x+1}{x^2}\div\frac{x^2-1}{x}=\frac{x+1}{x^2}\cdot\frac{x}{x^2-1}=\frac{x+1}{x^2}\cdot\frac{x}{(x+1)(x-1)}=\frac{1}{x(x-1)}$$

CHAPTER 6

Practice Exercises for Section 6.1, pages 247–250

1. $\dfrac{5}{4}$

$$t-\frac{3}{4}=\frac{1}{2}$$
$$t=\frac{1}{2}+\frac{3}{4}$$
$$t=\frac{2}{4}+\frac{3}{4}$$
$$t=\frac{5}{4}$$

2. $x=50$

$$\frac{7}{10}x=35$$
$$\frac{\overset{1}{\cancel{10}}}{\underset{1}{\cancel{7}}}\cdot\frac{\overset{1}{\cancel{7}}}{\underset{1}{\cancel{10}}}x=\frac{10}{\underset{1}{\cancel{7}}}\cdot\overset{5}{\cancel{35}}$$
$$x=50$$

3. $x=\dfrac{5}{12}$

$$4x-1=\frac{2}{3}$$
$$4x-1+1=\frac{2}{3}+\frac{3}{3}$$
$$4x=\frac{5}{3}$$
$$\frac{1}{4}\cdot 4x=\frac{1}{4}\cdot\frac{5}{3}$$
$$x=\frac{5}{12}$$

4. $k=15$

$$\frac{20}{36}=\frac{k}{27}$$
$$36k=20\cdot 27$$
$$k=\frac{\overset{5}{\cancel{20}}\cdot\overset{3}{\cancel{27}}}{\underset{1}{\cancel{36}}}=15$$

5. $5\frac{5}{8}$ gallons

Let n stand for the unknown number of gallons needed to travel 150 miles. There are several proportions that may be used to express the relationship between miles and gallons. For example,

$$\frac{150 \text{ miles}}{n \text{ gal.}} = \frac{80 \text{ miles}}{3 \text{ gal.}}; \quad \frac{n \text{ gal.}}{150 \text{ miles}} = \frac{3 \text{ gal.}}{80 \text{ miles}}; \quad \frac{80 \text{ miles}}{150 \text{ miles}} = \frac{3 \text{ gal.}}{n \text{ gal.}}$$

are all correct. If you cross multiply in any of these proportions the result is

$$80n = 3 \cdot 150$$

$$n = \frac{3 \cdot \overset{15}{\cancel{150}}}{\underset{8}{\cancel{80}}} = \frac{45}{8} = 5\frac{5}{8} \text{ gallons}$$

6. 7

$$\frac{x+9}{4} = \frac{2x-2}{3}$$
$$3(x+9) = 4(2x-2)$$
$$3x + 27 = 8x - 8$$
$$35 = 5x$$
$$7 = x$$

7. −10

$$\frac{5}{y} = \frac{8}{y-6}$$
$$5(y-6) = 8 \cdot y$$
$$5y - 30 = 8y$$
$$-30 = 3y$$
$$-10 = y$$

8. 2

$$\frac{5}{3z} + \frac{2}{3} = \frac{3}{z}$$

Multiply both sides by the *lcd*, 3z.

$$\left(\frac{5}{3z} + \frac{2}{3}\right) \cdot 3z = \frac{3}{z} \cdot 3z$$
$$5 + 2z = 9$$
$$2z = 4$$
$$z = 2$$

Practice Exercises for Section 6.2, pages 255–260

1. (a) .6 (b) .09 (c) .027 (d) .0059

In converting a rational number whose denominator is a power of 10, note that its decimal equivalent has as many decimal digits as there are zeros in the denominator.

(a) $\frac{6}{10} = .6$
(one zero) (one decimal digit)

(b) $\frac{9}{100} = .09$
(2 zeros) (2 decimal digits)

(c) $\frac{27}{1000} = .027$
(3 zeros) (3 decimal digits)

(d) $\frac{59}{10\,000} = .0059$
(4 zeros) (4 decimal digits)

2. (a) $\frac{2}{25}$ (b) $\frac{607}{1000}$ (c) $\frac{9}{10}$ (d) $\frac{4023}{1000}$

$$.08 = \frac{8}{100} = \frac{2}{25}$$

$$.607 = \frac{607}{1000}$$

$$.90 = \frac{90}{100} = \frac{9}{10}$$

$$4.023 = \frac{4023}{1000}$$

3. (a) .12 (b) .81 81 81... (c) .4375

To convert a rational number to a decimal, divide the numerator by the denominator.

$\frac{3}{25} = 3.00 \div 25 = .12$

```
      .12
 25 ) 3.00
      2 5
        50
        50
         0
```

(b) $\frac{9}{11} = 9.000 \div 11$

```
      .8181...
 11 ) 9.0000
```

The digits 8, 1 repeat infinitely.

(c) $\frac{7}{16} = 7.000 \div 16$

```
       .4375
 16 ) 7.0000
       6 4
         60
         48
         120
         112
           80
           80
```

4. (a) 11.45 (b) 16.274 (c) 7.33

Line up decimal points in a column and supply zeros in the columns where they are missing.

$$\begin{array}{r} 1.09 \\ 9.80 \\ .56 \\ \hline 11.45 \end{array} \qquad \begin{array}{r} 2.710 \\ 13.500 \\ .064 \\ \hline 16.274 \end{array} \qquad \begin{array}{r} 8.40 \\ -1.07 \\ \hline 7.33 \end{array}$$

5. x

Add like terms by adding their coefficients.

$$\begin{aligned} .4x + .8x - .2x &= (.4 + .8 - .2)\,x \\ &= (1.2 - .2)\,x \\ &= 1.0\,x \\ &= x \end{aligned}$$

6. (a) 650 (b) .0273

To multiply a decimal by 100, move the decimal point 2 digits to the right. Thus, $6.5 \times 100 = 650.$

$$\begin{array}{rl} 9.1 & \leftarrow \text{1 decimal digit} \\ \times .003 & \leftarrow +\text{3 decimal digits} \\ \hline .0273 & \leftarrow \text{4 decimal digits} \end{array}$$

7. (a) .000 21 (b) 1200

(a) To divide a decimal by 1000, move the decimal point 3 digits to the left.

$$.21 \div 1000 = .000\,21$$

(b) $8.4 \div .007 = \dfrac{8.4 \times 1000}{.007 \times 1000} = \dfrac{8400}{7} = 1200$

8. 30

$$\frac{.9 + .5 + .7}{.07} = \frac{2.1}{.07} = \frac{2.1 \times 100}{.07 \times 100} = \frac{210}{7} = 30$$

9. (a) .008 (b) .0001

(a) $(.2)^3 = (.2)(.2)(.2) = .008$

(b) $(.01)^2 = .01 \times .01 = .0001$

10. .246

$$\begin{aligned} \frac{x}{8.2} &= .03 \\ \frac{x}{8.2}(8.2) &= .03\,(8.2) \\ x &= .246 \end{aligned}$$

11. 1.27

$$\begin{aligned} 4y - 5 &= .08 \\ 4y - 5 + 5 &= .08 + 5 \\ 4y &= 5.08 \\ y &= \frac{5.08}{4} \\ y &= 1.27 \end{aligned}$$

12. (a) 12 (b) 5

(a)

$$\begin{aligned} .9x + 12 &= 22.8 \\ 10\,(.9x + 12) &= 10\,(22.8) \\ 9x + 120 &= 228 \\ 9x &= 108 \\ x &= 12 \end{aligned}$$

[Multiply both sides by 10 to eliminate decimals.]

(b) $.3y - 2.5 = .12y + 6.5$

The decimal in this equation with the greatest number of decimal digits is .12, or $\frac{12}{100}$. To eliminate all decimals, we must multiply both sides by 100.

$$\begin{aligned} 100\,(.3y - 2.5) &= 100\,(.12y + 6.5) \\ 30y - 250 &= 12y + 650 \\ 18y &= 900 \\ y &= 50 \end{aligned}$$

Practice Exercises for Section 6.3, pages 264–268

1. (a) .04 (b) .9 (c) 1.67 (d) .035

2. (a) 43% (b) 190 (c) .5% or $\frac{1}{2}$%

To change a decimal to a percent, move the decimal point two digits to the right (that is, multiply by 100) and insert the percent symbol.

$.43 = 43.\% = 43\%$ $\qquad 1.9 = 190.\% = 190\%$ $\qquad .005 = 00.5\% = .5\%$ or $\frac{1}{2}$%

3. 40.5

9% of $450 = .09 \times 450 = 40.5\not{0}$

4. \$285

$9\frac{1}{2}\% = 9.5\% = .095$

Rate × Principal = Interest

.095 × \$3000 = \$285.000

5. \$4900

Let P = principal (amount invested).

Using the formula R × P = I, we have

.07P = 343

$$P = \frac{343}{.07} = \frac{34\,300}{7} = 4900$$

6. 6%

Solving the equation R × P = I for R, we have $R = \dfrac{I}{P}$.

In this case, I = 30 and P = 500. Thus

$$R = \frac{30}{500} = \frac{6}{100} = 6\%$$

7. $67.20

$R = 8\% = .08; \quad P = \$1120; \quad T = \frac{3}{4}$ year

$R \cdot P \cdot T = I$

$I = .08 \times 1120 \times \frac{3}{4} = \frac{268.8}{4} = 67.2$

Thus, interest earned is $67.20.

8. $19.25

Because 1 month $= \frac{1}{12}$ of a year,

5 months $= \frac{5}{12}$ of a year. Thus,

$T = \frac{5}{12}; \quad R = 5\frac{1}{2}\% = .055; \quad P = \$840$

$$\begin{aligned} I &= R \cdot P \cdot T \\ &= .055 \times 840 \times \frac{5}{12} \\ &= .055 \times 70 \times 5 \\ &= .055 \times 350 \\ &= 19.25 \end{aligned}$$

The interest earned is $19.25.

9. $37.09

"Semiannually" means "every half year." $1\frac{1}{2}$ years $= 3 \times \frac{1}{2}$ year. Thus, in $1\frac{1}{2}$ years, interest will be compounded 3 times. Let P_1, P_2, and P_3 be the principals which earn interest for each of those half-year periods. Let I_1, I_2, and I_3 be the interest earned in those three periods. For the first half-year,

$$R \cdot P_1 \cdot T = I_1$$

$$.06 \times 400 \times \frac{1}{2} = 12$$

Now $12 interest is added to the original principal, P_1, to obtain P_2.

$$R \cdot P_2 \cdot T = I_2$$

$$.06 \times 412 \times \frac{1}{2} = 12.36$$

$$P_3 = P_2 + I_2 = 412 + 12.36 = 424.36$$

$$R \cdot P_3 \cdot T = I_3$$

$$.06 \times 424.36 \times \frac{1}{2} = 12.7308 = 12.73 \text{ (to the nearest cent)}$$

The *total* interest earned for $1\frac{1}{2}$ years is given by

$$I_1 + I_2 + I_3 = 12 + 12.36 + 12.73 = \$37.09$$

Practice Exercises for Section 6.4, pages 271–273

1. $2800

commission = rate of commission × sale price

Thus, $C = 3\frac{1}{2}\% \times 80\,000$

$= .035\,(80\,000) = 2800$

The commission on $80 000 at $3\frac{1}{2}\%$ is $2800.

2. 25

Let n = the number of vacuum cleaners.

The total sale price, S, of n vacuum cleaners at $200 each is $200n$ (dollars).

The commission, C, is $450. Thus, from the formula

$R \cdot S = C$, we can write

$$\begin{aligned} .09 \times 200n &= 450 \\ 18n &= 450 \\ n &= 25 \end{aligned}$$

3. $37

markup = rate of markup × cost.

markup = 85% × $20 = .85 ($20) = $17

Thus, $17 is the markup.

sale price = cost + markup

= $20 + $17

= $37

4. 250%

markup = sale price − cost = 210 − 60 = $150 (markup)

rate of markup $= \frac{\text{markup}}{\text{cost}} = \frac{150}{60} = \frac{5}{2} = 2.5$

Expressed as a percent, 2.5 = 250%.

5. $42.50

discount = rate of discount × original price

discount = .15 ($50) = $7.50

sale price = original price − discount

sale price = $50.00 − $7.50 = $42.50

The discounted price for a senior citizen is $42.50.

6. \$40

Let P be the original price of the skirt.
discount = rate of discount × original price
Thus, \$12 = 30% × P

$12 = .3P$

$$P = \frac{12}{.3} = \frac{120}{3} = 40$$

The original price of the skirt was \$40.

7. 20%

original price − sale price = discount
\$20 − \$16 = \$4

$$\text{rate of discount} = \frac{\text{discount}}{\text{original price}} = \frac{4}{20} = \frac{1}{5}$$

To change $\frac{1}{5}$ to a percent, multiply by 100 and insert the percent sign.

$$\frac{1}{5} \times 100\% = 20\%$$

Practice Exercises for Section 6.5, pages 277–279

1. $10y - 30$

$\frac{x}{5} + 6 = 2y$ [Isolates x by first adding -6 to both sides.]

$\frac{x}{5} + 6 - 6 = 2y - 6$

$\frac{x}{5} = 2y - 6$ [Multiply both sides by 5.]

$5\left(\frac{x}{5}\right) = 5(2y - 6)$

$x = 10y - 30$

Check:

To check this solution, substitute $10y - 30$ for x in the original equation to see if it produces a true statement.

$\frac{10y - 30}{5} + 6 \stackrel{?}{=} 2y$

$2y - 6 + 6 \stackrel{?}{=} 2y$

$2y \stackrel{\checkmark}{=} 2y$

2. (a) $\frac{9 + 4y}{3}$ (b) $\frac{3x - 9}{4}$

(a) To solve for x, isolate the "x-terms" on one side.

$x - 4y = 9 - 2x$

$x - 4y + 2x + 4y = 9 - 2x + 2x + 4y$

$3x = 9 + 4y$

$x = \frac{9 + 4y}{3}$

(b) To solve for y, isolate the "y-term."

$x - 4y = 9 - 2x$

$-4y = 9 - 3x$

$y = \frac{9 - 3x}{-4} = \frac{-9 + 3x}{4} = \frac{3x - 9}{4}$

3. $\frac{6r - 1}{r + 8}$

Bring all "k-terms" to the left side.

$8k + 1 = r(6 - k)$

$8k + 1 = 6r - rk$

$rk + 8k = 6r - 1$

$k(r + 8) = 6r - 1$

$k = \frac{6r - 1}{r + 8}$

4. $\frac{2A - hb}{h}$

$A = \frac{h(B + b)}{2}$

$2A = h(B + b)$

$2A = hB + hb$

$2A - hb = hB$

$\frac{2A - hb}{h} = B$

Here note that h, the height, is assumed to be nonzero.

Practice Exercises for Section 6.6, pages 282–285

1. (a) true (b) false (c) true

(a) $-12 \leq 0$ is true because -12 is *less than* 0.
(b) $-8 \geq 8$ is false because -8 is neither greater than 8 nor equal to 8.
(c) $20 \geq 20$ is true because 20 is *equal to* 20.

2. (a) $\geq$ (b) $\leq$ (c) both

(c) $-7 = -7$. Therefore, both $-7 \leq -7$ and $-7 \geq -7$ are true.

3. $x \leq 0$

By the Law of Transitivity, if $x \leq y$ and $y \leq 0$, then $x \leq 0$.

4. (a) closed interval (b) open interval (c) closed interval

In (c), the set of numbers that satisfy the condition $3 \leq x \leq 100$ can be denoted by [3, 100]. This set is a closed interval because the endpoints 3 and 100 are included in the set.

5. (a) true (b) true (c) false (d) false

(a) is true because the endpoint -9 is included in the closed interval $[-9, 5]$.

(b) is true because $0 < .002 < 1$.

(c) is false because the endpoint 5 is not included in the open interval $(-8, 5)$.

(d) is false because although -4 is less than 6, 6 is not less than 6. (*Both* statements, $-4 < 6$ and $6 < 6$, would have to be true in order for $-4 < 6 < 6$ to be true.)

6. (a) open right ray

(b) closed right ray

(c) closed right ray, denoted by $[2, \infty)$

(d) open left ray, denoted by $(-\infty, 0)$

7. (a) true, (b) false

Practice Exercises for Section 6.7, pages 288–291

1. (a) < (b) < (c) > (d) >

(a) $0 < 5$. When *any* number n is added to both sides of this inequality, the direction of the inequality is preserved. Thus,

$$0 + n < 5 + n, \text{ or } n < n + 5$$

(b) $-3 < 0$. Therefore, $-3 + n < 0 + n$, or $n - 3 < n$.

(c) $4 > 1$. Therefore, $n + 4 > n + 1$.

(d) $1 > -2$. Therefore, $1 + n > -2 + n$.

2. (a) > (b) <

(a) $n > n - 1$ If both sides of an inequality are multiplied by a *positive* number, the direction of the inequality is *preserved.* Thus,

$$10n > 10(n - 1)$$

(b) $n + 4 > n + 3$ If both sides of an inequality are multiplied by a *negative* number, the direction of the inequality is *reversed.* Thus,

$$-5(n + 4) < -5(n + 3)$$

3. (a) > (b) >

(a) $2 > -3$. Multiplication or division by a *positive* number *preserves* the sense of the inequality. Thus, if $k > 0$, then $\frac{2}{k} > \frac{-3}{k}$.

(b) $-5 < 2$. Multiplication or division by a *negative* number *reverses* the sense of the inequality. Thus, if $k < 0$, then $-5k > 2k$.

Practice Exercises for Section 6.8, pages 295–298

1. $(7, \infty)$

$$x - 10 > -3$$
$$x - 10 + 10 > -3 + 10$$
$$x > 7$$

The solution set is the open right ray $(7, \infty)$.

2. $(-\infty, 6)$

$$2y < y + 6$$
$$2y - y < y + 6 - y$$
$$y < 6$$

The solution set is $(-\infty, 6)$.

3. $(-20, \infty)$

$$\frac{x}{20} > -1$$
$$\frac{x}{20} \cdot 20 > (-1) \cdot 20$$
$$x > -20$$

The solution set is $(-20, \infty)$.

4. $(-6, \infty)$

$$-5x < 30$$
$$\frac{-5x}{-5} > \frac{30}{-5}$$
$$x > -6$$

The solution set is $(-6, \infty)$.

5. $(12, \infty)$

$$\frac{2}{3}x > 8$$
$$\frac{3}{2} \cdot \frac{2}{3}x > \frac{3}{2} \cdot 8$$
$$x > 12$$

The solution set is $(12, \infty)$.

6. $[-5, \infty)$

$$x + 1 \le 4(x + 4)$$
$$x + 1 \le 4x + 16$$
$$-15 \le 3x$$
$$\frac{-15}{3} \le x$$
$$-5 \le x \quad (\text{or } x \ge -5)$$

The solution set is $[-5, \infty)$.

7. $(-\infty, 180]$

$$18 - \frac{x}{10} \ge 0$$
$$18 \ge \frac{x}{10}$$
$$180 \ge x$$

The solution set is $(-\infty, 180]$.

8. $[1, 7]$

$$0 \le 3x - 3 \le 18$$
$$0 + 3 \le 3x - 3 + 3 \le 18 + 3$$
$$3 \le 3x \le 21$$
$$1 \le x \le 7$$

The solution set is the closed interval $[1, 7]$.

9. $\left(\frac{-1}{2}, 8\right)$

$$\begin{aligned} -10 &< 6-2x < 7 \\ -16 &< -2x < 1 \\ \frac{-16}{-2} &> \frac{-2x}{-2} > \frac{1}{-2} \\ 8 &> x > \frac{-1}{2} \end{aligned}$$

This inequality is equivalent to

$$\frac{-1}{2} < x < 8$$

The solution set is $\left(\frac{-1}{2}, 8\right)$.

10. [100, 130]

Let d = distance; r = rate; t = time = 2 hours.

$$d = r \cdot t = 2r$$

Because $50 \le r \le 65$, it follows that $100 \le 2r \le 130$.
Therefore, the distance traveled is between 100 miles and 130 miles.
That is, $100 \le d \le 130$, or "the distance interval" is given by [100, 300].

Practice Exam 1 on Chapter 6, page 303

1. 5

$$\begin{aligned} \frac{x+3}{4} &= \frac{2x}{5} \\ 5(x+3) &= 4(2x) \\ 5x+15 &= 8x \\ 15 &= 3x \\ 5 &= x \end{aligned}$$

2. 9

Let x be the number of melons that can be bought for \$2.25.
Use a proportion to compare melons to cost; then cross multiply.

$$\begin{aligned} \frac{4}{\$1.00} &= \frac{x}{\$2.25} \\ 1.00x &= 4(2.25) \\ x &= 9 \end{aligned}$$

3. 9

$$\frac{3}{t-3} + \frac{6}{t} = \frac{21}{2t}$$

This equation is not a proportion. Cross multiplying will not apply.
Instead, multiply each side by the *lcd* of these expressions; namely $2t(t-3)$.

$$\begin{aligned} 2t(t-3)\cdot\frac{3}{t-3} + 2t(t-3)\cdot\frac{6}{t} &= 2t(t-3)\cdot\frac{21}{2t} \\ 6t + 12(t-3) &= 21(t-3) \\ 6t + 12t - 36 &= 21t - 63 \\ 18t - 36 &= 21t - 63 \\ 27 &= 3t \\ 9 &= t \end{aligned}$$

4. 140

$$.01x - 1 = .4$$

Multiply both sides by 100 to eliminate decimals.

$$\begin{aligned} 1x - 100 &= 40 \\ x &= 140 \end{aligned}$$

5. \$3000

Let x be the amount invested (principal).

Rate $= 9\frac{1}{2}\% = .095$; Interest = \$285

$$\begin{aligned} R \cdot P &= I \\ .095x &= 285 \\ 95x &= 285\,000 \\ x &= \frac{285\,000}{95} \\ x &= 3000 \end{aligned}$$

6. \$47 000

6% of \$50 000 = .06 × 50 000 = \$3000 (agent's commission)
Subtract the agent's commission from \$50 000.

\$50 000 − \$3000 = \$47 000

(Alternately, you could show that she receives 94% of \$50 000, which equals \$47 000.)

7. $\frac{10-3x}{5}$

$$\begin{aligned} 3x+5y &= 10 \\ 5y &= 10-3x \\ y &= \frac{10-3x}{5} \end{aligned}$$

8. (a) $\frac{5}{3b+2}$ (b) $\frac{5-2a}{3a}$

(a)
$$\begin{aligned} 3ab+2a &= 5 \\ a(3b+2) &= 5 \\ a &= \frac{5}{3b+2} \end{aligned}$$

(b)
$$\begin{aligned} 3ab+2a &= 5 \\ 3ab &= 5-2a \\ b &= \frac{5-2a}{3a} \end{aligned}$$

9. (a) true (b) false (c) true

(a) $4 \le 4$ is true because $4 = 4$.

(b) The open left ray $(-\infty, 3)$ does not contain 3 because it consists only of those numbers x such that $x < 3$.

(c) -2 is in $[-3, 0]$ because $-3 \le -2 \le 0$.

10. (a) > (b) > (c) >

(a) $6 > -6$. Subtract c from both sides of this inequality and the sense is preserved.
$6 - c > -6 - c$

(b) If $x > y$, then subtracting y from both sides preserves the sense of the inequality.
Thus, $x - y > y - y$, or $x - y > 0$.

(c) $2 < 3$. If $c < 0$, then multiplying both sides by a *negative* number *reverses* the sense of the inequality. Thus, $2c > 3c$.

11. $(-\infty, 3]$

$$2y - 1 \le 8 - y$$
$$3y \le 9$$
$$y \le 3$$

The solution set is the closed left ray $(-\infty, 3]$.

12. $(7, 11)$

$$4 < \frac{x+1}{2} < 6$$
$$8 < x + 1 < 12$$
$$7 < x < 11$$

The solution is the open interval $(7, 11)$.

Practice Exam 2 on Chapter 6, page 303

1. 6

$$\frac{2}{3}x - 3 = \frac{x+3}{9}$$

Multiply both sides by the *lcd*, 9, to clear the equation of fractions.

$$9\left(\frac{2}{3}x - 3\right) = 9\left(\frac{x+3}{9}\right)$$
$$6x - 27 = x + 3$$
$$5x = 30$$
$$x = 6$$

2. 7

$$\frac{1}{x-5} = \frac{7}{2x}$$
$$1(2x) = 7(x - 5)$$
$$2x = 7x - 35$$
$$-5x = -35$$
$$x = 7$$

3. 175

Let x be the number of calories in 10 ounces of milk. Set up a proportion.

$$\frac{x \text{ calories}}{10 \text{ ounces}} = \frac{140 \text{ calories}}{8 \text{ ounces}}$$
$$8x = 10 \times 140$$
$$x = \frac{10 \times 140}{8}$$
$$x = 175 \text{ (calories)}$$

4. (a) .12 (b) 12%

$$75 \overline{)9.00} = .12$$
$$\begin{array}{r} 75 \\ \hline 150 \\ 150 \\ \hline \end{array}$$

$.12 = 12.\% = 12\%$

5. 52

$$.04(x + 3) = 2.2$$
$$100[.04(x + 3)] = 100(2.2)$$
$$4(x + 3) = 220$$
$$4x + 12 = 220$$
$$4x = 208$$
$$x = 52$$

Check:

$$.04(52 + 3) \stackrel{?}{=} 2.2$$
$$.04(55) \stackrel{?}{=} 2.2$$
$$2.20 \stackrel{\checkmark}{=} 2.2$$

6. \$147

Interest = Principal × Rate × Time
$I = P \times R \times T$

3 months $= \frac{3}{12}$ or $\frac{1}{4}$ of a year.

$P = \$8400$; $R = 7\% = .07$; $T = \frac{1}{4}$ year

$$I = 8400 \times .07 \times \frac{1}{4} = \frac{8400(.07)}{4} = \$147$$

7. \$540

discount = rate of discount × marked price
discount = 40% (\$900) = .4 (\$900) = \$360
sale price = marked price − discount
sale price = \$900 − \$360 = \$540

8. $\frac{2b+16}{3}$

$$3(a - 5) = 2b + 1$$
$$3a - 15 = 2b + 1$$
$$3a = 2b + 16$$
$$a = \frac{2b+16}{3}$$

9. (a) $\frac{4}{10-3y}$

$$10x - 3xy = 4$$
$$x(10 - 3y) = 4$$
$$x = \frac{4}{10-3y}$$

(b) $\frac{10x-4}{3x}$

$$10x - 3xy = 4$$
$$-3xy = 4 - 10x$$
$$3xy = 10x - 4$$
$$y = \frac{10x-4}{3x}$$

10. (a) true (b) false (c) true (d) false

11. (a) < (b) < (c) <

(a) $-4 + c < 7 + c$ because $-4 < 7$ and because adding c to both sides preserves the sense of the inequality.

(b) $-2 < 1$. Dividing both sides of this inequality by the *positive* number k *preserves* the sense of the inequality.
Thus, $-\frac{2}{k} < \frac{1}{k}$.

(c) $10 > -3$. Multiplying both sides of this inequality by a *negative* number *reverses* the sense of the inequality.
Thus, if $k < 0$, then $10k < -3k$.

12. $(-\infty, 7]$

$15 - 3x \geq 1 - x$

$-2x \geq -14$

$x \leq 7$

The solution set is $(-\infty, 7]$.

13. $[-3, 6]$

$6 \leq \frac{x}{3} + 7 \leq 9$

$-1 \leq \frac{x}{3} \leq 2$

$3(-1) \leq 3\left(\frac{x}{3}\right) \leq 3(2)$

$-3 \leq x \leq 6$

The solution set is $[-3, 6]$.

CHAPTER 7

Practice Exercises for Section 7.1, pages 306–309

1.

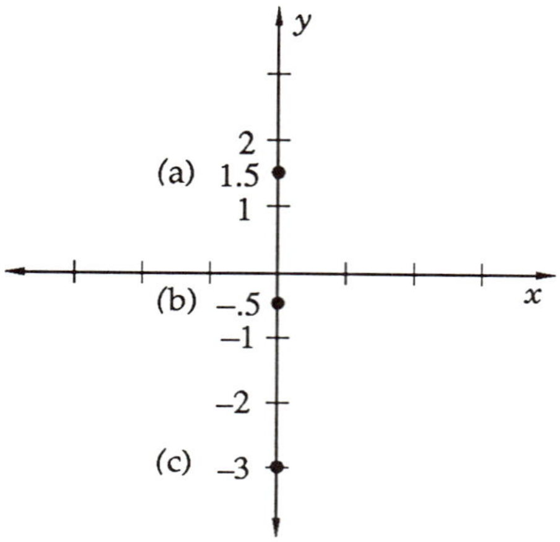

FIGURE 7A-S

2.

A:	(a) 1	(b) 3	(c) $(1, 3)$
B:	(a) 3	(b) 0	(c) $(3, 0)$
C:	(a) 1	(b) -3	(c) $(1, -3)$
D:	(a) 0	(b) -4	(c) $(0, -4)$
E:	(a) -2	(b) -4	(c) $(-2, -4)$
F:	(a) -3	(b) 2	(c) $(-3, 2)$

3.

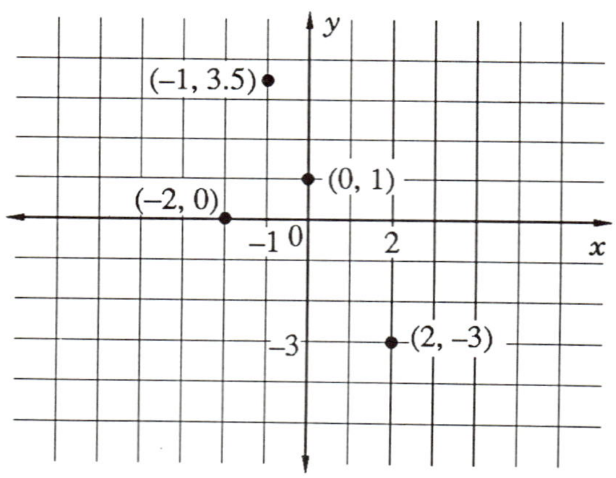

FIGURE 7B-S

Practice Exercises for Section 7.2, pages 312–318

1. (a), (d), and (e)

Substitute the first coordinate for x and the second for y in the equation $y = 3x - 4$ to see if it produces a true statement.

(a) $2 \stackrel{?}{=} 3(2) - 4$

$2 \stackrel{?}{=} 6 - 4$

$2 \stackrel{\checkmark}{=} 2$

(b) $1 \stackrel{?}{=} 3(1) - 4$

$1 \stackrel{?}{=} 3 - 4$

$1 \stackrel{\times}{=} -1$

(c) $2 \stackrel{?}{=} 3(6) - 4$

$2 \stackrel{?}{=} 18 - 4$

$2 \stackrel{\times}{=} 14$

(d) $-4 \stackrel{?}{=} 3(0) - 4$

$-4 \stackrel{\checkmark}{=} -4$

(e) $5 \stackrel{?}{=} 3(3) - 4$

$5 \stackrel{?}{=} 9 - 4$

$5 \stackrel{\checkmark}{=} 5$

2. (a) (0, 3) (b) (4, 1) (c) (6, 0)

Substitute the known coordinate to solve for the unknown in the equation $x + 2y = 6$.

(a) If $x = 0$, then $0 + 2y = 6$. Thus, $y = 3$.

(b) If $y = 1$, then $x + 2(1) = 6$ and $x = 6 - 2 = 4$.

(c) If $y = 0$, then $x + 2(0) = 6$ and $x = 6$.

3.

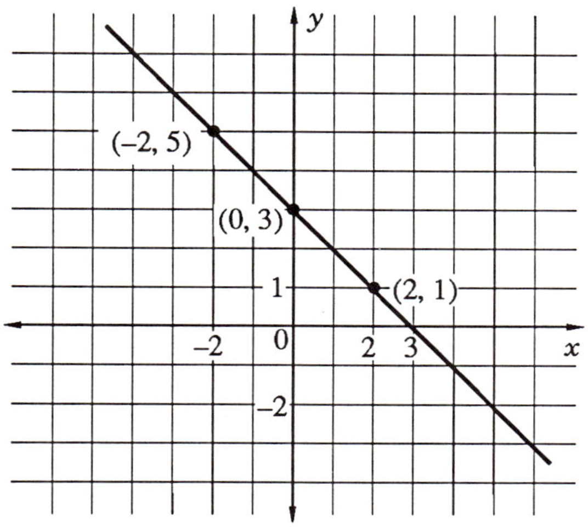

FIGURE 7C-S

x	$y = 3 - x$
−2	$3 - (-2) = 5$
0	$3 - 0 = 3$
2	$3 - 2 = 1$

The table indicates that the ordered pairs (− 2, 5), (0, 3), and (2, 1) are solutions of $y = 3 - x$.

Plot any two of these ordered pairs on a rectangular coordinate system and draw the line connecting them.

4.

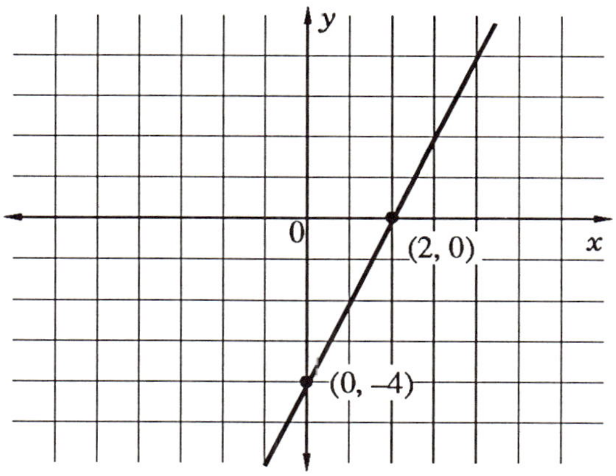

FIGURE 7D-S

x	y
0	−4
2	0

When $x = 0$, $2(0) - y = 4$ and $y = -4$.

When $y = 0$, $2x - 0 = 4$ and $x = 2$.

5.

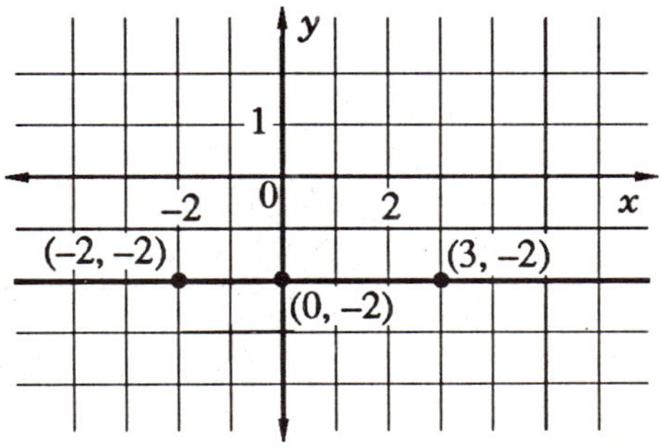

FIGURE 7E-S

6.

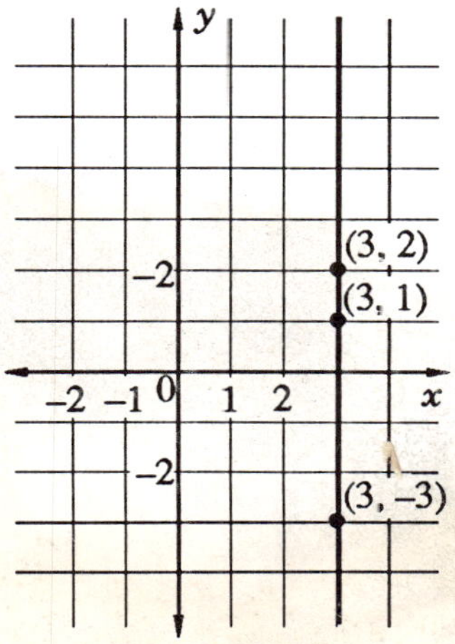

FIGURE 7F-S

Practice Exercises for Section 7.3, pages 324–327

1. (a) rise = 2 (b) run = − 3 (c) slope = $-\frac{2}{3}$ (d) downward

$P_1 = (1, 2)$, $P_2 = (-2, 4)$
rise $= y_2 - y_1 = 4 - 2 = 2$.
run $= x_2 - x_1 = -2 - 1 = -3$.
slope $= \frac{\text{rise}}{\text{run}} = \frac{2}{-3} = -\frac{2}{3}$.
The negative slope indicates that the line slopes downward from left to right.

2. (a) $\frac{1}{3}$ (b) $\frac{1}{3}$

(a) Using $P_1(-2, -1)$ and $P_2(1, 0)$,

$$\text{slope} = \frac{y_2 - y_1}{x_2 - x_1} = \frac{0-(-1)}{1-(-2)} = \frac{1}{3}$$

(b) Using $P_1(-2, -1)$ and $P_3(4, 1)$,

$$\text{slope} = \frac{y_3 - y_1}{x_3 - y_1} = \frac{1-(-1)}{4-(-2)} = \frac{2}{6} = \frac{1}{3}$$

3.

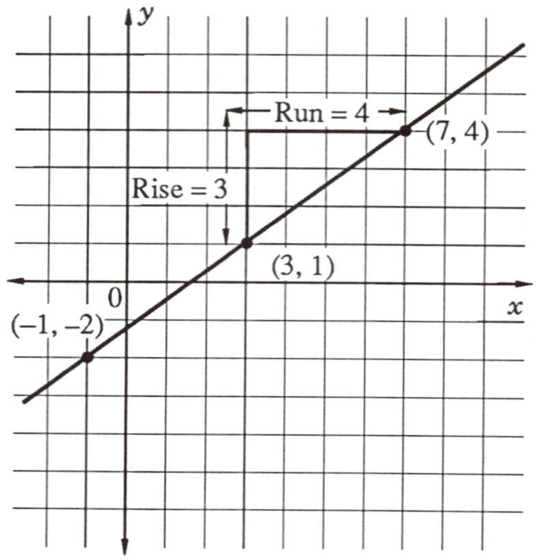

FIGURE 7G-S

4. (a) slope = 7 (b) (*iii*) upward

$$\text{slope} = \frac{6-(-8)}{8-6} = \frac{14}{2} = 7$$

A positive slope indicates that the line slopes upward to the right.

5. (a) undefined (b) (*ii*) vertical

Because the x-coordinates of $(9, 5)$ and $(9, -1)$ are the same, the change in x is 0. Thus, slope $= \frac{-1-5}{9-9} = \frac{-6}{0}$, which is undefined.
The line is a vertical line, and its slope is undefined.

6. (a) slope = 0 (b) (*i*) horizontal

The slope of the line through $(5, -4)$ and $(0, -4)$ is $\frac{-4-(-4)}{0-5} = \frac{0}{-5} = 0$. Because the y-coordinates are the same, the line is horizontal.

Practice Exercises for Section 7.4, pages 333–338

1. (a) slope = $\frac{1}{2}$ (b) y-intercept = − 5 (c)

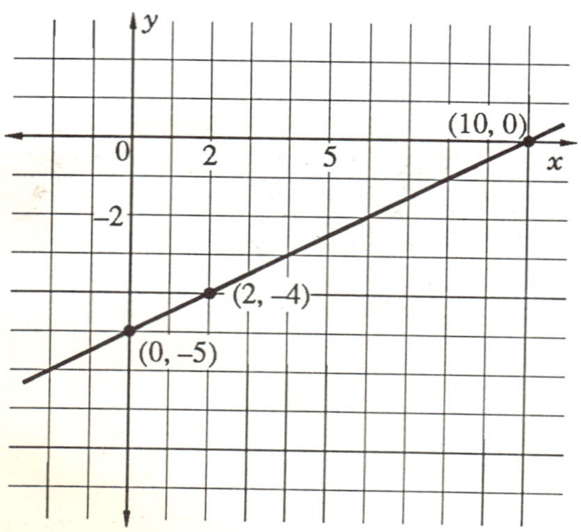

FIGURE 7H-S

2. $y = 4x - 2$

3. (a) $-\frac{x}{4} + y = 1$ (b) $-x + 4y = 4$

(a) The x-intercept is − 4; the y-intercept is 1.
The equation in intercept form is

$$\frac{x}{-4} + \frac{y}{1} = 1 \text{ or } -\frac{x}{4} + y = 1$$

(b) To clear of fractions, multiply by the *lcd*, 4.

$$4\left(-\frac{x}{4} + y\right) = 4(1)$$

$$-x + 4y = 4$$

4. (a) $\frac{x}{5}+\frac{y}{2}=1$ (b) $2x+5y=10$

(c)

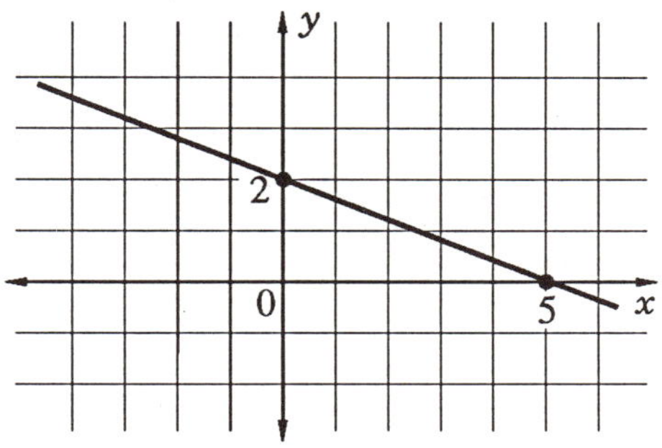

FIGURE 7I-S

5. $y=\frac{x}{2}$

Because the line passes through the origin, the y-intercept, b, is zero. Find the slope. Let $P_1=(0,0)$ and $P_2=(6,3)$.

$$m=\frac{3-0}{6-0}=\frac{1}{2}$$

The equation of the line is $y=\frac{1}{2}x+0$ or $y=\frac{x}{2}$.

6. (a) $y=7$ (b) $x=-2$ (c) $(-2,7)$

See Figure 7J-S.

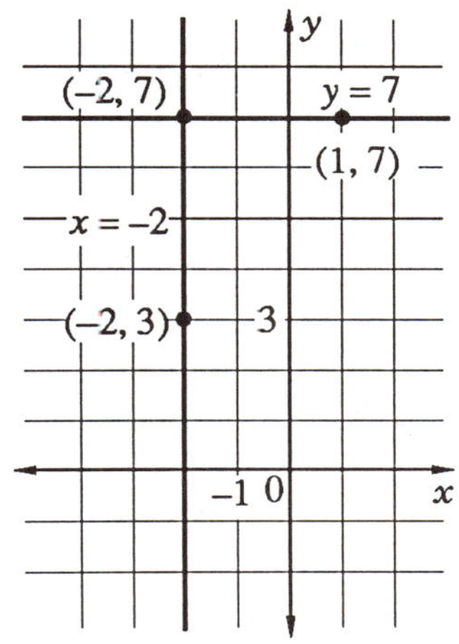

FIGURE 7J-S

Practice Exercises for Section 7.5, pages 344–345

1. The tables below are helpful in graphing the lines. In the first, we find the intercepts by substituting 0 for x, then 0 for y. In the second, we first solve for y in terms of x.

$L_1: 2y+x=4$ $L_2: y=-3x-3$

x	y
0	2
4	0

x	y
0	-3
-1	0

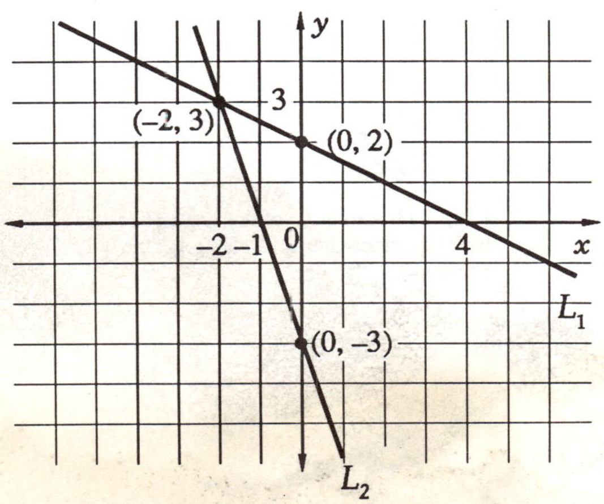

FIGURE 7K-S

The two lines intersect at the point $(-2,3)$. To check this solution, substitute -2 for x and 3 for y in the original equations.

Check:

L_1: $2y=4-x$

$2(3)\overset{?}{=}4-(-2)$

$6\overset{\checkmark}{=}6$

L_2: $y=-3(x+1)$

$3\overset{?}{=}-3(-2+1)$

$3\overset{?}{=}-3(-1)$

$3\overset{\checkmark}{=}3$

2.

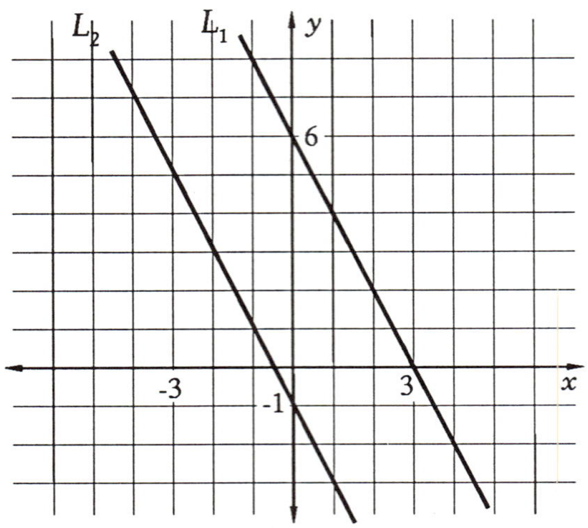

FIGURE 7L-S

Solving each equation for y in terms of x, we can determine the slopes and y-intercepts.

$L_1: 2x+y=6$

$y=-2x+6$

slope $=-2$, y-intercept $=6$.

$L_2: 2y+2=-4x$

$2y=-4x-2$

$y=-2x-1$

slope $=-2$, y-intercept $=-1$.

The lines L_1 and L_2 are parallel because they both have slope -2 and yet have different y-intercepts. They are therefore not the same line.

3. L_1 and L_3 are parallel.

Solve each equation for y to determine the slope of each line.

$L_1: \frac{y}{3}=x \quad y=3x \quad$ slope $=3$

$L_2: y=\frac{x}{3} \quad y=\frac{1}{3}x \quad$ slope $=\frac{1}{3}$

$L_3: y=3x+2 \quad$ slope $=3$

$L_4: \frac{x+y}{3}=3 \quad x+y=9 \quad y=9-x \quad$ slope $=-1$

L_1 and L_3 have the same slope, 3, but they have different y-intercepts. They are therefore not the same line, but are parallel.

Practice Exercises for Section 7.6, pages 348–356

1. $(10, 1)$ is a solution.

Substitute 10 for x and 1 for y in each equation.

$$x-4y=6 \qquad 2y=x-8$$
$$10-4(1)\stackrel{?}{=}6 \qquad 2(1)\stackrel{?}{=}10-8$$
$$10-4\stackrel{?}{=}6$$
$$6\stackrel{\checkmark}{=}6$$

$(10, 1)$ is a solution of the system because both equations check. (In fact, it is the only solution.)

2. $\left(4, \frac{-1}{2}\right)$

$$y=\frac{x-5}{2}$$
$$x+2y=3$$

In the first equation y is given explicitly in terms of x.

Substitute the expression $\frac{x-5}{2}$ for y in the second equation.

Thus,

$$x+2\left(\frac{x-5}{2}\right)=3$$
$$x+(x-5)=3$$
$$2x=8$$
$$x=4$$

Now substitute 4 for x in the first equation to find y.

Thus,

$$y=\frac{4-5}{2}=\frac{-1}{2}$$

The solution of the system is the ordered pair $\left(4, \frac{-1}{2}\right)$.

3. $(-9, -5)$

To use the substitution method, first solve one equation for one of the variables. The first equation is easily solved for x. Thus, $x-3y=6$ becomes $x=3y+6$. Substitute $3y+6$ for x in the other equation. Thus, $3y-2x=3$ becomes

$$\begin{aligned} 3y-2(3y+6)&=3\\ 3y-6y-12&=3\\ -3y&=15\\ y&=-5 \end{aligned}$$

Now, to find x, substitute -5 for y in the equation $x=3y+6$

$$x=3(-5)+6=-15+6=-9$$

The solution of the system is the ordered pair $(-9, -5)$.

Check:

$$\begin{aligned} x-3y&=6 \\ -9-3(-5)&\stackrel{?}{=}6 \\ -9+15&\stackrel{?}{=}6 \\ 6&\stackrel{\checkmark}{=}6 \end{aligned} \qquad \begin{aligned} 3y-2x&=3 \\ 3(-5)-2(-9)&\stackrel{?}{=}3 \\ -15+18&\stackrel{?}{=}3 \\ 3&\stackrel{\checkmark}{=}3 \end{aligned}$$

4. $\left(2, -\frac{1}{2}\right)$

$$\begin{aligned} x-2y&=3\\ 3x+2y&=5\\ 4x&=8\\ x&=2 \end{aligned}$$

Now substitute 2 for x in either equation. From the first equation, we have

$$\begin{aligned} 2-2y&=3\\ -2y&=1\\ y&=\frac{-1}{2} \end{aligned}$$

The solution of the system is $\left(2, \frac{-1}{2}\right)$.

5. $(3, 5)$

$$\begin{aligned} 5x-4y&=-5\\ 4x-y&=7 \end{aligned}$$

Simply adding the corresponding sides of these equations will not eliminate either variable. Multiply both sides of the second equation by -4 to obtain

$$-16x+4y=-28$$

Add this equation to the first equation.

$$\begin{aligned} 5x-4y&=-5\\ -16x+4y&=-28\\ \hline -11x\qquad&=-33\\ x\qquad&=3 \end{aligned}$$

Now substitute 3 for x in either equation. Thus, from the second equation, $4x-y=7$, we have

$$\begin{aligned} 4(3)-y&=7\\ -y&=-5\\ y&=5 \end{aligned}$$

The solution of the system is the ordered pair $(3, 5)$.

6. $(0, 3)$

In order to eliminate the x-terms, multiply both sides of the first equation by 3 and both sides of the second equation by -7.

$$\begin{aligned} -21x+30y&=90\\ 21x+28y&=84\\ \hline 58y&=174\\ y&=3 \end{aligned}$$

Substitute 3 for y in the second equation.

$$\begin{aligned} 3x+4(3)&=12\\ 3x+12&=12\\ 3x&=0\\ x&=0 \end{aligned}$$

The solution of the system is $(0, 3)$.

To check, substitute 0 for x and 3 for y in each equation.

Check:

$$\begin{aligned} -7x+10y&=30 \\ -7(0)+10(3)&\stackrel{?}{=}30 \\ 30&\stackrel{\checkmark}{=}30 \end{aligned} \qquad \begin{aligned} 3x+4y&=12 \\ 3(0)+4(3)&\stackrel{?}{=}12 \\ 12&\stackrel{\checkmark}{=}12 \end{aligned}$$

7. Ann: \$5.25 per hour; Bob: \$5.00 per hour.

Let a = Ann's hourly wage (in dollars) and b = Bob's hourly wage (in dollars). In 4 hours, Ann earns $4a$ dollars, and in 5 hours, Bob earns $5b$ dollars. Translate the given facts into algebraic equations.

Ann's hourly wage is \$.25 more than Bob's

$$a = .25 + b$$

The second equation says that the total earned by Ann in 4 hours and by Bob in 5 hours is \$46.

$$4a+5b=46$$

This system of equations is most easily solved by the substitution method because one equation is already solved for a in terms of b. We can substitute $.25+b$ for a in the second equation. Thus,

$$\begin{aligned} 4(.25+b)+5b&=46\\ 1+4b+5b&=46\\ 9b&=45\\ b&=5 \text{ (dollars, Bob's hourly wage)}\\ a&=.25+5\\ a&=5.25 \text{ (dollars, Ann's hourly wage)} \end{aligned}$$

Practice Exercises for Section 7.7, pages 360–361

1. (a) 8 (b) 56 (c) $\frac{3}{2}$ or 1.5

Algebraically, "y varies directly as x" can be written as $y = kx$, or as $\frac{y}{x} = k$, for $x \neq 0$, where k is the constant of variation.

(a) If $y = 2$ when $x = \frac{1}{4}$, then

$$k = \frac{2}{\frac{1}{4}} = 2 \div \frac{1}{4} = 2 \times \frac{4}{1} = 8$$

(b) When $x = 7$, $y = 8(7) = 56$.

(c) When $y = 12$, substitute 12 for y in the equation $y = 8x$. Thus,

$$12 = 8x \text{ and } x = \frac{12}{8} = \frac{3}{2} = 1.5$$

2. (a) $k = .4$ or $\frac{2}{5}$ (b) 18 (c) 22.5 or $\frac{45}{2}$

(a) $\frac{y}{x} = k$, for $x \neq 0$. If $y = 4$ when $x = 10$, then $k = \frac{4}{10} = .4$ or $\frac{2}{5}$. Thus, $y = \frac{2}{5}x$.

(b) When $x = 45$, then $y = \frac{2}{5}(45) = 18$.

(c) When $y = 9$, then $9 = \frac{2}{5}x$. Solve for x by multiplying both sides by $\frac{5}{2}$. Thus,

$$x = \frac{5}{2}(9) = \frac{45}{2} = 22.5$$

3. \$21

Let w = the amount withheld; let g = gross earnings.

Then w varies directly as g. Thus, $\frac{w}{g} = k$ (constant).

Thus, $k = \frac{12}{200} = \frac{3}{50}$, or .06. If $g = 350$, then $\frac{w}{g} = .06$ becomes $\frac{w}{350} = .06$.

$$w = .06(350) = 21 \text{ (dollars)}$$

Practice Exam 1 on Chapter 7, page 366

1. P: (a) 3 (b) 2 (c) $P = (3, 2)$
Q: (a) 3 (b) -2 (c) $Q = (3, -2)$
R: (a) -1 (b) -4 (c) $R = (-1, -4)$
S: (a) 0 (b) 5 (c) $S = (0, 5)$
T: (a) -5 (b) 1 (c) $T = (-5, 1)$

2.

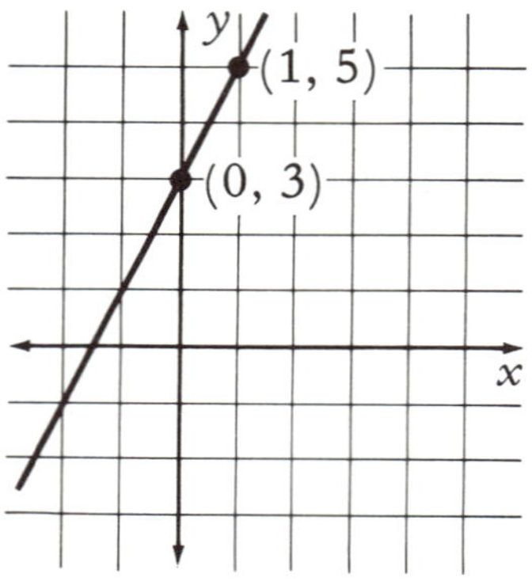

FIGURE 7M-S

x	$y = 2x + 3$
0	$2(0) + 3 = 3$
1	$2(1) + 3 = 5$

Plot the points (0, 3) and (1, 5). Draw the line connecting them.

3. (a) 3 (b) -3 (c) -1

(a) rise $= y_2 - y_1 = 6 - 3 = 3$

(b) run $= x_2 - x_1 = 3 - 6 = -3$

(c) slope $= \frac{\text{rise}}{\text{run}} = \frac{3}{-3} = -1$

(d) downward to the right (because the slope is negative).

4. (a) 4 (b) -3

In the form $y = mx + b$, m is the slope, b is the y-intercept. Thus, in the equation $y = 4x - 3$, the slope is 4, the y-intercept is -3.

5. $y = -2x + 3$

6. (3, 2)

By the addition method:

$$\begin{aligned} 2x+y&=8\\ \underline{x-y}&\underline{=1}\\ 3x+0&=9\\ x&=3 \end{aligned}$$

Now substitute 3 for x in either equation.

$$\begin{aligned} 2(3)+y&=8\\ 6+y&=8\\ y&=8-6=2 \end{aligned}$$

The solution is (3, 2).

Check:

$$\begin{array}{ll} 2x+y=8 & x-y=1\\ 2(3)+2\overset{?}{=}8 & 3-2\overset{?}{=}1\\ 6+2\overset{?}{=}8 & 1\overset{\checkmark}{=}1\\ 8\overset{\checkmark}{=}8 & \end{array}$$

7. (3, −2)

$$\begin{aligned} 6x+5y&=8\\ 5x+3y&=9 \end{aligned}$$

Using the addition method, we can eliminate the y-terms by multiplying the first equation by −3 and multiplying the second by 5.

$$\begin{aligned} -18x-15y&=-24\\ \underline{25x+15y}&\underline{=\ \ 45}\\ 7x\quad\ &=\ 21\\ x\quad\ &=\ 3 \end{aligned}$$

Now substitute 3 for x in either equation.

$$\begin{aligned} 6(3)+5y&=8\\ 5y&=-10\\ y&=-2 \end{aligned}$$

The solution is (3, −2).

8. $\frac{3}{4}$

$$\begin{aligned} y&=kx\\ 3&=k\cdot 2\\ k&=\frac{3}{2} \end{aligned}$$

When $x=\frac{1}{2}$, $y=\frac{3}{2}\cdot\frac{1}{2}=\frac{3}{4}$

Practice Exam 2 on Chapter 7, page 367

1. $A=(5, 1)$
$B=(3, -1)$
$C=(-4, -3)$
$D=(-4, 0)$
$E=(0, 3)$

2.

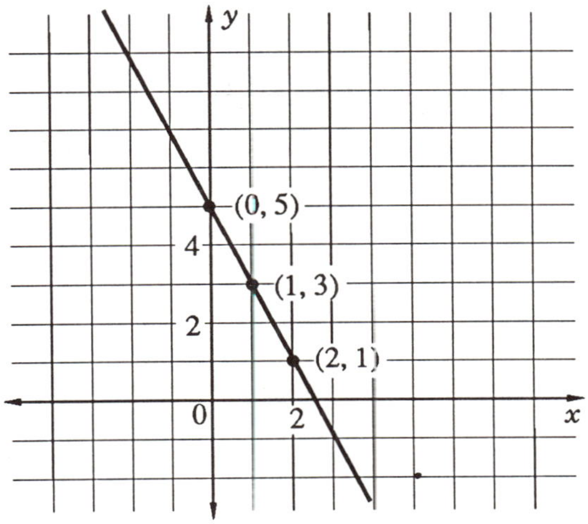

x	$y=5-2x$
0	$5-2(0)=5$
1	$5-2(1)=3$
2	$5-2(2)=1$

FIGURE 7N-S

Plot any two of the points, (0, 5), (1, 3), and (2, 1), and draw the line connecting them.

3. (a) $\frac{2}{3}$ (b) −1

(a) slope $=\frac{7-5}{3-0}=\frac{2}{3}$

(b) slope $=\frac{-2-10}{8-(-4)}=\frac{-12}{12}=-1$

4. (a) $\frac{-3}{2}$ (b) 2

Solve the equation for y in terms of x.

$$\begin{aligned} 3x+2y&=4\\ 2y&=-3x+4\\ y&=\frac{-3x+4}{2}\\ y&=\frac{-3}{2}x+2 \end{aligned}$$

This last equation is of the form $y=mx+b$.

The slope, m, equals $\frac{-3}{2}$ and the y-intercept, b, equals 2.

5. $y=2x+8$

$m=\text{slope}=\frac{0-8}{-4-0}=\frac{-8}{-4}=2$

$b=y\text{-intercept}=8$

$y=mx+b$

The equation of the line is $y=2x+8$.

6. (1, 3)

$$x+5=2y$$
$$3y-4x=5$$

Solve the first equation for x: $x=2y-5$.
Substitute $2y-5$ for x in the second equation and solve for y.

$3y-4(2y-5)=5$
$3y-8y+20=5$
$-5y=-15$
$y=3$

Now substitute 3 for y in the equation
$x=2y-5$
$x=2(3)-5$
$x=1$

The solution is (1, 3).
To check, substitute 1 for x, 3 for y in each equation.
Check:

$x+5=2y$	$3y-4x=5$
$1+5\stackrel{?}{=}2(3)$	$3(3)-4(1)\stackrel{?}{=}5$
$6\stackrel{\checkmark}{=}6$	$5\stackrel{\checkmark}{=}5$

7. (4, −2)
Multiply both sides of the second equation by −2 and add to the corresponding sides of the first equation. Thus,

$$\begin{aligned} 6x+5y &= 14 \\ -6x+4y &= -32 \\ \hline 9y &= -18 \\ y &= -2 \end{aligned}$$

Substituting −2 for y in the first equation, we have

$$6x+5(-2)=14$$
$$6x=24$$
$$x=4$$

The solution of the system is (4, −2).

8. (a) $\frac{4}{5}$ or .8 (b) 8 (c) $\frac{5}{4}$ or 1.25

Because y varies directly as x, we can write $y=kx$, or $\frac{y}{x}=k$ for $x\neq 0$, where k is the constant of variation.

(a) If $y=20$ when $x=25$, then $k=\frac{20}{25}=\frac{4}{5}$, or .8.

(b) When $x=10$, $y=\frac{4}{5}x=\frac{4}{5}(10)=8$.

(c) When $y=1$, $1=\frac{4}{5}x$

$$\frac{5}{4}(1)=\frac{5}{4}\left(\frac{4}{5}x\right)$$
$$\frac{5}{4}=x$$

CHAPTER 8

Practice Exercises for Section 8.1, pages 369–371

1. (a) 225 (b) 169 (c) 400 (d) 1 000 000

2. (a) 20 (b) −15 (c) 13

(a) $\sqrt{400}=20$ because $20^2=400$

(b) $\sqrt{225}=15$ because $15^2=225$; thus,
$-\sqrt{225}=-15$

(c) $\sqrt{169}=13$ because $13^2=169$

3. 9 inches

$\sqrt{81\text{in.}^2}=9$ in.

4. (a) −57 (b) $8a$ (c) a^3b^4

The radical, $\sqrt{\ }$, always indicates the *nonnegative* square root, unless it is preceeded by the minus sign, as in part (a).

(a) $\sqrt{57^2}=57$, and therefore,
$-\sqrt{57^2}=-57$

(b) $\sqrt{(8a)^2}=8a$. This result is nonnegative because $a>0$.

(c) For any nonnegative number p, $\sqrt{p^2}=p$.
Here $a^3b^4\geq 0$, so that $\sqrt{(a^3b^4)^2}=a^3b^4$.

5. (a) x^4 (b) 3^3 or 27 (c) a^{11}

(a) $\sqrt{x^8}=x^4$ because $(x^4)^2=x^8$ and because $x^4=(x^2)^2\geq 0$.

(b) $\sqrt{3^6}=3^3$ because $(3^3)^2=3^6$.

(c) $\sqrt{a^{22}}=a^{11}$ because $(a^{11})^2=a^{22}$.
($a^{11}>0$ because $a>0$.)

Practice Exercises for Section 8.2, pages 374–378

1. (a) rational (b) undefined (c) irrational (d) rational

In (d), $\dfrac{7\sqrt{3}}{8\sqrt{3}} = \dfrac{7}{8}$.

2. (a) $3 < \sqrt{12} < 4$ (b) $7 < \sqrt{50} < 8$
(c) $10 < \sqrt{120} < 11$ (d) $12 < \sqrt{150} < 13$

(a) $\sqrt{9} < \sqrt{12} < \sqrt{16}$. Thus,
$3 < \sqrt{12} < 4$

(b) $\sqrt{49} < \sqrt{50} < \sqrt{64}$. Thus,
$7 < \sqrt{50} < 8$

(c) $\sqrt{100} < \sqrt{120} < \sqrt{121}$. Thus,
$10 < \sqrt{120} < 11$

(d) $\sqrt{144} < \sqrt{150} < \sqrt{169}$. Thus,
$12 < \sqrt{150} < 13$

3. (a) 2.2 (b) 6.9

(a) $\dfrac{3+\sqrt{2}}{2} \approx \dfrac{3+1.41}{2} = \dfrac{4.41}{2} = 2.205$
To one decimal digit, $2.205 \approx 2.2$.

(b) $4\sqrt{3} \approx 4(1.73) = 6.92$.
Rounding to one decimal digit, $6.92 \approx 6.9$

4. Approximately 9 meters.

The area, A, of a square of side s is given by the formula $A = s^2$. Thus, $s = \sqrt{A}$. If the area is 80 square meters, then $s = \sqrt{80}$ meters. Because $\sqrt{81} = 9$, $\sqrt{80}$ is slightly less than 9.

5. $\sqrt{7}$ cm.

By the Pythagorean Theorem,
$c^2 = a^2 + b^2$,
where c is the hypotenuse. Thus,

$$4^2 = a^2 + 3^2$$
$$a^2 = 4^2 - 3^2$$
$$a^2 = 16 - 9$$
$$a^2 = 7$$
$$a = \sqrt{7} \approx 2.6 \text{ (cm.)}$$

Practice Exercises for Section 8.3, pages 381–385

1. (a) 120 (b) $4x^4$ (c) $7xy^3$

(a) $\sqrt{14\,400} = \sqrt{144 \cdot 100} = \sqrt{144} \cdot \sqrt{100} = 12 \cdot 10 = 120$

(b) $\sqrt{16x^8} = \sqrt{16} \cdot \sqrt{(x^4)^2} = 4 \cdot x^4 = 4x^4$

(c) $\sqrt{49x^2y^6} = \sqrt{49}\sqrt{x^2}\sqrt{y^6} = 7xy^3$

2. (a) $3\sqrt{3}$ (b) 5.2

$\sqrt{27}$ can be simplified because 27 has a factor that is a perfect square, namely 9.

(a) $\sqrt{27} = \sqrt{9 \cdot 3} = \sqrt{9} \cdot \sqrt{3} = 3\sqrt{3}$

(b) Using $\sqrt{3} \approx 1.73$,
$3\sqrt{3} \approx 3(1.73) = 5.19$
To one decimal digit, $5.19 \approx 5.2$.

3. (a) $x^2\sqrt{x}$ (b) $3x\sqrt{5x}$

To simplify a radical, find factors under the radical symbol that are perfect squares.

(a) x^4 is a factor of x^5 and $\sqrt{x^4} = x^2$. Thus,
$\sqrt{x^5} = \sqrt{x^4 \cdot x} = \sqrt{x^4} \cdot \sqrt{x} = x^2\sqrt{x}$.

(b) $\sqrt{45x^3} = \sqrt{9x^2 \cdot 5x} = \sqrt{9x^2}\sqrt{5x} = 3x\sqrt{5x}$

4. (a) 18 (b) 18 (c) −18 (d) undefined as a real number

5. (a) $\dfrac{4}{9}$ (b) $\dfrac{x^2}{5y^6}$

(a) $\sqrt{\dfrac{16}{81}} = \dfrac{\sqrt{16}}{\sqrt{81}} = \dfrac{4}{9}$

(b) $\sqrt{\dfrac{x^4}{25y^{12}}} = \dfrac{\sqrt{x^4}}{\sqrt{25}\sqrt{y^{12}}} = \dfrac{x^2}{5y^6}$

6. (a) $\dfrac{\sqrt{6}}{3x}$

Take the square root of each factor in the numerator and denominator.

$$\sqrt{\frac{2}{3x^2}} = \frac{\sqrt{2}}{\sqrt{3}\sqrt{x^2}} = \frac{\sqrt{2}}{\sqrt{3}\,x}$$

The irrational number $\sqrt{3}$ is left in the denominator. But we can then rationalize the denominator by multiplying by $\dfrac{\sqrt{3}}{\sqrt{3}}$. Thus,

$$\frac{\sqrt{2} \cdot \sqrt{3}}{\sqrt{3}\,x \cdot \sqrt{3}} = \frac{\sqrt{6}}{\sqrt{9}\,x} = \frac{\sqrt{6}}{3x}$$

(b) $\dfrac{x\sqrt{10x}}{10}$

$$\sqrt{\frac{x^3}{10}} = \frac{\sqrt{x^2} \cdot x}{\sqrt{10}} = \frac{x\sqrt{x}}{\sqrt{10}}$$

Rationalize the denominator:

$$\frac{x\sqrt{x} \cdot \sqrt{10}}{\sqrt{10} \cdot \sqrt{10}} = \frac{x\sqrt{10x}}{10}$$

7. (a) and (d) are errors.

(a) $\sqrt{100-64} = \sqrt{100} - \sqrt{64}$ is false because $\sqrt{100-64} = \sqrt{36} = 6$, whereas $\sqrt{100} - \sqrt{64} = 10 - 8 = 2$.

(d) $\sqrt{2} + \sqrt{3} = \sqrt{5}$ is false because unlike radicals cannot be added in this way. Using approximations, on the left side, $\sqrt{2} + \sqrt{3} \approx 1.41 + 1.73 = 3.14$, whereas on the right side, $\sqrt{5} \approx 2.24$.

8. (a) .07

$$\begin{aligned}\sqrt{.0049} &= \sqrt{49\,(.0001)}\\ &= \sqrt{49}\,\sqrt{.0001}\\ &= 7\,(.01)\\ &= .07\end{aligned}$$

(b) $1.1a^2$

$$\begin{aligned}\sqrt{1.21a^4} &= \sqrt{121\,(.01)\,a^4}\\ &= \sqrt{121}\,\sqrt{.01}\,\sqrt{a^4}\\ &= 11\,(.1)\,a^2\\ &= 1.1a^2\end{aligned}$$

(c) .009

$$\begin{aligned}\sqrt{.000\,081} &= \sqrt{\frac{81}{1\,000\,000}}\\ &= \frac{\sqrt{81}}{\sqrt{1\,000\,000}}\\ &= \frac{9}{1000}\\ &= .009\end{aligned}$$

Practice Exercises for Section 8.4, pages 387–388

1. (a) $5\sqrt{6}$ (b) $3\sqrt{5}$ (c) $a\sqrt{3a}$

(a) $\sqrt{6} + 10\sqrt{6} - 6\sqrt{6} = (1 + 10 - 6)\sqrt{6} = 5\sqrt{6}$

(b) Simplify each radical first.
$\sqrt{125} = \sqrt{25 \cdot 5} = \sqrt{25}\,\sqrt{5} = 5\sqrt{5}$
$\sqrt{20} = \sqrt{4 \cdot 5} = \sqrt{4}\,\sqrt{5} = 2\sqrt{5}$
Thus, $\sqrt{125} - \sqrt{20} = 5\sqrt{5} - 2\sqrt{5} = 3\sqrt{5}$

(c) $3a\sqrt{12a} = 3a\sqrt{4 \cdot 3a} = 3a\sqrt{4}\,\sqrt{3a} = 3a\,(2)\,\sqrt{3} = 6a\sqrt{3a}$
$\sqrt{75a^3} = \sqrt{25a^2 \cdot 3a} = \sqrt{25a^2}\,\sqrt{3a} = 5a\sqrt{3a}$. Thus,

$$\begin{aligned}3a\sqrt{12a} - \sqrt{75a^3} &= 6a\sqrt{3a} - 5a\sqrt{3a}\\ &= (6-5)\,a\sqrt{3a}\\ &= a\sqrt{3a}\end{aligned}$$

2. (a) $\frac{3+\sqrt{3}}{5}$ (b) $\frac{7\sqrt{5}}{4}$ (c) $\frac{2\sqrt{7}+3\sqrt{6}}{18}$

(a) $\frac{3}{5} + \frac{\sqrt{3}}{5} = \frac{3+\sqrt{3}}{5}$. This fraction cannot be simplified further.

(b) $lcd\left(\frac{2\sqrt{5}}{1}, \frac{\sqrt{5}}{4}\right) = 4$

Write $2\sqrt{5}$ or $\frac{2\sqrt{5}}{1}$ as a fraction with the common denominator 4 by multiplying numerator and denominator by 4.
Thus,

$$\frac{2\sqrt{5}\cdot 4}{1\cdot 4} - \frac{\sqrt{5}}{4} = \frac{8\sqrt{5}}{4} - \frac{\sqrt{5}}{4} = \frac{8\sqrt{5}-\sqrt{5}}{4} = \frac{7\sqrt{5}}{4}$$

(c) $lcm\,(9, 6) = 18$; $9 \cdot 2 = 18$; $6 \cdot 3 = 18$. Thus,

$$\frac{\sqrt{7}}{9} + \frac{\sqrt{6}}{6} = \frac{\sqrt{7}\cdot 2}{9\cdot 2} + \frac{\sqrt{6}\cdot 3}{6\cdot 3} = \frac{2\sqrt{7}}{18} + \frac{3\sqrt{6}}{18} = \frac{2\sqrt{7}+3\sqrt{6}}{18}$$

Practice Exercises for Section 8.5, pages 390–393

1. (a) 12 (b) $6x^2$ (c) $x\sqrt{21}$

(a) $\sqrt{6}\,\sqrt{24} = \sqrt{6 \cdot 24} = \sqrt{144} = 12$

(b) $\sqrt{2x^3}\,\sqrt{18x} = \sqrt{(2x^3)(18x)} = \sqrt{36x^4} = 6x^2$

(c) $\sqrt{3x}\,\sqrt{7x} = \sqrt{21x^2} = \sqrt{21}\,\sqrt{x^2} = \sqrt{21} \cdot x$, or $x\sqrt{21}$

2. (a) $8 - 2\sqrt{6}$ (b) $-5 + 9\sqrt{5}$

(a)
$$\begin{aligned}\sqrt{2}\left(\sqrt{32} - 2\sqrt{3}\right) &= \sqrt{2}\,\sqrt{32} - \sqrt{2}\left(2\sqrt{3}\right)\\ &= \sqrt{64} - 2\sqrt{2}\,\sqrt{3}\\ &= 8 - 2\sqrt{6}\end{aligned}$$

(b) $\left(5 - \sqrt{5}\right)\left(1 + 2\sqrt{5}\right)$

Multiply each term of the first factor by each term of the second.

$$\begin{array}{l} 5 - \sqrt{5}\\ \underline{1 + 2\sqrt{5}}\\ 5 - 1\sqrt{5}\\ \underline{\quad 10\sqrt{5} - 2\sqrt{5}\,\sqrt{5}}\\ 5 + 9\sqrt{5} - 2(5) = 5 - 10 + 9\sqrt{5} = -5 + 9\sqrt{5}\end{array}$$

3. (a) 2 (b) $\frac{1}{5}$ (c) $\frac{5}{2x}$

(a) $\frac{\sqrt{60}}{\sqrt{15}} = \sqrt{\frac{60}{15}} = \sqrt{4} = 2$

(b) $\frac{\sqrt{2}}{\sqrt{50}} = \sqrt{\frac{2}{50}} = \sqrt{\frac{1}{25}} = \frac{\sqrt{1}}{\sqrt{25}} = \frac{1}{5}$

(c) $\frac{\sqrt{75x}}{\sqrt{12x^3}} = \sqrt{\frac{75x}{12x^3}}$

[Divide numerator and denominator by the common factor, $3x$.]

$$= \sqrt{\frac{25}{4x^2}} = \frac{\sqrt{25}}{\sqrt{4x^2}} = \frac{5}{2x}$$

4. (a) $4\left(2-\sqrt{3}\right)$ (b) $\frac{2-\sqrt{3}}{5}$

(a) $\sqrt{48}=\sqrt{16\cdot 3}=\sqrt{16}\,\sqrt{3}=4\,\sqrt{3}$. Thus,

$$8-\sqrt{48}=8-4\,\sqrt{3}=4\left(2-\sqrt{3}\right)$$

(b) Using the factorization in part (a),

$$\frac{8-\sqrt{48}}{20}=\frac{4\left(2-\sqrt{3}\right)}{20}=\frac{2-\sqrt{3}}{5}$$

5. (a) $3\,(x-y)$ (b) $\frac{x-y}{3\,(x+y)}$

(a) $\sqrt{9x^2}-\sqrt{9y^2}=3x-3y=3\,(x-y)$

(b) Using part (a),

$$\frac{\sqrt{9x^2}-\sqrt{9y^2}}{9x+9y}=\frac{3\,(x-y)}{9\,(x+y)}=\frac{x-y}{3\,(x+y)}$$

6. $\frac{\sqrt{2}}{4}$

$$\frac{\sqrt{30}}{20}\cdot\frac{\sqrt{5}}{\sqrt{3}}=\frac{\sqrt{3}\,\sqrt{10}}{20}\cdot\frac{\sqrt{5}}{\sqrt{3}}=\frac{\sqrt{50}}{20}=\frac{5\,\sqrt{2}}{20}=\frac{\sqrt{2}}{4}$$

7. $\frac{\sqrt{5}}{5}$

Invert the divisor; then multiply.

$$\frac{\sqrt{8}}{10}\div\frac{\sqrt{2}}{\sqrt{5}}=\frac{\sqrt{8}}{10}\cdot\frac{\sqrt{5}}{\sqrt{2}}=\frac{\sqrt{2}\,\sqrt{4}}{10}\cdot\frac{\sqrt{5}}{\sqrt{2}}=\frac{2}{10}\cdot\frac{\sqrt{5}}{1}=\frac{\sqrt{5}}{5}$$

Practice Exercises for Section 8.6, pages 396–398

1. (a) $\frac{1}{49}$ (b) $\frac{1}{5}$ (c) $\frac{-1}{64}$ (d) 1

(a) $7^{-2}=\frac{1}{7^2}=\frac{1}{49}$

(b) $5^{-1}=\frac{1}{5^1}=\frac{1}{5}$

(c) $(-4)^{-3}=\frac{1}{(-4)^3}=\frac{1}{-64}=\frac{-1}{64}$

(d) $a^0=1$ for any nonzero number, a. Thus, $2.6^0=1$.

2. (a) $m=-2$ (b) $m=0$

(a) $8^m=\frac{1}{64}=\frac{1}{8^2}=8^{-2}$
Thus, $m=-2$.

(b) $6^m=1=6^0$
Thus, $m=0$.

3. (a) $\frac{4}{5}$ (b) $\frac{64}{9}$ (c) 125

(a) $\left(\frac{5}{4}\right)^{-1}=\frac{1}{\frac{5}{4}}=1\div\frac{5}{4}=1\times\frac{4}{5}=\frac{4}{5}$

(b) According to the rule $\left(\frac{a}{b}\right)^{-n}=\frac{b^n}{a^n}$,

$$\left(\frac{-3}{8}\right)^{-2}=\frac{8^2}{(-3)^2}=\frac{64}{9}$$

(c) Here are two possible solutions.

$$(.2)^{-3}=\left(\frac{1}{5}\right)^{-3}=\frac{5^3}{1^3}=\frac{125}{1}=125, \text{ or}$$

$$(.2)^{-3}=\frac{1}{.2^3}=\frac{1}{.008}=\frac{1000}{8}=125$$

4. (a) y^{-5} (b) 1 (c) y^3 (d) y^{-8}

(a) $\frac{y^1}{y^6}=y^{1-6}=y^{-5}$

(b) $y\cdot y^{-3}\cdot y^2=y^{1+(-3)+2}=y^0=1$

(c) $\frac{y^{-2}}{y^{-5}}=y^{-2-(-5)}=y^{-2+5}=y^3$

(d) $\frac{y^3\cdot y^{-4}}{y^7}=y^{3+(-4)-7}=y^{-8}$

Practice Exercises for Section 8.7, pages 401–402

1. (a) 6.307×10^7 (b) 9.02×10^{-4} (c) 8×10^{-3}

(a) In the number 63 070 000, the decimal point is understood to be to the right of the last zero. Move the decimal point 7 digits to the left and drop the unnecessary zeros. 6.307 is a number between 1 and 10. In moving the decimal point to the left, we have divided the number by 10^7. Make up for this by multiplying by 10^7. Thus, $63\,070\,000 = 6.307 \times 10^7$.

(b) In the number .000 902, place the decimal point after the first nonzero digit, 9. Note that 9.02 is a number between 1 and 10. We have moved the decimal point 4 digits to the right and, in effect, have multiplied the number by 10^4. Now balance that effect by multiplying by 10^{-4}. Thus, $.000\,902 = 9.02 \times 10^{-4}$.

(c) $.008 = 8 \times 10^{-3}$

Because we moved the decimal point 3 digits to the right to obtain 8, the number between 1 and 10, multiply by 10^{-3}.

2. (a) 40 000 (b) 805.3 (c) .0061

To convert from scientific notation to ordinary decimal notation, move the decimal point to the right if the exponent on base 10 is positive; move it to the left if the exponent is negative.

3. (a) 10^{-5} (b) 4×10^9

(a) $\dfrac{10^{-4}}{10^2 \times 10^{-1}} = 10^{-4-(2-1)} = 10^{-4-1} = 10^{-5}$

(b) $\dfrac{3600 \times 5000}{.0045} = \dfrac{(36 \times 10^2) \times (5 \times 10^3)}{45 \times 10^{-4}}$

$= \dfrac{\overset{4}{\cancel{36}} \times \overset{1}{\cancel{5}}}{\underset{1}{\cancel{\cancel{45}}}} \times \dfrac{10^2 \times 10^3}{10^{-4}}$ (45 reduced to 9, then to 1)

$= 4 \times 10^{2+3-(-4)}$

$= 4 \times 10^9$

Note that in this computation we wrote 3600 as 36×10^2 and .0045 as 45×10^{-4}. Strictly speaking, these are not in scientific notation; but the integers 36 and 45 are easier to work with than are the decimals 3.6 and 4.5.

Practice Exam 1 on Chapter 8, page 406

1. (a) 12 (b) $10x^2$ (c) $\frac{4}{7}$

(a) $\sqrt{144} = 12$ because $12^2 = 144$

(b) $\sqrt{100x^4} = \sqrt{100}\,\sqrt{x^4} = 10x^2$

(c) $\sqrt{\dfrac{16}{49}} = \dfrac{\sqrt{16}}{\sqrt{49}} = \dfrac{4}{7}$

2. $\dfrac{4\sqrt{2}\,a^2b^4}{c^4}$

$\sqrt{\dfrac{32a^4b^8}{c^8}} = \dfrac{\sqrt{16}\,\sqrt{2}\,\sqrt{a^4}\,\sqrt{b^8}}{\sqrt{c^8}} = \dfrac{4\sqrt{2}\,a^2b^4}{c^4}$

3. $\dfrac{\sqrt{2}}{4}$

$\dfrac{5\sqrt{2} - 3\sqrt{2}}{8} = \dfrac{(5-3)\sqrt{2}}{8} = \dfrac{\overset{1}{\cancel{2}}\sqrt{2}}{\underset{4}{\cancel{8}}} = \dfrac{\sqrt{2}}{4}$

4. $18x^3y^2z$

$\sqrt{9x^3y}\,\sqrt{3xy^3z}\,\sqrt{12x^2z} = \sqrt{9 \cdot 3 \cdot 12 \cdot x^3 \cdot x \cdot x^2 \cdot y \cdot y^3 \cdot z \cdot z}$

$= \sqrt{9}\,\sqrt{36}\,\sqrt{x^6}\,\sqrt{y^4}\,\sqrt{z^2}$

$= 3 \cdot 6x^3y^2z$

$= 18x^3y^2z$

5. $\dfrac{10}{7}$

$\dfrac{\sqrt{20}}{\sqrt{63}} \div \dfrac{\sqrt{7}}{\sqrt{45}} = \dfrac{\sqrt{20}}{\sqrt{63}} \cdot \dfrac{\sqrt{45}}{\sqrt{7}} = \dfrac{\sqrt{4}\,\sqrt{5}}{\sqrt{9}\,\sqrt{7}} \cdot \dfrac{\sqrt{9}\,\sqrt{5}}{\sqrt{7}} = \dfrac{2 \cdot 5}{7} = \dfrac{10}{7}$

6. (a) $3(\sqrt{2} - 3)$ (b) $\dfrac{\sqrt{2} - 3}{2}$

(a) $\sqrt{18} - 9 = \sqrt{9 \cdot 2} - 9 = 3\sqrt{2} - 9 = 3(\sqrt{2} - 3)$

(b) $\dfrac{\sqrt{18} - 9}{6} = \dfrac{3(\sqrt{2} - 3)}{3 \cdot 2} = \dfrac{\sqrt{2} - 3}{2}$

7. 4.4

$3\sqrt{2} + \dfrac{\sqrt{3}}{10} \approx 3(1.41) + \dfrac{1.73}{10}$

$\approx 4.23 + .173 = 4.403$

To one decimal digit, $4.403 \approx 4.4$

8. (a) $\frac{1}{64}$ (b) 1 (c) $\frac{25}{9}$

(a) $4^{-3} = \frac{1}{4^3} = \frac{1}{64}$

(b) $(-6)^0 = 1$

(c) $\left(\frac{3}{5}\right)^{-2} = \left(\frac{5}{3}\right)^2 = \frac{5^2}{3^2} = \frac{25}{9}$

9. a

$a^{-7} \cdot a^{10} \cdot a^{-2} = a^{-7+10+(-2)} = a^1 = a$

10. 4.87×10^4

11. .0287

$2.87 \times 10^{-2} = .0287$

12. 5.4×10^3

Express the factors as whole numbers multiplied by powers of 10.

$$\frac{.0081 \times 240}{.04 \times .009} = \frac{(81 \times 10^{-4}) \times (24 \times 10)}{(4 \times 10^{-2}) \times (9 \times 10^{-3})}$$

$$= \frac{\overset{9}{\cancel{81}} \times \overset{6}{\cancel{24}}}{\underset{1}{\cancel{4}} \times \underset{1}{\cancel{9}}} \times \frac{10^{-4} \times 10}{10^{-2} \times 10^{-3}}$$

$$= 54 \times 10^2$$

$$= 5.4 \times 10^3$$

Practice Exam 2 on Chapter 8, page 407

1. (a) 15 (b) $\frac{7x^3}{10y}$

(a) $\sqrt{225} = 15$ because $15^2 = 225$

(b) $\sqrt{\frac{49x^6}{100y^2}} = \frac{\sqrt{49}\,\sqrt{x^6}}{\sqrt{100}\,\sqrt{y^2}} = \frac{7x^3}{10y}$

2. (a) $10\sqrt{5}$ (b) $9x\sqrt{2x}$

Extract the factors that are perfect squares.

(a) $\sqrt{500} = \sqrt{100 \cdot 5} = \sqrt{100}\,\sqrt{5} = 10\sqrt{5}$

3. $\frac{3\sqrt{10}}{10}$ or $.3\sqrt{10}$

$$\sqrt{\frac{9}{10}} = \frac{\sqrt{9}}{\sqrt{10}} = \frac{3}{\sqrt{10}} = \frac{3 \cdot \sqrt{10}}{\sqrt{10} \cdot \sqrt{10}} = \frac{3\sqrt{10}}{10}$$

4. (a) $9x^2$ (b) $2(5 - \sqrt{15})$ (c) -2

(a) $\sqrt{3x}\,\sqrt{27x^3} = \sqrt{3x \cdot 27x^3} = \sqrt{81x^4} = 9x^2$

(b) $\sqrt{5}\left(\sqrt{20} - 2\sqrt{3}\right) = \sqrt{5}\,\sqrt{20} - \sqrt{5}\left(2\sqrt{3}\right)$

$= \sqrt{100} - 2\sqrt{5}\,\sqrt{3}$

$= 10 - 2\sqrt{15}$

$= 2\left(5 - \sqrt{15}\right)$

(c)
$$\begin{array}{l} \quad 1 - \sqrt{2} \\ \underline{\quad 2 + 2\sqrt{2}} \\ \quad 2 - 2\sqrt{2} \\ \underline{\qquad\quad 2\sqrt{2} - 2\sqrt{2}\,\sqrt{2}} \\ \quad 2 + 0 \quad -2(2) = 2 - 4 = -2 \end{array}$$

5. $\frac{1}{2}$

$$\frac{5}{\sqrt{8}} \div \frac{\sqrt{75}}{\sqrt{6}} = \frac{5}{\sqrt{8}} \cdot \frac{\sqrt{6}}{\sqrt{75}} = \frac{5}{2\sqrt{2}} \cdot \frac{\sqrt{2}\,\sqrt{3}}{5\sqrt{3}} = \frac{1}{2}$$

6. (a) $3(4 - \sqrt{5})$ (b) $\frac{4 - \sqrt{5}}{5}$

(a) $12 - \sqrt{45} = 12 - \sqrt{9 \cdot 5} = 12 - 3\sqrt{5} = 3\left(4 - \sqrt{5}\right)$

(b) $\frac{12 - \sqrt{45}}{15} = \frac{3\left(4 - \sqrt{5}\right)}{15} = \frac{4 - \sqrt{5}}{5}$

7. 9.23

$5 + 3\sqrt{2} \approx 5 + 3(1.41) = 5 + 4.23 = 9.23$

8. (a) $\frac{1}{121}$ (b) 1 (c) $\frac{100}{49}$

(a) $11^{-2} = \frac{1}{11^2} = \frac{1}{121}$

(b) $a^0 = 1$ for any nonzero number a. Thus, $(-5)^0 = 1$.

(c) $\left(\frac{-7}{10}\right)^{-2} = \frac{10^2}{(-7)^2} = \frac{100}{49}$

9. (a) n^{-5} (b) n^4

(a) $\frac{n^{-3}}{n^2} = n^{-3-2} = n^{-5}$

(b) $n^9 \cdot n^{-2} \cdot n^{-3} = n^{9+(-2)+(-3)} = n^4$

10. (a) 9.4×10^{-2} (b) 8710

11. 3.5×10^5

$$\frac{21\,000}{200 \times .0003} = \frac{21 \times 10^3}{2 \times 10^2 \times 3 \times 10^{-4}} = \frac{21}{2 \times 3} \times \frac{10^3}{10^2 \times 10^{-4}} = \frac{7}{2} \times 10^{3-2-(-4)} = 3.5 \times 10^5$$

12. 20 meters

$$c^2 = a^2 = b^2$$
$$c^2 = 12^2 + 16^2 = 144 + 256 = 400$$
$$c = \sqrt{400} = 20 \text{ (meters)}$$

CHAPTER 9

Practice Exercises for Section 9.1, pages 409–411

1. (a) $x^2 + x - 6 = 0$ (b) $2x^2 - 1 = 0$

(a) To clear the equation $\frac{3}{x+1} = \frac{x}{2}$ of fractions, multiply both sides by the *lcd*, $2(x+1)$. (Cross multiplying yields the same result.)

$$3 \cdot 2 = x(x+1)$$
$$6 = x^2 + x$$
$$0 = x^2 + x - 6, \text{ or } x^2 + x - 6 = 0$$

This is a quadratic equation of the form $Ax^2 + Bx + C = 0$, where $A = 1$, $B = 1$, and $C = -6$.

(b)
$$(x+1)^2 = x(2-x) + 2$$
$$x^2 + 2x + 1 = 2x - x^2 + 2$$
$$2x^2 - 1 = 0$$

Here $A = 2$, $B = 0$, and $C = -1$.

2. (a) (*ii*) is a first-degree equation. (b) -3

(*i*) $(x-3)2x = 4$ is equivalent to $2x^2 - 6x - 4 = 0$. This is a quadratic, or second-degree, equation in which $A = 2$, $B = -6$, and $C = -4$.

(*ii*)
$$4x^2 + 9x = (2x+3)^2$$
$$4x^2 + 9x = 4x^2 + 12x + 9$$
$$4x^2 + 9x - 4x^2 - 12x = 4x^2 + 12x + 9 - 4x^2 - 12x$$
$$-3x = 9$$
$$x = -3$$

Check:
$$4(-3)^2 + 9(-3) \stackrel{?}{=} (2(-3)+3)^2$$
$$4(9) + (-27) \stackrel{?}{=} (-6+3)^2$$
$$36 - 27 \stackrel{?}{=} (-3)^2$$
$$9 \stackrel{\checkmark}{=} 9$$

Practice Exercises for Section 9.2, pages 414–417

1. -4, 10

A product is zero whenever any one of its factors is zero. Thus, $(x+4)(x-10) = 0$ is a true statement whenever $x + 4 = 0$, or $x - 10 = 0$. Therefore, $x = -4$ or $x = 10$.

2. (a) -6, 5

$$x^2 + x - 30 = 0$$

To solve a quadratic equation in standard form (with zero on one side), first try to factor the expression that is equal to zero. $(x+6)(x-5) = 0$. Next, set each factor equal to zero and solve for x.

$x + 6 = 0$ or $x - 5 = 0$
$x = -6$ | $x = 5$

The roots are -6 and 5.

Check:

(for -6)	(for 5)
$(-6)^2 + (-6) - 30 \stackrel{?}{=} 0$	$5^2 + 5 - 30 \stackrel{?}{=} 0$
$36 - 6 - 30 \stackrel{?}{=} 0$	$25 + 5 - 30 \stackrel{?}{=} 0$
$0 \stackrel{\checkmark}{=} 0$	$0 \stackrel{\checkmark}{=} 0$

(b) -1, 2

$$\frac{10}{x-1} = 5x \text{ [Cross multiply.]}$$
$$5x(x-1) = 10$$
$$5x^2 - 5x = 10$$
$$5x^2 - 5x - 10 = 0$$
$$5(x^2 - x - 2) = 0$$
$$x^2 - x - 2 = 0$$
$$(x+1)(x-2) = 0$$

$x + 1 = 0$ | $x - 2 = 0$
$x = -1$ | $x = 2$

The roots are -1 and 2.

(c) 6, 1

$$1 + \frac{6}{x^2} = \frac{7}{x}$$
$$x^2 + 6 = 7x$$
$$x^2 - 7x + 6 = 0$$
$$(x-6)(x-1) = 0$$

$x - 6 = 0$ | $x - 1 = 0$
$x = 6$ | $x = 1$

The roots are 6 and 1.

3. (a) $\frac{1}{4}$, 1

$$4x^2-5x+1=0$$
$$(4x-1)(x-1)=0$$
$$4x-1=0 \quad\text{or}\quad x-1=0$$
$$4x=1 \qquad x=1$$
$$x=\frac{1}{4}$$

The roots are $\frac{1}{4}$ and 1.

(b) $\frac{1}{3}$, -1

$$12x^2+8x-4=0$$
$$4(3x^2+2x-1)=0$$
$$4(3x-1)(x+1)=0$$
$$(3x-1)(x+1)=0$$
$$3x-1=0 \quad\text{or}\quad x+1=0$$
$$3x=1 \qquad x=-1$$
$$x=\frac{1}{3}$$

The roots are $\frac{1}{3}$ and -1.

4. $\frac{\pm 5}{7}$

$$49x^2=25$$
$$49x^2-25=0$$
$$(7x+5)(7x-5)=0$$
$$7x+5=0 \qquad 7x-5=0$$
$$7x=-5 \qquad 7x=5$$
$$x=\frac{-5}{7} \qquad x=\frac{5}{7}$$

The roots are $\frac{5}{7}$ and $\frac{-5}{7}$, or $\frac{\pm 5}{7}$.

5. (a) 0, $\frac{\pm 1}{2}$

$$8x^3-2x=0$$
$$2x(4x^2-1)=0$$
$$2x(2x+1)(2x-1)=0$$
$$2x=0 \qquad 2x+1=0 \qquad 2x-1=0$$
$$x=0 \qquad 2x=-1 \qquad 2x=1$$
$$x=\frac{-1}{2} \qquad x=\frac{1}{2}$$

The roots are 0, $\frac{1}{2}$, and $\frac{-1}{2}$.

(b) 0, 1, 3

$$x^4-4x^3+3x^2=0$$
$$x^2(x^2-4x+3)=0$$
$$x^2(x-1)(x-3)=0$$
$$x^2=0 \qquad x-1=0 \qquad x-3=0$$
$$x=0 \qquad x=1 \qquad x=3$$

The roots are 0, 1, and 3.

6. $3x^2+7x-6=0$

If the roots of an equation are given by

$x=\frac{2}{3}$ or $x=-3$, then

$x-\frac{2}{3}=0$ or $x+3=0$.

The product of these two binomials must also be zero.

Multiply:

$$\begin{array}{r} x+3 \\ x-\frac{2}{3} \\ \hline x^2+3x \\ -\frac{2}{3}x-2 \\ \hline x^2-\frac{7}{3}x-2 \end{array}$$

Thus,

$$x^2-\frac{7}{3}x-2=0$$

is a quadratic equation with the required roots, $\frac{2}{3}$ and -3. An equivalent equation *without fractions* is found by multiplying both sides by 3. Thus,

$$3(x^2-\frac{7}{3}x-2)=3(0)$$
$$3x^2-7x-6=0$$

Practice Exercises for Section 9.3, pages 421–422

1. (a) $\frac{\pm 2}{5}$ (b) $\pm 3\sqrt{5}$ (c) $\frac{3}{5}$, $\frac{-1}{5}$ (d) $\frac{-7\pm\sqrt{6}}{2}$

(a) $x^2=\frac{4}{25}$

$$x=\pm\sqrt{\frac{4}{25}}$$
$$x=\pm\frac{2}{5}$$

(b) $x^2-45=0$

$$x^2=45$$
$$x=\pm\sqrt{45}=\pm\sqrt{9\cdot 5}=\pm\sqrt{9}\sqrt{5}=\pm 3\sqrt{5}$$

(c) $(5u-1)^2=4$

This equation is of the form $x^2=a$, where $x=5u-1$.

$$5u-1=\pm\sqrt{4}$$
$$5u=1\pm 2$$
$$u=\frac{1\pm 2}{5}$$

One root is $\frac{1+2}{5}$, or $\frac{3}{5}$; the other root is $\frac{1-2}{5}$, or $\frac{-1}{5}$.

Check: (for $\frac{3}{5}$)

$$\left[5\left(\tfrac{3}{5}\right)-1\right]^2\stackrel{?}{=}4$$
$$(3-1)^2\stackrel{?}{=}4$$
$$2^2\stackrel{?}{=}4$$
$$4\stackrel{\checkmark}{=}4$$

(for $\frac{-1}{5}$)

$$\left[5\left(-\tfrac{1}{5}\right)-1\right]^2\stackrel{?}{=}4$$
$$(-1-1)^2\stackrel{?}{=}4$$
$$(-2)^2\stackrel{?}{=}4$$
$$4\stackrel{\checkmark}{=}4$$

(d) $(2u+7)^2=6$

$$2u+7=\pm\sqrt{6}$$
$$2u=-7\pm\sqrt{6}$$
$$u=\frac{-7\pm\sqrt{6}}{2}$$

One root is $\frac{-7+\sqrt{6}}{2}$; the other root is $\frac{-7-\sqrt{6}}{2}$.

2. (a) $\frac{5}{3}$ (b) $0, \frac{-3}{2}$ (c) 0, 1

(a)

$$(5-3x)^2=0$$
$$5-3x=0$$
$$5=3x$$
$$\frac{5}{3}=x$$

$\frac{5}{3}$ is the only root.

(b)

$$x(2x+3)^2=0$$
$$x=0 \quad \text{or} \quad (2x+3)^2=0$$
$$2x+3=0$$
$$x=\frac{-3}{2}$$

The roots are 0 and $\frac{-3}{2}$.

(c)

$$y^3-2y^2+y=0$$
$$y(y^2-2y+1)=0$$
$$y(y-1)^2=0$$
$$y=0 \quad \text{or} \quad (y-1)^2=0$$
$$y-1=0$$
$$y=1$$

The roots are 0 and 1.

3. (a) one real root (b) no real roots (c) two real roots

Each equation is of the form $x^2=a$.

(a) $(2u+9)^2=0$. Here, $x=2u+9$, $a=0$. There is only one real root.

(b) $(3u)^2=-9$. Here, $x=3u$; $a=-9$. There are no real roots because $a<0$.

(c) $(u-4)^2=4$. Here, $x=u-4$; $a=4$. Because $a>0$, there are two different real roots.

Practice Exercises for Section 9.4, pages 425–427

1. (a) $\frac{3}{4}, -2$ (b) $2\pm\sqrt{3}$

(a) $4x^2+5x-6=0$

This is a quadratic equation of the form $Ax^2+Bx+C=0$, where $A=4$, $B=5$, and $C=-6$.

Substitute A, B, and C in the formula, $x=\frac{-B\pm\sqrt{B^2-4AC}}{2A}$. Thus,

$$x=\frac{-5\pm\sqrt{5^2-4(4)(-6)}}{2(4)}=\frac{-5\pm\sqrt{25+96}}{8}=\frac{-5\pm\sqrt{121}}{8}=\frac{-5\pm 11}{8}$$

One root is given by $\frac{-5+11}{8}=\frac{6}{8}=\frac{3}{4}$; the other root is given by $\frac{-5-11}{8}=\frac{-16}{8}=-2$.

Check:

(for $\frac{3}{4}$)

$$4\left(\frac{3}{4}\right)^2+5\left(\frac{3}{4}\right)-6\stackrel{?}{=}0$$
$$4\left(\frac{9}{16}\right)+\frac{15}{4}-6\stackrel{?}{=}0$$
$$\frac{9}{4}+\frac{15}{4}-6\stackrel{?}{=}0$$
$$\frac{24}{4}-6\stackrel{?}{=}0$$
$$6-6\stackrel{?}{=}0$$
$$0\stackrel{\checkmark}{=}0$$

(for -2)

$$4(-2)^2+5(-2)-6\stackrel{?}{=}0$$
$$4(4)+(-10)-6\stackrel{?}{=}0$$
$$16-10-6\stackrel{?}{=}0$$
$$0\stackrel{\checkmark}{=}0$$

(b) $x^2-4x+1=0$. Here, $A=1$; $B=-4$; $C=1$.

$$x=\frac{-B\pm\sqrt{B^2-4AC}}{2A}$$

$$x=\frac{-(-4)\pm\sqrt{(-4)^2-4(1)(1)}}{2(1)}=\frac{4\pm\sqrt{16-4}}{2}=\frac{4\pm\sqrt{12}}{2}=\frac{4\pm\sqrt{4\cdot 3}}{2}=\frac{4\pm 2\sqrt{3}}{2}=\frac{2(2\pm\sqrt{3})}{2}=2\pm\sqrt{3}$$

2. (a) $\frac{3}{2}, \frac{2}{3}$

$6x^2 - 13x + 6 = 0$

$A = 6,\ B = -13,\ C = 6$

$$x = \frac{-B \pm \sqrt{B^2 - 4AC}}{2A}$$

$$x = \frac{-(-13) \pm \sqrt{(-13)^2 - 4(6)(6)}}{2(6)} = \frac{13 \pm \sqrt{169 - 144}}{12} = \frac{13 \pm \sqrt{25}}{12} = \frac{13 \pm 5}{12}$$

One root is given by $\frac{13+5}{12} = \frac{18}{12} = \frac{3}{2}$; the other root is given by $\frac{13-5}{12} = \frac{8}{12} = \frac{2}{3}$.

(b) $\frac{3}{2}, \frac{2}{3}$

We can also solve this equation by the factoring method.

$6x^2 - 13x + 6 = 0$

$(3x - 2)(2x - 3) = 0$

$3x - 2 = 0 \quad | \quad 2x - 3 = 0$

$x = \frac{2}{3} \quad | \quad x = \frac{3}{2}$

3. (a) 0 (b) 1 real root (c) $\frac{3}{2}$

$4x^2 - 12x + 9 = 0$ Here, $A = 4;\ B = -12;\ C = 9.$

(a) The discriminant, $B^2 - 4AC = (-12)^2 - 4(4)(9) = 144 - 144 = 0$

(b) Because the discriminant is zero, there is only one real root.

(c) $x = \frac{-B \pm \sqrt{0}}{2A} = \frac{-(-12)}{2(4)} = \frac{12}{8} = \frac{3}{2}$

Practice Exercises for Section 9.5, pages 429–432

1. 5, 6, and 7

Let x = the smallest of the three consecutive integers;
$x + 1$ = the next consecutive integer;
$x + 2$ = the largest of the three.
The sum of the squares of these integers is 110. In symbols,

$x^2 + (x+1)^2 + (x+2)^2 = 110$

$x^2 + x^2 + 2x + 1 + x^2 + 4x + 4 = 110$

$3x^2 + 6x + 5 = 110$

$3x^2 + 6x - 105 = 0$

$x^2 + 2x - 35 = 0$

$(x+7)(x-5) = 0$

$x + 7 = 0 \quad | \quad x - 5 = 0$

$x = -7 \quad | \quad x = 5$

Because the problem states that the integers are *positive*, choose $x = 5$. Then

$x + 1 = 6$

$x + 2 = 7$

(To check, note that $5^2 + 6^2 + 7^2 = 25 + 36 + 49 = 110$.)

2. $w = 4$ feet, $l = 9$ feet

Let w = the width of the rug; l = the length. The length is one foot longer than twice the width. Thus, $l = 2w + 1$.

Area $= l \cdot w = 36$ [Substitute $2w + 1$ for l.]

$(2w + 1)w = 36$

$2w^2 + w = 36$

$2w^2 + w - 36 = 0$

$(2w + 9)(w - 4) = 0$

$2w + 9 = 0 \quad | \quad w - 4 = 0$

$2w = -9 \quad |$

$w = -\frac{9}{2} \quad | \quad w = 4$

The width must be a positive number. Thus, $w = 4$ (feet); $l = 2w + 1 = 2(4) + 1 = 9$ (feet).

3. (a) 336 feet (b) 5 seconds

(a) If $t = 2$ seconds, then $h = 400 - 16(2)^2 = 400 - 64 = 336$ (feet)

(b) When the object reaches the ground, its height above the ground is zero. Thus, solve the equation $0 = 400 - 16t^2$ for t.

$16t^2 = 400$

$t^2 = 25$

$t = \pm 5$

Discard the negative solution because t must be positive. Thus, $t = 5$ seconds.

4. 8 machines or 12 machines

In the formula, P is measured in *thousands* of dollars. Thus, let $P = 196$.

$196 = 100 - x^2 + 20x$

$x^2 - 20x - 100 + 196 = 0$

$x^2 - 20x + 96 = 0$ [Factor.]

$(x - 8)(x - 12) = 0$

$x - 8 = 0 \quad | \quad x - 12 = 0$

$x = 8 \quad | \quad x = 12$

Producing either 8 machines or 12 machines will derive the same profit of \$196 000.

Practice Exam 1 on Chapter 9, page 437

1. $\frac{1}{4}$

$$(2x+1)^2=4(x-1)^2$$
$$4x^2+4x+1=4(x^2-2x+1)$$
$$4x^2+4x+1=4x^2-8x+4$$
$$12x=3$$
$$x=\frac{3}{12}$$
$$x=\frac{1}{4}$$

2. 7, −2

$$x^2-5x-14=0$$
$$(x-7)(x+2)=0$$
$$x-7=0 \quad | \quad x+2=0$$
$$x=7 \quad | \quad x=-2$$

The roots are 7 and −2.

3. $\frac{1\pm\sqrt{5}}{2}$

$$(2y-1)^2=5$$
$$2y-1=\pm\sqrt{5}$$
$$2y=1\pm\sqrt{5}$$
$$y=\frac{1\pm\sqrt{5}}{2}$$

4. $\frac{-9\pm\sqrt{69}}{2}$

$x^2+9x+3=0$ $[A=1,\ B=9,\ C=3]$

$$x=\frac{-B\pm\sqrt{B^2-4AC}}{2A}$$
$$x=\frac{-9\pm\sqrt{9^2-4(1)(3)}}{2(1)}=\frac{-9\pm\sqrt{69}}{2}$$

5. $\frac{5}{2}$, −1

By the factoring method:

$$2y^2-3y-5=0$$
$$(2y-5)(y+1)=0$$
$$2y-5=0 \quad | \quad y+1=0$$
$$2y=5 \quad | \quad y=-1$$
$$y=\frac{5}{2} \quad |$$

The roots are $\frac{5}{2}$ and −1.

Check: $\left(\text{for } \frac{5}{2}\right)$

$$2\left(\frac{5}{2}\right)^2-3\left(\frac{5}{2}\right)-5\stackrel{?}{=}0$$
$$2\left(\frac{25}{4}\right)-\frac{15}{2}-5\stackrel{?}{=}0$$
$$\frac{25}{2}-\frac{15}{2}-5\stackrel{?}{=}0$$
$$5-5\stackrel{?}{=}0$$
$$0\stackrel{\checkmark}{=}0$$

(for −1)

$$2(-1)^2-3(-1)-5\stackrel{?}{=}0$$
$$2(1)+3-5\stackrel{?}{=}0$$
$$2+3-5\stackrel{?}{=}0$$
$$5-5\stackrel{?}{=}0$$
$$0\stackrel{\checkmark}{=}0$$

6. $\frac{1\pm\sqrt{19}}{2}$

$$(2u-1)^2-19=0$$
$$(2u-1)^2=19$$
$$2u-1=\pm\sqrt{19}$$
$$2u=1\pm\sqrt{19}$$
$$u=\frac{1\pm\sqrt{19}}{2}$$

7. $-1\pm\sqrt{3}$

$x^2+2x-2=0$

Using the quadratic formula, $A=1,\ B=2,\ C=-2$.

$$x=\frac{-B\pm\sqrt{B^2-4AC}}{2A}$$
$$x=\frac{-2\pm\sqrt{2^2-4(1)(-2)}}{2(1)}=\frac{-2\pm\sqrt{12}}{2}=\frac{-2\pm\sqrt{4\cdot 3}}{2}=\frac{-2\pm2\sqrt{3}}{2}$$
$$=\frac{2(-1\pm\sqrt{3})}{2}=-1\pm\sqrt{3}$$

8. $x(17-x)=60$; 5 and 12

Let x be one of the two numbers whose sum is 17. The other number can be expressed by $17-x$. Their product is 60. Thus,

$$x(17-x)=60$$
$$17x-x^2=60$$
$$x^2-17x+60=0 \quad \text{[Factor.]}$$
$$(x-12)(x-5)=0$$
$$x-12=0 \quad | \quad x-5=0$$
$$x=12 \quad | \quad x=5$$
$$17-x=5 \quad | \quad 17-x=12$$

In either case, the two numbers are 5 and 12.

Practice Exam 2 on Chapter 9, page 437

1. (a) −1 (b) first-degree

(a)
$$3(x+1)^2=(3x+4)(x+1)$$
$$3(x^2+2x+1)=3x^2+7x+4$$
$$3x^2+6x+3=3x^2+7x+4$$
$$-1=x$$

(b) This is a first-degree equation because it reduces to an equation with no second-degree terms; the remaining terms are either first-degree terms or numbers.

2. 13, −2

$$x^2-11x-26=0$$
$$(x-13)(x+2)=0$$
$$x-13=0 \quad | \quad x+2=0$$
$$x=13 \quad | \quad x=-2$$

Check: (for 13)

$$(13)^2-11(13)-26\stackrel{?}{=}0$$
$$169-143-26\stackrel{?}{=}0$$
$$0\stackrel{\checkmark}{=}0$$

(for −2)

$$(-2)^2-11(-2)-26\stackrel{?}{=}0$$
$$4+22-26\stackrel{?}{=}0$$
$$0\stackrel{\checkmark}{=}0$$

3. 0, 5

$\frac{x}{3} = 2 - \frac{2}{x+1}$ [Multiply both sides by the *lcd*, 3 (x + 1).]

$3(x+1)\frac{x}{3} = 3(x+1)\cdot 2 - 3(x+1)\frac{2}{x+1}$

$x^2 + x = 6x + 6 - 6$

$x^2 - 5x = 0$

$x(x-5) = 0$

$x = 0 \quad | \quad x - 5 = 0$

$x = 5$

The roots are 0 and 5.

4. $\frac{4 \pm \sqrt{3}}{5}$

$(5x-4)^2 = 3$

$5x - 4 = \pm\sqrt{3}$

$5x = 4 \pm \sqrt{3}$

$x = \frac{4 \pm \sqrt{3}}{5}$

One root is $\frac{4+\sqrt{3}}{5}$; the other is $\frac{4-\sqrt{3}}{5}$.

5. (a) 4 (b) 2 real roots (c) $-1, \frac{-1}{3}$

(a) $3x^2 + 4x + 1 = 0$ is of the form $Ax^2 + Bx + C = 0$.

$A = 3,\ B = 4,\ C = 1$. The discriminant is given by:

$B^2 - 4AC = 4^2 - 4(3)(1) = 16 - 12 = 4$

(b) There are 2 distinct real roots because

$B^2 - 4AC = 4 > 0$.

(c) $x = \frac{-4 \pm \sqrt{4}}{2(3)}$

$x = \frac{-4 \pm 2}{6}$

$x = \frac{-4+2}{6} = \frac{-2}{6} = -\frac{1}{3}$ or $x = \frac{-4-2}{6} = \frac{-6}{6} = -1$

6. $\frac{1}{5}$, 1

$10x^2 + 2 = 12x$

$10x^2 - 12x + 2 = 0$

$5x^2 - 6x + 1 = 0$

$(5x-1)(x-1) = 0$

$5x - 1 = 0 \quad | \quad x - 1 = 0$

$5x = 1 \quad | \quad x = 1$

$x = \frac{1}{5}$

The roots are $\frac{1}{5}$ and 1.

7. 1, −5

$(x+2)^2 - 9 = 0$

$(x+2)^2 = 9$

$x + 2 = \pm 3$

$x = -2 \pm 3$

One root is $-2 + 3$, or 1; the other is $-2 - 3$, or -5.

8. $w = 3\frac{1}{2}$ feet, $l = 4$ feet

Let w = width, l = length.

Perimeter is 15 feet; area is 14 square feet.

$P = 2l + 2w = 15;\quad A = l \cdot w = 14$

Solving the second equation for l, $l = \frac{14}{w}$.

Substitute this expression for l in the first equation.

$2\left(\frac{14}{w}\right) + 2w = 15$ [Multiply both sides by w.]

$28 + 2w^2 = 15w$

$2w^2 - 15w - 28 = 0$

$(2w-7)(w-4) = 0$

$2w - 7 = 0$	$w - 4 = 0$
$w = \frac{7}{2} = 3\frac{1}{2}$ (feet)	$w = 4$ (feet)
$l = 14 \div \frac{7}{2} = 14 \times \frac{2}{7} = 4$ (feet)	$l = \frac{14}{4} = \frac{7}{2} = 3\frac{1}{2}$ (feet)

With either solution, the dimensions are $3\frac{1}{2}$ feet by 4 feet.

Index